NOBEL SYMPOSIUM 25

Medicine and Natural Sciences

Chemistry in Botanical Classification

Proceedings of the Twenty-Fifth Nobel Symposium
held August 20–25, 1973
Södergarn, Lidingö (near Stockholm), Sweden

Editors
GERD BENDZ and JOHAN SANTESSON

Administrative editor
VERA RUNNSTRÖM-REIO

NOBEL FOUNDATION · STOCKHOLM

ACADEMIC PRESS · NEW YORK AND LONDON

A Subsidiary of Harcourt Brace Jovanovich, Publishers

Academic Press ISBN 0-12-086650-1
Library of Congress Catalog Card Number 74-967

Printed in Sweden by
Almqvist & Wiksell, Uppsala 1974

Foreword

This volume contains the papers presented at Nobel Symposium 25, Chemistry in Botanical Classification. Together these papers give a comprehensive review of the position of the research frontier and the present state of knowledge in one of the most active and promising interdisciplinary areas of the natural science of today. However, only to a minor extent can the articles—excellent as they are—reflect the intellectual activity during the Symposium. To those present the exchange of ideas during the discussions, not only during the sessions but almost continuously during the week we lived together at Södergarn, may well be regarded as the most memorable and important part of the Symposium.

Quite a number of the discussions were put in writing by the participants. These written remarks, often different from the original presentation, are included in this volume.

Some of the comments reflect upon the communication-barrier still existing between plant taxonomists and chemists—but also show that "we have begun an important and useful dialogue between chemists and systematists" as Professor Takhtajan said in the general discussion.

The program was not planned to give the most systematic presentation of the papers contributed, but rather to promote an intermixing of ideas. In this volume, however, the papers are presented in a slightly different order. Rapid publication has been made possible through the kind cooperation of the members of the Symposium whom we wish to thank for adhering to the very short deadline set for delivery of their contributions.

A Nobel Symposium cannot be accomplished without help from many sources and many persons. It is a pleasure to have this opportunity to express our thanks to those who helped to make the Symposium successful.

On this occasion we would like to express our gratitude to the late Professor Arne Tiselius, once a member of the Nobel Symposium Committee, for initiating this Symposium. Our thanks are also due to The Tri-Centennial Fund of the Bank of Sweden, to the Nobel Foundation, and its Nobel Symposium Committee and to the Royal Academy of Sciences and its Nobel Institute for Chemistry for the generous grants which have made this Symposium possible. For highly appreciated help and advice during the preparation of the Symposium we wish to thank the Nobel Foundation and its staff. We are very grateful to Svenska Handelsbanken for placing at the disposal of the Symposium the ideal place for such an activity, Södergarn, the beautiful conference hotel where the hostess Miss Kerstin Ekblad did so much to make ours a very pleasant stay.

The Symposium was held at Södergarn except for two occasions. On Tuesday, August 21, the afternoon session was held at the Royal Academy of Sciences where Professor C.-G. Bernhard welcomed the participants of the Symposium and gave an orientation of the activities of the Academy. We want to thank the Academy for the hospitality and for the dinner given in connection with our visit there. On Thursday, August 23, the morning session was held in Uppsala to give the members an opportunity to visit Hammarby, the Summer Home of Carolus Linnaeus. We are grateful for the hospitality of the University of Uppsala on this occasion.

Finally we wish to express our thanks to the other members of the organizing committee for pleasant collaboration and to all who contributed to the success of the Symposium for their valuable help.

Uppsala, October 1973

Gerd Bendz   Johan Santesson

Contents

8

Sponsors

The Nobel Foundation
The Tri-Centennial Fund of the Bank of Sweden
The Royal Academy of Sciences

Nobel Symposium Committee

Ramel, Stig, Chairman, Executive Director of the Nobel Foundation
Hulthén, Lamek, Professor, Member of the Nobel Committee for Physics
Fredga, Arne, Professor, Chairman of the Nobel Committee for Chemis-
  try
Gustafsson, Bengt, Professor, Secretary of the Nobel Committee for Medi-
  cine
Gyllensten, Lars, Professor, Member of the Swedish Academy
Greve, Tim, Director of the Norwegian Nobel Institute (Peace)
Svensson, Nils-Eric, Director of the Tri-Centennial Fund of the Bank
  of Sweden

Organizing committee

Professor Arne Fredga
Ass. Professor Gerd Bendz
Professor Olov Hedberg
Professor Hans Runemark
Professor Rolf Santesson
Docent Johan Santesson

List of Participants

Bate-Smith, E C, Institute of Animal Physiology, Agricultural Research Council, Babraham, Cambridge CB2 4AT, UK

Bendz, G, Institute of Chemistry, University of Uppsala, POB 531, S-751 21 Uppsala, Sweden

Bergström, G, Ecological Station, University of Uppsala, S-380 60 Färjestaden, Sweden

Birch, A J, Research School of Chemistry, Australian National University, POB 4, Canberra, ACT 2600, Australia

Boulter, D, Department of Botany, University of Durham, South Rd, Durham DH1 3LE, UK

Bu'Lock, J D, Department of Chemistry, University of Manchester, Manchester Ml3 9PL, UK

Cronquist, A, New York Botanical Garden, Bronx, N.Y. 10458, USA

Farnsworth, N R, Department of Pharmacognosy and Pharmacology, College of Pharmacy, University of Illinois at the Medical Center, POB 6998, Chicago, Ill. 60680, USA

Fredga, A, Institute of Chemistry, University of Uppsala, POB 531, S-751 21 Uppsala, Sweden

Geissman, T A, Department of Chemistry, University of California, Los Angeles, Calif. 90024, USA

Grant, W F, Genetics Laboratory, Macdonald Campus of McGill University, Ste Anne de Bellevue 800, Queb., Canada

Harborne, J B, Phytochemical Unit, Plant Science Laboratories, University of Reading, Reading RG6 2AS, Berks, UK

Hedberg, O, Institute of Systematic Botany, University of Uppsala, POB 541, S-751 21 Uppsala, Sweden

Hegnauer, R, Laboratorium voor Experimentele Plantensystematiek, Rijksuniversiteit, Leiden, The Netherlands

Herout, V, Institute of Organic Chemistry and Biochemistry, Czechoslovak Academy of Science, Praha 6, Czechoslovakia

Herz, W, Department of Chemistry, The Florida State University, Tallahassee, Fla 32306, USA

Heywood, V H, Department of Botany, Plant Science Laboratories, University of Reading, Reading RG6 2AS, Berks, UK

Jensen, U, Botanisches Institut der Universität Köln, 5000 Köln-Lindenthal 41, Gyrhofstrasse 15, W. Germany

Kjær A, Institute of Organic Chemistry, Danmarks Tekniske Højskole, Bldg 201, 2800 Lyngby, Denmark

Kullenberg, B, Department of Entomology, University of Uppsala, POB 561, S-751 22 Uppsala, Sweden

Lavie, D, Department of Organic Chemistry, Weizmann Institute of Science, Rehovot, Israel

Mabry, T J, Department of Botany, University of Texas at Austin, Austin, Tex. 78712, USA

Merxmüller, H, Botanische Staatssammlung, D 8000 München 19, Menzinger Strasse 67, W. Germany

Natori, S, National Institute of Hygienic Sciences, Kamiyoga-l-chome, Setagaya-ku, Tokyo, Japan

Ourisson, G, Institut de Chimie, Université Louis Pasteur, 1, rue Blaise Pascal, POB 296/R 8, 67008 Strasbourg, France

Reichstein, T, Institut für organische Chemie der Universität, CH-4056 Basel, Switzerland

Runemark, H, Institute of Systematic Botany, University of Lund, Ö. Vallgatan 18–20, S-223 61 Lund, Sweden

Sandberg, F, Department of Pharmacognosy, Faculty of Pharmacy, Lindhagensgatan 128, S-112 51 Stockholm, Sweden

Santesson, J, Institute of Chemistry, University of Uppsala, POB 531, S-751 21 Uppsala, Sweden

Santesson, R, Institute of Systematic Botany, University of Uppsala, POB 541, S-751 21 Uppsala, Sweden von Schantz, M, Farmaceutical Institute, University of Helsinki, Fabianink. 35, Helsinki 17, Finland

Schwarting, A E, School of Pharmacy, University of Connecticut, Storrs, Conn. 06268, USA

Shibata, S, Faculty of Pharmaceutical Sciences, University of Tokyo, Bunkyo-ku, Tokyo, Japan

Swain, T, Royal Botanical Gardens, Kew, Richmond, Surrey, UK

Sørensen, N A, Norges Tekniske Høgskole, Trondheim, Norway

Takhtajan, A L, Komarov Botanical Institute of the USSR Academy of Sciences, Popov ul. 2, Leningrad P-22, USSR

Tétényi, P, Institute for Research of Medical Plants, Dániel u. 38–42, Budapest XII, Hungary

Turner, B L, Department of Botany, University of Texas at Austin, Austin, Tex. 78712, USA

Wagner, H, Institut für pharmazeutische Arzneimittellehre der Universität, 8 München 2, Karlstr. 29, W. Germany

Weimarck, G, Department of Plant Taxonomy, University of Lund, Ö. Vallgatan 20, S-223 61 Lund, Sweden

Runnström-Reio, V, The Nobel Foundation, c/o Karolinska Institutet, S-104 01 Stockholm 60, Sweden

Author Index

Opening speech

Ladies and Gentlemen,

On behalf of the Organization Committee I have the honour and the pleasure to bid you a most hearty welcome to this Nobel Symposium No. 25.

I think we have all a feeling that pure research, pure science is not very popular today. Authorities and public opinion ask for practical results; investments in research should repay within a few years. Under such conditions we are very happy and perhaps a little proud that it was possible to organize this Symposium; the money invested here will certainly not repay in 2 or 3 years. This was made possible by the existence of some organizations here in Sweden, founded and built up under more favourable conditions. I shall give you their names: The Nobel Foundation, The Swedish Royal Academy of Sciences and—last but not least—The Tri-Centennial Fund of the Bank of Sweden, which in fact has spent the greater part of the money. I shall not waste time by explaining how these organizations operate and co-operate, but they are all involved in the symposium.

Perhaps I should say a little about the general stipulations for organizing Nobel Symposia. The purpose is to bring together a number of scientists from related fields to discuss and lecture on some important problem of current interest. There are some other interesting stipulations: it is said that controversial subjects should not be avoided. There are also rather strict rules as to the number of participants. There should be 20–30 from abroad—even the generosity of the Nobel Foundation has its limits—and not more than 10 from Sweden. Then we can have a limited number of observers.

A few words on the origin of our special Symposium. We have here in Sweden a feeling that contacts between botanists and chemists are not what they ought to be: we know too little about each other and about problems of mutual interest. Some years ago, Professor Bendz, aided by our friend Holger Erdtman and his wife, organized a meeting between botanists and chemists, a so-called Nobel Workshop, at Väversunda in Östergötland.

It was a small and mainly local symposium, but there came a few guests from abroad, Professor Turner among them. It was sponsored by The Nobel Institute for Chemistry and I think it was rather successful. A report was delivered to the head of the Nobel Institute, Professor Arne Tiselius, who proposed that Professor Bendz should organize a Nobel *Symposium* in 1973. A committee was formed and they asked me to act as president, that is as some kind of figure-head.

A few months later Professor Tiselius passed away, which was, of course, a severe blow, but it was possible to realize the plans and raise the money even without his support and aid. In the organization we have also had good help from Holger Erdtman; we are sorry that he is not here to-day but very grateful for his help and advice.

It is of course a difficult task to select the participants, to decide which people should be invited. We have asked a number of prominent scientists for advice and at last we had a list of more than 50 persons. Thus we really had to make a selection. Of course this selection can be criticized. Let me say that the merit here is due to the advisers: the responsibility and the blame are to be taken by the Committee.

What are we going to do here?

The number of participants is restricted not only for economical reasons, but also because we want the Symposium to have an intimate and informal character. All people present should have the opportunity to meet each other, get acquainted and discuss matters of mutual interest. We have asked each lecturer how long time he needs for his lecture and we are glad if they keep the allotted time, but we have no red lights or other signals to stop you if you talk too much. Ample time has been set apart for discussions and we hope that everything will turn out for the best.

Of course I should also say something about

Chemistry in Botanical Classification, but I am sure that all our distinguished guests know more about that topic than I do. I am an organic chemist and I have always been much interested in botany, but I have never performed any research work on natural products. And so I think, the less said by me the better.

However, even an outsider must find Chemistry in Botanical Classification very fascinating. Some kinds of substances like proteins and nucleic acids are found in every living organism but they may be of different kinds. There are types of compounds, whose presence or absence is significant for large domains in the vegetable kingdom; anthocyanins are for example not found in fungi. On the other hand we have the chemical races: apparently homogeneous species may include groups clearly differing in chemistry.

The modern methods of organic chemistry have furnished a wealth of phytochemical information. This information is increasing every day, but as to the interpretation of the data there are surely different opinions. Which chemical characters are really significant and when is a character of real importance?

We are sure that the Symposium volume will give a good picture of the present state of things. We hope that the Symposium will be a powerful stimulant for the future development. In this hope I declare the Symposium opened.

Arne Fredga

General Aspects on Plant Chemotaxonomy

The Chemical Approach to Plant Classification with Special Reference to the Higher Taxa of Magnoliophyta

A. Takhtajan

Komarov Botanical Institute of the USSR Academy of Sciences, Leningrad, USSR

Summary

The chemical approach to plant classification started with micromolecular chemistry and many works are still devoted to the secondary constituents. Like all other methods micromolecular studies have their own advantages and limitations. The basic defect of the micromolecular approach is connected with the fact that as a result of convergent evolution unrelated taxa may have the same metabolic end product and even the same biosynthetic pathway of its origin, especially when the pathway is relatively simple. Consequently, the micromolecular approach to plant classification, though useful at the generic and lower levels is of very limited value at the higher taxonomic levels. It may prove of some significance mainly in helping the systematists to decide between alternative views as to relationships and taxonomic position of problematical taxa.

The amount of evolutionary and taxonomic information highly increases from non-polymeric secondary constituents to proteins and nucleic acids. There is therefore a definite tendency towards the study of these two major groups of polymeric molecules rather than to less informational end products of metabolic processes.

While there are numerous cases of convergence known in the evolution of secondary compounds, it seems highly unlikely that proteins of the same primary structure can be produced by evolutionary convergence. Therefore the comparative studies on the primary structure of proteins provide much more objective criteria of the actual degree of genetic relatedness existing among present day organisms, and far more valuable information on evolutionary classification.

Though proteins are but functional translations of operational units of DNA, at present nucleic acid chemistry seems less reliable as a source of evolutionary insight and taxonomic study. Though the comparative study of the degree of pairing between complementary strands of nucleic acids of different organisms may be eventually very useful, there are so many difficulties both in the technique and interpretation that at present molecular hybridization is of relatively limited value.

It is becoming more and more obvious that no single method can provide fully confident information on evolutionary classification. No one single source of information can substitute for integrative approach to plant classification based on the correlation and synthesis of the evidence taken from all available sources of knowledge. Therefore without wide comparative morphological studies the chemical approach alone cannot introduce any fundamental changes into the present day general system of classification of Magnoliophyta or any other major branches of higher plants.

The construction of the evolutionary classification of organisms is one of the most important aims and objects of the evolutionary biology and the one which is most difficult to attain. Present day evolutionary systematics tries to construct a classification of organisms which is co-ordinated with their cladistic and patristic affinities [27, 28, 47, 48, 64, 78, 79, 94]. All taxa of an evolutionary classification must be "monophyletic" and each taxon can contain only the organisms that show the greatest relationships with each other (a relationship of common ancestry or an ancestor–descendent relationship). "A monophyletic group is one whose most recent common ancestor is cladistically a member of that group" [5].

Evolutionary classification must be based on all available differences and similarities among homologous characters and on the correlation and synthesis of the evidence taken from all obtainable sources of knowledge. Such an evolutionary (phylogenetic) classification of plants, the flowering plants in particular, has not been constructed yet, though much has already been done in this field. In dealing with flowering plants only horizontal relationships are directly involved, which makes their evolutionary classification so difficult. In order to improve the existing systems of classification, which are inevitably only partly evolutionary and in great part still more or less "phenetic", we need larger amounts and more accurate data, and the data should not be limited to comparative morphology only. Other approaches are necessary, including the chemical approach.

Living beings are unique in having a great amount of information on their past history inscribed at any level of their organization— from the molecular level to the level of the organism. Each successive level of organization is a new translation of the basic molecular message. And though each level is important for classification of organisms, at this Symposium we naturally have to deal mainly with the molecular level. We will consider molecules as documents of evolutionary history [113] which can be utilized as important additional data for evolutionary classification of plants.

Chemical data are being used either by recording the presence or absence of various compounds in different taxa, or by comparing structural features and biosynthetic pathways of a common compound or chemically related compounds. The two large classes of compounds which are being used for taxonomic and evolutionary purposes are the high mol. wt and essentially polymeric molecules of primary constituents such as nucleic acids and proteins and relatively low mol. wt non-polymeric metabolic end products such as non-protein amino acids, alkaloids, flavonoids, betalains, glycosides, terpenoids etc. commonly referred to as secondary constituents.

Historically the chemical approach to plant classification started with the secondary constituents. Some rudimentary plant chemistry was utilized for plant classification as early as the first quarter of the last century. In the latter part of the XIX century, when a period of intense development of chemistry of natural products and the elucidation of their structure and biosynthesis had begun, the chemical approach to plant classification received a new impetus. But several decades passed before there was a real start of the chemical approach to plant classification.

Beginning around 1960 a stream of publications started, including a series of books of fundamental importance [3, 7, 11, 49, 57, 59, 72, 99, 100, etc.] which greatly stimulated fur-

ther progress in the chemical approach to systematics and there arose an outburst of new investigations. Now it is an "explosively developing field" [96] and the very organization of this Symposium proves its growing importance.

Secondary Constituents

Many works are still devoted to the secondary constituents. Many of these compounds are widespread among numerous unrelated taxa and therefore have little value as taxonomic markers, but quite a few of them have a more or less restricted distribution and are found in two or more taxa, both related and unrelated. Sometimes they are found only in one taxon. Among the most attractive chemical compounds for taxonomic purposes are alkaloids [11, 60, 61], which can easily be detected and are characterized by a great variety of structures and biosynthetic complexity [10, 63]. The alkaloids of the benzylisoquinoline type provide a good example of the potentialities and limitations of alkaloid chemistry.

Members of the Berberidaceae s.l. contain berberine, palmatine and related alkaloids, which are found not only in the genera *Berberis* and *Mahonia*, but also in *Epimedium*, *Bongardia*, *Caulophyllum*, *Leontice* and *Gymnospermium*, as well as in *Podophyllum* and *Nandina* [59, 82]. Alkaloids of the berberine type occur also in the ·Annonaceae (palmatine in *Enantia*), Hydrastidaceae (berberine and hydrastine), Menispermaceae, Ranunculaceae (*Aquilegia*, *Thalictrum*, *Coptis*, *Zanthorhiza*), as well as in the Papaveraceae and Fumariaceae. These families belong to 3 related orders: Magnoliales, Ranunculales and Papaverales. The Ranunculales and Papaverales belong to the same subclass Ranunculidae, which most probably derived from the Magnoliidae. Thus all these taxa are more or less related and the occurrence of alkaloids of berberine type in their members raises no problems. The Papaverales are rather closely related to the Ranunculales and it is interesting that bicuculine and adlumine, found in the Fumariaceae are very closely related to hydrastine, found in *Hydrastis canadensis* L.

But alkaloids of the berberine group occur also in the Rutaceae and *Evodia glauca* Miq.,

Orixa japonica Thunb., the genus *Phellodendron, Toddalia asiatica* (L.) Lam. and *Zanthoxylum clava-herculis* L. contain berberine [86], which is most unexpected. Many years ago Hallier [56] mainly on phytochemical grounds (alkaloid berberine and polyphenole rutin) derived "more or less directly" the order "Terebinthines", which includes the Rutaceae as the most primitive family, from the Berberidaceae s.l. He greatly overvalued the taxonomic significance of the chemical data. The order Rutales in all probability derived from the Saxifragales s.I. [101, 103] and belongs to the subclass Rosidae, which is only distantly related to the Ranunculidae. The occurrence of berberine, jatrorrizine, magnoflorine, menispermine, palmatine etc. in one species, *Phellodendron amurense* Rupr., is certainly surprising, but as Price [86] points out, in respect to the alkaloids it produces, the family Rutaceae is probably the most versatile of all the families of the higher plants.

Betalains and the Delimination of Caryophyllales

In recent years much attention has been paid to plant pigments called betalains, which are found in all investigated families of the order Caryophyllales except the Bataceae, Molluginaceae and Caryophyllaceae [68, 73, 111, 112]. Certainly the presence of betalains in some families, especially in the Cactaceae and Didiereaceae, is good additional evidence for their inclusion into the order Caryophyllales. The affinities between the Cactaceae and the Aizoaceae, Portulacaceae and Phytolaccaceae [23–25, 55, 56, 110], as well as the affinity of the Didiereaceae with the Cactaceae, Portulacaceae and related families [39, 56, 89, 91, 101, 103] were established on a wide basis of comparative morphology. Chemical data came only as a further support. As regards the Bataceae, Molluginaceae and Caryophyllaceae, which contain no betalains, the situation is different. Whereas the systematic position of the Bataceae within the order Caryophyllales is to a certain extent still open to question and Cronquist [27] prefers to maintain *Batis* as a separate order, both of us will not agree to separate the Molluginaceae and Caryophyllaceae from the betalain-containing group

of families. The Molluginaceae is closely allied to the Phytolaccaceae and Aizoaceae, and is connected to both by way of the intermediate palaeotropical genus *Gisekia*. Many authors do not even recognize the Molluginaceae as a separate family and include its genera into the Aizoaceae [52, 71, 85, 90, 107, 111]. Cronquist [27] derives the Molluginaceae directly from the Phytolaccaceae. The Caryophyllaceae has an analogous origin and is closely related to the Aizoaceae, Molluginaceae and Portulacaceae and most probably derived directly from the Phytolaccaceae [27, 84, 90, 101, 103, 104, 110]. Both the Molluginaceae and Caryophyllaceae are so closely related to the betalain-containing Phytolaccaceae, Aizoaceae and Portulacaceae, that excluding them from the order would require that one ignores all the accumulated evidence from various botanical disciplines.

As regards the Bataceae, which differs from all other caryophyllalean families, including also the Caryophyllaceae, in having sieve-tube plastids with ring-shaped bundles of proteinaceous filaments [6], I would rather keep it within the Caryophyllales than to maintain *Batis* in a separate monotypic order, accepted by Wettstein [110], Lawrence [71], Eckardt [38], Cronquist [27] and Thorne [107], and even less prefer it in a separate subclass Batidae [6]. *Batis* has some important similarity both with the Phytolaccaceae (paracarpous gynoecium in the Microteoideae) and Gyrostemonaceae (very small stipules and pollen morphology). Erdtman [39] came to the conclusion that *Batis* pollen has some morphological characters in common with *Gyrostemon* on the one hand and with the Polygonaceae on the other. According to Kuprianova [70] pollen grains of *Batis* have some characters in common with the Phytolaccaceae, but much more resemble those of the Gyrostemonaceae. Electron micrographs also show that the exine stratification and fine structure in *Batis* are similar to the corresponding features in the Gyrostemonaceae though the Bataceae differ from the Gyrostemonaceae by having compound apertures without opercules [87].

Though the presence of betalains cannot serve for strict delimination of the order Caryophyllales, it is one of the good diagnostic characters for the delimination of the family Caryophyllaceae within the order, as shown by an example of the monotypic North American genus *Geocarpon,* which occupies a somewhat intermediate position between the Aizoceae and Caryophyllaceae and has several times been changed in its position between these two families. Recently Bogle et al. [12] on the grounds of both morphological and phytochemical data have come to the conclusion that *Geocarpon* must be placed in the Caryophyllaceae rather than in the Aizoaceae to which it was originally assigned. Like all other members of the Caryophyllaceae *Geocarpon* contains no betalains. This example shows different diagnostic significance of the same chemical compound on different taxonomic levels.

The Role of Some Other Secondary Metabolites

If we turn now to coumarines, flavonoids, glycosides and other secondary compounds we will see that as taxonomic guides they have more or less the same value as alkaloids. Glycosides have probably less value. There appears to be little direct relationship between the presence of glycosides and the systematic position of genera and higher taxa. Apart from certain special cases, analogous glycosides are found both in dicots and monocots and only a few families contain a single predominant type of glucoside. Moreover, usually several categories are found, not only in the same family, but also in the same genus or species [83]. As Hegnauer [58] and Paris [83] point out, from the taxonomic point of view glycosides appear to be interesting at the level of species or variety. The same is true regarding flavonoids, coumarines and other secondary constituents.

There are, however, a few good examples of possible correlation between the presence of some peculiar types of coumarines or glycosides and systematic position at a generic and higher levels. One of the best examples is a monotypic Japanese genus *Glaucidium* which is usually placed in the family Ranunculaceae, but which differs markedly from the Ranunculaceae in the structure of the gynoecium, ovule and seed, in the mode of dehiscence of the follicles and in its caryology and is therefore separated into a new family Glaucidiaceae [103, 105, 106]. Re-

cently in the rhizome of *Glaucidium* there has been found a new substance, glaupelol, with a coumarine skeleton which is unknown in any other plant [65, 81]. It is an additional evidence supporting the separation of the genus *Glaucidium* into a family Glaucidiaceae. In fact the differences between the Glaucidiaceae and the Ranunculaceae and related families are so remarkable, that Tamura [106] finds it difficult to assume a close relationship between the Glaucidiaceae and any of the ranunculalean families. He even excludes the Glaucidiaceae from the Ranunculales and puts it in the Hypericales, which is not at all acceptable.

As a result of convergent or parallel evolution unrelated taxa may have not only the same metabolic end products but even the same biosynthetic pathways of their origin. As Erdtman [40] points out many complex compounds, even those of quite complex nature, may be formed by relatively simple biosynthetic processes. It has been repeatedly emphasized by a number of workers, that the longer the chain of biosynthetic processes the less probable are the phenomena of their parallel evolution and convergence. Thus, the taxonomic significance of secondary constituents depends on the complexity of their biosynthetic pathways. However, even rather complex biosynthetic pathways may arise independently and repeatedly in unrelated taxa.

Whereas comparative plant morphology is one of the most advanced branches of botany, comparative phytochemistry is still in its early stage, and a detailed classification of secondary products based on their biosynthetic pathways is not worked out yet. But even if we knew the biosynthetic origin of all the existing secondary compounds it would be much more difficult to detect their evolutionary convergence than that of whole organisms. The convergent similarity of organisms is never deep and usually is limited to the organs and tissues which are directly connected with similar environmental factors. Therefore when given taxa are morphologically well studied the convergent features do not create any serious obstacle to evolutionary classification. As Mayr [79] rightly says, "actually the difficulties caused by convergence have been exaggerated". In cases like that of the genus *Batis* (Bataceae), which morphologically is very insufficiently studied, we may not be sure that the similarity with some other halophytic representatives of the Caryophyllales is not a convergent adaptation to saline habitats. Unfortunately there are still many gaps in our knowledge. But with the progress of comparative morphology cases of this kind become rarer. On the other hand, to detect the facts of convergence of both the secondary compounds and their biosynthetic pathways on purely chemical grounds is much more difficult. The frequent occurrence of the same compounds in unrelated taxa proves the frequence of their independent origin along the different lineages of flowering plants. Therefore the micromolecular approach to classification of flowering plants seems to be of limited value. However it may prove of some significance particularly where the systematists do not agree among themselves, as in the case of the genus *Geocarpon* mentioned above, as well as "to point out problems which may not occur to them" [40]. "Its merit would seem rather to lie in helping the taxonomist to decide between alternative views as to relationship that have already been put forward by taxonomists", says Gibbs [46] about the micromolecular approach to plant classification.

Primary Constituents

1. *Proteins*

The amount of evolutionary and taxonomic information increases from secondary constituents to proteins and from proteins to messenger RNA molecules and to genes. Molecules of nucleic acids and proteins have been named semantophoretic molecules or semantides—molecules that carry the information of the genes or a transcript thereof [113]. "In macromolecules of this type", say Zuckerkandl & Pauling, "there is more history in the making and more history preserved than at any other single level of biological integration." Consequently, secondary constituents are not nearly as useful for evolutionary and taxonomic studies as polymeric molecules—proteins and nucleic acids. There is therefore a definite tendency towards these more informational polymeric molecules rather than to the relatively simple and much less informational non-polymeric micromolec-

ular final products of metabolic processes (episemantic molecules in Zuckerkandl & Pauling's terminology).

2. *Amino acid*

In the past two decades we have witnessed a rapidly increasing use of proteins for evolutionary and taxonomic purposes. While numerous cases of convergence are known in the evolution of secondary compounds, it seems unlikely that proteins of the same primary structure can be produced by evolutionary convergence [51]. What is more, at present it is possible to determine whether two groups of similar proteins have a common ancestor or are of independent origin [42].

Each amino acid in a protein corresponds to a triplet of nucleotides in the DNA and therefore it is possible to compute the mutation distances between two homologous proteins, which is defined as the minimal number of nucleotide changes required for one polypeptide to code for another [37, 67]. Consequently protein chemistry can be used as one of the most objective criteria of the actual degrees of genetic affinities existing among extant organisms.

There are several levels at which comparative studies of proteins can be made, including partial or complete amino acid sequence determination, peptide maps, electrophoretic mobility, chromatographic characteristics, serological cross-reactivity ("immunodiffusion systematics"), etc. It is generally agreed, however, that the comparative studies on primary structures of proteins, that is amino acid sequence determination, provide the most valuable information on evolutionary classification [45]. As Dobzhansky [35] points out, "the amino acid sequence in certain proteins of different organisms are far too similar for these likenesses to be ascribed to chance. It is only reasonable to suppose that these proteins are homologous, and the genes coding for them have descended by gradual accumulation of mutational changes from the same gene in a common ancestor."

Unfortunately the technical difficulties in investigating the primary structure of proteins considerably limits the choice of proteins for comparative studies. Therefore at present only a few proteins which have small molecules and are available in adequate amounts can be used for comparative studies, cytochrome c and fer-redoxins [26, 54, 77] among them.

Cytochrome c, a small and easily purified protein which is very suitable for amino acid sequence determination, has recently received much attention [14, 15, 19, 29, 33, 34, 75, 88]. A constant average rate of evolutionary change of cytochrome c over at least 800 million years indicated that it is a suitable protein both for determining the relationships between major taxa of organisms and the geological times of their divergence [15, 18, 29, 31, 43, 44, 88]. The application of the ancestral sequence method [19, 31, 32] strongly suggests that all 3 kingdoms of eucaryotic organisms—animals, fungi and plants—derived from a common ancestor, since almost one third of the residues in all species so far examined are invariant [17, 31]. The results of calculation of the times of divergence of 3 eucaryotic kingdoms using the so called unit evolutionary period [74] for cytochrome c, which is greater than the period for enzymes of comparable size [34], indicate they emerged at about the same time [88]. There are also some indications of approximately equal evolutionary distances between higher animals, fungi and higher plants.

Another conclusion they reach was that the higher plants have a common ancestor of their own [16] and the monocots are a natural (monophyletic) group which arose from dicotyledonous ancestor [16, 19]. They also conclude that the flowering plants originated at least several geological periods before the Cretaceous [88], which is in the agreement with the views of some palaeobotanists. Thus, many years ago Seward [93] wrote on the "antiquity of angiosperms antedating by many millions of years, probably by several geological periods" their first appearance in the Cretaceous deposits. Several later authors postulated a Triassic, Permo-Triassic or even Permo-Carboniferous origin of the flowering plants.

The cytochrome c amino acid sequence can provide some useful information on the relationships of higher taxa. The available evidence however is so scanty and fragmentary, that attempts to derive a "phylogenetic tree" of the flowering plants from the cytochrome data are premature. Boulter's [15] "flower-plant phylogenetic tree" relating the cytochrome c se-quence of 21 species, though interesting in itself, cannot satisfy the systematists. It would be difficult indeed to say anything definite about the relationships of the Caryophyllales and Polygonales without their being compared with the Ranunculales, or to discuss the origin of the Asteraceae without available data on the cytochrome c amino acid sequence of some representatives of the Gentianales and Campanulales. I cannot agree with Stevens et al. [98], that monocots diverged from the main dicotyledonous stock at about the same time as the Asteraceae. We know nothing yet about the primary structure of cytochrome c of such key orders as the Magnoliales, Hamamelidales and Ranunculales, Dilleniales, Violales and Saxifragales, as well as the Nymphaelales, Alismales and Liliales.

Like any other method cytochrome c chemistry has some important limitations. Because of mosaic evolution a count of mutations can be used to construct a dendrogram of cytochrome c evolution rather than a phylogenetic tree. The method tells us the amount of difference between given cytochromes c but not between the corresponding organisms. Two taxa might be very similar or even identical in cytochrome c structure, but very different in all other characters. The evolutionary rate of cytochrome c is very slow [30] and therefore one might expect that the evolutionary distances in high-rate (tachytelic) lines are much greater than the cytochrome c mutation distances. Cytochrome c chemistry can detect only ancient divergence, but not rapid evolutionary divergence of tachytelic groups.

3. Nucleic acids

Along with the rapidly increasing protein approach to plant classification we have witnessed also an equally rapid increase of the use of the nucleic acids. There are different methods of studying nucleic acids for evolutionary and taxonomic purposes. One of them is the determination of the proportions of the four bases in DNA, that is the ratio of $C+G$ to $A+T$. Because of the possibility of the convergent origin of the same $C+G/A+T$ ratio in different taxa this method has a rather limited value [4, 28]. The best possible method would be nucleotide sequence study, but at present there is no sufficient technique yet.

So called "molecular hybridization" is used to gain some insight into the genetic relationships between organisms that cannot be hybridized [4, 7, 8, 9, 13, 36, 50, 69, 76, 80, 92, 95]. Although the degree of pairing, i.e. the degree of relative homology between complementary strands of nucleic acids of different organisms, have been shown to be useful for taxonomic and evolutionary study, the method of molecular hybridization has also some important limitations [1, 2, 16, 20, 41].

One of the greatest difficulties of the molecular hybridization technique is the fact that in evolution inversions, translocations and repetitions of DNA sequences have occurred which may obscure evolutionary relationships when this method is used [16]. The redundancy in genetic material causes especially great difficulties. In the genome of eucaryotic organisms from 15 to 80% of the DNA consists of sequences repeated from 100 to 1 000 000 times and consequently from 85 to 20% of unique sequences, which have much lesser chance to come into contact with homologous ones [22]. Besides, redundant and unique genes within the same genome may play different roles [21, 35, 97]. As Dobzhansky [35] points out, "most structural genes are found among the unique genetic materials, while the redundant fraction of the genetic endowment may consist mainly of regulatory or controlling genetic elements. The highly complex developmental processes in higher organisms may require a minority of relatively stable structural genes outnumbered by more liable controlling genes."

But even if we could overcome all these difficulties both of technique and in interpretation there would remain the question of the degree of correlation between "genetic relatedness" as revealed by the nucleic acid homologies of various organisms and the realized nature of the organism [20]. "We should remember, says Boyden [20], that gene mutations can occur so that the phenotype is drastically altered without affecting the homology of the genes concerned. There must therefore be instances where there is some considerable disagreement between the nature of the organisms as revealed by their genetically determined phenotypes and their

genetic relatedness as measured by DNA homology. There may also be epistatic genes which though active and presumably represented in RNA, will prevent the appearance of the characters determined by the hypostatic and non-allelic genes. We do not see therefore that there is likely to be any simple and direct relatedness as measured by DNA-RNA homology and a classification which should reveal the natures of the 'goods' as finally delivered" (p. 115).

Another great difficulty in the interpretation of the molecular hybridization data is the phenomena of neoteny (paedomorphosis, or juvenilisation), that is the genetically controlled persistance of the earlier stage of ontogeny and thus their evolutionary transformation into the adult stages of later generations. Neoteny may be understood genetically in terms of mutations of genes controlling the speed of ontogenetic development, and involves the reduction in activity of certain genes [53]. In this respect it is important to note that the slowing-down activity of even a single gene would affect the time of action of many genes and, as a consequence, the whole developmental processes [109]. Thus we may conclude that to explain neoteny there is no need to admit any radical genetic change. From the evolutionary point of view the significance of neoteny lies in the attainment of maximum phenotypic effect by means of minimal genetic change [102]. It is interesting that to exactly the same conclusion has independently come Boyden [20], who says: "We might expect that evolutionary changes resulting in paedomorphosis would alter the genetic code little, while affecting the phenotypes markedly" (p. 115). There is therefore no simple proportionality between the degree of pairing and the degree of real genetic affinity between the organisms and consequently no direct conversion in classification. It certainly greatly limits the potentialities of the molecular hybridization method and diminishes its reliability.

It is almost a paradox that though proteins are but functional translations of operational units of DNA, at present nucleic acids seem less reliable as a source of evolutionary and taxonomic information. Though molecular hybridization may be eventually very useful, there are so many difficulties both in the technique and interpretation that at the moment it is of relatively limited value. "DNA homology provides simply one more special test of affinity, comparable, say, to genetic hybridization, and does not offer a definite assessment of 'similarity' of transcendental taxonomic importance", says Heslop-Harrison [62].

It is obvious that natural selection operates on reproducing individuals (phenotypes) and only indirectly on genome coding (genotype). It is the phenotype throughout its development which is exposed to environmental factors and selection pressure. Since phenotypes are the joint resultant of genotype and environment and the complex product of the interaction and collaboration of many genes during the long chains of epigenetic pathways, they contain much more evolutionary information than macromolecules including nucleic acids. The frequently expressed idea of "systematics of genotypes" as an ultimate goal of systematics is therefore basically wrong. I cannot agree with Dobzansky [35] who says: "In the fullness of time, the sequence of the genetic 'letters', the nucleotides, in all genes of all organisms, may imaginably become known. It would then be possible to quantify the similarities and the differences between organisms and to erect a classification of living things, based on precise information about their genetic endowments" (p. 414). I would rather agree with Heslop-Harrison [62], who says, that "the commonly accepted aim of classifying 'genotypes' is not necessarily the most appropriate one for taxonomists, even could it be achieved". It is improbable indeed that genetic codes will ever provide more guides to evolutionary distances and affinities than the phenotypes themselves. Even if we were able to read directly the code of genetic information there would still be the necessity of studying all phenotypic levels of the translation of the basic molecular message.

Conclusions

It is becoming more and more obvious that no single method can provide fully confident information on evolutionary classification. No one single source of information can substitute for the integrative approach to plant classification based on the correlation and synthesis of the evidence from all available sources of knowl-edge. Therefore without the wide basis of comparative morphological studies the chemical approach alone cannot introduce any fundamental changes into the present-day systems of classification of Magnoliophyta or any other major branches of higher plants. The future of systematics lies both in the expansion of approaches and their integration. Biochemical systematics will be integrated into knowledge accumulated by generations of systematists, but as Heslop-Harrison [62] notes, "It is improbable in the extreme that comparative morphology should ever be superseded as the principle source of taxonomic criteria for the higher organisms."

I wish to express my cordial thanks to Dr Janice C. Coffey for reading this manuscript and making some necessary English corrections.

References

1. Alston, R E, Lloydia 1965, 28, 300.
2. — Biochemical systematics, in Evolutionary biology (ed Th Dobzansky, M K Hecht & W C Steere) vol. 1, p. 197. Appleton-Century-Crofts, New York, 1967.
3. Alston, R E & Turner, B L, Biochemical systematics. Prentice-Hall, Englewood Cliffs, N.J., 1963.
4. Antonov, A C, Miroshnichenko, G P & Slursarenko, A G, Adv modern biol (Moscow) 1972, 74, 247. (In Russian.)
5. Ashlock, P K, Syst zool 1971, 20, 63.
6. Behnke, H D & Turner, B L, Taxon 1971, 20, 731.
7. Belozersky, A N & Antonov, A S (ed), DNA structure and the position of organisms in the system. Moscow University Press, Moscow, 1972.
8. Bendich, A J & Bolton, A T, Plant physiol 1967, 42, 959.
9. Bendich, A J & McCarthy, B J, Genetics 1970, 65, 545.
10. Bentley, K W, The alkaloids. Interscience, New York and London, 1957.
11. Blagoveschensky, A V, Biochemical evolution of flowering plants. Nauka, Moscow, 1966. (In Russian.)
12. Bogle, A L, Swain, T, Thomas, R D & Kohn, E D, Taxon 1971, 20, 473.
13. Bolton, E T & McCarthy, B J, Proc natl acad sci US 1962, 48, 1390.
14. Boulter, D, Sci progr 1972, 60, 217.
15. — Pure & appl chem 1973, 34, 539.
16. Boulter, D, Laycock, M V, Ramshaw, J & Thompson, E W, Amino acid sequence studies of plant cytochrome c, in Phytochemical phylogeny (ed J B Harborne) p. 179, Academic Press, London, 1970.
17. Boulter, D, Thompson, E W, Ramshaw, J A & Richardson, M, Nature 1970, 228 (5251), 552.

18. Boulter, D, Laycock, M V, Ramshaw, J A M & Thompson E W, Taxon 1970, 19, 561.
19. Boulter, D, Ramshaw, J A M, Thompson, E W, Richardson, M & Brown, R H, Proc roy soc 1972, B 181, 441.
20. Boyden, A, Bull natl inst sci India (Symp new trends in taxonomy) 1967, 34, 108.
21. Britten, R J & Davidson, E H, Science 1969, 165, 349.
22. Britten, R J & Kohne, D E, Science 1968, 616, 529.
23. Buxbaum, F, Bot Arch (Leipzig) 1944, 45, 190.
24. — Jahrb Schweiz kakt Ges 1948, 2, 3.
25. — Kakt & Sukk 1962, 13, 194.
26. Cammack, R, Hall, R D & Rao, K, New scientist & sci j 1971, 51, 696.
27. Cronquist, A, The evolution and classification of flowering plants. Houghton Mifflin, Boston, Mass., 1968.
28. Crowson, R A, Classification and biology, Heinemann, London, 1970.
29. — J molec evol 1972, 2, 28.
30. Dayhoff, M O, Atlas of protein sequence and structure, vol 4. National Biomedical Research Foundation, Silver Spring, Md, 1969.
31. — Ibid vol. 5, 1970.
32. Dayhoff, M O & Eck, R V, Atlas of protein sequence and structure, vol 2. National Biomedical Research Foundation, Silver Spring, Md, 1966.
33. Dickerson, R E, J mol evol 1971, 1, 26.
34. — Sci Am 1972, 4, 58.
35. Dobzansky, Th, Genetics of the evolutionary process, Columbia University Press, New York and London, 1970.
36. Dutta, S K, Richman, N, Woodward, V & Mandel, M, Genetics 1967, 57, 719.
37. Eck, R V & Dayhoff, M O, Atlas of protein sequence and structure, vol 2. National Biomedical Research Foundation, Silver Spring, Md, 1966.
38. Engler, A, Syllabus der Pflanzenfamilien, 12 Aufl. (ed H Melchior) vol. 2, Angiospermen. Bornträger, Basel, 1964.
39. Erdtman, G, Pollen morphology and plant taxonomy, angiosperms, Almqvist & Wiksell, Stockholm, 1952.
40. Erdtman, H, Some aspects of chemotaxonomy, in Chemical plant taxonomy (ed T Swain) p. 89. Academic Press, London and New York, 1963.
41. Fairbrothers, D E, Chemosystematics, in Modern methods in plant taxonomy (ed V H Heywood) p. 141. Academic Press, London and New York, 1968.
42. Fitch, W M, Syst zool 1970, 19, 99.
43. Fitch, W M & Margoliash, E, Science 1967, 155, 279.
44. — Biochem genet 1967, 1, 65.
45. — The usefulness of amino acid and nucleotide sequences in evolutionary studies, in Evolutionary biology (ed Th Dobzansky, M K Hecht & W C Steere) vol. 4, p. 67. Appleton-Century-Crofts, New York, 1970.
46. Gibbs, R D, Lloydia 1965, 28, 279.
47. Ghiselin, M T, The triumph of the Darwinian method, Berkeley, 1969.
48. — The principles and concepts of systematic biology, in Systematic biology (ed C D Sibley) p. 45. Proc natl acad sci US 1969.
49. Florkin, M, A molecular approach to phylogeny. Elsevier, Amsterdam, 1966.
50. Gibbson, I, Evolution 1968, 22, 398.
51. Goodman, M, Bull natl inst sci India (Symposium on new trends in taxonomy) 1967, 34, 119.
52. Gundersen, A, Families of dicotyledons. Chronica Botanica, Waltham, Mass., 1950.
53. Haldane, J B S, Am naturalist 1932, 66, 5.
54. Hall, D O, Cammack, R & Rao, K K, Nature (Lond.) 1971, 233, 136.
55. Hallier, H, New phytol 1905, 4, 151.
56. Hallier, H, Arch néerl sci 1912. Ser. 2, 1, 146.
57. Hawkes, J G (ed), Chemotaxonomy and serotaxonomy. Academic Press, London and New York, 1968.
58. Hegnauer, R, Pharm weekbl 1957, 92, 860.
59. — Chemataxonomie der Pflanzen. Birkhäuser, Basel, 1962–1966.
60. — The taxonomic significance of alkaloids, in Chemical plant taxonomy (ed T Swain) p. 389. Academic Press, London and New York, 1963.
61. Hegnauer, R, Comparative phytochemistry of alkaloids, in Comparative phytochemistry (ed T Swain) p. 211. Academic Press, New York, 1966.
62. Heslop-Harrison, J, Chairman's summing-up, in Chemotaxonomy and serotaxonomy (ed J G Hawkes) p. 279. Academic Press, London and New York, 1968.
63. Heywood, V H, Phytochemistry and taxonomy, in Comparative phytochemistry (ed T Swain) p. 1. Academic Press, London, 1966.
64. Hull, D L, Ann rev ecol syst 1970, 1, 19.
65. Irie, H, Uyea, S, Yamamoto, K & Kinoshita, K, Chem commun 1967, 547.
66. Jay, M, Taxon 1968, 17, 136.
67. Jukes, T H, Molecules and evolution. Columbia University Press, New York, 1966.
68. Kimler, L, Mears, J, Mabry, T J & Rösler, H, Taxon 1970, 19, 875.
69. Kohne, D E, Taxonomic applications of DNA hybridization techniques, in Chemotaxonomy and serotaxonomy (ed J G Hawkes) p. 117. Academic Press, London and New York, 1968.
70. Kuprianova, L A, The palynology of the Amentiferae. Nauka, Moscow-Leningrad, 1965.
71. Lawrence, G H, Taxonomy of vascular plants. Macmillan, New York, 1951.
72. Leone, C A (ed), Taxonomic biochemistry and serology. Ronald Press, New York, 1964.
73. Mabry, T J, The betacyanins and betaxanthins, in Comparative phytochemistry (ed T Swain) p. 231. Academic Press, London and New York, 1966.

74. Margoliash, E & Fitch, W M, Ann NY acad sci 1968, 151, 359.
75. Margoliash, E, Fitch, W M & Dickerson, R E, Brookhaven symp biol 1968, 21, 259.
76. Marinova, E I, Antonov, A S & Belozersky, A N, Dokl USSR acad sci 1969, 184, 483.
77. Matsubara, H, Jukes, T H & Cantor, C R, Brookhaven symp biol 1968, 21, 201.
78. Mayr, E, Syst zool 1965, 14, 73.
79. Mayr, E, Principles of systematic zoology. McGraw-Hill, New York, 1969.
80. McCarthy, B J & Bolton, E T, Proc natl acad sci US 1963, 50, 156.
81. Murakami, T, Mikami, Y & Itokawa, H, Chem pharm bull 1967, 15, 1817.
82. Panov, P P, Panova, L N & Mollov, N M, Compt rend acad bulgare sci 1971. 25, 55.
83. Paris, R, The distribution of plant glycosides, in Chemical plant taxonomy, p. 337. Academic Press, London and New York, 1963.
84. Pax, F, Bot Jahrb 1927, 61, 223.
85. Pax, J & Hoffmann, K, Aizoaceae, in Die naturlichen Pflanzenfamilien (ed A Engler & K Prantl) vol. 16c, p. 179. Leipzig, 1934.
86. Price, J R, The distribution of alkaloids in the Rutacease, in Chemical plant taxonomy (ed T Swain) p. 429. Academic Press, London and New York, 1963.
87. Prijanto, B, Batidateae, in World pollen flora (ed G Erdtman) 1970, 3, 1.
88. Ramshaw, J A M, Richardson, D L, Meatyard, B T, Brown, R H, Richardson, M R, Thompson, E W & Boulter, D, New phytol 1972, 71, 773.
89. Rauch, W & Reznik, H, Bot Jahrb 1961, 81, 94.
90. Rendle, A B, The classification of flowering plants. Cambridge, 1938.
91. Schölch, H F, Ber deut bot ges 1963, 76, 49.
92. Searcy, D G, Evolution 1970, 24, 207.
93. Seward, A C, Plant life through the age, 2nd edn. Cambridge, 1933.
94. Simpson, G G, Principles of animal taxonomy. Columbia University Press, New York and London, 1961.
95. Slusarenko, A G, The primary structure of DNA as taxonomic character in higher plants, in DNA structure and systematic position of organisms (ed A N Belozersky & A S Antonow) p. 196. Moscow University Press, Moscow, 1972. (In Russian.)
96. Stafleu, F A, Taxon 1969, 18, 318.
97. Stebbins, G L, The basis of progressive evolution. University North Carolina Press, Chapel Hill, N.C., 1969.
98. Stevens, F C, Glazer, A N & Smith, E L, J biol chem 1967, 242, 2764.
99. Swain, T (ed), Chemical plant taxonomy. Academic Press, London and New York, 1963.
100. — Comparative phytochemistry. Academic Press, London and New York, 1966.
101. Takhtajan, A, Die Evolution der Angiospermen. Gustav Fischer Verlag, Jena, 1959.
102. — Foundations of the evolutionary morphology of angiosperms. Nauka, Moscow-Leningrad, 1964. (In Russian.)
103. — A system and phylogeny of the flowering plants. Nauka, Moscow-Leningrad, 1966. (In Russian.)
104. — Flowering plants, origin and dispersal. Oliver & Boyd, Edinburgh, 1969.
105. Tamura, M, Sci repts, Osaka University 1963, 11, 115.
106. Tamura, M, Bot mag Tokyo 1972, 85, 29.
107. Thorne, R F, Aliso 1968, 6(4), 57.
108. Turner, B L, Taxon 1969, 18, 134.
109. Wardlow, D W, Phylogeny and morphogenesis. Macmillan, London, 1952.
110. Wettstein, R, Handbuch der systematischen Botanik. Franz Deuticke, Leipzig und Wien, 1935.
111. Willis, J C, A dictionary of the flowering plants and ferns, 7th edn (revised by H K Airy Shaw). The University Press, Cambridge, 1966.
112. Wohlpart, A & Mabry, T J, Taxon 1968, 17, 148.
113. Zuckerkandl, E & Pauling, L, J theoret biol 1965, 8, 357.

Discussion

Ourisson: (1) Chemists will agree that chemical characters are not better than any other ones; but they are different, independent (to a large extent) of morphological information. Therefore, they are useful. And any chemist will agree that "no single method can provide fully confident information on evolutionary classification".

(2) It may be true that convergent morphology causes no real problem, but I do not agree that convergence in chemical characters is more difficult to spot. There are of course cases where this may be true (identical compounds formed in related taxa by different biosyntheses would be one such case). But even chemists will not be fooled by the isolation of identical sesquiterpenes in Hepaticae and in Compositae, for example.

(3) Most chemists will agree with Professor Takhtajan when he concludes that proteins or nucleic acids are, after all, not "better" characters than micromolecular metabolites. The last ones are after all only reflections of the first ones.

(4) If one agrees with Professor Takhtajan in believing that taxonomy must take a series of characters into account, then it would be very important to use this agreement to solve important problems. We should use this sym-

posium to have some of us—botanists—select problems of general interest, and others—chemists—agree to examine them by a variety of methods.

Turner: I, for one, would be willing to state that at least a few chemical characters are superior to morphological characters for phyletic purposes. For example, cytochrome c is a much more useful character for determining suprafamilial relationships among angiosperms than say trichomes, ovary construction, etc. In fact, there is no comparable structure among flowering plants to permit any kind of objective insight into probable cladistic assemblages, which Prof. Takhtajan so eloquently pleads for as a systematic objective.

Now to his remarks regarding our treatment of the Caryophyllales: We do not believe that he has any better objective insights into the *cladistic* relationships among this order than do others. We do not disagree with the position accorded the Caryophyllaceae and Molluginaceae, merely the rank accorded these two families, for it appears to us that the betalains arose in the Chenopodiales prior to the development of anthocyanins in the angiosperms generally. Hence, we prefer to accept the Caryophyllales as composed of only two families—reflecting its parallel but subsequent development out of a yet more ancestral line which has persisted as the order Chenopodiales.

Cronquist: I think Professor Takhtajan's point about the greater difficulty of recognizing chemical than morphological convergence may not have been fully understood by all. If there is convergence in some one morphological feature or correlated set of features, this can be discovered through a comparison of other morphological features. The single term morphology is here used to cover a broad range of characters; it is not one character, but many. But if one had to discover convergence by a study of only the character in which there had been convergence, one would probably fail, and the convergence would go undetected. No such broad range of chemical characters has been studied. We know about one chemical character in one set of taxa, and some other character in some other set of taxa, and so on. Thus, if there is convergence in a chemical character,

it may well be difficult to detect from the chemistry alone without reference to the morphology.

Heywood: I cannot agree with Mayr [78, 79] that the difficulties caused by convergence have been exaggerated. It would be truer to say that the extent of morphological convergence has not yet been appreciated—sufficient studies have not yet been made. If a single character shows convergence in different groups it is likely to be easily detected, but difficulties arise when the genetic relatedness of the groups showing convergence is close. It may well then be difficult to detect as it will occur in many characters. Examples can be sought in the Gramineae, for example, and in other "natural" families, and indeed throughout the plant kingdom and fungi (e.g. Basidiomycetes). I disagree with Takhtajan's statement that all taxa of an evolutionary classification must be monophyletic. For certain kinds of relationships, it may be more useful or as useful to work with polyphyletic (anagenetic) grades of evolution rather than with monophyletic clades. Within the content of a traditional or conventional definition monophyletic groups are desirable but not necessarily so in other contexts, where a more flexible approach may be needed.

Cronquist: The question of the monophyletic or polyphyletic nature of taxa deserves further consideration. I have written about this in another place, so I will comment only briefly here.

In my experience, so far as I can interpret the evidence, few if any taxa, other than some amphiploid species, are truly monophyletic in the sense that there was an original member of the taxon, properly included in the taxon, from which all other members are descended. Instead we find that before you can trace all the members of a taxon to a common ancestor, you are just outside the limits of the group.

This is true in animals as well as in plants. There is a very good fossil record of mammals, and as G. G. Simpson pointed out about 1945, no matter how you define the mammals, there was never an original kind of a mammal that was ancestral to all other mammals. Before you get to a common ancestor, you have not a mammal but a reptile. Members of one group of reptiles (not just any reptile, but a group of closely related reptiles) underwent a series of parallel evolutionary changes, leading to the formation of mammals. In the same paper Simpson propounded the rule-of-thumb that if all the members of a taxon of a particular rank are derived from a taxon of lesser rank, that is sufficient monophylesis for taxonomic purposes. In general, I agree, although I would not want to take the rule literally in all cases.

Just as there was never an original mammal, so there was probably never an original angiosperm. Before the angiosperms can be traced back to a single common ancestor, you will have not an angiosperm but some kind of gymnosperm.

Taxonomy should be in general consistent with phylogenetic relationships, but anyone seeking absolute monophylesis in taxonomic groups is chasing a will-o-the-wisp.

Bu'Lock: I think it is important to avoid what may turn out to be misleading analogies between the data accumulated by botanists and those from biosynthetic organic chemists. For example, the resemblance between cases of "convergence" in botanical evolution and the so-called convergence of biosynthetic pathways is often very superficial, and we may even doubt whether the latter has any real significance for phylogeny. Again, when phytochemists construct a "tree" of biosynthetic reactions, its resemblance to a phylogenetic construct may be wholly superficial and we should be aware of the confusions this may lead to. In particular, given a biosynthetic "tree", there is no reason to suppose that the genes for what we can see in biochemical terms as basic reaction steps have any different status than those for peripheral reactions in the diversification phase of biosynthesis. However, the consequences of a single-gene defect in a "basic" step are the absence of a whole class of compounds whereas for a peripheral defect only one or two may be affected. This is one justification—additional to the experimental reason usually given—for disregarding isolated negative data in making chemotaxonomic groupings.

Birch: An added dimension in chemical considerations of small molecules is that the chem-

ist can express some views on the probability of occurrence of a biochemical process. This emphasises the importance of changes of synthetic steps, rather than the structures of the products as such. The isoquinoline alkaloids, for example, arise so readily from natural amino acids, one at least of tyrosine type, that it is merely surprising that they do not occur more frequently. The only distinguishing character in berberine mentioned, which might be of some significance because of biosynthetic improbability, is the "berberine-bridge" carbon inserted by a rather unusual oxidation process.

Merxmüller: There is a certain danger in the product of this remarkable Americo-Russian collaboration, the Takhtajan-Cronquist System, namely that everybody considers it (consciously or even subconsciously) the nearest approach to the truth at the present. So, chemical characters are mostly judged only with respect of fitting or not-fitting into this system. In this context the Cactaceae, e.g., are now said to fit very well into the Centrospermae, chemically and morphologically—in spite of the fact that still nobody can explain the perianth structure of Cactaceae in terms of Centrospermae. On the other hand, one says that the chemical similarity of Rutaceae to the Ranunculales cannot mean anything (because the Rutaceae "belong" to the Rosidae)—in spite of the fact that some members of Rutaceae like *Boeninghausenia* show very strange morphological similarities to the Ranunculales. I am convinced that a surprising chemical evidence should always be a starting point for morphological re-investigations, especially if it is not in agreement with the above-named systems. Another point to be stressed is the exaggerated generalization which at least sometimes is made from extremely scanty chemical evidence. So everybody now is convinced that the Molluginaceae possess anthocyanins; as far as I can see all this information goes back to an investigation in one species of *Hypertelis* made by Beck, Wagner and myself some 10 years ago. As long as chemical data are so few in contrast to the enormous amount of morphological ones, the preponderance of morphology explains itself quite simply.

Chemical Plant Taxonomy: A Generalist's View of a Promising Specialty

A. Cronquist

The New York Botanical Garden, Bronx, N.Y. 10458, USA

Summary

Morphology has traditionally been the major source of information on which taxonomic schemes are based. This is true both because of the relative ease of getting the information, and because *Homo sapiens* is psychologically adapted to the use of visual pattern data. Morphological characters of course have a chemical foundation; in a sense all characters are chemical. What we call chemical characters in taxonomy are features that do not have an obvious morphological expression. Chemical characters may on the average be a little closer to the gene and a little less subject to environmental modification than morphological ones, but the difference is only one of degree.

The taxonomic system is supposed to reflect the totality of similarities and differences among organisms. Therefore good taxonomists try to use all kinds of data, including chemical data, insofar as is reasonably possible. The relative difficulty of acquiring chemical information does not make it inherently any more or less important than information on classical morphological features.

The betalains provide a good example of both the use and the misuse of chemical characters. There is now a growing consensus that all betalain-containing families belong in the same order. The Cactaceae and Didiereaceae, whose affinities have previously been controversial, are now seen to fit well with the other betalain-containing families. The proposed exclusion of the non-betalain families Caryophyllaceae and Molluginaceae from the group is another matter. The weight of the evidence, including the newly available information on sieve-tube plastids, is that these two families belong where they have traditionally been put, with the other families of the traditional order Centrospermae. Here we have an example of the principle that taxa should be defined on the basis of all the available evidence, rather than by assigning heavy a priori weight to a particular character; and also the principle that for sound genetic reasons the presence of a character is more likely to be taxonomically important than its absence.

There is not necessarily any relationship between the biological importance of a character and its taxonomic importance. The C_4 pathway in photosynthesis, for example, as opposed to the C_3 pathway, must surely be highly important to the plant. At the same time it appears to be of relatively little taxonomic importance, varying sometimes even from one species to another in the same genus.

Serological reactions have been touted as providing an independent check on the morphologically based taxonomic system, and they do have some value. Serological data, combined with information on the kinds of alkaloids and other chemical constituents have played a strong role in the now generally accepted exclusion of *Paeonia* from the Ranunculaceae, and in the dissolution of the traditional order Rhoeadales into two only rather distantly related orders, the Papaverales and Capparales. The usefulness of serological reactions is limited, however, by uncertainty as to what is actually being measured, by the occurrence of "antisystematic" reactions (commonly associated with lectins), and especially by the fact that they exist only as one to one comparisons rather than as concrete data. The number of tests required to check a large number of items reciprocally against each other is astronomical, and so we must make a subjective choice of what to check against what.

A kind of chemical character that shows great promise is the amino-acid sequence of specific proteins. The structure of cytochrome *c* in various kinds of organisms is now being intensively investigated. We do not yet have enough data to do more than show that this approach can be expected to correlate reasonably well with the more traditional ones. Hopefully, the amino-acid sequences may eventually provide an objectively measurable body of data that are truly independent of the more traditional taxonomic characters.

Some Taxonomic Philosophy

Morphology has traditionally been the major source of information on which taxonomic schemes are based. It provides a wealth of relatively easily observed differences among individuals and groups of individuals, and lends itself readily to the use of pattern as well as point data. Furthermore, *Homo sapiens* in general characteristically uses his eyes as the major tool for gathering data; his ears come second, and other tools such as his fingers, nose, and tongue are far behind in this regard. A blind man is under a much greater handicap in learning about the nature of the world, and getting along in it, than a deaf man, or one who cannot smell. Even our ears are used primarily to receive information from other humans, rather than to gather data about the world directly. Our ears may indeed provide the easiest way to distinguish between certain species of swan (*Cygnus*) or meadowlark (*Sturnella*), but obviously this sort of information is not available to *botanical* taxonomists. The use of hand lenses, light microscopes of various power, and now various sorts of electron microscopes are natural extensions of the use of the naked eye —extensions made possible by the progressive growth of technology, but not fundamentally different, psychologically, from the familiar use of the naked eye at the macroscopic level.

All morphological characters of course have a chemical foundation. Life itself is essentially a chemical phenomenon. The leaves, flowers, fruits, or whatever morphological structures we observe are the results of complex chemical interactions. Likewise the physiological, ecological, and any other kind of characters we might wish to consider are eventually chemical. To a very large extent, differences in morphological characters depend on differing amounts and proportions of vital chemicals at various times, rather than on simple presence or absence.

How, then, do what we call chemical characters in taxonomy differ from morphological characters, which as we have noted are also essentially chemical? In two ways: (1) the chemical characters do not have an obvious morphological expression; (2) chemical characters commonly reflect the presence or absence of specific chemicals or classes of chemicals, or very gross differences in their quantity.

As might be expected, there is in fact no sharp line between chemical and morphological characters. When 19-century botanists first distinguished the green, blue-green, red, and brown algae from each other they were thinking primarily in terms of the colors they could see, but it turns out that these color differences are best understood in terms of the chemistry of the pigments concerned.

One might also try to distinguish chemical characters from morphological ones by saying that the chemical characters are simpler, since the morphological characters depend on interactions among many chemicals. There is a degree of truth in this, but it is indeed only a matter of degree. Each of the chemicals that might be used as a taxonomic character is at or near the end of a long biosynthetic chain that stretches all the way back to the gene and is influenced by various chemical processes occurring in the cell.

Along with the idea that chemical characters are simpler than morphological ones might go the idea that chemical characters are less subject to direct modification by the environment. Here again there is some substance to the thought, but not so much as might at first seem. It appears that no amount of environmental manipulation will induce anything outside the order Caryophyllales [8, 28] to produce betalains, but in many or most species of angiosperms a shortage of sugar will cut off the production of anthocyanins. When chemical characters are considered in terms of amounts and proportions, they are often subject to environmental control, as for example in the production of HCN and cyanogenetic substances by certain grasses, and tetrahydrocannabinol by *Cannabis*. Among the kinds of plants that accumulate aluminium or selenium, this accumulation is obligate in some species and facultative in others. The relationship of chemical characters to the natural environment may not be so very different from that of morphological characters, in which the size and sometimes the shape of various organs are much more subject to environmental control than is their basic structure.

Taxonomists are fond of saying that the taxonomic system is supposed to reflect the totality of the similarities and differences among organisms, and that individual morphological characters are merely handy guides toward the general goal. I say it myself. It is implicit in this concept that the more information we can use in producing the taxonomic system, the less the system is subject to the vagaries of sampling, and the more nearly we can approach the desired end.

Here we have the basic rationale for the use of chemical characters in taxonomy. The taxonomist, ever avid for new data to use in refining or if necessary rebuilding his system, is happy to make use of chemical characters as they become available. At least good taxonomists are; we shall consign the others to outer darkness.

There are of course certain problems in the application of chemical data to taxonomy. The greatest problem is that the data are relatively hard to get. Even when the necessary laboratory procedures are relatively simple, they still take much longer than simple observation with the naked eye or a low-power lens, and a different set of specialized and often expensive equipment is necessary for each of many different kinds of chemical approaches. Furthermore, the chemical analysis tends to use up the material in the process of getting the information. The world's herbaria, which are great repositories of more or less permanently available morphological information, cannot effectively be used as a major source of supply of samples for chemical analysis.

The natural psychological result of the difficulty of getting the chemical information is to consider that it must be inherently more important than classical morphological features. We all tend to value things by their scarcity or cost. Who gave a thought to the importance of clean air before industrial pollution became a problem? Through long and often sad experience taxonomists have learned that all morphological characters are subject to failure, and must be evaluated in each instance in terms of their correlation with other characters. Chemical characters are in fact no different in this regard, although not everybody has yet recognized it. Twenty years ago it was thought that the so-called anthochlor pigments (chalcones and aurones collectively) were largely restricted to the subtribe Coreopsidinae in the tribe Heliantheae of the family Asteraceae, and that the presence of anthochlors provided a chemical marker for this subtribe. Now we know that they are much more widespread both within and without the Asteraceae, and their taxonomic significance has accordingly been considerably downgraded.

Another problem with chemical data is that they often do not lend themselves to pattern recognition. Taxonomists necessarily deal to a large extent in pattern data, which can be perceived as a whole by the naked eye [27]. The parts that make up the pattern can be individually analyzed and recorded, but it is the whole pattern that is perceived and remembered. That is the way our minds work, not just because we are taxonomists, but because we are humans. Our minds are much more receptive to data that fit together to make a pattern, especially a visual pattern, than to the same number of uncorrelated point data. If one has to remember which ones of a list of 50 or 100 chemicals occur in which species of plants, one may have to resort to a computer to keep track of them. It is no accident that some bacteriologists, who have to deal with such point data as response to various media and environmental conditions, have become early backers of computer taxonomy.

Visual patterns can be obtained by paper chromatography, and it is relatively easy for the eye to recognize similarities and differences in the patterns. When the chemicals making up the spots are characterized, the data become much more significant, but also much harder to keep in mind.

Betalains

The betalains provide a good example of both the usefulness and the limitations of chemical characters. A relatively large amount of data has been obtained, and much has been written about them. Unfortunately, the taxonomic evaluation of the data has in my opinion often had the flavor of advocacy rather than dispassionate analysis.

At least as long ago as 1945 [12], more than 20 years before the term betalain itself was coined, it was recognized that this group of pigments is widespread in the Centrospermae, and unknown in families that have not been considered at least by some botanists to be allied to the Centrospermae. (I use the familiar but irregular name Centrospermae here, instead of the nomenclaturally preferable name Caryophyllales, for reasons which will become apparent.) Subsequent studies by various authors, notably Mabry and his associates [18–23, 30] have clarified our knowledge of the nature of the betalains, and expanded our information about their taxonomic distribution.

Among the items we should note at this point are that the betalains are chemically very different from the flavonoid pigments, and that in each of these groups we can recognize a blue to purple to crimson series and a yellow to orange to scarlet series of colors. The blue to purple to crimson flavonoids are called anthocyanins, and the yellow to orange to scarlet ones are called anthoxanthins, although this latter term has not been much used in recent years. The visually comparable betalain groups are called betacyanins and betaxanthins, in deliberate linguistic parallel to anthocyanins and anthoxanthins. The color difference has a chemical marker in each case: A double bond 0 attached to the central ring distinguishes anthoxanthins from anthocyanins [1]; a somewhat more complex difference at one end of the molecule distinguishes betaxanthins from betacyanins [30].

There is a seeming antipathy between anthocyanins and all betalains. The two have never been found to occur together; the single recorded exception proves, on reinvestigation, to be erroneous [18]. On the other hand, there is no such antipathy between anthoxanthins and at least the betaxanthin group of betalains, which do frequently occur together [1, 18].

Mabry et al. proposed in 1963 [19] to define the order Centrospermae on the presence of betacyanins. On this basis they added two families (Cactaceae and Didiereaceae) that had not generally been included in the group in the past, and excluded the Caryophyllaceae, which had usually been considered central to the order. In this paper I find no mention of the Molluginaceae, which had generally been included in the Centrospermae either as a separate family or as a part of the family Aizoaceae. In 1962, however, Beck et al. [3] had reported anthocyanin rather than betacyanin in the genus *Hypertelis* of the Molluginaceae, a fact noted by Eckardt [10] in his treatment of the Centrospermae for the 12th edition of the Engler Syllabus in 1964. By 1966 Mabry [21] was routinely excluding the Molluginaceae from the Centrospermae, on the basis of the pigments. Mabry and his associates, such as Turner, have continued to define the Centrospermae essentially on the basis of the present of betalains [vide 5].

Beginning with the 1963 paper [19], the Mabry & Turner group have maintained that the presence of betacyanins (more recently, betalains) and the absence of anthocyanins in the Centrospermae as defined by them is of inherently great taxonomic significance, because of the completely different nature of the two sets of pigments and their mutual exclusiveness.

The usefulness of betalains as a marker of affinity has become widely recognized especially in the last decade. For example, the systems of Takhtajan [28], Thorne [29] and Cronquist [8] all include all the betalain-containing families within a single order. This developing consensus depends partly on Mabry's argument that a correlated set of genes is necessary for the production of betalains, and partly on the fact that none of the families involved (including the Cactaceae and Didiereaceae) has any very obvious major sets of allies outside the order. As more species in diverse families are examined, without the finding of betalains in any additional families, their importance in the taxonomic system continues to rise. Furthermore, assigning a high significance to the presence of betalains resolves some long-standing problems without creating any new ones. Thus the betalains provide the additional evidence that was needed to clarify the taxonomic position of the Cactaceae and Didiereaceae.

The exclusion of the non-betalain families Caryophyllaceae and Molluginaceae from the Centrospermae is another matter. Takhtajan [28], Thorne [29], and Cronquist [8] all include both the Caryophyllaceae and the Molluginaceae in the same order with the betalain families. Thorne also includes the non-betalain family Polygonaceae in this order, and Takhtajan includes the monotypic non-betalain family Bataceae. Each of these families is by the other two authors assigned to a monotypic order adjacent to the Centrospermae (under whatever name). The position of the Polygonaceae and Bataceae is not essential to the present discussion, and is mentioned only to avoid the possible appearance of too carefully choosing the information to present.

I cannot speak for other authors in this regard, but my reasons for including both betalainic and anthocyanic families in the same order are several.

First, as Ownbey & Aase [25] have pointed out, the presence of a character is in general likely to be more important than its absence. It has been amply demonstrated that individual phenotypic characters are commonly governed by a complex system of genes, rather than by a single gene unaffected by others. The appearance of a particular character in the phenotype requires the whole genetic system, and if two individuals or allied taxa share the same structure (or complex chemical substance) then they probably also share the same genetic system governing its development. If any one of the essential genes in the system is lost, then the structure or compound fails to develop. Absence of a character may therefore reflect the absence of any part or all of the necessary set of genes; in closely related groups it may reflect the absence of a single gene.

Ownbey's principle is of considerable importance with respect to the absence of anthocyanin from betalain-containing families. As Alston [1] and subsequently Kimler et al. [18] have noted, anthoxanthins and betaxanthins frequently occur together. Furthermore, throughout the Centrospermae these anthoxanthins very commonly include flavonols [13, 15]. According to Grisebach's scheme of the biogenesis of flavonoids [13] both flavonols and anthocyanins are derived directly from dihydroflavonol, which is a compound well along in the flavonoid biosynthetic chain. Thus the Centrospermae in general lack only the terminal link in the biosynthetic chain for the production of anthocyanin. This could easily be a matter of a single gene.

Second, although in this case it is a matter of minor importance, we have no assurance that further investigation will not disclose the coexistence of anthocyanin and betalain in the same species. Since anthoxanthins and betaxanthins occur together, and since anthocyanins are chemically so similar to anthoxanthins, a complete mutual exclusion seems inherently unlikely. According to Wohlpart & Mabry [30] only about 85 genera and 200 species of some 650 genera and 8 000 species in the Centrospermae (as defined by them) had been examined for these pigments by 1968, and today

the number has grown to only about 100 genera and 250 species (Turner, personal communication). If anthocyanins and betalains commonly occurred together we would know it by now, but at this stage we should not feel confident about their complete mutual exclusion. If the gynoecium of only 15% of the genera of Rubiaceae had been examined, we might very well be unaware that *Gaertnera* and *Pagamea* have a superior instead of inferior ovary. Given the concomitant circumstances, however, even if the complete mutual exclusion of anthocyanins and betalains were eventually demonstrated I would consider it merely an interesting anomaly rather than a matter of major taxonomic importance.

Third, and most important, exclusion of the Caryophyllaceae and Molluginaceae from the Centrospermae goes contrary to the great bulk of the evidence. The close relationship of these several families has been so obvious to so many taxonomists for so long that no single newly discovered difference can effectively dissociate them. There is a set of embryological characters that unites the bulk of these families and separates them from all other orders. Furthermore, about half the species of the group (including the Caryophyllaceae) characteristically have free-central or basal placentation in a compound ovary, an otherwise rather uncommon character. At least 8 of the 11 families of the order, including the Caryophyllaceae, sometimes have anomalous secondary thickening of the stem, and in 3 of the families this feature is so consistent as to be almost a family character. Other features that tend to unite the traditional Centrospermae could also be mentioned, but I will content myself with a striking new one.

All families of the Centrospermae, including the Caryophyllaceae and Molluginaceae, have a characteristic type of sieve-tube plastid that is completely unknown in other families of angiosperms [4]. This feature has led Behnke & Turner [5] to beat a partial retreat with regard to the taxonomic significance of betalains as opposed to anthocyanins. They now divide the small subclass Caryophyllidae in the sense of Takhtajan [28] and Cronquist [8] into 3 smaller subclasses. Their more narrowly defined subclass Caryophyllidae, characterized by the special type of sieve-tube plastid, contains only two orders. Of these two orders, the Caryophyllales (with anthocyanin but without betalain) contain only the families Caryophyllaceae and Molluginaceae; and the Chenopodiales (with betalains but without anthocyanin) contain all the remaining families of the Caryophyllales as defined by Cronquist [8]. In this arrangement the affinity of the Caryophyllaceae and Molluginaceae to the betalain-containing families is in effect admitted, even though the assignment to different orders is maintained.

The Behnke & Turner proposal of 1971 [5] is in some respects an improvement ·over the Mabry, Taylor & Turner proposal of 1963 [19], but it is still unsatisfactory. Both proposals reflect a reliance on one character taxonomy, which is scientifically about as dependable as a one mouse experiment.

Good taxonomists have known for more than 200 years that taxonomic characters acquire their importance through their correlation with other characters, rather than having a fixed, inherent, a priori importance. I have expounded on this matter at some length in another place [8], so I will speak about it only briefly here. I will point out, however, that if the special type of sieve-tube plastid that characterizes the Centrospermae (i.e. the Caryophyllales as I have previously defined them) occurred helter-skelter in a wide range of families and orders, none of us would be likely to consider it of much taxonomic importance. Likewise the betalains. Taxonomic decisions are properly based on the sum of the available evidence, and a character is only as important as it proves to be in marking groups that have been perceived in this way. Linnaeus' maxim, "Scias characterem non constituere genus, sed genus characterem" (Philosophia Botanica), still holds, at all taxonomic levels.

To recapitulate: The betalains have been very useful to taxonomists in helping to show the relationship of the Cactaceae and Didiereaceae to the traditional Centrospermae, but efforts to use the presence or absence of betalains as the sole character to define a major taxonomic group have done taxonomy a disservice.

C₄ Photosynthesis

Another general principle that taxonomists have

had to learn, and are continually having to relearn, is that there is not necessarily any relationship between the biological importance of a character and its taxonomic importance. A recent, very striking example of this principle is provided by C_4 photosynthesis. I think no one will deny that C_4 photosynthesis, with its associated features of low photorespiration, high optimum temperature, and high tolerance to dissolved intracellular oxygen, is of great importance to the plant. The list of families in which C_4 photosynthesis is known includes Poaceae, Cyperaceae, Aizoaceae, Amaranthaceae, Chenopodiaceae, Nyctaginaceae, Portulacaceae, Euphorbiaceae, Zygophyllaceae, and Asteraceae [9]. It is easy to see that 5 of the 10 families on the list belong to the Caryophyllales (sensu meo), and that two others, the Cyperaceae and Gramineae, have often been considered closely related inter se. The other 3 families, Asteraceae, Euphorbiaceae, and Zygophyllaceae, belong to 3 different, additional orders in anybody's system. One might be tempted to say at least that this feature has originated only a limited number of times, and thus might prove to have considerable, even if limited, taxonomic importance. Unfortunately for this thought, the photosynthetic pattern is not consistent within a family or even always within a genus. Some species of *Atriplex* follow the C_4 pathway, others the taxonomically more widely distributed C_3 pathway [6]. Furthermore, the list of families in which C_4 photosynthesis is known to occur will almost certainly be expanded as more studies are made. Fortunately, no one has yet attempted a major reorganization of the taxonomic system of higher plants on the basis of photosynthetic pathways. At the same time, all algologists will admit that the photosynthetic pigment system is of major taxonomic importance in the algae.

Serology

Whether serological reactions should be considered as chemical characters depends on one's definitions. In any case, they have their own special set of strengths and weaknesses.

The great virtue of serological data is that they are objectively measurable and absolutely independent of all other characters. One gets a result that can be measured in terms of percentages, or counted in terms of numbers of bands, without the need to decide which character is more important than which other character. Furthermore, as the late Marion Johnson was fond of pointing out, they can be used as either a microscope or a telescope, depending on how strongly you sensitize the rabbit.

One of the great weaknesses of serological data is that we do not really know what we are measuring. It may be the number of proteins in common (but which ones?), or the number of similar (but not identical) proteins, or the amount of difference between similar but not identical proteins, or the relative amounts of some one or more kinds of proteins, or probably some combination of these. Furthermore, individual rabbits may give somewhat different results, and who knows how different the results might be if we used cows or alligators or Barbary apes instead of rabbits. Anyone who has noted the varying sets of kinds of pollen to which different human sufferers from hay fever are sensitive must have some doubts about the taxonomic reliability of allergic reactions.

Another very serious problem with serological data is that they exist only as one to one comparisons gathered at the expense of much time and effort. Other kinds of characters, both chemical and morphological, exist independently of comparison. Once the data are gathered for the various taxa or individuals under investigation they can be cross-compared in all sorts of ways by the mind or with the aid of the computer. Each serological comparison, in contrast, requires a separate laboratory test.

It takes only one test to compare two items, but 3 tests for 3 items, 6 tests for 4 items, 10 tests for 5 items, 15 for 6 items, and so on at an ascending rate. The formula for the number of tests required to compare each member of a group individually with all other members is $n(n-1)/2$, if n is the number of members of the group. When n is more than about half a dozen, this figure approaches one half of the square of n. If we consider that there are about 350 families of flowering plants, then it would take more than 60 000 tests to compare one sample of one species in each family with one sample of one species in each of all other families. And if we consider that there are about 230 000 species of angiosperms, then it would

take more than 2.6×10^{10} tests to check one sample of each species against all other species.

Obviously such large numbers of tests are out of the question. We must content ourselves with a very limited number, and choose our test items carefully. But this choice introduces the very element of subjectivity that the serological method is supposed to avoid.

The startling results of Simon [26, 27] suggest that many surprises might await us if larger numbers of tests were made. He found a strong serological affinity between *Nelumbo* and several members of the Agavaceae, but little or no affinity between *Nelumbo* and various other monocots, including members of the family Liliaceae and the order Alismatales. The sporadic occurrence of strong serological reactions between taxa not considered to be closely related led Moritz [24] and his associates to speak of antisystematic reactions. They consequently tried to devise means of dealing with such reactions, by precipitating or otherwise disposing of whatever was causing them, so that the remaining material could be used to give a proper measure of affinity. Here again the subjective element is introduced into the assessment of results.

Dr D. Fairbrothers advises me that these antisystematic reactions (a term he prefers to avoid) commonly involve a group of carbohydrate-protein-lipid complexes called lectins, and that when the lectins are chemically removed from the serum in advance the test is much more significant and reliable. But here again we are being just as subjective as when we use more ordinary kinds of characters. Characters that correlate with others, that give us results we are prepared to accept, are considered to be useful, and the ones that don't correlate are dismissed as unimportant. This sort of outlook is in fact necessary in taxonomic work, and can be minimized only by the computer approach. My point here is that the serological method does not escape the problem to which other kinds of characters are subject.

In spite of these problems with serological data, I believe that in proper context they are taxonomically useful. They so often produce results in conformity with other information that as a practical matter they must be given some weight in reaching conclusions, especially when other data are ambiguous. Hammond's studies [14], which show *Paeonia* to be serologically isolated from all tested genera of the Ranunculaceae, are among the factors that have led me to put *Paeonia* in its own family, and indeed in a different subclass than the Ranunculaceae. I hasten to add that several other features also point to this disposition of the group. Serological studies likewise support the dissolution of the traditional order Rhoeadales into two smaller orders, the Capparales and Papaverales, and the assignment of the Papaverales to the subclass Magnoliidae, near the order Ranunculales [11].

In the case of the Rhoeadales we have another set of chemical characters pointing in the same direction as the serological data. The Papaverales, but not the Capparales, have the characteristic isoquinoline (phenylalanine) alkaloids of the Magnoliidae [16, 17]. On the other hand, the Capparales, but not the Papaverales, characteristically produce mustard oil, which is almost but not entirely restricted to this one order. If we had neither the chemical nor the serological data, there would surely be a continuing strong body of taxonomic opinion supporting the traditional broad definition of the order Rhoeadales.

Amino Acid Sequences

I could go on with a grocery list of chemical characters that have been or might be used in plant taxonomy, and a laundry list of the particular places where I have used them in my own work, but I will spare you the ordeal. Rather I would like to comment briefly about a kind of chemical character that is going to be increasingly important to taxonomy in the future. That is the amino acid sequence of proteins. The procedure is complex and difficult, but it can probably be made easier. Data on amino acid sequences are still too sparse and scattered to be directly useful to taxonomists, but surely the promise for the future is there, and here indeed we have a kind of data that can be evaluated independently of all other data, providing an independent check on the system.

Enough has been done on cytochrome *c*

sequences to demonstrate the prospects. I will pass over the work that has been done on animals and consider instead the preliminary work of Boulter et al. [7] on 14 species in 13 genera flowering plants. They found that *Spinacia* (Chenopodiaceae) and *Fagopyrum* (Polygonaceae) are on the same major phyletic branch, apart from other things tested, but still fairly far removed from each other. This is in harmony with existing systems that treat the two families as belonging to different orders of the subclass Caryophyllidae. No other members of the Caryophyllidae were included in their study. They further found a close relationship between *Abutilon* and *Gossypium,* both in the family Malvaceae, and they found no difference at all in the cytochrome *c* of two species of *Brassica.* So far so good. Their finding of a seemingly close relationship between *Brassica* and *Cucurbita* is not quite so good. These genera, both in the same subclass but in different orders [8, 28], should be about as far apart as *Spinacia* and *Fagopyrum.* Their further finding of a fairly close relationship between *Phaseolus* (Fabaceae) and the *Cucurbita-Brassica* group is completely out of harmony with our other information.

Thus it appears that the amino acid sequence in cytochrome *c*, like other taxonomic data, is partly but not wholly correlated with other data. The information will certainly prove to be very interesting and very useful to taxonomists, but it is not going to solve all our problems. The same will surely be true of other proteins whose intimate structure will be elucidated in time to come. Everything is grist for the taxonomist's mill, but neither the grist nor the mill contains a unique Rosetta-stone.

I wish to thank Kenneth Becker, David Fairbrothers, David Giannasi, and B. L. Turner for help with particular details of the manuscript. I am indebted to my wife, Mabel A. Cronquist, for editorial help and consultation. All interpretations and opinions are my own.

References

1. Alston, R, Biochemical systematics, in Evolutionary biology (ed T Dobzhansky, M K Hecht & W C Steere) vol. 1, p. 197. Appleton-Century-Crofts, New York, 1967.
2. Anderson, E, Am j bot 1956, 43, 882.

3. Beck, E, Merxmüller, H, & Wagner, H, Planta 1962, 58, 200.
4. Behnke, H-D, Bot rev 1972, 38, 155.
5. Behnke, H-D & Turner, B L, Taxon 1971, 20, 731.
6. Björkman, O, Nobs, M A & Berry, J A, Carnegie inst Wash yearbook 1971, 70, 507.
7. Boulter, D, Ramshaw, J A M, Thompson, E W, Richardson, M & Brown, R H, Proc roy soc Lond 1972, B 181, 441.
8. Cronquist, A, The evolution and classification of flowering plants. Houghton Mifflin, Boston, 1968.
9. Downton, W J S, Check list of C₄ species, in Photosynthesis and photorespiration (ed M D Hatch, C B Osmond & R O Slayter) p. 554. Wiley, New York, 1971.
10. Eckardt, T, Reihe Centrospermae, in A Engler's Syllabus der Pflanzenfamilien, vol. 2, Angiospermen (ed H Melchior) p. 79. Bornträger, Berlin, 1964.
11. Frohne, D, Planta med 1962, 10, 283.
12. Gibbs, R D, Trans roy soc Can III, 1945, 39 (5), 71.
13. Grisebach, H, Recent investigations on the biosynthesis of flavonoids, in Recent advances in phytochemistry (ed T J Mabry, R E Alston & V C Runeckles) vol. 1, p. 379. Appleton-Century-Crofts, New York, 1968.
14. Hammond, H D, Serol mus bull 1955, 14, 1.
15. Harborne, J B, The evolution of flavonoid pigments in plants, in Comparative phytochemistry (ed T Swain) p. 271. Academic Press, London, New York, 1966.
16. Hegnauer, R, Planta med 1958, 6, 1.
17. — Ibid 1961, 9, 37.
18. Kimler, L, Mears, J, Mabry, T J & Rösler, H, Taxon 1970, 19, 875.
19. Mabry, T J, Taylor, A & Turner, B L, Phytochemistry 1963, 2, 61.
20. Mabry, T J & Turner, B L, Taxon 1964, 13, 197.
21. Mabry, T J, The betacyanins and betaxanthins, in Comparative phytochemistry (ed T Swain) p. 231. Academic Press, London and New York, 1966.
22. Mabry, T J & Dreiding, A S, The betalains, in Recent advances in phytochemistry (ed T J Mabry, R E Alston & V C Runeckles) vol. 1, p. 145. Appleton-Century-Crofts, New York, 1968.
23. Mabry, T J, Kimler, L & Chang, C, The betalains: structure, function, and biogenesis, and the plant order Centrospermae, in Recent advances in phytochemistry (ed V C Runeckles & T C Tso) vol. 5, p. 105. Appleton-Century-Crofts, New York, 1972.
24. Moritz, O, Some special features of serological work, in Taxonomic biochemistry and serology (ed C A Leone) p. 275. Ronald Press, New York, 1964.
25. Ownbey, M & Aase, H C, Res stud state coll Wash monog, suppl. 1, 1956.
26. Simon, J-P, Aliso 1970, 7 (2), 243.
27. — Ibid 1971, 7 (3), 325.
28. Takhtajan, A L, Sistema i phylogenia tsvetkovich rasteniji. Soviet Sciences Press. Moscow and Leningrad, 1966.
29. Thorne, R F, Aliso 1968, 6, 57.
30. Wohlpart, A & Mabry, T J, Taxon 1968, 17, 148.

Discussion

Mabry: You have incorrectly referred to the Mabry et al. [19] suggestion that a distinct order be reserved for the betalain families with a closely related order for such anthocyanin families as the Caryophyllaceae as "one character taxonomy" and the "misuse of chemical characters". My own contribution will indicate the way our suggestion was, in fact, based upon multiple characters. However, I might point out one of the ways that the betalain distribution data has been misused by you. In your 1968 work [8], you unfortunately downgrade the systematic significance of the remarkable distribution of the betalains by emphasizing an old, obscure and, to a chemist, unreliable report of the co-occurrence of betalains with cyanidin. Reinvestigation quickly established that only betalain occurred naturally in the plant, the cyanidin being an artifact of the workup procedure.

Ourisson: (1) Chemical studies can very often be run on minute specimens, certainly not larger than those sometimes needed for morphological studies. Yet, it is true that curators of Herbaria usually refuse to give samples to chemists, whereas they do give them (with care) to morphologists. This is one more reason why it is important for us chemists to convince botanists that we can help them seriously, and for botanists to suggest to us problems of real importance to them.

(2) The numerical limitations of chemical studies mentioned by Cronquist are of course not specific of chemical data. For instance, are there many more genera studied for the presence of sieve-tube plastids than chemically?

(3) As regards the Caryophyllaceae-Molluginaceae problem: is the whole biosynthetic chain missing, or (a) do the species of these groups contain precursors of betalains (but lack the later enzymes required)? (b) are they able to transform into betalains added precursors

(and do they lack the enzymes required for the earlier biosynthetic steps)?

Jensen: You have explained that there are serious limitations in the use of protein characters. However, it is generally accepted by those who use protein characters that these alone cannot be used to construct natural groups; indeed you did not mention some important limitations and these I shall consider in my own lecture. You said that there was uncertainty with respect to what was being measured in serological reactions. Whilst this is so for the quantitative precipitation reaction of Boyden, etc., this limitation need not apply to immunodiffusion, immunoelectrophoresis and adsorption tests where analysis of the proteins involved is possible. Cronquist also cited "antisystematic" reactions as a further limitation of serological methods. This term was first used by Dr Moritz to cover cross-reactions involving widely separated taxa. We now know that such reactions are normal and could be explained by the presence of proteins such as fraction I protein which is known to have similar determinants in plants as different as *Chlorella* and *Nicotiana*. Cronquist said that antisystematic reactions are due to lectins, but this is not so in our experiments since a sample of serum is first tested with the plant extract, and is shown not to give a precipitation reaction prior to immunisation of the rabbits.

Cronquist: Dr Fairbrothers and his associates have recently proposed some rather drastic changes in concepts of relationships among certain families of angiosperms, on the basis of serological studies. My discussion of some of the problems with the serological approach relates in large part to my efforts to evaluate the significance of the information on which he proposed these changes.

Kjær: Absence of a given compound in a taxon is—though in principle, less significant than its presence—frequently more difficult to establish experimentally. The inherent difficulty in utilizing the absence of a compound as a biochemical marker hence calls for caution.

Cronquist (reply to Harborne): I do not smoke. One of my students does smoke, however, and

he applied the cigarette test to many Asteraceae which we collected together in Mexico. On that basis we recorded on the labels the presence of anthochlors in the ray flowers of a number of species. Now Dr David Giannasi, a chemo-systematist working at the New York Botanical Garden, tells me that some things other than anthochlors will respond to the cigarette test, so possibly some of our label data are incorrect.

Turner: Concerning your remarks regarding the Centrospermae: there is more to the betalain question than the mere presence or absence of a secondary compound. There is the absence, presumably, of an entire metabolic pathway—further, the order is sufficiently large and diverse to have evolved out of a relative complex lineage. This being so, it is remarkable that there should be the mutual exclusion of betalains and anthocyanins. In fact, why have not betalains arisen independently in yet other groups? Presumably because selection of anthocyanin pigments militated against the presumably more primitive betalain pigments (angiospermically speaking); at least that is our surmize. If, as Cronquist suggests, the absence of betalains is due to single-gene blocks in a metabolic pathway leading to their production, then such rigid fixation and mutual exclusion with anthocyanin would not be expected. We entertain no hope that an exception to mutual exclusion will be found, for widespread sampling suggests that such does not exist.

Regarding his comments on C_4 metabolism: it is clear that this is a very useful character for systematic purposes, at least at the familial level or lower. The fact that it has arisen independently in a number of plant groups is not relevant to the question of systematic utility. Many characters, including morphological ones, arise independently and yet find useful application in the groups within which they arise. Thus, Smith & Brown, in a recent article (1973) has shown C_4-plants in the Poaceae to be related, at least in part, and my own surveys (Smith & Turner, unpubl.) among the Asteraceae show that C_4 plants are confined to only two very natural taxa, and serve to relate a group of genera whose relationships here-to-fore have been quite problematical.

Hegnauer: Regarding the presence of trace amounts of individual compounds in taxa (example: azetidine carboxylic acid): Many secondary metabolites are toxic to cells. Plants which contain them in easily detectable amounts have evolved measures enabling accumulation. Therefore, the presence of trace amounts of compounds in apparently not related taxa does not invalidate the taxonomic meaning of the accumulation of the same compounds in other taxa.

Birch: In connection with this paper, I would like to make another example of the importance of pathways rather than structures as such. In 1953 I pointed out on theoretical grounds that chalcones must be the precursors of all of the flavonoid and anthocyanin pigments. They must be formed as intermediates, and can therefore be potentially accumulated. Their significance may be rather small. Aurones, which mark another oxidative step, may be more significant as evolutionary markers.

Geissman: You suggested that a taxonomic significance had been attached to aurones, which was not claimed in the early work. Although all of the Coreopsidinae contained aurones, the presence of aurones was not regarded as meaning that the plant was a member of this group.

It is also to be remarked that "chalcones" occur free in nature simply because they lack a nuclear hydroxyl group which, if present, induces their almost complete cyclization to a flavanone. Flavanones are extremely widely distributed but are simply cyclic isomers of chalcones.

Cronquist: Certainly you did not suggest any changes in the taxonomic system on the basis of the occurrence of anthoclors. My recollection is that in the early 1950's, the great number of reports of anthochlors in the subtribe Coreopsidinae, and the paucity of such reports from outside this subtribe, automatically suggested the possibility that the presence of anthochlors might be a significant chemical marker of the group. The information that has become available since that time tends to reduce rather than increase the taxonomic significance of anthochlors.

Bate-Smith: You remarked that "all morphological characters have a chemical foundation". Over the years, chemists working on the systematic distribution of the different classes of secondary constituents have been getting into the habit of looking for association between their occurrences and particular morphological characters, and I think that the recognition of such associations, and the establishment of possible causative relationships between the one and the other will be an important contribution from chemistry to plant taxonomy. I myself have quite a number of such possibilities in mind, but none which can at present be positively specified.

Chemosystematics—an Artificial Discipline

V. H. Heywood

Department of Botany, Plant Science Laboratories, University of Reading, UK

Borné dans sa nature, infini dans ses vœux,
l'homme est un dieu tombé qui se souvient des cieux.
Lamartine, Méditations poétiques

Summary

The extensive use of chemical data in systematics over the past ten years has led to the recognition of a field widely known as chemosystematics. Now, it is time to evaluate the nature of the research undertaken under this name, and assess its aims, advantages and limitations within the context of modern systematics as a whole.

While the methodology of the chemical component of chemosystematics is in the main rigorous and scientific, the same cannot be said about the systematic or taxonomic component. Although research in numerical taxonomy has focussed attention on the need to break down information about organisms into characters, followed by strategies designed to synthesize these characters into classifications, it is by no means certain that this approach is valid, at least as at present practised. Traditional classifications, in almost universal use today, have not been produced by such a procedure but rather by a complex process of imprecise model formation, followed by testing, reformulation and retesting, many times over. Characters explicitly recognized as such do not come into this complex series of processes at the beginning but at various later stages and in particular when there is a need to describe precisely the resultant classifications and provide keys to the component groups.

The nature of chemical information is such that it is unlikely to be used in this procedure at the stage of initial model formation and is therefore to that degree secondary, although its ultimate value in the final product may be high. If used in numerical taxonomic procedures, however, chemical data can be treated on a par with other kinds of data.

The actual handling of chemical data in either approach presents many problems which have not yet been adequately studied, such as homology, comparability, variability, etc. quite apart from the difficulties posed by their "indirect" nature.

Perhaps the most useful introduction to a symposium involving two very different disciplines is to attempt to define the key words that near-ly every speaker will employ and to clarify some of the major concepts involved. I shall attempt to do this for the botanical side, following the brief I was given by the President. That there is a need for clarification is very evident from a perusal of some of the papers of those who have recently written on chemosystematics. For example, Birch [1] in a paper delivered at the IUPAC symposium in Strasbourg in July 1972 on Chemistry in evolution and systematics writes: "Attempts have been made to assist that essentially artificial classification, taxonomy, by considering the structures of plant constituents as markers on a level with morphological characteristics." Ignoring the fact that I do not understand what this comment means as written, two of the key words used are classification and taxonomy, and other terms to be considered include systematics, relationships, phylogeny, phenetic, phyletic/phylogenetic, evolutionary (as applied to systems of or approaches to classification and relationships).

I assume that the title of this symposium is not to be understood literally, as classification is but one aspect of the general field of systematics and consequently of biochemical systematics (if such a term is admitted, see below for discussion).

Definitions

(a) *Systematics*. In the broadest sense, systematics is concerned with the scientific study of the diversity and differentiation of organisms and the relationships (of any kind) that exist between them [2, 3]. Without further qualification (other than botanical/zoological/fungal/microbial, etc.) it embraces phylogenetic, evolutionary, phenetic, morphological, traditional, classical, etc. approaches. It can cover such studies at all levels in the hierarchy, ranging from the individual and population level to the family, order, class and even higher levels. At the lower levels of the hierarchy, it includes what is often called biosystematics and indeed Solbrig's recent text book entitled "Plant biosystematics" [4] is, in effect, a contribution to plant systematics. His definition of biosystematics is "the application of genetics, cytology, statistics, and chemistry to the solution of systematic questions in order to provide explanations about the diversity of organisms within the framework of evolution". This is essentially comparable with the broad definition of systematics given above although it is unnecessarily restrictive in listing only genetics, cytology, etc. as disciplines to be applied: in fact, any discipline that provides data or evidence may be employed, including, of course, morphology, anatomy, palynology, electron microscopy, etc. It is the comparative approach which characterises systematics, and the various disciplines contributing to systematics are often termed comparative or systematic, e.g. comparative/systematic anatomy, comparative/systematic phytochemistry, etc.

(*b*) *Taxonomy*. Although this term has been widely misapplied in the past, there has been a tendency in recent years to restrict it to its original meaning, namely that part of systematics which deals with the study of classification, its bases, principles, procedures and rules. It therefore covers such highly complex areas as character-selection and handling, key-making, hierarchies, definitions of taxa and all such techniques as are involved in the mechanics of actually preparing classifications and keys of any kind from the raw data of whatever nature. This is an aspect of systematics that is not, normally, well understood by nonprofessionals who are usually concerned with the results of these activities, not with the processes themselves.

(*c*) *Classification*. Classification is the act or process of arranging plants into classes or groups (taxa). In biological classification the classes usually form a nested hierarchy—the so-called box-within-box arrangement based on a simplified form of set theory. It is important to note that in this form of classification, the groups are non-overlapping so that a single

group (taxon) cannot belong to more than one next higher group (taxon) in the hierarchy. For example, a species can only belong to one genus at any one time. Although alternative non-hierarchical classification is possible it is of little general application at the present time. Classification of chemical compounds is not strictly or at all hierarchical as is discussed below.

The word classification is also used to describe the product of the process of classifying. A more explicit name for the product is arrangement or system of classification. These are intended to express the relationships of the plants and groups and serve as a filing system. The distinction between the act of classification and the arrangement of the consequent groups relative to one another is not unimportant. For example, the classification of the angiosperms into 250–300 families implies an arrangement of them only to the extent that the hierarchy imposes one on them—whether or not they are grouped into orders and if so, how many. The same classes, without any change in their content or circumscription, can be arranged in many different ways as in the various natural, phylogenetic or evolutionary systems proposed by, for example, Engler & Prantl, Emberger, Ehrendorfer, Cronquist, Takhtajan, Hutchinson, Soó, etc. [5–11]. It is, I think, important to stress this since by and large there is a very wide amount of agreement amongst taxonomists as to which angiosperm families should be recognized (although there are uncertainties still in some areas as to the level to be accorded to some groups as in the Leguminosae (Fabaceae) where the Papilionoid, Mimosoid and Caesalpinoid groups may be treated as families, sub-families or tribes).

(*d*) *Relationships or affinity*. Relationships between plants or taxonomic groups may be of many different kinds and should be specified to avoid ambiguity. Although the use of the term, unqualified, may be taken to mean evolutionary or phylogenetic relationships, this is not necessarily a justified assumption.

Relationships may be overall which is usually interpreted as meaning in respect of all the kinds of evidence available or employed. In phenetic relationships any sources of information other than phylogenetic ones may be used, such as chemistry, cytology, embryology, morphology, anatomy, ecology, but there is no implication of evolutionary significance in such relationships. (The term phenetic has been widely used in recent years as meaning numerical but although phenetic classifications are often produced using numerical techniques, they can also be produced by non-numerical methods. Conversely, numerical classifications need not be phenetic as techniques are available for making computer-based cladistic schemes.) Phylogenetic or phyletic relationship, on the other hand, does refer to various evolutionary components which in turn should be specified so as to avoid misunderstanding. These components are: *patristic* relationship which refers to common ancestry of the groups involved. The usual explanation of resemblances between groups is that they are the consequence of their having had a common ancestor whose features have been retained in the descendants or have been altered independently in each group in the same way from the ancestral condition. These two types of patristic relationship may be referred to as *primitive* or *derived,* respectively [12]. *Cladistic* relationship, on the other hand, refers to evolutionary pathways by which observed resemblances have been achieved. This may also be termed branching or genealogical relationship. A third and important element in phylogeny is *chronistic* relationship which refers to the time scale during which the evolutionary events take place. The time element is often indicated in phylogenctic diagrams (especially "trees") by the vertical dimension, although all too often it is either implied or ignored rather than being explicitly stated. It is clearly important to consider not just the ancestry and pathways in evolution but the timing of events. Because of the difficulties involved in establishing these kinds of relationships from the evidence usually available, some authors advocate the use of the general term evolutionary relationship when the kind of phylogenetic components involved cannot be clearly specified.

A further kind of relationship is homoplastic which is applied when similarity is due not to inheritance from a common ancestor but to convergent or parallel evolution.

In addition to these major kinds of relationship, it is perfectly legitimate to talk about genetic relationship, chemical relationship, cyto-

genetic relationship, genomic relationship, etc., etc., when one wishes to single out the evidence from a particular discipline or sub-discipline. The term affinity is often used as synonymous with relationship.

Several of the terms discussed above are comparatively new, having been introduced as a result of recent attempts to clarify the principles and methodology of systematics and classification. It is incumbent on chemists working in the field of chemosystematics to familiarize themselves with the correct application of this technical vocabulary just as the botanist needs to understand the comparable chemical vocabulary. Reference to recent literature will show the need for mutual comprehension as chemistry begins to take its place in plant systematics and evolutionary studies. This point is discussed further in the next section.

Chemistry and Systematics

1. *General considerations*

I have discussed in a recent paper [13] the unusual, even surprising nature of the hybrid subject chemosystematics. As I pointed out, chemistry, as one of the physical sciences, is what Pantin has called a restricted science in the sense that it is concerned with a limited range of phenomena and it is this restriction which has permitted it to become precise, exact and mature. Unlike biology, and in particular, systematics, it does not have to venture into nearly every other discipline. Chemistry is able to define with a high degree of precision the molecules with which it works, the parameters involved in any experiment, and therefore the techniques to be used. As Daniels [14] has written, in the context of computers and botany: "Classical taxonomy has, since the time of Linnaeus, been rather in the nature of an art, in spite of all protestations as to its scientific status. The questioning of such status comes not from the end result which has stood the test of time well but from the manner in which the decision is arrived at. In the physical sciences, an experiment can theoretically be conducted by anyone if the conditions and the equipment involved in the experiment are stated, and if this set of conditions and equipment can be duplicated. Given the same data and the same procedures, the same results should be obtained. What has been lacking in botanical taxonomy and, in fact, in taxonomy in general, has been the specific designation of the parameters involved in any given experiment.''

Because of the diverse and complex nature of systematics, it involves both rigorous experiments as when disciplines such as genetics, cytology and, of course, chemistry are employed, and imprecise areas such as the actual production of classifications with the data obtained from these other disciplines. In other words, although the data may be obtained with full scientific rigour the taxonomic processes whereby these data are selected, processed and arranged remain to a large degree imprecise and it is for this reason that taxonomy has often been called a scientific art.

The test of a classification is largely a pragmatic one—does it work, i.e. does it satisfy the needs of the user, does it permit identifications to be made, will it withstand the addition of further information without major recasting, does it allow a high degree of predictivity, does it store large amounts of information, etc? While it is generally admitted that the majority of biological classifications in use today do meet these requirements, otherwise alternative classifications would have been sought, it is extremely difficult to analyse the exact processes involved in their production. It is equally difficult to obtain from the general body of taxonomists, or for that matter from the users of their work, a clear statement as to the aims and methodology to be employed to implement these goals, except in the most general terms which are not, to use modern jargon, operational [15].

The reasons for this situation are to be found by analysing the various taxonomic processes, the need for which has not arisen as a serious issue until relatively recently, especially when attempts were made to use computers in taxonomy. The computer is a hard taskmaster and requires the rigid statement of the logical processes to be applied in a given procedure. This is shown very clearly in both attempts to produce classifications by using computer programmes and by using computer systems for classification keys [16]. As Daniels [14] succinctly puts it: "If the taxonomist is not able

precisely to define his terms and with equal precision to define the steps by which he arrives at a given conclusion, then he is unable to define his experiment. If such parameters cannot be precisely determined and defined, then it is not possible for other individuals to consistently arrive at the same conclusion, given the same starting point.''

Why is it, one might ask, that these steps cannot be precisely determined, or at least, agreement reached as to what the steps are, since various attempts have been made recently by so-called "numerical taxonomists" only to be met with distrust, disagreement and controversy? Part of the answer, at least, lies in the extent to which traditional methods of classification have involved subtle processes such as Gestalt perception and a related phenomenon, typology, which bring into play complex human mental processes which it is difficult to appreciate and analyse, let alone machine copy.

It is, I believe, important to discuss these problems in the context of chemistry and classification since they have a fundamental bearing on what use can be made of chemical data in taxonomy. To give but one example, it appears that in constructing classifications we do not start off with an array of pieces of information or characters and then set about to seek correlations between these so as to produce groups. On the contrary, classes or groups are perceived mentally (or aesthetically as Pantin puts it [17], or neurally as I have described it myself [18]) and then our impression can be analysed and given precision by extracting characters from it. This "unanalysed entities approach", as I have termed it elsewhere [18], see also Cullen [19], implies that to a very large extent we by-pass characters in making classifications and have recourse to them a posteriori when we need to describe and communicate the results to other people. There may, in fact, be little direct connection between the two processes—classification and character selection —as I have discussed on a previous occasion [20]. What is more, some taxonomists may be good at the one but ineffectual at the other which explains the apparent paradox one often comes across in taxonomy whereby one is convinced as to the validity of a classification but cannot see how to make it work.

There is nothing new or original about these views—they were in fact clearly expressed by Whewell [21] in 1847. It is an increasing preoccupation with the methodology of taxonomy and with the needs to handle large amounts of new data that has brought them to the fore. Two points emerge: the first, which is discussed below, is that with few exceptions, the information derived from chemistry, cannot be used in traditional methods of classification as primary data; the second is that the procedures of numerical taxonomy may be misconceived to the extent that they involve detailed character analysis as a first step in classification unless it can be shown that such analysis is capable of selecting the correct data for establishing acceptable pattern perceptions in a manner comparable with the tried methods of human Gestalt perception. It can, of course, be argued that numerical taxonomy can produce better (i.e. more acceptable, and, of course, repeatable) results than traditional classifications but there is little evidence that this is so and the onus is on numerical taxonomists to establish this, if it is so. I suspect that the eventual answer will lie in a combination of some of the techniques of numerical taxonomy with some of those of traditional taxonomy.

A related phenomenon to Gestalt perception is that of typology. Although crude typology in the pre-evolutionary sense of establishing an invariant ground-plan for a taxonomic group has rightly been discredited (see discussion in Davis & Heywood [11, pp 8–12]), typological procedures are an essential part of present day taxonomy. None of our taxonomic groups are precisely defined—they are built up by a process of accretion. When we talk about the Compositae (Asteraceae), none of us has ever considered each species, each genus, each tribe and achieved a complete oversight of the family. None of the descriptions of the family allow for all the variations in the so-called family characters displayed by the constituent genera and species. Every family description is typological in the sense that it conveys a generalized impression or type-picture of the family (not, of course, the nomenclatural type which is quite irrelevant). Every taxonomist or person familiar with the family has his own typological picture of it but it is not capable of precise

definition. It is well known that in terms of so-called diagnostic features, the higher (in the hierarchy) the fewer (the characters) and reliance is placed more on tendencies. We often identify plants as belonging to a particular group despite their possession of individual features which do not agree with those listed in the published typological descriptions of the group. We are all familiar with situations whereby we "know" that a plant is a member of a particular family without being able to say why. The nature of the "type" in typology is, as Pantin says, ". . . the machinery of our trained perception. It is that modification of our perceptive faculty which has been produced by the integration of all our past experience connected with objects of the class".

Although to a large extent the concept of a perceptual type applies to the classification of molecules, the classes in such cases are susceptible to precise definition in most cases. The classes of compounds such as alkaloids, terpenoids, amino acids, etc. were established by quite different mental processes from the classes (taxa) in biological classification, and they can be defined rigorously.

It might be thought (and I wonder if this is in fact implicit in the approach of some chemists to biological classification) that the application of rigorous, definable chemical information to plant taxonomy might put it on a more precise and scientific basis. Unfortunately this is not so, due to basic differences between chemical and biological classification which are discussed in the next section.

2. *Classification in biology and chemistry*

I have already discussed the way in which biological classifications in general use today are nested hierarchies. Our ability to classify plants and animals in a linear branching system of successive subordinate classes stems from the evolutionary descent of species from common ancestry. If one takes a horizontal biological classification of families with included genera with included species, this can be drawn in the form of a vertical dichotomous branching system resembling an evolutionary tree (fig. 1). To the extent that the similarities shown by the constituent members of each group can be shown to be due to common ancestry, such a

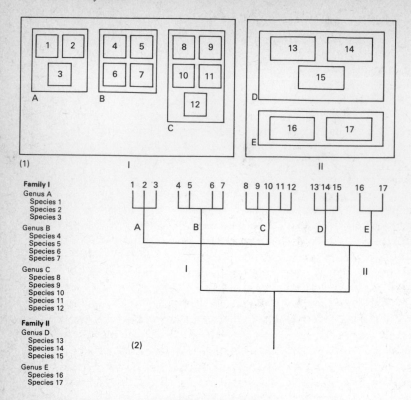

Fig. 1. Two forms of presenting the same hierarchical classification of two families, five genera and seventeen species listed: (1) a box-within-box arrangement; (2) a tree-like phenogram.

representation of a classification is a partly phylogenetic diagram but not fully so as discussed above. In other words, it is the fact of evolution that causes characters to hang together or show correlations although there may be other reasons for such clustering. Such a type of classification appeals to the logical machinery of our minds which explains why biological classification has made such progress. On the other hand, the objects to be classified, i.e. various kinds of taxa, cannot, as we have seen be precisely defined which is a serious drawback in terms of the ways our minds work.

If we compare the classification of molecules by the organic chemist in the light of these requirements we find that the different kinds of molecule can be specified with a much higher degree of precision than can organisms, by using structural formulae and extended names. On the other hand, there are no clearly defined higher classes of compound, comparable to the genera, of the families, etc. of the biological

taxonomist, so that simple brief mnemonic names cannot be used. This is not because organic chemistry still awaits a Linnaeus but because the molecules to be classified have not evolved by natural selection. What classes there are cannot be subordinated in a linear dichotomous manner. The result is, as we all know, a chaotic nomenclature without genera, terms coined by individuals with few ground rules and no official co-ordination.

Another purpose of classifications is to express the relationships of the component members. Again a comparison of chemical and biological classification is illuminating. The latter do permit us to assign a position to all known organisms but they certainly do not display all the relationships between organisms as this would demand a multi-dimensional system. The relationships in any chemical classification are, on the other hand, of necessity multi-dimensional, not linear-dichotomous, even though they include relationships of a higher order. The classes overlap and a lower order class can belong to more than one higher order class at any one time.

Although it is not always admitted by taxonomists, biological classifications, even phylogenetic systems, are very limited in the kinds of relationship they actually express. A fully evolutionary or natural classification would require $n+1$ dimensions which is, of course, quite beyond the capacity of the human brain to conceive; in no way do we even pretend to aim at such an ideal. It will be recalled that I discussed earlier the various kinds of relationship that may be referred to in systematics. In considering subjects such as genetics, ecology, physiology, ultrastructure, biochemistry, enzymology, etc. we discuss many important kinds of relationships between organisms that may cut across taxonomic or evolutionary relationships expressed by a classification. In the particular context of chemistry, it is well known that the same complex molecules may occur in organisms which are widely separate from a systematic/evolutionary viewpoint. The common occurrence of such compounds in diverse organisms—even in plants and animals—may be explained by their formation by different biosynthetic pathways and in a more basic way by reflecting the limited number of possible kinds of molecule which are available for their construction. Chemical similarity does not imply taxonomic identity or common ancestry to anything like the degree that morphological similarity does. When there is little structural or morphological diversity shown, as in bacteria, thus preventing unique character combinations, taxonomic classification runs into difficulties and this is not solved by recourse to chemical attributes such as the presence or absence of a biochemical process which is in turn due to the presence or absence of a particular enzyme.

Biological classifications rely on similarities in homologous features. They do not express relationships provided by analogous features, nor chemical molecular relationships which cut across taxonomic/evolutionary relationships. Nor do they express dynamic functional relationships, including those that are essentially biochemical.

3. Chemical data in classification

In the light of the discussion on the nature of classification and relationships in chemistry and biology, I now want to consider the application of chemical data to biological classification.

As I have described on previous occasions [23] because of man's visual sense, the biological classifications he makes are primarily morphological. He needs to see what he is doing. The Gestalt approach and typological appreciation discussed in the last section stem from this characteristic of man. In practice the raw materials of classification—the plants in the field growing in populations—have to be looked for before information of any sort about them can be used or extracted from them. Thus there is a morphological bias built into the system before any of the taxonomic processes can begin.

Again, as we have seen, since biological classifications are limited in the kinds of relationships they can express because of their linear dichotomous nature, which is caused by and reflects their underlying evolutionary basis, only those data which lend themselves to this kind of relationship will tend to be selected. In higher plants, because of the enormous diversity of form and structure which permits unique combinations to evolve through natural selection, morphological and anatomical data tend to be employed and show the restricted kinds of relationship imposed by the form of the classification. Chemical data whose relationships are essentially multi-dimensional do not lend themselves so well for this purpose. When we consider the fact that in constructing biological classifications today we are faced with an unprecedented repertoire of disciplines from which data can be drawn, it is obvious that we are forced in most cases to select those kinds of information which we have reason to believe will suit our purpose. This would appear to run counter to the precepts of modern biological classification that all possible information should be taken into account, a viewpoint stemming from John Lindley, the father of natural classification. Apart from the practical grounds which require us to limit our search for data, there are obvious procedural reasons which help to explain this paradox. Since our traditional method of classification depends to a large extent on Gestalt perception, those features which participate in the Gestalt, i.e. morphological, clearly are at an advantage. Non-visual information, or to be more exact, visual information that cannot be perceived without the intermediacy of machines such as microscopes (e.g. anatomical, cytological, micromorphological features) and non-visual information, can only be added in after the morphologically-biased Gestalt groups are produced, and if they agree (i.e. correlate) they are regarded as supporting the tentative classification; if they disagree, and especially if several independent lines of non-visual or non-immediate information all disagree in the same direction, the tentative classification will be examined with a view to revision along the lines of the additional evidence, or the non-visual and non-immediate information will simply be stored in the classification even though it does not support it.

This latter aspect of biological classifications as data-storage and retrieval systems is extremely important and provides part of the explanation of the paradox mentioned above. It is by using classifications as data-storage systems that other kinds of data and relationships can be handled, i.e. they will contain information that does not of itself diagnose the constituent classes. This is one of the major services of systematics to biology and in our present context to chemistry: it alone provides an effective information system based on the taxonomically circumscribed classes within a mnemonically workable structure even though it was not designed specifically for these other kinds of data or relationship.

This means, in effect, that all kinds of relationship can be extracted from biological classifications although the classification does not in itself express them. I believe that this is the answer to the problem posed by the production of chemical data that is not immediately useful taxonomically. Accumulation of chemical information which is one of the results of phytochemical studies is given sense by our ability to relate it to a biological classification, store it and retrieve it.

As our knowledge and experience increases, we should be able to obtain some assessment as to the kinds of compound which, because of their diversity and systematic distribution, are more likely to be of value in assisting in the production of biological classifications.

If one considers non-traditional methods of classification, chemical data can be handled initially on a par with other classes of data as in taximetric programmes. There are, however, very few instances where chemical data have been employed on an appreciable scale in this way, largely due to the difficulties of obtaining a sufficiently wide sampling of taxa. Too many blank or "non-available" entries in a numerical matrix lead to distortions in the end product. Our work at Liverpool and Reading on the multivariate systematics of the Umbelliferae is one of the few projects in which a broad spectrum of micro- and macromolecular data will be included [24–26]. Very little attention has been paid so far to the question of what constitutes a representative chemical sampling of taxa. As a taxonomist I do not feel able to discuss this question with any confidence and I would like to see a dialogue between botanists and chemists on this.

I should like now to pursue the question of whether the experience of chemists to relationships as displayed by the ways in which molecules can be classified, affects their attitude to the role of such molecules in biological classification. This may well be important in that our attitudes to what are ostensibly the same process, i.e. classification, will differ widely according to our training. Systematists and chemists are trained to think differently to a large degree because the kinds of phenomena they deal with are different, the range and diversity of phenomena is different, their scientific aims are different and the vigour and precision of their techniques and definitions is different.

Although in the general sense the basic aim of all scientific research is the classification of attributes and events [17], at the particular level classification is not the major preoccupation of the chemist whereas it is quite explicitly so for the systematist. The balance, therefore, between the two components in chemosystematics—chemistry and systematics—is a very uneven one. Although the prefix chemo- might suggest that the stress is in favour of chemistry, on the contrary chemistry is not much more than a data-provider and the systematics side of the hybrid handles, arranges and interprets these data along with all kinds of other data and it is the systematist who decides which are important for classification or interpretation, not the chemist. Even the evolution of particular classes of compound is dependent on working out the evolution of the organisms or classes of organisms in which they occur, either directly or indirectly, not just on our knowledge of the progressive complexities or changes in the structure of molecules on chemical grounds. For example, Harborne's recent analysis of evolutionary trends in the flavonoid patterns in higher plants were established by plotting the different structural modifications against the more or less agreed level of advancement of the families in which they occurred so as to establish the sequence, as in the case of the substitution in the 6- or the 8-position in the yellow flavonols [27]. Kubitzki [28], on the other hand, argued against some current views on the origin of the Rosalean and Guttiferalean orders vis à vis the Ranales on the basis of the "primitive" flavonoid patterns found in the former which are different from that found in the Rosales, these decisions of chemical primitiveness and derivations being based on earlier established correlations, not on independent chemical grounds.

The establishment of evolutionary trends in characters (semophyleses) is nearly always difficult in the absence of adequate historical sequences. One is forced to try and arrange the different expressions of the character or organ concerned in contemporary organisms in a series of increasing derivation or complexity based on logic and commonsense as to the ways in which change is most likely to have happened, bearing in mind mechanical or physiological constraints and evidence from other comparable known situations. Often the decisions rely heavily on extrapolation from knowledge of trends which have been established in other situations where there is reasonable fossil evidence. It should also be mentioned here, however, that some phylogenetic schools, especially the Hennig-type cladistic one [29], have devised methods for the reconstruction of cladogenesis from contemporaneous organisms, although there is considerable debate as to the validity of such approaches.

It might have been expected that it would have been easier to establish trends of complexity in molecular structure but this is not generally so because so few biosynthetic stud-

ies have been made of the kinds of compounds most frequently used in chemosystematic work. It seems to be generally agreed that increasing attention should be paid by phytochemists to comparative biosynthesis and enzyme studies so that not only structural configurations of compounds will be known but their relationships in terms of variations, transformations, and underlying genetic basis. Most of the comments by chemists and biologists on the priorities for research in phytochemistry during the coming years given in Mabry's paper "Major frontiers in phytochemistry" [40] express views along these lines. Such dynamic information will not only contribute to evolutionary studies but greatly assist taxonomists in assessing the value and significance of chemical information. The absence of fossil chemical information will continue to be a major drawback in working out the time-scale of chemical changes even though the sequences can be elucidated so that it is unlikely that it will be possible to arrive at evolutionary assessments of chemical information which are completely independent from biological classifications.

Current Work in Chemosystematics

Several excellent reviews of the role of chemistry in plant systematics have been published during the past few years [30–35]. These vary somewhat in their appreciation of the field, depending largely on their interests and training of their authors—whether they are chemists, biochemists or botanists. Some stress the value of micromolecular features, some stress the potential of macromolecular approaches; others give preference to particular techniques such as serology or to particular classes of compound such as terpenoids or flavonoids. Emphasis may be given to systematic screening or to biosynthetic studies. It is perhaps a sign of incipient maturity that the exaggerated claims made some ten years or so ago by some authors have been replaced by more realistic assessments.

It is clear from such reviews, and from a general perusal of the literature, that chemical information may be of value in most aspects of plant systematics from the population to the family level and above, in elucidating phenetic and evolutionary relationships and variation patterns, hybridization situations, evolutionary trends, and so on. I do not feel it is profitable to publish any further general reviews for some time: the needs of the student, the interested non-specialist chemist and botanist are already well served by the essays published by Alston, Erdtman, Harborne, Mabry, Swain, Turner and others. What we now need are critical assessments of particular problems or topics such as those of Boulter on cytochrome c [36], Harborne on flavonoids [37], Weimarck on statistical methods [38], Holmquist & Jukes on ambiguities in sequence data [39], etc.

Mabry [40] in his assessment of major frontiers in phytochemistry defines biochemical systematics as "a field which crosses many scientific barriers, and employs knowledge from all areas of chemistry and biology in an effort to detect and understand the processes of evolution". He echoes Alston's [32] earlier assessment that "there is still an aura of immaturity about the field and that an overall perspective is badly needed". The very fact that the term biochemical systematics (or the less acceptable chemical taxonomy) is used is an indication of this immaturity since in a proper biochemosystematic study, as Mabry indicates, biochemistry would simply be one of the contributory disciplines to an investigation in systematics or biosystematics. In the pioneer days of the use of cytology, especially karyology, in systematics, the term cytotaxonomy was frequently used, but nowadays chromosome information is such a normal tool that the term is seldom used as it would indicate an unacceptable bias of approach. At the 1972 IUPAC symposium on Chemistry in Evolution and Systematics, several distinguished chemists commented to me that in the light of the discussions, it seemed that there was no such subject as chemosystematics, just systematics in which chemistry had a role to play like many other disciplines.

In making these comments, I do not wish to suggest that we should not hold such symposia. On the contrary, this one is correctly conceived as what we need to establish is a dialogue between a wide spectrum of scientists with very different training so as to identify goals, discuss the limitations of the techniques being supplied in pursuit of these goals, establish priorities, etc. I should like to see more exchange of opinion on what chemists are aiming at in these studies, whether the present structure of systematics is capable of handling these objectives, what sort of information or explanations chemists would like the systematists to provide them with, etc. I should like to see discussions on such topics as the relevance of homology and analogy in chemistry, and other such fundamental topics of systematics and classification which were established in a context of comparative morphology and anatomy.

Having attended symposia on the relations between taxonomy and such closely cognate fields as ecology, I am all too conscious of the need for closer inter-communication [41]. It is easy to assume that one is talking the same language because one is handling the same terms (such as species, population, sampling variation, etc.) but closer analysis soon reveals how differently these terms may be conceived. In such widely different fields as chemistry and systematics, how much more necessary is the dialogue. Too much emphasis has been placed, I believe, on considering detail rather than general principles. Some of these matters are discussed in the next section on phylogenetic schemes.

Phylogenetic Schemes

The construction of phylogenetic schemes, usually in the form of tree diagrams, was an obvious consequence of the acceptance of the theories of Darwinian evolution. As I have discussed earlier, the linear-dichotomous form of biological classification pre-adapted pre-evolutionary classifications to expression in the form of an evolutionary tree, without any alteration in the circumscription of the taxa in these classifications. In very many instances, it was simply assumed that similarity was an indication of common ancestry. Unlike parts of the animal kingdom where sufficiently good fossil sequences are available to permit the construction of reasonably factual trees, most parts of the plant kingdom, including the angiosperms, are insufficiently represented in the fossil record at key times.

Evolutionary reconstruction is not, of course, entirely dependent on a good fossil record and numerous techniques have been devised for the extrapolation backwards of evolutionary relationships from the characters shown by contemporary organisms, supported by what fossil evidence there is and information derived from palaeogeography, palaeoclimatology, etc. Recently statistical models employing computer programmes have been devised to handle such problems but a great deal of controversy surrounds such techniques (numerical phyletics) and many of the basic assumptions have been challenged. Part, but by no means all of the difficulty arises from the need to express the information used in terms of unit characters as I have discussed above in connexion with numerical taxonomy, and it has even been suggested that there is a considerable element of circularity involved, the so-called phyletics being in some cases simply a re-arrangement of a phenetic classification, the same data being used for both. This is a most controversial area and cannot be thoroughly explored here. Moreover it is a subject in its infancy and it is perhaps unfortunate that the hard-earned data on amino-acid sequences should be involved in the fray. My purpose here is simply to draw attention to the fairly primitive state of the art at the present day.

For reasons which I do not entirely comprehend, the work of many chemical or biochemical contributions to systematics seems to be aimed at assisting in the elucidation of phylogenetic relationships within the angiosperms. It is largely the fault of systematists that a false picture may have been presented to the outside world of the state of angiosperm phylogeny. Reference to handbooks tends to reveal a small number of competing systems, none of which is properly justified. Not many systematists concern themselves directly with the overall schemes proposed, such as those of Cronquist, Engler & Prantl, Hutchinson, Takhtajan, Thorne, Soó, etc., although they may contribute, through their own research, information which could modify some particular aspect of these schemes. They are useful for teaching purposes— taxonomy without phylogeny can be boring. My impression after consulting with colleagues in various parts of the world is that angiosperm phylogeny is not a subject of major concern to them although they are quite happy for

other people to engage in such studies. It has little practical value from the point of view of identification, arranging collections or extracting information as opposed to a well prepared taxonomic classification. I would go so far as to suggest that if it is generally felt that biology would advance significantly if agreement could be reached on the probable evolutionary disposition of the families of angiosperms, then we ought to make more serious efforts to this end, rather than rely on the devoted efforts of a small number of individuals. The task would demand a major programme of co-ordinated research.

Within the overall set of relationships shown by the components of the organic world, a thread of ancestor-descendent relationships exists and it is this which we purport to show in phylogenetic trees. It is understandable that we should attempt to retrace this thread since its very existence poses a challenge and at the same time provides a backbone to the kinds of diversity and relationship we work with. Unfortunately the skeleton provided by these ancestor-descendent relationships, usually represented in the form of a tree, is both difficult to reconstruct in most groups and of limited value if successfully produced. It is the biological and chemical processes involved in the organisms (and their organ systems, organs and attributes) which make up the trees that most biologists are interested in, not the architecture of the tree itself. It is as well, therefore, that we remind ourselves why we attempt such reconstructions.

As regards the angiosperms, the position is well summarised by Emberger [6]—"De magnifiques arbres ont été conçus: leur frondaison nous enchante, et on voudrait se reposer à leur ombre. Hélas, ils sont peu solides, car ils sont nourris uniquement de théories, ce qui, on le sait, n'est pas très substantiel". In practice most angiosperm systems depend to a large extent on a preconceived notion as to the idea of the primitive flower. I suspect we have underestimated the extent to which such notions have influenced and biased phylogenetic thinking. Admittedly, there is a good deal of evidence from various sources which tends to support in a general way the most widely accepted view of the Magnoliad-type of plant as "primitive" or ancestral. On the other hand there is a tendency to select those lines of evidence which support a particular theory, especially one which in one form or another goes back to De Candolle and Goethe and appeals to our sense of logical progression. Despite recent clarifications, there is little in the fossil record of the angiosperms which contributes convincingly to one theory or another. Even the monocolpate grains associated with leaf-fossils of a type found in the Magnoliads today, in the Albian deposits, can in no way be taken to indicate that the plant bearing them possessed other features that would place it in this complex. Even if this were to be established it would tell us nothing about what other groups might have existed at that epoch and until extensive fossil series at a critical early level are found, we are forced to rely very heavily on conjectures.

It would be wrong to overlook the impressive body of evidence building up which suggests that more than one line may have been involved in the evolution of angiosperms. By seeking a single basal group we may be placing ourselves in a conceptual straightjacket. Evidence from the origin of other groups of organisms would lend support to the viewpoint that the angiosperms constitute a pleiophyletic grade, rather than a monophyletic clade. The painstaking morphological work of Meeuse [42–46], the gonophyll theory of Melville [47], the chemical evidence of Kubitzki, the cytochrome c evidence of Boulter, and the arguments of Croizat, Emberger, and others, together with the fact that our knowledge of homologies in the angiosperm flower are very limited, all add up to an impressive challenge to what we might consider traditional views.

The most promising novel approach to reconstructing angiosperm phylogeny is the method applied by Boulter and co-workers for relating observed amino acid differences in pairs of homologous proteins in contemporary organisms to the minimum number of DNA base-pair substitutions that would be necessary to account for present day amino acid sequences if they were all derived from a single ancestral DNA structural gene [36]. Boulter's group works with cytochrome c and more recently with plastocyanins.

Even if it is accepted that (1) the technical biochemical basis of this work is sound (as would indeed appear to be the case), (2) the logic of the procedure is valid and (3) the concept of evolutionary parsimony in reconstructing trees is tenable, the results would appear to have a limited value in suggesting the relative sequence in which the representatives of the groups sampled evolved. It does not tell us which group as such is ancestral to which nor does it tell us what characters, other than those of the cytochrome c any ancestral group may have possessed. No modification of this approach can circumvent these limitations.

It is not profitable at this stage to attach too much importance to the results of the method: the sampling, however impressive from a technical point of view, is far too limited to bear detailed discussion. If it proves possible to extend this work and greatly increase and refine the sampling in terms of organisms and proteins, the results may well establish a case for detailed re-thinking of our concepts of evolutionary derivation of existing angiosperms. This in turn would force systematists to adopt a much more vigorous and scientific approach to the handling of the general evidence on which phylogenetic schemes of the angiosperms are based, as suggested above.

It is somewhat ironic that one of the newest biochemical approaches should be applied to one of the oldest problems of systematics and highlight the strengths and weaknesses of the classical method.

References

1. Birch, A J, Pure & appl chem 1973, 33, 17.
2. Sylvester-Bradley, P C, Heywood, V H & Sneath, P H A, Nature 1964, 203, 358.
3. Davis, P H & Heywood, V H, Principles of angiosperm taxonomy. Oliver and Boyd, Edinburgh and London, 1963.
4. Solbrig, O T, Plant biosystematics. The MacMillan Company, London, 1970.
5. Engler, H G A & Prantl, K A E, Die natürlichen Pflanzenfamilien. Leipzig.
6. Emberger, L, in Traité de Botanique systématique, Chadefaud, M & Emberger, L. Masson, Paris, 1960.
7. Melchior, H, A. Engler's Syllabus der Pflanzenfamilien, vol. 2. Bornträger, Berlin-Nikolasser, 1964.
8. Cronquist, A, The evolution and classification of flowering plants. Nelson, London, 1968.

9. Takhtajan, A, Flowering plants—origin and dispersal. Oliver and Boyd, Edinburgh, 1969.
10. Hutchinson, J, The families of flowering plants, 2nd edn. Oxford University Press, Oxford, 1959.
11. Soó, R von, Fejlödéstörténeti Növényrendszertan. Tankönyvkiadó. Budapest, 1963.
12. Sokal, R R & Camin, J H, Syst zool 1965, 184.
13. Heywood, V H, Pure & appl chem 1973, 34, 355.
14. Daniels, G S, Arnoldia 1973, 33, 26.
15. Heywood, V H, Bull jard bot nat belg 1967, 37, 31.
16. Morse, L E, Taxon 1971, 20, 269.
17. Pantin, C F A, Relations between the sciences. Cambridge University Press, London, 1968.
18. Davis, P H & Heywood, V H, Principles of angiosperm taxonomy, p. 113. Oliver and Boyd, Edinburgh and London, 1963.
19. Cullen, J, Botanical problems of numerical taxonomy, in Modern methods in plant taxonomy (ed V H Heywood) p. 175. Academic Press, London and New York, 1968.
20. Heywood, V H, Ecological data in practical taxonomy, in Taxonomy and ecology (ed V H Heywood) p. 330. Academic Press, London and New York, 1973.
21. Whewell, W, Philosophy of the inductive sciences, 2nd edn. John Parker, London, 1847.
22. Davis, P H & Heywood, V H, Principles of angiosperm taxonomy. Oliver and Boyd, Edinburgh and London, 1963.
23. Heywood, V H, The characteristics of the scanning electron microscope and their importance in biological studies, in Scanning electron microscopy (ed V H Heywood) p. 1. Academic Press, London and New York, 1971.
24. McNeill, J, Parker, P F & Heywood, V H, A taximetric approach to the classification of the spinyfruited members (tribe Caucalideae) of the flowering plant family Umbelliferae, in Numerical taxonomy (ed A J Cole) p. 129. Academic Press, London and New York, 1969.
25. Crowden, R K, Harborne, J B & Heywood, V H, Phytochemistry 1969, 8, 1963.
26. Heywood, V H, Boissiera 1971, 19, 289.
27. Harborne, J B, Recent adv phytochem 1972, 4, 107.
28. Kubitzki, K, Taxon 1969, 18, 360.
29. Hennig, W, Phylogenetic systematics (transl. D Dwight Davis & R Zangerl). University of Illinois Press, Urbana, Chicago and London, 1966.
30. Turner, B L, Pure & appl chem 1967, 14, 189.
31. Harborne, J B, Progress in phytochemistry 1968, 1, 545.
32. Swain, T, Plants in the development of modern medicine. Harvard University Press, Cambridge, Mass., 1972.
33. Alston, R E, Evolutionary biology 1967, 1.
34. Heywood, V H, Pure & appl chem 1973, 34, 355.
35. Erdtman, H, Pure & appl chem 1963, 6, 679.
36. Boulter, D, Pure & appl chem 1973, 34, 539.
37. Harborne, J B, Recent adv phytochem 1972, 4, 107.
38. Weimarck, G, Taxon 1972, 21, 615.
39. Holmquist, R & Jukes, T H, J mol evol 1972, 2, 10.
40. Mabry, T, Recent adv phytochem 1972, 4, 273.
41. Heywood, V H (ed), Taxonomy and ecology. Academic Press, London and New York, 1973.
42. Meeuse, A D J, Advanc front pl sci 1962, 1, 105.
43. — Ibid 1965, 11, 1.
44. — Fundamentals of phytomorphology. New York, 1966.
45. — Acta biotheoretica 1972, 21, 167.
46. — Acta bot neerl 1972, 21, 113, 235, 351.
47. Melville, R, Kew bull 1962, 16, 1; 1963, 17, 1.

Discussion

Geissman: Would you care to comment on the disconcerting observation that there can exist qualitative differences in chemical composition of plants with, for example, age and polyploidy with respect to the use of chemical characters in taxonomy.

Heywood: Although chemical variants within species can be described and named, it is more usual simply to accept the taxonomic units, such as species, as containers within which much variation of the kinds you have mentioned can be found, but which is not directly expressed by the classification. Recognition of chemical ecological and other kinds of variation within species would often lead to a multidimensional overlapping system of intraspecific categories which few people would advocate today: Seasonal or organ to organ chemical variation mainly poses a sampling problem.

Sørensen: I should not like to join in the discussion between Professor Heywood and Professor Geissman, but I should like to ask why you state that chemical data should—or perhaps could not—be handled in the same way as morphological data. Since the 1880's the botanists working with lichens have used chemical tests—in reality extremely primitive chemical tests—in their keys for lichen determinations. And since the start of this century the pattern of fermentation tests has played a steadily increasing role in the classification of bacteria and other microorganisms. I quite agree that you do not have the same need for chemical tests on the flowering plants, but what I like to know is why you give the chemical tests a lower taxonomical value in the angiosperms.

Heywood: I was referring to the fact that in angiosperm taxonomy a morphological bias is introduced into the process of classification by using our visual approach. Chemical characters do not participate in the Gestalt perception, but, if used, have to be added in afterwards. This traditional method of taxonomy, which has been employed for nearly all higher organism classifications in use today, does not consider variation in terms of separate characters initially but as visual perception patterns. Non-visual or indirect characters, which need auxiliary means for their appreciation, simply cannot be taken into account in this primary stage of classification. Indeed seeking out the organisms with which one is going to work by looking for them introduces a morphological emphasis. Unless one uses taximetric or numerical techniques, chemical information has to occupy a secondary role. Chemical tests may be used as a means of diagnosis as in lichens, but this is not the same as constructing the classes one is going to diagnose.

Hedberg: The intraspecific phytochemical variation referred to by Professor Geissman is matched in many cases by the large and ±continuous variability in morphological characters —often much larger in tropical than in temperate areas.

Runemark: (Reply to Geissman). In a plant species with a reasonable amount of cross-fertilization all individuals are genetically different and also chemically different. This makes recognition of chemical races in taxonomy impracticable.

Birch: In discussing the general question of relations between chemists and botanists, I note that chemists on the whole have had in this area rather restricted interests. They have defined interesting structures and have often proceeded to use these to try to define biosynthetic reactions. Only in a very secondary way have they tried to use the data to contribute to classification, a fact attested by the vagueness of the botanical nature of the material, examined chemically in an exact manner. The botanist probably has the major responsibility to define the critical areas, and then to persuade or even employ a chemist to carry out appropriate work.

The availability of new practical techniques: t.l.c., g.l.c. and m.s., etc. makes this approach increasingly feasible.

Cronquist: I want to comment briefly on Dr Geissman's problem with Dr Constance's identification of two chemically distinct things as belonging to the same species. It is of course possible that Constance has simply made a mistake. But it is also perfectly possible to have two chemically or morphologically distinct local races which cannot be treated as specifically different, because the differences between them are bridged by other local races of the species. Such local races may maintain themselves as distinct because of some barrier to interbreeding such as polyploidy, self-pollination, or apomixis; yet in the context of the total population, they cannot reasonably be treated as representing different species.

Heywood: Perhaps, as an example of the kind of research suggested by Professor Birch, I should mention the Umbelliferae project on which Dr Harborne and I and many collaborators have been engaged for the last 10 years (see Heywood, V H (ed), Biology and chemistry of the Umbelliferae. Academic Press, London, 1972). The group we are studying—the tribe *Caucalideae*—contains about 20 genera and 100 species and despite a great deal of research only about 50% have been sampled for a range of compounds. It is difficult to maintain the interest of a team of workers on a single project after a number of years. In addition to the phytochemical research, all other kinds of information about the genera and species concerned has to be accumulated—anatomical, cytological embryological, palynological, etc., and the handling of the mass of data produced poses major problems for construction of a classification.

A Chemical Compound as a Taxonomic Character

V. Herout

Institute of Organic Chemistry and Biochemistry, Czechoslovak Academy of Sciences, Prague, Czechoslovakia

Summary

From the definition of a taxonomic character it follows that a chemical character may also serve for the classification of living organisms. In this article an attempt is made to define the extent of its validity, and its importance is demonstrated on examples by comparison with characters of other types.

In this article a modest attempt has been made to define the importance of chemical compounds as taxonomic characters.

If we accept the definition recently formulated by Heywood [1] for a taxonomic character we can see that today the use of chemical characters has become indisputable: ". . . A character may be generally defined in taxonomy as any attribute referring to form, structure, physiology of behaviour which is considered separately from the whole organism for a particular purpose such as comparison, identification or interpretation." . . . "Defined in this way biochemical, physiological, cytological and other kinds of data are covered as well as traditional morphological features."

The fact that the chemical character has been included in the arsenal of taxonomists does not change the problem concerning what exactly may be taken safely from the field of phytochemistry, for example, as utilizable for classification, and what significance the evaluation of such a character has. An over- or underestimation of this relatively newer character still remains a problem.

This question fundamentally concerns the principle itself: to what extent chemical compound in general may be utilized for classification purposes? It is clear that substances that are ubiquitous in living organisms are valueless for such a purpose, as for example glucose and other sugars, about 20 common amino acids which form the basis of all proteins, simple fatty acids (cf [2]) etc. In contrast to this the accumulation of silicic acid is usually considered as a character typical of Equisetopsida [3]. Another character, investigated for a long time and typical of some phylogenetically very young families, is the accumulation of polysaccharides based on fructose which differ in their chemical composition from starch (for example inulin in Compositae, Campanulaceae, etc.).

However, attention has more recently been focused on substances which are directly connected by their structure with the genetical equipment of the cell. Such substances are mainly desoxyribonucleic acids the parts of which directly constitute the genetic material of the organism. Progress in the knowledge of these substances, however, is not yet at the stage when the utilization of the collected knowledge on the chemical structure of the genetic material might become directly useful for the purpose of determining the differences of species or of contributing to a similar extent to our knowledge of the phylogenesis of single taxons of organisms existing today.

From this point of view the knowledge of the primary structure of proteins, the structure of which in each organism is determined just by the grouping of the desoxyribonucleic acids present, seems more promising not only for today but more especially for the near future. Further, the activities of the complements of proteins form the basis of all functions of any organism and give rise to its morphological characteristics. This theme has recently been treated for example by Boulter [4] who already uses the comparison of amino acid sequence data for the determination of evolutionary relations of higher plants. Up to now, amino acid sequence data of some simple and ubiquitous proteins, such as ferredoxins and cytochrome *c*, have proved suitable: the same is also true, for example, of hemoglobin in animals. The advantage of the study of the structure of a single protein consists in the fact that it can be considered in phylogenetic schemes as a reflection of the direct relation of protein and gene.

From this point of view one may also judge the earlier attempts to use catalytic properties [5] and particularly serological methods for comparative taxonomic studies (for example see reviews [6, 7] or Jensen, p. 217). These somewhat complex approaches are at present substituted by precise separation methods, as for example gel or disc electrophoresis, which permit the recognition of the identity or the difference of the investigated protein complex in the compared species; cf e.g. [8, 9]. Hence, it seems hopeful that in the future more reliable phylogenetic relations will substantially contribute to the accumulation of data on the primary structures of different proteins from the same organism [4].

However, at present the investigation of micromolecular compounds which are commonly summarised under the somewhat unsuitable name of "secondary metabolites" still remains the dominant approach. We do not know the true function of these substances, comprising alkaloids, terpenoids, flavonoids, glycosides, non-protein amino acids, polyacetylenes, etc. It is clear, however, that their incidence is a result of a coactivity of a series of enzymatic systems present in single species; thus their direct relationship with their genetic equipment again appears. In the case of these substances their practical importance for systematics has been summarized many times and some examples have become classical; (for example the incidence of red and yellow betalain pigments in the group of families classified as Centrospermae where they completely replace the pigments of the anthocyanine type (e.g. [10, 11]). However, it should be stressed that not all substances, concerning their occurrence, can be regarded as typical for a restricted circle of genera, families or higher categories. Therefore many attempts have been made which base this clas-

(*1*) Daucane (*2*) Pinguisane (*3*) Bazzanane

sification on substances whose occurrence is by no means so narrowly limited. Here I may mention a few examples from my own field of work. The occurrence of some sesquiterpenoids based on typically cyclised skeletons is quite typical and it seems a good character for their classification in a certain family. For example, the daucane skeleton (*1*) has until now been found with certainty only within several tribes of Umbelliferae (see [12, 13]); with the pinguisane skeleton (*2*) substances have only been isolated from some liverworts [14, 15], just like the bazzanane skeleton (*3*), see [16]. The occurrence of other skeletal types is by no means so narrow and is therefore not at all interesting from the point of view of systematics (for example cadinane, selinane, caryophylane, humulane, and many others). Therefore it is possible to regard with scepticism a whole series of papers in which inferences on systematics are based on the occurrence of these compounds whose diffusion in nature is such a poor characteristic. This may also be stated with greater justification of the occurrence of many monoterpenes (common α- and β-pinenes, myrcene, or limonene) and oxygenated derivatives (e.g. cineol) which are very common in the plant kingdom. In particular quantitative differences in the content of compounds of this type may, at first sight, seem interesting, but the ascertained difference in quantity is far from being a true "taxonomic character" and it cannot be evaluated without taking environmental factors into account. Even then statistical and computer programs are indispensable but, of course, only at the species level or below it and never in higher taxons (see also Turner, p. 123). The effects of external media very often significantly influence the quantitative composition of essential oils as their producers are well aware. Similarly the general use of the composition of *n*-alkanes for taxonomic conclusions is rather dubious, as we discovered in one of our studies on the dependence of the occurrence and the composition of cuticular n-alkanes of various plants on the season of the year, microclimatic conditions, and the origin (difference in composition on the upper and the lower side of the leaves [17]). We consider that the tendency to overestimate these and similar correlations for the purposes of systematics is predominant among chemists who lack a more detailed knowledge of the value of taxonomic characters.

In contrast to this, specialists in biology, to whom more detailed information on the nature, chemical structure, and the biosynthesis of the compounds studied lies somewhat outside their scope, may have the tendency to overestimate the importance of chromatographic spots or peaks of whose chemical nature nothing substantial is known. However, it would certainly not be correct—in this case—to base the relationship or its remoteness on a larger or smaller number of "identical" coloured spots only. The number of "spots", for example, in thin-layer or paper chromatography is a function of the amount of the extract under investigation, because the threshold amounts are not identifiable. Nevertheless chromatographic methods and their evaluation in the hands of an experienced investigator are a valuable aid in the study of the relationship of genera, for example, within the frame of tribes or families and are moreover very useful in the study of interspecific hybrids (cf e.g. the numerous examples of comparisons of paper-chromatographic maps of flavonoids from *Baptisia sp.*; e.g. [18]; the comparison of albaspidins from different *Dryopteris sp.* [19] or, last but not least, the articles in this volume by Bate-Smith, p. 93, or Grant, p. 293.)

Similarly, gas chromatography of volatile substances present in the glandular hairs of some species of the *Rosa* genus may contribute in checking the accuracy of the classification into corresponding sub-sections [20]. For example the records of chromatograms of these substances from the genera *R. albiflora* Opiz, *R. gizellae* Bork., and *R. inodora* Fries (all from the subsection Sepiaceae; number of chromosomes $2n=42$) carried out under the same conditions differ very little among themselves. *R. tomentosa* Sm. and *R. sherardii* Davies (subsection Vestitae; number of chromosomes $2n=36$) are also very similar; of course, both types of re-

(*4*) Furanoeremophilane

cords differ from each other substantially. Similar relationships have so far been studied up to now in 14 species of the genus *Rosa* and in most instances good agreement was found with the classification based on morphological and cytological characters. Hence, although the nature of the substances producing the perfume of these glandular species of roses is not known —it is under study even this year—a gas-chromatographical record may serve as an aid in following the previously determined relationships. Similarly, other accurately defined records may also be employed, as has been shown in the paper by Schechter & de Wet [8] dealing with the comparison of disc-gel electrophoretic patterns of different proteins, isolated from seeds of cultivars *Sorghum,* or by Fairbrothers [9] of seed proteins and isoenzymes of *Danthonia sericea*.

The utilizability of a chemical character, especially in the case of the so-called secondary metabolites, may sometimes be reduced because the occurrence of the substances of the expected type is not universal, i.e. some species do not contain these substances in a detectable concentration. Let me mention as an example our studies of several genera of the Senecioneae tribe (Compositae family): for the intrageneric classification of the plants of *Petasites* and for the tracing of the relationships of this genus we successfully used in our laboratories [21–23] the occurrence of sesquiterpenic substances of furanoeremophilane type (basic compound furanoeremophilane (*4*)). These substances were also found by us (cf e.g. [24]) abundantly in a number of other related genera, for example *Senecio, Homogyne, Adenostyles,* and by other working groups also in *Ligularia* [25], *Cacalia* [26] etc. To these genera the species *Tussilago farfara* is very close, but no single trace of eremophilane derivatives has so far been discovered in this species. However, it is quite possible that some error is involved, because we have not yet studied all the parts of this plant, or have not chosen the proper season. In a recent paper we demonstrated the variability of the content of furanoeremophilane derivatives in single parts of *Petasites hybridus, P. albus* and *P. kablikianus* [27]. It is also possible that in the afore-mentioned species the blocking of some biosynthetic routes occurred and the synthesis of derivatives of this type became "dormant", as supposed by Erdtman for similar secondary plant products (see, e.g. [28]), i.e. a "chemical divergence" is involved. This existence of a divergence, however, is not typical of the chemical character alone but is also common in morphological characters, whether the frequency or merely the simple occurrence of flower organs, etc. is involved.

One of the advantages of the chemical character is the fact that it can suitably complete the lack of criteria for the assessing of the accuracy of the classification of some taxon.

Thus we have found that *Adenostyles alliariae* [29] contains in its vegetative parts the same or very similar derivatives of the furanoeremophilane type as a series of the members of Senecioneae tribe. But, with the exception of a negligible number of cases, according to the present view on the main morphological character, the genus *Adenostyles* is classified in the Eupatoriae tribe (the flowers are tubular, not ligulate). The problem arose as to whether the occurrence of this chemical character is not anomalous, because it has not yet been observed in any plant of the Eupatoriae tribe. On the basis of an analysis of a complex of morphological and anatomical characteristics from the biology of flowers, and on the basis of phytogeographic and cytological analysis our collaborators have demonstrated that the classification of the *Adenostyles* genus into the Senecioneae tribe is justified by all the characters mentioned. The absence of ligulate flowers may be explained satisfactorily by the disappearance of the border rows of the heads and thus also the reduction in the number of flowers.

Another example [30, 31] in which the difficulties of classification have been caused mainly by the lack of good characters is the liverwort *Riccardia*, very rarely forming sporangia. An anatomically different species, *R. pinguis*, is sometimes classified in the genus *Aneura*, as *A. pinguis*. In the majority of classifications it is usually left in the genus *Riccardia* (the taxon

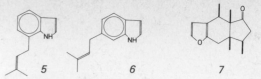

(5), (6) Alkaloids from *Riccardia incurvata* Lindb. and *R. sinuata* (Hook.) Trev. (7) pinguisone from *Aneura pinguis* (L.)

Aneura mostly considered as a subgenus). We have been able to compare the chemical composition of this species with the composition of *R. incurvata* Lindb. and *R. sinuata* (Hook.) Trev.; they are morphologically close types which anatomically correspond to the original concept of the *Riccardia* genus. While both *Riccardia* contained two simple indole alkaloids (5, 6), we were unable to find these alkaloids in *Aneura* at all, while its typical and unique component was the sesquiterpenic furanoketone pinguisane (7). Hence, in this case the chemical characters show themselves by their objective attributes, to be extremely valuable, and in this debatable case even as decisive criteria.

In spite of these indisputable results we cannot regard the occurrence of single separate micromolecules representing fortuitously isolated secondary metabolites as the final aim on which biosystematics could be exclusively based. It is becoming increasingly evident that our knowledge of structure-specific biosynthetic pathways is more likely to be of significance in specific evolutionary development. It has long been known that from the chemical point of view the same metabolites could have arisen as a result of very different biosyntheses based on the effect of specific enzyme systems. Thus the presence of the same substance may have quite a different biological meaning [28, 32]. As an illustration we may mention the very elegant study by Zenk [33], who proved that natural derivatives of naphthoquinone are formed by in at least four different biosynthetic pathways (at least in the plant kingdom), and he also demonstrated that phylogenetically differently related groups of plant families also make different use of these possibilities.

Knowledge of a specific biosynthetic pathway should therefore be considered as more important than the mere knowledge of the specific structure of the compound during bio-

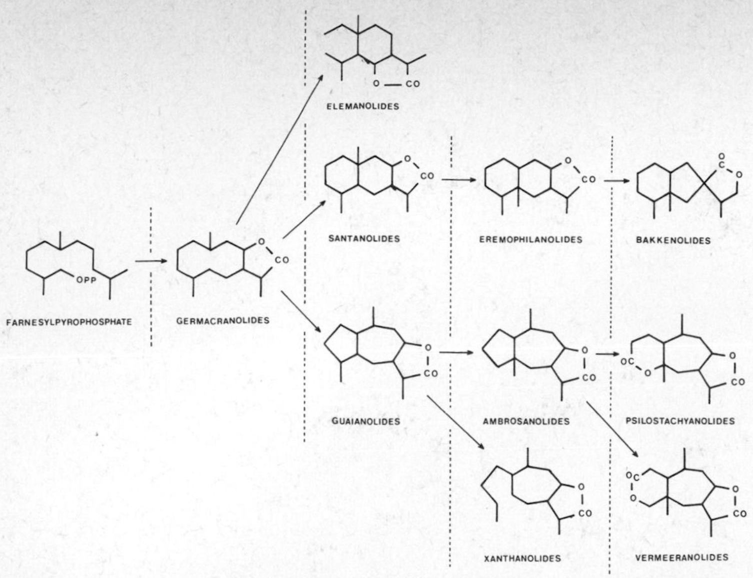

Fig. 1. Scheme of the supposed biosynthesis of sesquiterpene lactones of various types (on the basis of carbon skeletons only).

systematic and phylogenetic investigations. Hence, information should become available from studies of the enzyme systems involved and from their substrate specificities and catalytic activities, especially in the future.

Our study of the relations between the single tribes, in which the Asteraceae (Compositae) family is generally classified, is based on these premises. On the basis of biogenetic relationships during the formation of sesquiterpenic γ-lactones (fig. 1)—at present merely conjectured but in many respects very probable— we have attempted [34] to create a "phylogenetic tree" when comparing them with the deductions deriving from morphological characters (especially the structures of flower organs); the concept, though not in entirely common form, is shown in fig. 2. For the biosynthesis of such stereo- and structure-specific cyclisations or oxidations of the original farnesol chain, the presence of very characteristic enzyme systems should be supposed. It is also useful to remember again that on the basis of our present knowledge of the occurrence of lactones of exactly this type, the existence of probably very analogous systems may be supposed, not only in Umbelliferae (which, by the way, are considered by some authors to be not

too far from Compositae), but also in relatively primitive families of Angiospermae, as for example Magnoliaceae and Lauraceae, and especially newly in liverworts. These, like one of the phyla of Bryophyta, represent plants much lower from the evolutionary point of view, which seems to prove that even very complex biosynthetic pathways may be independently repeated in the course of evolution (or even forgotten and then utilised again in unrelated organisms).

In connection with the loss of the ability to synthetize some biosynthetic stage it is known that a mutation causing a similar loss is usually a drawback rather than an advantage for the mutant [cf 32]. Although this fact was observed in many artificially obtained mutants, especially in micro-organisms, it may be considered that for the evolutionary process even those—probably relatively few—mutations have a decisive importance in consequence of which the acquirement of new, and (for the existence of the individual) positive properties take place. For example, the production of a new metabolite the presence of which manifests itself positively in the "struggle for survival" (the formation of a substance whose presence in the plant protects it more or less from voracious predators) means a positive acquisition. The progeny of such a mutant in comparison with other individuals has an enhanced chance to survive.

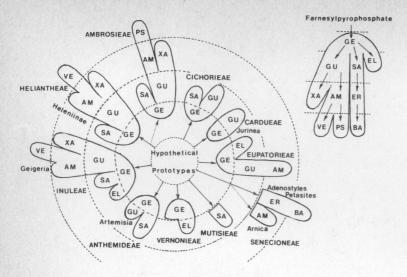

Fig. 2. Scheme of the supposed phylogenetic relations between the tribes of Compositae family, accounting for the ability to biosynthetize sesquiterpenic lactones of different skeletons (each dashed circle represents a step of the biosynthesis of the skeletons of these substances supposingly involving a single enzymatic system). *GE,* germacronolides; *EL,* elemanolides; *SA,* santanolides; *ER,* eremophilanolides; *BA,* bakkenolides; *GU,* guaianolides; *AM,* ambrosanolides; *PS,* psilostachyanolides; *XA,* xanthanolides; *VE,* vermeeranolides.

This view does not contradict Birch's recently published assumption [35] according to which the more advanced biogenetic sequence must necessarily be the latest in evolution. It should be admitted that such more advanced sequence (from the biogenetical point of view) might lead to "less advanced" compounds—from a purely chemical viewpoint—if the last biogenetic step leads, for example, to chemical degradation of a part of the intermediate. In fact, by this a chemically simpler end-product is obtained. In such a case a mutation occurring with loss of such a biogenetic stage would therefore cause the product to appear to be more complex. Therefore the fact should be remembered again that in the evolution of organisms the existence of reductive processes may be followed not only in morphological but certainly in chemical characters as well. Of course, this shows that a simplified view, as mentioned above, often becomes too simple.

References

1. Heywood, V H, Plant taxonomy, 4th edn, p. 29. The English Language Book Society and E. Arnold, London, 1970.
2. Erdtman, H, in Chemical plant taxonomy (ed T Swain) p. 89. Academic Press, London and New York, 1963.
3. Hegnauer, R, Chemotaxonomie der Pflanzen, vol. 1, p. 244. Birkhäuser, Basel and Stuttgart, 1962.
4. Boulter, D, in Progress in phytochemistry (ed L Reinhold & Y Liwschitz) vol. 3, p. 199. Interscience-Wiley, London, 1972.
5. Blagověschenskij, A V, Die biochemischen Grundlagen des Evolutionsprozesses der Pflanzen (in German). Akademie-Verlag, Berlin, 1955.
6. Leone, C A, Taxonomic biochemistry and serology. Ronald Press, New York, 1964.
7. Gibbs, R D, in Chemical plant taxonomy (ed T Swain) p. 64. Academic Press, London and New York, 1963.
8. Schechter, Y & de Wet, J M J, in Abstracts of plenary lectures, p. 24. IUPAC Symp. Chemistry in evolution and systematics, Strasbourg, 1972.
9. Fairbrothers, D E, in Abstracts of plenary lectures, p. 25. IUPAC Symp. Chemistry in evolution and systematics, Strasbourg, 1972.
10. Dreiding, A S, in Recent developments in the chemistry of natural phenolic compounds (ed W D Ollis) p. 194. Pergamon, Oxford, 1961.
11. Mabry, T J, in Comparative phytochemistry (ed T Swain) p. 231. Academic Press, London and New York, 1966.
12. Hegnauer, R, in The biology and chemistry of the Umbelliferae, suppl. 1. to the Bot j of the Linnean soc. 1971, 64, p. 267.
13. Williams, C A & Harborne, J B, Phytochemistry 1972, 11, 1981.
14. Benešová, V, Samek, Z, Herout, V &Šorm, F, Coll Czech chem comm 1969, 34, 582.

15. Krutov, S M, Samek, Z, Benešová, V & Herout, V, Phytochemistry 1973, 1405, 1973.
16. Hayashi, S & Matsuo, A, Experientia 1969, 25, 1139.
17. Streibl, S, Stránský, K & Herout, V, Coll Czech chem comm 1967, 32, 3213.
18. Alston, R E & Hempel, E F, J heredity 1964, 55, 267.
19. Widén, C-J & Britton, D M, Can j bot 1971, 49, 247.
20. Kolátorová, E, Konečný, K & Streibl, M, Acta musei Silesiae, ser. Dendrologia 1972, 133 (in Czech).
21. Novotný, L, Toman, J, Starý, F, Marques, A D, Herout, V & Šorm, F, Phytochemistry 1966, 5, 1281.
22. Novotný, L, Toman, J & Herout, V, Phytochemistry 1968, 7, 1349.
23. Herout, V & Šorm, F, in Perspectives in phytochemistry (ed J B Harborne & T Swain) p. 158. Academic Press, London and New York, 1969.
24. Novotný, L & Šorm, F, Beiträge zur Biochemie und Physiologie von Naturstoffen, p. 377. G Fischer, Jena, 1965.
25. Ishii, H, Tosyo, T & Minato, H, Tetrahedron 1965, 21, 2605.
26. Rodríguez-Hahn, L, Guzmán, A & Romo, J, Tetrahedron 1968, 24, 477.
27. Novotný, L, Kotva, K, Toman, J & Herout, V, Phytochemistry 1972, 11, 2795.
28. Erdtman, H, in Chemical plant taxonomy (ed T Swain) p. 89. Academic Press, London and New York, 1963.
29. Harmatha, J, Samek, Z, Novotný, L, Herout, V & Šorm, F, Coll Czech chem comm 1969, 34, 1739.
30. Benešová, V, Samek, Z, Herout, V & Šorm, F, Coll Czech chem comm 1969, 34, 582.
31. — Ibid 1969, 34, 1807.
32. Birch, A J, Biosyntetic pathways in chemical plant taxonomy (cd T Swain) p. 141. Academic Press, London and New York, 1963.
33. Zenk, M H, in Abstracts of plenary lectures of IUPAC symposium on chemistry in evolution and systematics, p. 15. Strasbourg 1972.
34. Herout, V, Chemotaxonomy of the family Compositae (Asteraceae) in pharmacognosy and phytochemistry (ed H Wagner & L Hörhammer) p. 93. Springer, Berlin, Heidelberg, New York, 1971.
35. Birch, A J, Pure and appl chem 1973, 35, 17.

Discussion

Merxmüller: In my opinion some of these problems are not as controversial as they are thought to be. Especially in the large families it has become extremely difficult to survey the whole recent taxonomic literature concerned. In the Compositae it is simply misleading to go back to the last overall treatment by O. Hoffmann. In the case of *Adenostyles,* e.g., nobody will continue to treat it as a member of Eupatorieae, at least if he knows Vierhapper's convincing paper of 1925. The incompatibility of *Arnica* with all other Senecioneae has already been demonstrated by the serological work of Schumacher in the early sixties. The remarkable chemical inhomogeneity of the so-called tribe Inuleae may be explained by the morphological findings of Leins and other collaborators of my Institute that the Inulinae/Plucheinae and the Gnaphaliinae and related groups are separated to such an extent that both groups merit tribal recognition. As far as *Geigeria* is concerned it also holds in Leins' morphological groupment of Inuleae quite an isolated position.

Herout: We have just found by using sesquiterpenoid lactones as a useful character for establishing possible relations between the Compositae tribes that the genera mentioned by Professor Merxmüller do not fit with our concept; they stay, minimally speaking, aside.

Herz: What is the affinity of *Geigeria?* Is it related to all to the tribe Heliantheae? The poisonous constituents you have referred to are also found in some representatives of one genus of Heliantheae, i.e. in *Hymenoxys.*

Mabry: I note that you suggest independent origin for the structurally similar sesquiterpene lactones which occur in the Compositae and the Magnoliaceae and nowhere else in the angiosperms. Is it not possible that the Compositae is much older than the fossil records indicate and has direct links to the Magnoliaceae?

Cronquist: Who knows what is possible? I can say that in the fossil record, the Asteraceae can be traced back as far as the Miocene-Oligocene boundary. It is of course possible that they really go much farther back, but I would be very much surprised to find any early fossil connection directly to the Magnoliales.

Turner: I wish to respond to the observation by Prof. Merxmüller that *Arnica* perhaps belongs to the tribe Heliantheae instead of the Senecioneae. I disagree, preferring to retain it in that tribe; in addition I would place near this genus a number of morphologically isolated, North American genera. In short, the European view of the Senecioneae appears to have been too provincial; at least it would make easier the disposition of a number of North American genera *if* the tribe's circumscription was extended to accommodate these (Powell & Turner. In press). Thus, observe the eternal taxonomic shifting that must go on amongst the more classically oriented systematicists using purely morphological characters. Chemical characters should do much to resolve such controversy, hence the significance of the synthesis you have presented here.

Herout: (reply to Bu'Lock). In the past we tried to encourage Professor Bohlmann, University of Berlin, to summarize his rich material on polyacetylenes in Compositae with regard to possible classification of this family.

Sørensen: According to investigations of Schulte and later on by Bohlmann some *Arnica* species contain acetylenes in all parts of the plants. As far as I know, acetylenes have only been detected once in the flowers of one single *Senecio*; none in the root part which is mostly the best source. So from a chemists point of view it is very nice if *Arnica* has been removed from the Senecioneae.

As to Professor Turner's question how many Senecioneae have been investigated, I would guess some 50, mostly European and Australian species. The only American species I remember is again an *Arnica* species from Alaska, which behaved as the European *Arnicas*.

The Chemistry of Disjunct Taxa

T. J. Mabry

The Cell Research Institute and Department of Botany, The University of Texas at Austin, Austin, Tex. 78712, USA

I will not review here (in the written version of my lecture) the natural products chemistry of disjunct taxa, since a review paper on this topic just appeared in print [1]. Moreover, additional aspects of the chemistry of disjunct taxa were touched upon in several of a series of papers presented at a symposium entitled "Disjunctions in plants", which were collectively published [2].

Here, I wish to note only one or two of the trends which have emerged from our recent chemical investigations of disjunct taxa which occur in North and South America. Most of these studies are being conducted as one aspect of a project entitled "Origin and structure of ecosystems". The latter program, which is one of a group of related projects involving scientists from different disciplines, is under the auspices of the International Biological Program (IBP). The results of our chemical investigations of disjunct taxa will be published in detail in IBP publications now in preparation.

It appears that in many instances more advanced taxa (populations and species derived from other populations of the same and different ploidy levels) have fewer and structurally simpler compounds relative to those found in the more primitive members of the same genus. Thus, the chemical patterns of derived taxa often include new compounds with simpler structural features probably involving fewer enzymatic steps. Such trends have been observed for some of the studies with a number of genera, including, for example, *Parthenium, Vernonia, Hymenoxys* and *Ambrosia*. Thus, it appears that loss mutations occur more frequently than do gain mutations in the course of speciation within a genus. It should be stressed that despite what appears to be a tendency for more loss than gain mutations in terms of the production of natural products chemistry, we and others have recorded examples of what may be exceptions to this trend. We are presently trying to assemble from the literature, as well as our own investigations, data which will bear upon this question. If such a trend can be established it would, in some cases, provide information which would assist in the determination of the origin of disjunct taxa.

In this connection, it might be noted that small, isolated island populations of mainland taxa usually exhibit a chemical pattern which contains both fewer and simpler compounds (see [1]), supporting the view that these island populations were derived from sources on the mainland.

In addition, we have found that taxa known to be recently introduced from one ecosystem into a new but similar one, often exhibit qualitatively and even quantitatively the same natural products chemistry. When, however, the new ecosystem is different and requires some physiological adaptation of the taxon, the plants that survive usually exhibit some modified natural products chemical pattern, albeit involving minor structural modifications.

Finally, I wish to invite other investigators to forward to me their own observations with regard to the chemistry of disjunct taxa, especially when these data can be correlated with other features of the ecosystems; in addition, information which bears upon primitive versus advanced species should be noted. Perhaps by pooling our observations significant trends can be established.

The research mentioned here was supported by NSF, Grants GB-27152 and 29576X, the Robert A. Welch Foundation, Grant F-130, and NIH, Grant HD-04488.

References

1. Mabry, T J, Pure appl chem 1973, 34, 377.
2. — Ann Missour bot garden 1972, 59, 105.

Discussion

Herout: Can the occurrence of some similar (or identical) plants in North and South America be explained by migrations which may have occurred at the start and end of the glacial periods?

Mabry: The occurrence of disjunct taxa in North and South America resulted in some cases from long distance dispersal mechanisms; in other instances, the disjunct taxa represent populations which became isolated in earlier times as the deserts and mountain ranges formed and the continents drifted. The glacial periods also played a role in the formation of some disjunct populations.

Cronquist: This is something that cannot be answered quickly and easily. There are diverse opinions on it.

Mabry: I agree; I am suggesting that it is worthwhile to examine the available data in light of the possibility that loss mutations more frequently characterize an advanced species relative to other members of the genus than do gain mutations. Even if the trend I suggest is supported by future investigations, we will, I think, find numerous exceptions.

Weimarck: Do you claim that chemical complexity generally decreases during speciation?

Mabry: Yes; certainly some of our observations support this view. However, as I have already mentioned, I am suggesting this as a working hypothesis. Additional studies are required to determine if the phenomenon is widespread.

Turner: Your hypothetical statement that within a given genus the more specialized members tend to lose compounds, is bound to create a paradox. How can a genus, with specialization, generally lose compounds and at the same time build up a chemical complexity within the species that comprize that genus?

Mabry: I suppose the latter question may be answered in part by reference to noted systematists who have suggested that during a burst of evolution peripheral taxa may exhibit considerable reduction in certain morphological characters. It appears that the same may be true for chemical characters. Regarding the origin of the genus *Larrea*, my statement would still hold if the hypothesis is correct, that is, *Larrea* originated in South America if the South American taxa are more chemically complex relative to those in North America. These chemical studies are in progress.

Cronquist: In response to your question whether *Parthenium hysterophorus* is an advanced species in the genus I would say: "Provisionally, yes".

Mabry: Parthenium hysterophorus has a simpler chemistry than other more primitive *Parthenium* species; similarly, in many other genera, the species considered advanced are also chemically less complex than other members of the same genus. Moreover, introduced populations of a species tend to have a simpler chemistry than the source population.

Reichstein: The assumption that near the center of origin of a genus more members will be found than on the periphery of its distributions is of course widely accepted. But a member growing on the periphery need not necessarily be one of the oldest because it had more time to spread. More important may be its spreading power. I can mention one example, i.e. the well known "male fern" *Dryopteris filix-mas* (L.) Schott. It is a tetraploid species and its two ancestors are growing together today only in the Caucasus and adjacent areas. It is therefore fairly safe to assume that it is on or near this limited area where *D. filix-mas* originated. As an allotetraploid it must be younger than its diploid parents. Nevertheless its greater spreading power has allowed it to cover today the greater part of the northern hemisphere. The phloroglucinols of *D. filix-mas* are well known, they correspond fairly well to the sum of the components present in both putative ancestors.

Geissman: At levels above the generic, it appears that the simplest compounds of a class are the most widely distributed among species, genera, tribes and families. Costunolide, the simplest sesquiterpene lactone, is found not only throughout the tribes of Compositae, but in another family (Magnoliaceae) as well. At the more complex structural levels, structures tend more to be unique and reflect genetic individuality and thus appear to reflect an evolutionary trend toward the addition of genetic factors for chemical (structural) elaboration.

Mabry: With regard to the situation at the generic level, it appears from our studies that advanced species within a genus contain fewer and simpler compounds. No doubt many exceptions will be found to this trend. I also expect exceptions to your observations will be found at higher taxonomic categories.

Merxmüller: Many of such conclusions rely upon the firm conviction that the evolution of different characters or character groups must be always strongly correlated. But we all admit the existence of heterobathmy (mosaic evolution), i.e. different rates of evolution in different morphological characters. Likewise there is ample evidence for a frequent lack of correlation between karyological and morphological differentiation (very uniform karyotypes in morphologically extremely diversified groups and vice versa). I cannot see the slightest reason why there should be always a clear correlation between the evolution of chemical characters and that of morphological ones. As long as the morphological and geographical evidences have not cleared up beyond any doubt, I certainly would not like to use the chemical one for decisions. At any case, each example ought to be treated individually, as general conclusions certainly cannot be reached.

Mabry: You may be correct. Nevertheless since much of our data can be correlated with the trend for advanced species in a genus to have a simpler chemistry, we suggest that other investigators examine their data with this observation in mind.

Weimarck: A general trend towards decreasing complexity seems illogical. I believe that the examples you have chosen which illustrate possible cases of plants having been introduced to a new continent rather could be examples of

the Sewall Wright effect, or bottle-neck effect, and not of speciation in general.

Mabry: Perhaps, but I don't believe the Sewall Wright effect could be applied to all the examples upon which my views are based.

Mears: In my own research on the flavonoids and other phenolics of *Parthenium* (Compositae-Helianthoideae), I find that indeed those species suggested as most similar to primitive types have very complicated phenolic patterns. However, certain clearly "derived" species have much less phenolic pattern complexity, with such patterns composed of parts of the pattern of the primitive species; other clearly derived species have quite complex phenolic patterns. In *Parthenium* an environmental factor seems to distinguish among clearly derived species with either complex or simplified phenolic patterns.

Bu'Lock: In biosynthetic terms, the addition of new processes to a system may make it either more complex or less complex so far as the pattern of products indicates. The addition of an extra transformation step, if it is not highly substrate-specific, will perhaps double the number of products; such an addition would lead to evolution with increasing complexity. Contrariwise, the evolution of tighter control mechanisms or of higher grades of substrate-specificity would reduce the complexity of the range of products seen. Whether one type of development corresponds to genus development and another with speciation within a genus is something we must leave to the botanical authorities, but I would suggest that development of a complex range of products could conceivably provide a range of possibilities upon which selective chemo-ecological mechanism might go on to work in a reductive direction.

Mabry: Yes, either increased or decreased chemical complexity could occur. Although it appears from our examples that at the generic level loss mutations represent an advanced species, I must emphasize that these few observations by no means constitute a firm evolutionary rule. I feel that each case must still be examined individually.

Cronquist: I would like to speak to the question that several other speakers have referred to, of how you can always reduce chemical complexity in the course of evolution within a group, without some means of getting the complexity in the first place. If I understand you correctly, you are trying to meet the problem by saying that in the origin of genera or perhaps higher taxa, the complexity is established, and then in the course of evolution within the genus it is reduced. If I understand correctly the people who are concerned with evolutionary mechanisms per se, transspecific evolution is essentially similar to specific evolution; it consists of a series of speciations. Thus your proposed explanation seems unlikely. I think it will turn out that chemical evolution is essentially similar to morphological evolution, which of course is basically chemical anyway. In the same groups you have progressive evolutionary reduction as in the flowers of the subclass Hamamelidae. In others you have progressive evolutionary elaboration, as in the perianth of the family Orchidaceae. You have come upon some examples of progressive chemical reduction, which I would not question, but the reverse must also occur.

Mabry: Additional studies may, of course, reveal other trends or refine the one proposed here.

Homology of Biosynthetic Routes: the Base in Chemotaxonomy

P. Tétényi

Research Institute for Medicinal Plants, Budapest, Hungary

Summary

Discussion. The possibility to pass from chemistry to biology is only given by keeping in mind the history and the increasing individuality during phylogeny. The differences of phylogeny are recapitulated both in metabolism and in the structure of living beings during their ontogeny.

The chemical taxa represent at the same time the past and the differentiation, so they could be the base for taxonomic evaluation.

From the existing different biochemical levels there were proved most facts of phylogenetic differentiation of the metabolism. Neither the biosynthetised materials, nor their frequency can serve as a base for taxonomy, but the solution is given by the observed homology of biosynthetic routes. These can be detected only step by step and can lead to a newly recognized phylogenetical tree, founded on metabolism.

Two examples. The biosynthetic routes of the essential fatty acids were elucidated by the fundamental research of photosynthesis and by the experimental work on prostaglandins. The utilization of these new results in taxonomy will be presented by homology of some species of Boraginaceae and by analogy of their fatty acid routes and those of *Sinapis alba* L., *Medicago sativa* L., *Linum usitatissimum* L., *Oenothera* sp.

The homology and differentiation of terpenoid metabolism at the infraspecific chemical taxa of *Chrysanthemum vulgare* (L.) Bernh. are possible to evaluate.

Conclusion. It is demonstrated, how and why the newly recognized facts have come to be included in botany and chemistry from "chemo"-, to chemotaxonomy.

The principle of taking into account the genetical factor—the historical development—is gaining ground increasingly even beyond the sciences which are concerned with the living world, e.g. in chemistry and in soil science. The recent pre-biological research on the development of organic matter into a living substance is also of a historical character. This goes to show that the transition from chemistry to biology can be comprehended only when we consider the historical aspect since the descent, the phylogeny are fundamental characteristics of living beings.

Living organisms—beyond the above-mentioned features of their descent—are differentiated, various and individual. Their individuality and specialization are even due to differences of descent and this is reflected in the whole course of their ontogeny, of their life. This relationship between phylo- and ontogeny has been proved already in the past century for animals (and man) and in the case of plants it has been explained in 1955 by the modern phylembryogenetic theory of Takhtajan [1]. The recapitulation of phylogeny is at the same time structural and functional but also represents a process of growing independence which usually occurs within the maternal organism. The descendant conserves the genetic determination, the historical adaptation of the parent but it also represents a new living individual which is obliged to uphold independently the interior dynamic metabolic equilibrium which is the condition of its life and development and which is unknown in the world of chemistry although its elements may be traced there. The continuous metabolism of living beings varies with each moment but is also historically, i.e. genetically determined for the whole period of its ontogeny. The existence of chemical taxa within a species is a highly significant proof of actual divergences in metabolism and inherited differentiation. This is evidently due to the fact that chemical taxa are the most pregnant expressions of the above-mentioned two characteristics of life, namely of the dependence on time and of individuality.

Since the individual is always a member of a population, a species, a taxon, its existence

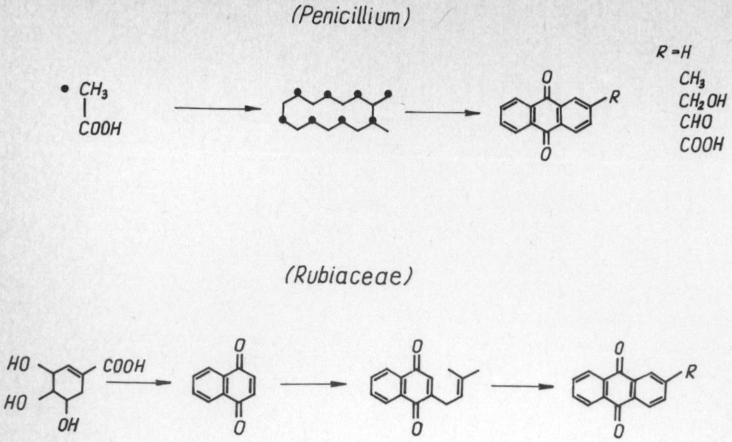

(Penicillium)

(Rubiaceae)

Fig. 1. Two fundamentally different ways of antra-quinone-biosynthesis (Zenk, M H, Ber deut bot Ges 1967.)

and metabolism are mass phenomena. The chemistry of a taxon manifests itself in many places and collectively at the same time. Therefore the chemical properties, the chemical differentiation, may serve as a basis of classification. However, the inherited chemistry of a living individual cannot be classified according to the macro- and microstructure of substances, since the performed vital functions are always bound to complex structures, although these may be traced back to simple compounds. It would be strictly correct to state that an amino acid is simpler or more complex than a terpene, even though a volatile oil-complex of terpene-compounds may differ substantially—with respect to the degree of organization and historical determination—from a protein-complex of amino acids. According to our present knowledge DNA-RNA codification is not very useful in classification and the enzyme-protein level is applicable only to larger taxonomic units (aside from isozymic diversification). Metabolism, however, is known very closely and in much detail, due to the research on the pathways of biosynthesis. Therefore it is in principle suitable for detecting individual specialisation and the common phylogeny, i.e. it may serve as a basis for classification according to chemistry.

Not all metabolites possess the same significance from the point of view of classification. For instance the essential amino acids cannot be missing from the organism and therefore their presence cannot be characteristic; however, if the pathways of biosynthesis are different—and sometimes even differing in intensity and accumulation—the differences in essential amino acids may serve as a basis for systematics. On the other hand, the statement made by Rumphius [2] ca 250 years ago, that anthraglycoside-containing plants are of related descent, does not represent a similarity in chemism, since it is well-known that fungi form their anthraglycosides in a manner different from that of higher plants (fig. 1). This is also true of several other substances, e.g. for alkaloids, since anabasine or nicotine among others are produced in different ways by different plant species. Neither is the frequency of occurrence of the biosynthesis of a certain substance to be taken as decisive (fig. 2); rarity is rather an indication of the aptitude to serve as a taxonomic marker.

Therefore, it is not the produced substances but the homology of their biosynthetic routes which is the historical proof of phylogeny and differentiation. This is how I can interpret the hypothesis of Vavilov [3] on the morphologically homologous series which was published in the 1920's, extended a decade later to the chemical substances of plants by Nilov [4] and Gurvich [5] and discussed also at the conference in Wageningen [6] in 1957. I should like to stress, however, that inferences on phylogeny —as a temporal process—may be drawn from the determined temporal arrangement and order of pathways of biosynthetic routes, but not of the substances themselves; in this way the bio-

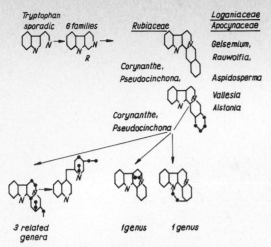

Fig. 2. Tryptophan derivative alkaloids suitable for detecting the relationship between Rubiaceae and Loganiaceae. (Hänsel, R, Arch Pharm 1956, 289, 619.)

chemical process realized in the living of today reflects the past. This may prove whether a chemical character, a typical feature of the chemism—in consideration of the analogous and homologous above-mentioned cases—may be regarded as taxonomically significant or merely accidental. In this evaluation of chemical characters one really seeks for the homologous series of biosynthesis, treating chemistry as a basis for the botanical classification.

Because of the magnitude of the hiatus, the study of the homology of biosynthesis and that

of the divergences should not be attempted in the largest taxonomic units—although here they are self-evident—but at the level of the species or even within the species. Here the divergences in biosynthesis are very small and it is possible to detect step by step and to prove the laws which are essential and evident. It is from these slight divergences that one must reconstruct—on the basis of the homology of chemistry—the systematic tree. This is a beautiful and exciting task because one must detect the really related similarities; in addition to homology one must also show the analogous cases, although it is much more difficult to detect the identity of metabolites and biogenesis than their differences. All this will not result in a reproduction of the previous systematic tree—even though in several cases they will prove to be identical—but in a well proved and more thorough knowledge of phylogeny according to chemistry. Such a system will acquire a similar significance in biology as the periodic system in chemistry, opening an immense vista of genetic transformation.

I should like to present two models as examples for using the homology and analogy of the biosynthetic pathways in taxonomy; one of the models refers to fatty acids, the other to the biosynthesis of terpenes.

The fact that α-linolenic acid is the main fatty acid component of galactolipids in photosynthetizing tissues of all higher plants, while

Fig. 3. Phylogenetic classification of algae based on their lipid metabolism.

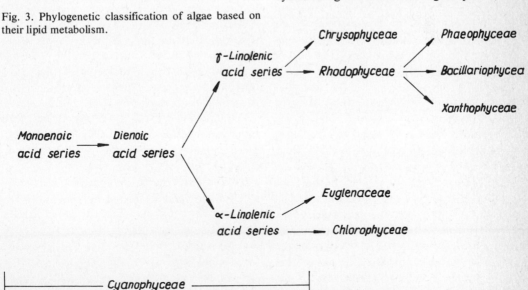

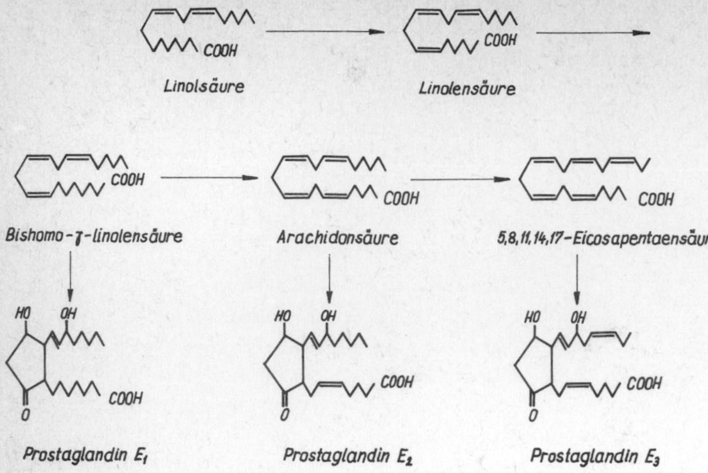

Linolsäure   Linolensäure

Bishomo-γ-linolensäure   Arachidonsäure   5,8,11,14,17-Eicosapentaensäure

Prostaglandin E_1 Prostaglandin E_2 Prostaglandin E_3

Fig. 4. Biosynthesis of prostaglandins. (Luckner, M, Sekundäre Stoffwechsel in Pflanze und Tier, p. 73. Fischer, Jena 1969.)

linolic acids are incorporated in the phospholipids, shows that the formation of unsaturated fatty acids in leaves is controlled by very fine but different enzyme systems [7]. These systems —although influenced by the environment— operate principally according to their genetic determination, depending on the phase of development. This genetic determination of the formation of unsaturated fatty acids furnishes a reliable basis for comparative biochemical studies and for a more accurate chemotaxonomical reflection of phylogeny, as has been demonstrated e.g. by Nichols [8] in his diagram referring to algae (fig. 3).

The unsaturated fatty acids play a significant role in the hormone equilibrium of higher animals and man being the building blocks of prostaglandins (fig. 4). At the beginning of prostaglandin research in Hungary attention was directed to the occurrence of unsaturated fatty acids in plants and we joined in this work. Our investigations were directed towards such species of the Boraginaceae family which occur in Hungary and seemed specially interesting.

Table 1 is based on our studies on the composition of fatty acids (according to the method published by Tétényi [9]). Let us consider, first of all, the simpler kind of homology of acetate units which is shown by the elongation of the carbon chain (fig. 5). While the oleic acid 18:1 occurs in significant percentages in all taxa of the table (11–43%) the formation of

the unsaturated fatty acid 22:1—erucic acid— is not general: it cannot be detected at all, e.g. in Lithospermeae species. However, all investigated species of Cynoglosseae contain considerable amounts (4–15%) of this fatty acid. In this respect the 18 samples of the investigated 7 species are uniform and this phenomenon may be considered as proof of the homology of biosynthetic routes.

It is obvious that oleic acid is present in lower percentages in the fatty acid composition of mustard or lucerne than in that of the Cynoglosseae showing that there the enzymes bring about the elongation of the chain with a greater intensity. Another difference consists in the fact that the 20:1 component percentage is 8.5% in mustard; this level is never attained in Cynoglosseae, and the component cannot be detected at all in lucerne because of its rapid transformation. The end result of these processes is an unusually high erucic acid content. All this goes to show that in these two species the metabolism of carbon chain elongation shows an analogy to that of Cynoglosseae but is not homologous to it, although both processes lead to 22:1.

In contrast to the chain elongation biosynthesis of Cynoglosseae, the other tribes of Boraginaceae are characterized by an enzymatic mechanism which produces 18 : 1 oleic acid and the further double bonds. The biosynthesis in Heliotropioideae is the simplest and evidently also the most ancient because in the seeds of the species the unsaturated fatty acids are char-

Table 1. *Homology and analogy in fatty acid composition*

Chain lengthening	Fatty acids as % of methyl esters				
	18:1	*20:1*	*22:1*		
CYNOGLOSSEAE					
Cynoglossum hungaricum Simonkai	37.9	3.9	14.6		
C. amabile Stapf et Drum.	34.0	5.1	11.7		
C. officinale L.	35.7	3.3	10.1		
Paracaryum coelestinum (Lindl.) Voss.	31.4	3.9	9.3		
Omphalodes linifolia (L.) Moench.	36.1	3.3	5.7		
Solenanthus appenninus (L.) Fischer & G. A. Meyer	30.3	1.9	4.0		
Sinapis alba L.	26.1	8.5	42.5		
Medicago sativa L.	14.9	O	28.0		
Further doubled bonds					
(a) in postposition to △9				*18:2*	*18:3α*
HELIOTROPIOIDEAE	25.7	Tr.	O	51.5	0.3
Papaver somniferum L.	9.9	O	O	74.2	O
P. bracteatum Lindl.	13.5	O	O	72.6	Tr.
ERITRICHIEAE	29.2	2.0	1.0	23.2	19.4
LITHOSPERMEAE	20.1	1.5	O	28.1	23.9
ECHIEAE	24.7	Tr.	O	17.5	34.9
Caccinia strigosa Boiss.	38.5	O	O	16.7	34.6
Alkanna graeca Boiss. and Spurner in Boiss.	19.2	O	O	27.9	34.2
A. orientalis (L.) Boiss.	15.5	O	O	26.0	33.0
Linum usitatissimum L.	20.3	O	O	17.2	53.4
(b) in praeposition to △9				*18:3γ*	*18:4*
Asperugo procumbens L.	28.6	Tr.	O	38.4	14.0	0.1
ANCHUSEAE					
Symphytum asperum Lepech.	17.2	0.7	0.9	38.8	28.1	6.4
S. orientale L.	27.0	1.6	0.9	38.1	22.2	0.1
S. caucasicum Bieb.	24.5	1.0	1.6	38.4	20.1	1.4
Borago officinalis L.	20.3	O	O	36.2	20.0	2.5
Anchusa riparia DC.	17.7	3.7	3.3	28.8	17.6	3.7
A. arvensis (L.) Bieb.	21.8	1.2	O	25.7	14.3	7.7
Oenothera biennis L.	7.5	O	O	82.3	3.9	O
O. lamarckiana Vries.	9.7	O	O	81.2	2.7	O
LITHOSPERMEAE					
Buglossoides arvensis (L.) J. M. Johnston	12.2	O	O	18.1	7.8	18.8
Lithospermum purpureo caeruleum L.	11.7	O	O	20.4	18.2	11.9
L. officinale L.	11.2	O	O	22.8	17.2	10.8
Myosotis silvatica Hoffm.	15.8	4.7	O	25.2	9.0	10.2
M. arvensis (L.) Hill.	26.1	O	O	25.9	3.6	8.9
Onosma visianii Clem.	26.4	O	O	23.0	8.3	6.0

O, Absent; Tr, traces.

acteristically represented only by oleic acid (*18 : 1*) and linolic acid (*18 : 2*). Other Boraginaceae species which also contain substantial amounts of *18:3α*-linolenic acid show a more complicated enzyme activity because—with the exception of *Caccinia*—they also contain other characteristic unsaturated fatty acids. Even so, it is a characteristic homologous feature that the sum of linolic acid and α-linolenic acid is as high as 40–70% in 30 samples of the studied 17 species.

As regards the analogy of unsaturation, the

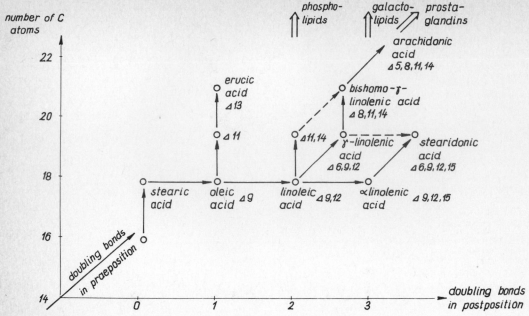

Fig. 5. Biosynthetic routes of formation of some unsaturated fatty acids.

percentage of linolenic acid is at most 4 times that of oleic acid in the *Heliotropium* species and 7 times that of oleic acid in the poppy. Linseed oil also contains two thirds of linolic and α-linolenic acid in combination but the latter component alone accounts for 50–70%; this excess has been found only in a single case in the tested Boraginaceae samples, therefore this is a case of analogy.

The third enzymatic activity observed in Boraginaceae—the formation of γ-linolenic acid and stearidonic acid (*18:3* and *18:4*)—is also connected with unsaturation. This is char-

acteristic for *Asperugo procumbens* L. and Anchuseae species as on the other hand mainly of the Lithospermeae. The former homology was observed in 17 samples of 13 species with the formation of 11–28% of γ-linolenic acid and the latter in 21 samples of 9 species with the formation of 4–17% stearidonic acid, although really significant infraspecific differences were found for both fatty acids in samples of one-one species.

The slight amount (3–4%) of γ-linolenic acid formed in the *Oenothera* species included in our investigations, merely presents an analogy since the divergent direction of enzymatic activity is pointed out even by the maximal percentage (81–82%) of linolic acid (*18:2*).

Fig. 6. Classification of Boraginaceae in accordance with their fatty biosynthetic routes.

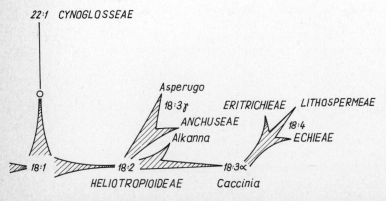

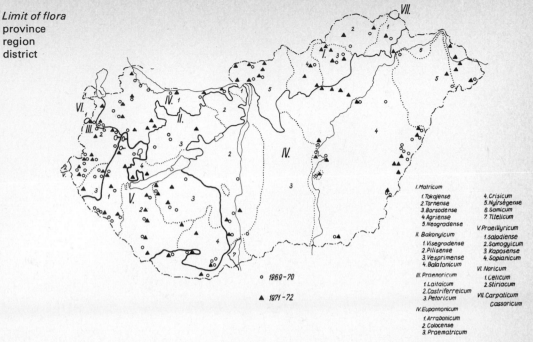

Fig. 7. Sites of collection of *Chrysanthemum vulgare* L. samples on the map of the Hungarian Flora.

As to the homology of biosynthetic routes the family system of Boraginaceae is shown in fig. 6. This is based on our own investigations with respect to the 3 types of characteristic enzymatic activity in the formation of fatty acids.

The homology and analogy of the biogenesis of terpene components will be presented with the aid of our investigations on *Chrysanthemum vulgare* (L.) Bernh.

The study of the volatile components which are products of biosynthetic pathways based on sugar, i.e. the first product of photosynthesis—as has been proved by the most recent research—has led to two different chemotaxonomical views. According to one view the occurrence of a substance is in itself of a certain taxonomical value; according to the other the occurrence of the main components should be regarded above all. However, it is much more difficult to prove the absence of a terpene component—due to the limited sensitivity of detection—than its presence in a certain proportion, in excess of a certain threshold value. Therefore it is more acceptable to elucidate the successive steps of biosynthesis and simultaneously the divergence and the polychemism with

the aid of components present in high percentages. The consideration of components present in high percentages which signify divergent biosynthetic routes also reduces the problem of our only partial knowledge of the substances. The lack of knowledge is a very real obstacle to an arrangement based on the homology of biosynthesis, but we cannot establish a contrast between exact comparative phytochemistry which only employs known substances and the numerical chemotaxonomy of substances based on spot and peak chromatography. This absolute formulation of known and unknown is not reasonable because our knowledge is always relative, lying between these two extremes.

This is also the case in our studies on *Chrysanthemum* (method see Tétényi et al. [10]); the determination of the ratio of volatile oil components was taken as our point of departure. So far in this species and with respect to the main terpene components 14 taxa have been found with different volatile oil compositions as chemical "races" or "chemotypes" (summary see Tétényi [11]).

This chemical differentiation of the species proves its adaptability which is also proved by the large distribution. Since the species is living, it adapts itself repeatedly to the variable condi-

74 P. Tétényi

Table 2. *Essential oil composition of some Chrysanthemum vulgare samples* (*Hungary 1969–72*)

Area of collection	Components (in % of oil)										
	1	2	3	4	5	6	7	8	9	10	11
Mosonsomorja	54.4					××				××	
401/ (coll.)	×	42.2	××	15.7	23.4						
Röjtökmuzsaj	××		33.3	46.2		×	×				
Drégelypalánk				82.2		×	×	×			
Kerkakutas						94.0					
Moson							74.4	22.0			
Sarkad				××			×	77.2			
Kőszeg	×	×	×	×					42.2		×
Szarvas				××		×				53.0	
Bodajk	×	×		×							41.2
Murakeresztur											
413/3 (coll.)											
Tiszafüred	×						×				
Kőszeg	××	××						××	×		×
Mersevát				××			×	×			
435/2 (coll.)											
401/4 (coll.)											
Borsodbóta	×			××							×

Notes: × = 1–5%; ×× = 5–10%.

1. Pinenes+camphene; 2. cineol; 3. γ-terpinene; 4. arthemisiaketone; 5. yomogi-alcohol; 6. thujone; 7. unknown 1; 8. unknown 2; 9. campher; 10. chrysanthenylacetate; 11. bornylacetate; 12. unknown 3; 13. thujylalcohol; 14. unknown 4; 15. borneol; 16. umbellulone; 17. unknown 5; 18. piperitone; 19. unknown 6.

tions, varying in its inheritance and in this way others among the numerous volatile oil components may become preponderant. Evidently, however, new chemism can result only from a further development of the former routes of biosynthesis or from their inhibition which are produced by very fine micro-changes. That is why here—and also in the case of other species —the polychemism should be studied in its dependence on the area, considering the conditions which may induce a divergence of the standard reaction.

On the map of Hungarian Flora (fig. 7) it has not yet been possible to define everywhere the occurrence of different chemical taxa of *Chrysanthemum vulgare*. According to our studies taxa containing the following main components are found (table 2): pinenes+camphene, chrysanthenyl acetate, thujone, campher, 1,8-cineol, γ-terpinene, artemisiaketone, borneol-bornylacetate and umbellulone. A new taxon with piperitone as the main component has been identified in cooperation with Lawrence (Canada). One taxon with thujyl-alcohol as main component has not been described so

far and we have been unable to identify 6 other main components. The investigations are not concluded yet but the main components occurring with the highest frequency are artemisia ketone and umbellulone. These components, single or in combination, represent the main percentage of the volatile oil in nearly 40% of the samples and in combination with a third component they form the main percentage in a further 25% of the samples. This combined occurrence in two thirds of the samples is just as characteristic of Hungarian populations of *Chrysanthemum vulgare*, as the campher-thujone component combination in Finland, the iso-thujone in Canada or the monoterpene hydrocarbons in Western Europe. Very slight variations have been found so far in the plants collected from Hungary or in their clones with respect to the volatile oil composition; the polychemism of the collection was conserved vegetatively.

Fig. 8 illustrates the relationships of the biosynthesis of the components. Within the species no connections of descent might be found but the differentiation of biosynthesis points to

12	13	14	15	16	17	18	19
						×	
						×	
	×						
						××	
×			23.6		××	×	
			×	×	×	×	
			30.2				
95.0							
××	84.0						
		77.5	×	×	×	×	
			54.5			×	
				81.5		×	
					94.0		
	××		×			64.2	
××				××	××		55.0

some distance or closeness of the relationship. This diagram shows that the taxa with artemisia ketone, or 1,8-cineol and those formed according to Banthorpe [12] by the menthenyl-4-carbonium ion way seem to be the most ancient; the sesquiterpene branching, the isopinocamphone and the chrysanthenyl acetate taxa seem of later origin.

We hope that further research on biosynthetic routes will elucidate the infraspecific relationships, e.g. the fact that chemical taxa with mainly pinene content do not occur in Hungary but simpler ketones do. The chemical taxa forming an oxidized hydrocarbon (ketone) as their main component represent a homologous series within the species, as do also the infraspecific units which form mono- and sesquiterpene hydrocarbons. On the other hand the infraspecific chemical taxa of *Cinnamomum camphora* (L.) Sieb., which according to Hegnauer [13] have arisen by parallel mutations and repeatedly form the same main components—are only analogous, because their biosynthetic routes are different (fig. 9).

These two examples—important from the point of view of plant physiology—prove that the inclusion of chemistry into botanical classification, i.e. chemotaxonomy, is justified, due to the high genetic and phylogenetic signifi-

cance of the biosynthetic routes of the species or of higher taxonomic units. Since, due to the progress of biochemistry, the laws of formation of the various metabolites represent a more satisfactory basis for chemotaxonomy than was possible before now, the level of exactitude attainable in chemistry and other natural sciences will be realisable also in biology. Chemotaxonomic research will open the road to a more detailed comprehension of the nature of plants, of their chemism and morphological characteristics.

How can this novel group of knowledge be included in the present material of chemistry and biology, what problems should this new science help to solve? The main problem consists in defining the development or more profound interpretation of the old technical terms and of the concepts they represent. Based on my proposal made in 1957, to use the prefix "chemo"- [14], the nomenclature of chemotaxonomy, chemosystematics for this new science, as used by Hegnauer [15], is gaining ground. I have repeatedly presented and discussed in detail (in Wageningen and Zurich and later in "Taxon" [16]) the classification and nomination of infraspecific chemical taxa according to the rules of botanical nomenclature. These new technical terms have been generally accepted and we have passed from "chemo" to the chemotaxonomist [17].

The novel nomenclature is merely a more expressive professional tool but my fundamental statements on ontogenetic hypothesis [18], on the laws of chemical differentiation [19], the discovery that the quantitative and qualitative chemical changes in living individuals occur with the same probability [20] and the proof of the general occurrence of polychemism [21] have been more or less incorporated in botanical papers and books. I may mention the works of Soó in Hungary [22], Stahl and Merxmüller in Germany [23, 24], Lambinon in Belgium [25], Hillis in Australia [26] and Gurvich in the USSR [27]. This may be taken as an indication that chemotaxonomy has been accepted in botany.

The relationship between this new science and chemistry is not very consolidated. Sometimes in discussing biochemical systematics or comparative phytochemistry the novel and in-

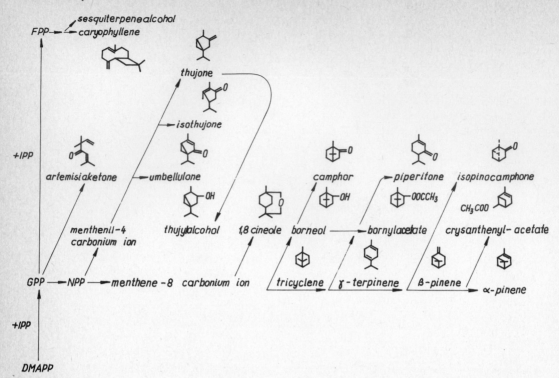

Fig. 8. Biosynthetic routes of some terpenes in *Chrysanthemum vulgare*.

dependent character of this new science does not stand out clearly enough as representing a new concept reaching beyond our knowledge up to date. Still, an approach from the point of

view of chemistry has been deemed necessary: this is indicated by the system of Mentzer [28]. That is why we welcome this conference as a great help in elucidating the role of chemistry in botanical taxonomy.

I have already discussed in detail one method of approaching chemistry, the possibilities of the grouping of metabolites in my lecture in Vienna in 1970 and in ref. [29]. Chemistry

Fig. 9. Biosynthetic routes of formation of some monoterpene compounds in *Cinnamomum*. (Fujita, Y, J Jap bot 1967, 42, 278.)

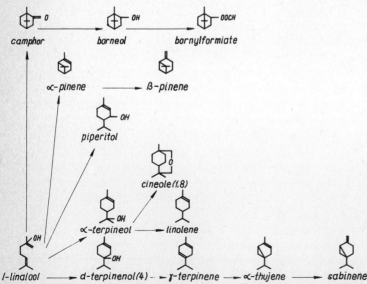

should accept the historical aspect which characterizes the botanical object and should reflect the differentiation which is the real basis for the raising of chemotaxonomy. This enlarged knowledge and this novel aspect—we can find them in the article by Birch [30] on chemical phylogeny—will serve as a new bridge between the separated sciences, chemistry and biology, transforming them into an organic whole.

Starting from the nature of living and evaluating chemotaxonomy as the science of the homology of the biosynthetic routes, we have arrived again, but on a higher level, at the course of time and of specialization, which are simultaneous active factors in chemistry and in botanical systematics. Consequently we took a significant step forward on the endless spiral of the acquisition of knowledge.

References

1. Takhtajan, A L, Bot zhur 1955, 40, 789.
2. Rumpius, G E. Herbarium amboinense. Elsevier, Amsterdam, 1750.
3. Vavilov, N I, Zakoni homologicheski rjadov, in Trudi III. siezda po selekcii, p. 82. Saratov, 1920.
4. Nilov, V I, Trudi prikl bot gen sel 1936, 3, 5.
5. Gurvich, N L, Dokl akad nauk SSSR 1936, 12, 171.
6. Mothes, K, Pharm weekbl 1957, 15, 74.
7. Trémolières, A, Phytochemistry 1972, 11, 3453.
8. Nichols, B W, Comparative lipid biochemistry, in Phytochemical phylogeny (ed J B Harborne) p. 105. Academic Press, London and New York, 1970.
9. Tétényi, P, Acta bot acad sci Hung 1974, 19 (1).
10. Tétényi, P, Kaposi, P & Héthelyi, I, Herba hung 1974, 13 (2).
11. Tétényi, P, Infraspecific chemical taxa of medicinal plants, p. 125. Akadémiai Kiadó Chemical Publisher, New York, 1970.
12. Banthorpe, D K & Turnbull, K W, Chem comm 1966, 166.
13. Hegnauer, R, Chemotaxonomie der Pflanzen, vol. 4, p. 358. Birkhäuser, Basel 1966.
14. Tétényi, P, Taxon 1957, 7, 40.
15. Hegnauer, R, Pharm acta Helv 1958, 33, 287.
16. Tétényi, P, Taxon 1968, 17, 261.
17. — Ibid 1967, 16, 178.
18. — Bot Közl 1964, 51, 187.
19. — Bot zhur 1962, 47, 1731.
20. — Bull soc bot France 1963, 110, 177.
21. Tétényi, P, Hung agric rev 1965, 9, 10.
22. Soó, R, Acta bot acad sci Hung 1970, 16, 445.
23. Stahl, E, Pharm weekbl 1971, 106, 242.
24. Merxmüller, H, Ber deut bot Ges 1967, 80, 608.
25. Lambinon, J, Bull soc roy bot Belge 1965, 98, 295.
26. Hillis, W E, Phytochemistry 1967, 6, 259.
27. Gurvich, N L, Moskow Obsh Isp prirod 1970. Thesis IV. sovesh. filogenii rast, p. 88.
28. Mentzer, C, Bull soc chim France 1960, 203.
29. Tétényi, P, Herba Polon 1972, 18, 131.
30. Birch, A P, Pure appl chem 1973, 33, 17.

Discussion

Wagner: One comment to your first example: The occurrence of unsaturated fatty acids with four double bonds in higher plants seems to be very rare. It is remarkable that these were found only in some families and here it is noteworthy, that under them are many "aquatic" plants. Since, on the other hand, these fatty acids are wide spread in lower plants, e.g. in algae, mosses and ferns, which have a biosynthetic pattern in synthesizing unsaturated f.a. very similar to that of animals, there arises the question, whether the availability to introduce a fourth double bond into a linolenic acid is correlated with the evolutionary step of the f.a. biosynthesis or whether there are other factors (e.g. ecological influences) which have created this availability.

Tétényi: The ecological conditions have little influence on the fatty acid composition of seeds in the investigated Boraginaceae species. This was proved by the samples collected in our botanic garden, which are all affected by the same environmental factors.

More can be due to the genetical factor: the infraspecific differences were remarkably high in stearidonic acid component (*18 : 4*) at *Myosotis silvatica* Hoffm. which is a complicated botanical taxon, on the other side as well, due to morphological variability proved by its rich synonymic naming.

Sørensen: As to Professor Wagner's remark that C_{18}-fatty acids with 4 double bonds are very unusual, I may add that this may be true with Professor Tétényi's acid, which as far as I could realize from the slide, had isolated double bonds. The parinaric acid, with 4 conjugated

double bonds is the dominating fatty acid in
the seeds of all members of the Balsaminaceae
so far investigated, and it is widely distributed
in the Rosaceae.

Wagner: The parinaric acid with conjugated
double bonds is such an exception. Therefore
the statement is valid that these f.a. are relatively rare in higher plants.

Birch: I would like to ask if the terpenes mentioned have known absolute configurations?
Since the steric course of transformations is
well understood, the configurations add another
dimension to structure in deciding whether compounds may be sequentially formed from each
other, or only rather remotely related through
open-chain precursors.

Tétényi: I agree with Prof. Wagner's remark
that parinaric acid has restricted occurrence and
I do not know the absolute configuration of
the terpenes isolated from *Chrysanthemum vulgare* L.

Applications of Special Classes of Compounds

Flavonoids as Evolutionary Markers in Primitive Tracheophytes

T. Swain

Royal Botanic Gardens, Kew, Surrey, UK

Summary

The distribution of flavonoids is discussed in relation to the phylogenetic affinities of classes of lower plants.

It is relatively easy, using both morphological and anatomical criteria, to arrange the major divisions of the present-day plant kingdom in hierarchal sequence from primitive to advanced [9, 14]. To a large extent, such arrangements have played a major role in determining our ideas about plant evolution. The fossil record, which in the case of vascular plants dates back to Silurian times (400 million years ago), undoubtedly has also helped to unravel the interrelationships of many of the phylogenetic sequences proposed and to demonstrate the magnitude of the time scale involved in the overall change [1]. But the record is unfortunately too scanty and incomplete to allow us to deduce the early history of land plant evolution with any degree of certainty [1, 10]. The discovery of several major groups of long extinct plants which bear no close resemblence to any now living, however, indicates that there may be many other transitional forms in the overall evolution of the plant kingdom which have yet to be found.

Although the morphological and anatomical data from both present-day and fossil plants have given us an important insight into evolutionary relationships, there are a number of difficulties which are not readily overcome. For example, it is not always easy to decide whether apparently equivalent morphological features in different taxa indicate a close relationship or have arisen as a result of parallel or convergent evolution [16]. Coupled with the fact that few intact pre-Cenozoic fossils are found [15], this makes comparison of ancient and modern taxa relatively difficult. Secondly, there is no way of knowing what magnitude of genetic change is needed for the development of any new advanced feature and hence what importance should be ascribed to it in the evolutionary sense. As a corollary, it might be noted that the genome of even the most primitive of today's plants has undoubtedly undergone many changes since the time of their presumed ancestral forms. It is useful, therefore, to consider whether other information, for example, biochemical data, might be used to overcome some of these difficulties. Since such information is independent of the classical data, it might be expected to lead to new insights into phylogenetic problems. Of course, biochemical features may be equally subject to convergent or parallel evolution. In this case, however, it is usually possible to decide the true situation with some degree of certainty, since it is unlikely that the genetic information required for the production of a given biochemical moiety has arisen more than once. For example, the close similarity of either the sequence of amino acids in various cytochromes [24] or of the individual steps in the biosynthesis of aromatic amino acids in different organisms [37] is unlikely to be due to chance or convergence.

Even though such data have only recently become available, they have already led to a rational rethinking of some existing ideas. For example, the fungi are traditionally believed to be associated with the plant kingdom, although sometimes recognised as having equal status [45]. Cytochrome data, however, show that they may be more closely related to the animal kingdom [24]. A second example, which is dealt with in detail elsewhere (p. 93), is the reconsideration of angiosperm evolution based on chemical data [4, 22] which indicates that the Rosidae, Hamamelidae and Dilleniidae are probably coeval with the Magnoliidae rather than being derived from them as is generally believed [13, 39]. It seemed useful, therefore, to see if biochemical data might throw any light on the evolution of lower tracheophytes. Since much data already exists on the distribution of flavonoids and related compounds in these non-flowering plants [21 a, b] it was decided to accumulate new comparative information on these substances specifically to examine their use as evolutionary markers [12] and this is discussed in subsequent sections of this paper.

Evolution of Lower Tracheophytes

It is usually believed, on the basis of similarities in the chemistry of the cell walls and of the chloroplast pigments, that land plants arose from an heterotrichous green alga, perhaps related to present-day Chaetophorales [1, 18]. Several species of this order are adapted to damp rather than the truly aqueous habitats of other algal groups. There is, however, great difficulty in deciding the nature of any of the now extinct transitional forms or even that of the most primitive true land plant. On morphological grounds, the mosses and liverworts, although they exist predominantly as gametophytes (haploids) with a very much reduced sporophyte (diploid) generation and are non-vascular and lack true roots, might be regarded as one link, and as progenitors of other terrestrial plants [17]. For example, like all other land plants, the bryophytes have a cuticle and stomata, a differentiated thallus (vegetative stage), show heteromorphic alternation of generations, produce aerial spores, have enclosed sex organs and, in some mosses, even have conducting tissues in the stem for both food and water, somewhat like a primitive non-lignified vascular system. The bryophytes also show several characters in common with the green algae, particularly in their protonemal phase (the early phase of spore germination), in having biflagellate spermatozoids and in storing starch as a food reserve [17]. It might thus appear that the bryophytes are by far the best candidates as intermediates between the Chlorophyta and the vascular plants, but the fossil record suggests that the most primitive land plants (ca 420 million years ago) were, in fact, the Psilopsida with a well developed vascular system [1, 10]. It is possible,

of course, that the most primitive bryophytes, like many algae, did not yield good fossils and that the earliest known members of the group (ca 360 million years ago) may represent the result of the development of lignin and other evolutionary features [44]. It must also be recognised that the change from an heterotrichous alga to a primitive liverwort or psilophyte is too great to be encompassed in a single step and it seems highly probable that well over 100 million years involving several, perhaps "amphibian", transitional stages preceded the evolution of the first land plant.

The most primitive vascular plants are usually regarded as members of the Psilopsida, ancient fossils of which are found in sedimentary rocks from Devon in England, Canada, and other parts of the world, the most celebrated being found in the Rhynie Chert in Scotland [1]. It is possible that these early Psilopsida arose directly from algal stock independently from the ancestors of other vascular plants since, in spite of their anatomical characteristics, they were never successful and, indeed, there is no continuous fossil record which connects the earliest member of the taxa, *Cooksonia,* of late Silurian age (400 million years ago) [10] with those of the presumed present day representatives, *Psilotum* and *Tmesipteris* [5]. Many authorities [6] have pointed out that these latter two genera possess a number of advanced characters and their lack of roots, like *Cooksonia* and the bryophytes, is not necessarily indicative of a primitive status. In fact, Bierhorst [6] places them close to the primitive fern genera *Stromatopteris* (Stromatopteridaceae) and *Actinostachys* (Schizaeaceae) which also have rootless embryonic forms. However, it should be noted that the young embryo of *Tmesipteris* bears a striking resemblance to the young sporophyte of present-day horned liverworts (*Anthoceros*), which may indicate a closer link between them and the bryophytes [5].

Most authorities consider that the next most advanced phyla are the lycopods which were abundant in the Carboniferous period, many as substantial trees. Only small herbaceous forms are extant today. The three present orders, although having many features in common, including endoscopic embryogeny (the embryo develops from the inner cell of the zygote) and sporangia in the axils of the microphylls (leaf-like organs) which have a single vein leaving no gap in the stele (vascular tissue of the stem), nevertheless possess a number of important differences [5, 17, 34]. Thus, the Lycopodiales are homosporous, with eligulate leaves (without scales) whilst the Selaginellales and Isoetales are heterosporous with ligulate leaves. To group the latter two orders together also appears to be a matter of convenience, since the two modern genera of the Isoetales have a remarkable rush-like aquatic habit with monolete (having a linear scar) spores rather than being branched and often heterophyllous with trilete spores as are most *Selaginella,* the sole genus of Selaginellales. Furthermore, the Isoetales have multiflagellate rather than biflagellate spermatozoids, thus resembling the ferns. The relationship of the lycopods as a whole to other living plants is obscure and they probably represent three separate offshoots of some ancient endosporous Devonian line which have not evolved much further [5].

Another group of primitive vascular plants, the Sphenopsida, is also represented in the earliest plant records [1]. However, this sub-division of the plant kingdom, once as successful as the lycopods, cannot be regarded as any more than a parallel development which has only barely survived the competition from higher plants. They may show relationships with the "proto ferns" [6, 31], but the affinities of the only existing genus, *Equisetum,* with other vascular plants is difficult to assess.

The ferns are the most primitive class of the Pteropsida which are distinguished from lower subdivisions of the Tracheophyta (vascular plants) by having a much more complex and efficient vascular system [5, 34]. This has allowed the development of tree-like forms able to exist in relatively dry habitats, unlike the earlier aborescent lycopods and horsetails which needed to live in saturated swamps.

The living ferns, which like other lower tracheophytes have a well-defined gametophyte generation, are usually divided into three orders, Ophioglossales (sometimes regarded as a distinct class), Marattiales and Filicales. The first order is distinguished from the other two by the absence of circinate vernation (uncurling of young leaves) and the fact that the fertile region of

the frond is in the form of a spike clearly set off from the vegetative portion [34]. Although there is no well-established fossil record, this arrangement points to their being of ancient lineage [5]. The Marattiales are usually regarded as being more primitive than the Filicales on account of their having sporangia which, like the Ophioglossales, bryophytes and lower vascular plants, develop from a *group* of initial cells (eusporanginate) rather than as in the Filicales from a *single* cell (leptosporanginate). However, there are several members of the Filicales which are thought to be more primitive than any species of the Marattiales [6]. For example, as mentioned earlier, *Stromatopteris* and *Actinostachys* have many features akin to *Psilotum.* The Osmundaceae are also a contentious group, being regarded by some authorities as primitive members of the Filicales and by others as a separate Order [6].

On the grounds of morphological complexity and their position in the fossil record there are several groups of arborescent plants, all now extinct, which are regarded as intermediate between the ferns and the seed bearing gymnosperms and angiosperms. The position and relationship of the main taxa of these Mesozoic plants, Pteridospermales, Bennettitales, Caytoniales and Cordaitales, is still open to argument, although it is generally believed that the first three orders gave rise to the present day Cycadales and later the angiosperms, while the last led to the second main gymnosperm line, Ginkgoales, Coniferales and Gnetales [33]. In all these Gymnospermae, the gametophyte is reduced to a condition of being parasitic on the sporophyte and never attains independence. In the cycads and in *Ginkgo,* the spermatozoids are motile as in the ferns, but they are not released directly from an antheridium (male sex producing organ). Instead the male cell is enclosed in a pollen grain which consists of a three-celled structure to aid in transporting the spermatozoids to the archegonium (female sex organ) [33]. In the other two orders, Coniferales and Gnetales, there are no flagellated spermatozoids. There are several other important differences which set these last two orders apart. The Gnetales, consisting of the three rather diverse genera *Ephedra, Gnetum* and *Welwitschia,* are similar to the other gymnosperms

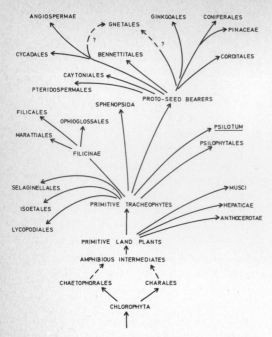

Fig. 1. Evolution of the major divisions of the lower tracheophytes.

1. VITEXIN

2. LUNULARIC ACID

by examining the overall distribution of different types of flavonoid compounds in the present representatives of the various classes of these lower plants.

Distribution of Flavonoids in Lower Vascular Plants

The most primitive plants in which true flavonoids are found are the advanced green algae, Characeae, which contain C-glycoflavones (*1*) [21]. It should be noted, however, that the dihydrostilbene growth inhibitor lunularic acid (*2*), a compound closely related to the flavonoids, is found in all algae from the procaryotic blue-greens upwards, having been originally isolated from a liverwort [28]. It is not found in mosses or most higher plants, although it may be present in trace amounts in those gymnosperms and angiosperms which contain the related stilbenes.

The distribution of lunularic acid, C-glycoflavones and all other types of flavonoids found in the lower vascular plants is outlined in table 1. It can be seen from this data that with few exceptions, there is a progressive increase in complexity of flavonoid compounds from the Charales to the Gymnospermae. This progression starts with C-glycoflavones in the green algae and the bryophytes, 3-deoxyanthocyanidins in the mosses, biflavones in *Psilotum*, flavonols and proanthocyanidins (leucoanthocyanins or condensed tannins) with trihydroxy B-rings in *Equisetum*, *O*-methylation in the ferns and C-methylation in gymnosperms. The latter class shows almost the full flowering of flavonoid evolution in containing all classes with the exception of neoflavonoids and aurones. The angiosperms, of course, contain a number of other flavonoid derivatives, not found in lower taxa, but these do not involve any radically new ring system.

in having a naked seed in which the female gametophyte acts as the nutritive tissue for the embryo, in contrast to the angiosperms where the embryo is surrounded by a triploid endosperm derived from fusion of male and female nuclei. However, they were previously regarded as showing close affinities with, and perhaps ancestral to, the angiosperms on account of their having vessels instead of tracheids in the secondary vascular tissue, and also in having a perianth-like structure in the male flowers. Today, these features are generally believed to be evidence of parallel evolution rather than of close affinity [33]. Indeed, the angiosperms are usually thought to have arisen from the Pteridospermalean-Cycadalian line [6, 33]. In the Gnetales, *Ephedra* stands apart from the other two genera, especially in pollination and ovule development [47], but the interrelationships of all three members of the Gnetales are unclear.

The overall evolution of the lower vascular plants discussed above is shown in fig. 1. Obviously there are a number of doubts and uncertainties about the relationships outlined which cannot be resolved using traditional taxonomic criteria. The rest of this paper is concerned to see if these difficulties can be resolved

Table 1. *Distribution of flavonoids in plants*

Division or sub-division	Class or Order	Lunularic acid	Glycoflavones	Biflavones	Flavones	Deoxyanthocyanidins	Flavonols	Anthocyanidins	Proanthocyanidins	Trihydroxy B-ring	C- or O-methylation	Remarks (see [21, 38])
Rhodophyta	Nemalionales	+	−	−	−	−	−	−	−	−	−	Caffeic-like acids present
Phaeophyta	Fucales	+	−	−	−	−	−	−	−	−	−	Vanillin+ve compounds
Chlorophyta	Chaetophorales	+	−	−	−	−	−	−	−	−	−	Possible cinnamic acids
	Charales	+	+	−	−	−	−	−	−	−	−	p-hydroxy-benzoate (?)
Bryophyta	Hepaticae	+	+	−	+	−	(+)	−	−	(+)	(+)	8-methoxy-luteolin & tricin
	Anthoceratae	?	−	−	−	−	−	−	−	−	−	
	Musci	−	+	(+)	+	+	(+)	−	−	−	−	*Dicranium* has a biflavone
Psilopsida	Psilotales	−	−	+	−	−	−	−	−	−	−	p-coumaryl lignin
Lycopsida	Lycopodiales	−	−	−	+	−	−	−	−	−	(+)	Vanillyl lignin
	Selaginellales	−	−	+	−	−	−	−	−	−	−	Syringyl lignin
	Isoetales	−	−	−	+	−	−	−	−	−	−	Syringyl lignin
Sphenopsida	Equisetales	−	−	−	+	−	+	−	+	+	−	
Pteropsida	Filicinae	−	+	−	+	+	+	+	+	+	+	
	Ophioglossales	−	−	−	−	−	+	−	−	−	+	Q-3-O Me in *Ophioglossum* not *Botrychium*
	Marattiales	−	−	−	−	−	+	−	+	+	−	
	Filicales	−	+	−	+	+	+	(+)	+	+	+	*Davallia* has pelargonidin
	Stromatopteris	−	−	−	−	−	+	−	+	−	−	
	Osmunda	−	−	−	+	−	+	−	+	−	−	
	Gymnospermae	−	−	+	+	−	+	+	+	+	+	
	Cycadales	−	−	+	−	−	−	−	+	−	+	
	Coniferales	−	−	+	+	−	+	+	+	+	+	
	Pinaceae	−	−	−	+	−	+	−	+	+	+	
	Ginkgoales	−	−	+	−	−	+	−	+	+	+	
	Gnetales	−	−	−	+	−	(+)	(+)	(+)	(+)	−	
	Ephedra	−	−	−	+	−	+	+	+	+	−	White fluorescent
	Welwitschia	−	−	−	+	−	−	−	−	−	−	Compound in common
	Gnetum	−	−	−	+	−	−	−	−	−	−	
	Angiospermae	(+)	+	(+)	+	+	+	+	+	+	+	Biflavonyls in few primitive families

− Not reported.
(+) Present in one or two species.
+ Present in several species.

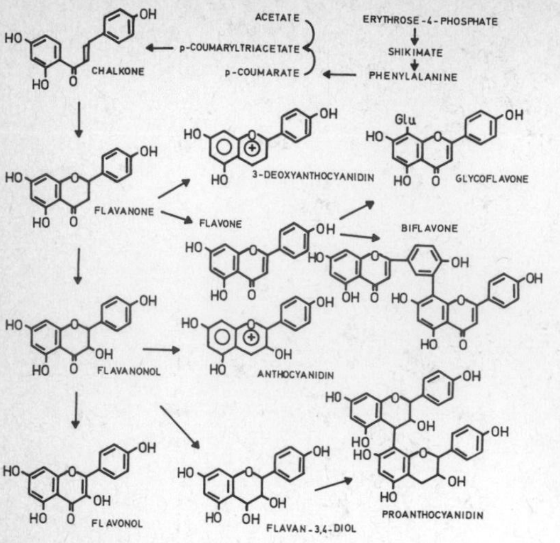

Fig. 2. The biosynthesis of flavonoids.

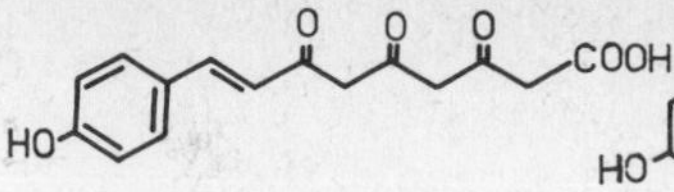

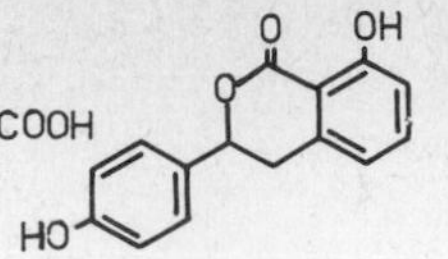

3. p-COUMARYLTRIACETIC ACID   4. HYDRANGENOL

The Genetic Implications of Biochemical Diversity

What do these changes mean in terms of increased complexity of the genome? In other words, how much extra genetic information is required to establish a new biosynthetic pathway such as that to the flavonoids [20] (fig. 2)? It may be assumed, since phenylalanine occurs in the proteins of all organisms, that the only variation in that part of the chromosomal DNA which codes for the synthesis of this amino acid when it is used as a starting point for flavonoid synthesis, will be concerned with possible slight changes in the amino acid sequences of the relevant enzymes which catalyse each step in its synthesis from erythrose-4-phosphate and pyruvic acid and in the overall control mechanisms [25, 42]. Again, since lunularic acid (2) is common to all lower plants, it can also be postulated that the steps from phenylalanine to p-coumaryltriacetic acid (3) are already encoded in the nucleus of all taxa up to and including liverworts. The only initial difference in the biosynthesis of the dihydrostilbenes (2) and the flavonoids (1) is the mode of ring closure of this mixed polyketide intermediate [7]. It would appear, therefore, that the only new enzyme system required by Chara to form the flavonoid nucleus is the requisite precursor-chalkone cyclase [20 a]. It is probable that this enzyme may have an active centre, and indeed an entire primary structure, very similar to that of

the corresponding stilbene cyclase. It is generally agreed that new enzymes arise by an original doubling of the chromosome coding for an analogous protein, followed by some minor mutation [43], and this could well be the case here.

The next step in flavonoid biosynthesis, the isomerisation of the chalcone to the flavanone [20], also has its analogies in the biosynthesis of lunularic acid (2). Pryce [27] has shown that hydrangenol (4), which contains an isocoumarin ring, is a good precursor of the growth inhibitor. If the route to lunularic acid in all organisms involves the prior formation of such an oxygen heterocyclic as in (4), involving a link between the terminal COOH of the mixed polyketide (3) and the C atom adjacent to the cinnamoyl moiety, this is akin to flavanone formation. The next step in the route (fig. 2) is the conversion of a flavanone to a flavone, and this resembles several other biochemical dehydrogenations of double bonds such as those involved in unsaturated fatty acid metabolism.

The major steps leading to the production of the hitherto unknown flavone precursor of the C-glycoside type found in Chara are, therefore, in the main, extensions of known biochemical reactions and presumably did not involve a series of interlocked multiple mutations of the DNA of the ancestral organism. This conclusion probably applies to most of the other steps involved in the elaboration of the flavonoid molecule, such as the introduction of the 3-hydroxy group and the reduction of the 4-carbonyl group. Instead, as mentioned above, it appears that all the new enzymes which were necessary for flavonoid biosynthesis could have arisen by chromosome doubling followed by minor mutations to alter the tertiary structure of the original enzyme and so allow either the approach to the reactive centre of somewhat different substrates and/or some slight change in the position of activation of a substrate which could lead to a varied product.

5. R = H, AMENTOFLAVONE
6. R = CH$_3$, SOTETSUFLAVONE

7. R=H, LEUCOCYANIDIN
8. R =OH, LEUCODELPHINIDIN

The Formation of Carbon-Carbon Bonds

One biosynthetic feature which does appear to be completely new is the formation of C–C bonds in the biflavonyls (5, 6), proanthocyanidins (7, 8) and, more especially, C-glycosyl flavones (1). Since the latter group of flavonoid compounds were presumably first to appear in an evolutionary sense, it is obvious that the capabilities of plants to form such bonds is, indeed, a very ancient one. It has been suggested [36, 38] that the evolution of flavonoids in *Chara* had the evolutionary advantage of allowing the more advanced green algae to exist near the surface of their aquatic environment, and eventually invade the land, thus improving photosynthetic activity, by acting as a screen to prevent UV light causing excessive mutations in the nucleic acids. If this is true it might be necessary for the compound to be transported about the cell, and since the flavones are very water insoluble the attachment of a sugar or other hydrophilic group would be desirable. The attachment of sugars to phenolic compounds is quite common in higher plants, the derivatives usually occurring as O-glycosides [26]. However, it seems probable that the attachment of a sugar, or sugar derivative, to the acidic phenolic group may have been an advanced step in evolution as O-glucosylation has been found to be lacking in all algae examined [26]. Since nucleophilic reactivity of both resorcinol- and phloroglucinol-like rings is high, an electrophilic addition of a sugar might be expected to take place directly, providing the deactivating effect of the 4-carbonyl can be overcome. It must be presumed that the enzyme responsible for the addition reduced to obviated this carbonyl de-activation in some way.

The formation of biflavones may have been an extension of this reaction. However, it should be noted that, with one exception, in the moss *Dicranium* ([23], see p. 117) they occur first in *Psilotum*, the most primitive plant with lignified tracheids. It is known that *Sphagnum* moss lignin is quite unlike higher plant lignin in consisting mainly of a polymer of p-hydroxy-cinnamyl alcohol units which are linked via the positions ortho to the hydroxyl group [8]. According to Siegel [32] the lignin in the sporophyte of *Psilotum* is similar to that of two other mosses, *Dawsonia* and *Dendroligotrichum*, in solubility, colour reactions and chemical spectroscopical properties and, if it is assumed that these lignins are of a similar type to that in *Sphagnum*, it seems plausible to conclude that the phenolic coupling reaction involved in lignin formation has been extended in *Dicranium* to give the 5′,8″-biluteolin [23] and in *Psilotum* to form the simple biflavone, amentoflavone (5).

The other major group of C–C linked compounds are the proanthocyanidins (7, 8) [42] which are first found in *Equisetum*. These compounds are ubiquitous in the ferns and gymnosperms and in the more woody angiosperms ([2], p. 93). Again it is possible that the formation of the C–C link is catalysed by enzymes modified from those responsible for the formation of C-glycosyl flavones. It seems more probable, however, that the reaction is quite different since as soon as a 4-carbonium ion is formed from the presumed 3,4-diol precursor, the reaction takes place spontaneously. For example, polymerisation rapidly ensues when a flavan-3,4-diol is treated with dilute mineral acid in the cold. It seems probable also that the formation of such compounds required the simultaneous elaboration of the flavan-3-ols (catechins) to act as "end groups" and prevent further polymerisation [35]. The paucity of flavans without hydroxyl substitution in the heterocyclic ring presumably prevented compounds analogous to the proanthocyanidins from being formed from flavones. Flavan-4-ols are known but are of limited occurrence [3].

The importance of the formation of these C–C linked and other flavonoids can now be considered in relation to the evolution of the lower tracheophytes.

The Evolutionary Advantage of Flavonoids

If we accept the fact that any group of compounds will only be produced if they confer on the taxa concerned an evolutionary advantage, it can be seen that the flavonoids offer an interesting example of structural change which has led to a range of varying functions, all of which have been important in differing situations. It is generally accepted that the flavonoid colouring matters of the angiosperm flower, such as the anthocyanins and flavonols, are important in attracting pollinators of one sort or another [20 b, 41]. However, one could not suggest the same role for the 3-deoxyanthocyanidins (fig. 2) in mosses. Here, it seems probable, their role is to act as a light screen which perhaps affects a phytochrome-controlled or similar response. Similarly, the chalcone colouring matters of the ray florets of several members of the Compositae undoubtedly act as pollination attractants or guides, whereas the chalcones found as pigments in the sori underneath the fronds of the fern *Pityrogramma chrysophylla* [20 b] must have another role, again perhaps controlling some light-catalysed reaction. The development of more complex flavonoids in the angiosperms also shows a second important role for this group of compounds, which is their ability to inhibit the growth of saprophytic organisms. The biflavones are known to be fungistatic in gymnosperm heartwoods and perhaps played a similar role in the primitive *Cooksonia* from which the present day *Psilotum* may have evolved. However, these compounds are relatively poor as antibiotics because their insolubility precludes transport to the site of attack. The possibility of producing soluble antifungal compounds based on flavones seems to have been "overlooked" during the early course of evolution, and the next group of satisfactory substances awaited the enzyme systems necessary for the introduction of the 3-hydroxyl group and the elaboration of the proanthocyanidins. These compounds, which occur in plants as oligomers

having from 2 to 10 flavonoid units [35, 40] are capable of inhibiting a large number of enzymes by their "tanning" action [19]. As such, they can be regarded as broad-spectrum antibiotics [12]. Their disadvantage lies in the fact that since they are present in relatively high concentration in the plant, any wounding is likely to release them and inhibit the plant's own enzymes. This could prevent the mobilization of other defences to infection. On the positive side, however, the production of relatively large amounts of proanthocyanins in leaves does make them much less attractive to herbivores of one sort or another. The proanthocyanidins are reinforced in angiosperms by ellagitannins (p. 93) and by more selective antibiotics, mainly based on the isoflavones, which are only produced when infection actually occurs [20 b, 37, 38].

Flavonoids as Evolutionary Markers

The more or less exclusive ability of the plant divisions outlined in fig. 1 to produce only certain classes of flavonoid (table 1) allows us to answer a number of important phylogenetic questions. First of all, it should be noted that although condensed tannins have been reported in the brown algae *Ascophyllum nodosum* and *Sargassum* species, these compounds appear to be quite unrelated to the proanthocyanidins found in vascular plants [12]. Furthermore, as can be seen from table 1, the Rhodophyta and members of the Chaetophorales do not contain any flavonoids, although the former appears to have relatively high quantities of caffeic acid-like compounds. We can presume, therefore, that the ability to produce true flavonoids in the algae is limited to the Charales. The internal distribution of flavonoids in individual plants of *Chara* and *Nitella* has also been examined to see whether the upper part of the plant contains more flavonoids than the region near the holdfast, as might be expected if the flavonoids were playing a role as UV screens. However, no difference was found [12]. This result might be expected in view of the fact that there are no true cellular differences between the upper and lower regions of this heterotrichous algae.

A re-examination of *Psilotum* species has confirmed that no flavonols or proanthocyanidins occur but, whereas Voirin [46] reported the presence of amentoflavone (5) only, it appears that a second biflavone, probably sotetsuflavone (6) is also present. An examination of the localization of these compounds in *Psilotum,* indicates that they are most probably present in, or attached to, the walls of the tracheids. Here they could serve as UV screens and inhibit the hydrolysis of polysaccharides by saprophytic hydrolases. The presence of biflavones (5) in *Psilotum* tells strongly against Bierhorst's [6] suggestion that they may be regarded as primitive ferns, since the latter contain no trace of these compounds. Indeed, close examination of both *Stromatopteris* and *Gleichenia* showed that these two, obviously primitive, fern genera nevertheless contain the same group of flavonoids generally present in other members of the Filicales [12].

The absence of biflavones in the ferns is paralleled in the other lower tracheophytes with the exception of Selaginellales which contain no other flavonoid. This sets *Selaginella* away from other Lycopsida which contain simple flavone-*O*-glycosides [46] and suggests the need for further work on the relationship of this plant division. The one report of a luteolin-based biflavone in the moss *Dicranium scoparium* [23] is also of great interest and, like that of 6-hydroxyapigenin in *Bryum* (see p. 117), 8-methoxy luteolin in *Monoclea* [48], and flavonols in the liverwort *Corsinia coriandrina* [29], requires further consideration to be given to the overall evolution of the bryophytes.

It may be noted in respect of the ferns that Voirin [46] had reported the absence of proanthocyanins from the Marattiales, with the exception of *Angiopteris evecta* which contained traces of both leucodelphinidin (8) and leucocyanidin (7). Since the latter genus is regarded as an advanced member of the order, Voirin suggested that these results gave general support to the division of the ferns into eusporangiate and leptosporangiate series [11]. Further examination, however, showed that 7 out of 8 members of the Marattiales contain relatively high quantities of proanthocyanidins including leucodelphinidin (8) [12]. Indeed, there is virtually no difference between the species examined in terms of their flavonoid, hydroxycinnamic acid and hydroxybenzoic acid content,

and that of the vast bulk of the Filicales [12]. The Ophioglossales, however, do not contain any proanthocyanidins and the consistent presence of 3-*O*-methylquercetin [46] indicates that they may be chemically quite distinct from the other fern orders.

The Sphenopsida as a class show a number of advanced features, including the production of the tri-hydroxy B-ring and proanthocyanidins (7, 8), and extra hydroxylation of the A-ring [30, 46]. One can thus regard these plants as having reached a relatively advanced state of flavonoid evolution on par with that of the ferns, thus supporting the proposals of Bierhorst [6] and Schweitzer [31] mentioned earlier.

The ability to produce proanthocyanidins as in the ferns and horsetails is maintained in all members of the gymnosperms except the Gnetales. In this order only *Ephedra* produces flavonols, proanthocyanidins and compounds with a trihydroxy B-ring (e.g. 8). The gymnosperms as a whole are noted for producing a wide variety of biflavones which are present in all families with the exception of Pinaceae and the three genera of the Gnetales, *Gnetum, Ephedra* and *Welwitschia* [12, 20 b]. As mentioned above, *Ephedra* differs from *Welwitschia* and *Gnetum* in having proanthocyanidins, which is interesting in view of the similar split made by Eames [47]. However, both *Ephedra* and *Welwitschia* contain a compound, absent from *Gnetum* and other gymnosperms, which fluoresces white under UV light and occupies an identical position on 2-dimensional chromatograms; its structure is at present being investigated. It is hoped, therefore, that a closer examination of the chemistry of these three genera will throw more light on their relationship with each other and with other Divisions of the plant kingdom.

Conclusions

It can be seen from the above brief outline that the presence of certain types of flavonoid compounds can be useful in examining phylogenetic affinities of classes of lower plants. Obviously, a large number of other different classes of compounds could be used in a similar way, and it is to be hoped that such information will be forthcoming in the near future. The flavonoid data, however, shows that, as pointed out earlier, there is a progressive evolutionary increase in the structural complexity of this group of compounds from glycoflavones in *Chara* to proanthocyanidins oligomers in the ferns and gymnosperms. It appears likely that these changes reflect the different advantages which the flavonoids confer on the plants in question. It is to be expected that this trend would continue into the angiosperms, whatever their immediate ancestors. On these grounds, the most primitive angiosperm, besides having the attributes summarised by Cronquist [13], would contain proanthocyanidins and lack alkaloids based on the aromatic amino acids, phenylalanine and tryptophan, since these are conspicuously absent in all lower vascular plants [37]. This supports the suggestions of Bate-Smith [4] and p. 93 and Kubitsky [22] outlined in detail elsewhere (p. 93), that the Magnoliidae, which more or less lack proanthocyanidins and are richly endowed with phenylalanine-derived alkaloid are not ancestral, in the chemical sense at least, to other plant groups. The problem of angiosperm evolution is a difficult one, but it is to be hoped that when we have more information on the biochemistry of lower plants we shall have some better ideas about their possible progenitors.

I wish to thank Mrs G. A. Cooper-Driver for useful discussions and allowing to quote unpublished data and Mrs S. Saunders for help in the preparation of this manuscript.

References

1. Banks, H P, Evolution and plants of the past. Wadsworth, Belmont, Calif, 1970.
2. Bate-Smith, E C, J Linn soc (bot) 1962, 60, 325.
3. — Phytochemistry 1969, 8, 1803.
4. — Nature 1972, 235, 157.
5. — Bell, P & Woodcock, C, The diversity of green plants, 2nd edn. Arnold, London, 1972.
6. Bierhorst, D W, Morphology of vascular plants. MacMillan, New York, 1971.
7. Birch, A J, Proc chem soc 1962, 3.
8. Bland, D E, Logan, A, Menshun, M & Sternhell, S, Phytochem 1968, 7, 1373.
9. Brown, W H, The plant kingdom. Ginn, Boston, 1935.
10. Chaloner, W G, Biol rev 1970, 45, 353.
11. Christensen, C, Filicinae, in Manual of pteridology, p. 522. Nijhoff, La Haye, 1938.
12. Cooper-Driver, G A & Swain, T, 1973. In preparation.
13. Cronquist, A, The evolution and classification of flowering plants. Houghton-Mifflin, Boston, 1968.
14. — Introductory botany, 2nd edn. Harper & Row, New York, 1970.
15. Crowson, R A, Classification and biology. Heinemann, London, 1970.
16. Davis, P H & Heywood, V H, The principles of angiosperm taxonomy. Oliver & Boyd, Edinburgh, 1965.
17. Doyle, W T, The biology of higher cryptogams. MacMillan, London, 1970.
18. Fritsch, F E, Ann bot (N S) 1945, 9, 1.
19. Goldstein, J L & Swain, T, Phytochemistry 1965, 4, 185.
20 a. Grisebach, H, Pure & appl chem 1973, 34.
20 b. Harborne, J B, Comparative biochemistry of the flavonoids. Academic Press, London, 1967.
21. — Recent adv in phytochem 1972, 4, 107.
22. Kubitski, K, Taxon 1969, 18, 360.
23. Lindberg, G, Österdahl, B-G & Nilsson, E, Chemica scripta (Sweden) 1973. In press.
24. McLaughlin, P J & Dayhoff, M J, J molec evolution 1973, 2, 99.
25. Miflin, B J, in Biosynthesis and its control in plants (ed B V Milborrow). Academic Press, London, 1973.
26. Pridham, J, Phytochem 1964, 3, 493.
27. Pryce, R J, Phytochem 1971, 10, 2679.
28. — Ibid, 1972, 11, 1759.
29. Reznik, H & Wierman, R, Naturwiss 1966, 53, 530.
30. Saleh, N A M, Majak, W & Towers, G H N, Phytochem 1972, 11, 1095.
31. Schweitzer, H J, Linnean meeting on paleobotany. Bristol, 1973.
32. Siegel, S M, Am j botany 1969, 56, 175.
33. Sporne, K R, The morphology of gymnosperms. Hutchinson, London, 1965.
34. — The morphology of the pteridophytes. Hutchinson, London, 1970.
35. Swain, T, in Plant biochemistry (ed J D Bonner & J E Varner). Academic Press, New York, 1965.
36. — in Biochemical evolution and the origin of life (ed E Schoffeniels) vol. 2. North Holland, Amsterdam, 1970.
37. — in Comprehensive biochemistry (ed M Florkin & E H Stotz) vol. 29 A. Elsevier, Amsterdam, 1973.
38. — in The flavonoids (ed J B Harborne & T Mabry). Chapman & Hall, London, 1973.
39. Takhtajan, A, Flowering plants. Oliver & Boyd, Edinburgh, 1969.
40. Thompson, R S, Jacques, D, Haslam, E & Tanner, R J N, J chem soc (Perkin trans) 1972, 1387.
41. Thompson, W R, Meinwald, J, Aneshansley, D & Eisner, T, Science 1972, 177, 528.
42. Tristam, H, in Biosynthesis and its control in plants (ed B V Milborrow) p. 21. Academic Press, London, 1973.
43. Watts, R L, in Biochemical evolution and the

origin of life (ed E Schoffeniels). North Holland, Amsterdam, 1971.

44. Watson, E V, The structure and life of the Bryophytes, 2nd edn. Hutchinson, London, 1967.
45. Whittaker, R H, Science 1969, 163, 150.
46. Voirin, B, Thèse, Docteur Science, University of Lyon.
47. Eames, A J, Phytomorph 1952, 2, 79.
48. Markham, K, Phytochem 1972, 11, 2047.

Discussion

Herout: Two other glycoflavonoids, saponarin and saponaretin, have been found in *Porella* (=*Madotheca*) *platyphylla*(Hepaticae).

Ourisson: The role of flavonoids has been presented as being to act as a UV-screen. Surely, this cannot be accepted as *the* reason for their presence. I doubt if we can really discern the positive, neutral, and negative effects of *one* character in an evolutionary sense.

Swain: I said that the glycoflavones in *Chara* could not possibly fulfill the same roles that flavonoids have in the flowers and fruit of angiosperms and I suggested that they became fixed in these algae because of their ability to act as UV screens. I firmly believe that all characters originally become fixed in the populations because they do confer one or more advantages to the organisms. We have shown that when solutions of glycoflavones are layered over *Alternaria* colonies, they do act as good UV screens and prevent denaturation, so I do not share your pessimistic view.

Ourisson: As regards your mention of the repellence of tannins to tortoises, it remains to be proved whether predation by reptiles was on the whole a negative factor in the evolution of plants: it may have been compensated by positive effects, such as more favourable dispersal means, or positive modifications of some niches, etc. The proved existence of one favourable (or unfavourable) effect of one factor does not prove that this factor is on the whole favourable (or unfavourable). One well-known case is that of the sickle-cell anemia: an unfavourable feature in general, but a favourable one in a population exposed to endemic malaria.

Swain: I do not believe that any major evolutionary change can be ascribed to a single fac-

tor acting alone; nevertheless, we must consider the advantages each factor might confer, both independently and together with others, if we are to discern their importance. Thus I believe that protection against herbivore predation is an advantage to those plants concerned. We have merely demonstrated that reptiles are like mammals in avoiding tannin-containing plants, and I suggest this is *one* important factor in plant–animal co-evolution. The example of the sickle-cell haemoglobin indeed shows that when it confers an advantage it is found in that population; where not, it is lost, or would be eventually.

Sørensen: In table 1, you gave an outline of the phenolic compounds occurring in lower plants. The Phaeophyceae were shown to contain lunularic acid and nothing else. I agree that lunularic acid is a very important compound, but the amount present in the brown algae is about 0.01%. As most of you know, the Phaeophyceae contain rather large amounts of other phenolics —5.25%. Although no chemist has so far been able to deduce the constitution of these compounds, I am rather worried when these main phenolics are completely ignored.

Wagner: Recently the compounds in *Fucus* species have been shown to be a new "tannin type" whose structures have been partially elucidated by Professor Glombitza from Bonn (see his forthcoming publication. In press). Obviously they are important when considering the dimeric and polymeric/phenolic compounds with C–C linkages in lower plants.

Swain: We have also examined two Phaeophyta "tannins", and they do not correspond to proanthocyanidins, and so were not included in the table. It will certainly be nice to know their structure. My intention in drawing attention to lunularic acid was to point out that the biosynthetic route to the mixed shikimate-polyketide intermediate of stilbenes and flavones was extremely ancient.

Bu'Lock: What about the C-alkylated phloroglucinols which occur in *Dryopteris*?

Swain: These compounds appear to be very restricted in distribution, but the ability to form

these *C*-alkylated phloroglucinols perhaps follows the same sort of mechanism as for the *C*-
substituted flavonoids.

Reichstein: Phloroglucinol derivatives occur in
nearly all species of the genus *Dryopteris* and
also in some members of the related genus
Chenitis. There are a few members of *Dryopteris* which do not contain phloroglucinols.

Birch: The biosynthetic evolution of flavonoids
(fig. 2, p. 86) indicates once again that Nature
is a good organic chemist, since the sequence is
identical with that which I predicted in 1957 on
purely mechanistic chemical grounds. Lunularic
acid is a very interesting case, since it is a de-
oxyresorcylic acid derivative, and must arise
from a reduced polyketide intermediate. This
lends some weight to the speculation that it
may have evolved from the fatty acid route,
substituting a cinnamoyl coenzyme-A for acetyl
coenzyme-A as a starter-unit, and leaving two
oxygens in the chain. The ring-closure by aldol-
condensation to a stilbene would be predicted
by any organic chemist to be easier than the
C-acylation required for flavonoids; it is therefore logical that it should evolve first. If a "reason" is sought for 3-oxygenation in anthocyanins, it may be that the 3-oxyglycosides are
particularly stable to light, and Nature may
have "decided" to take the extra trouble for
this reason.

Systematic Distribution of Ellagitannins in Relation to the Phylogeny and Classification of the Angiosperms

E. C. Bate-Smith

Agricultural Research Council Institute of Animal Physiology, Babraham, Cambridge CB2 4AT, UK

Summary

Ellagitannins are exclusively present in dicotyledonous angiosperms, but they are not present in all of them. The fundamental cleavage is between the subclass Magnoliidae (Takhtajan) from which they are absent, and the Hamamelidae-Dilleniidae.

The commonest flavonoid constituents in vascular plants are the flavonols myricetin, quercetin and kaempferol, and the leucoanthocyanidins leucodelphinidin and leucocyanidin. In all classes of vascular plants evolutionary change has proceeded by progressive elimination of hydroxyl function in these constituents, either by complete loss of the constituent, by loss of one particular OH group, or by methylation of one or more particular OH groups, processes which can be described as "degeneration" of phenolic function. The same process of degeneration can be discerned in the ellagitannins, where *O*-methylation has, in particular groups such as the Myrtiflorae, modified the functional properties of the ellagitannin, whilst in other groups the tannin has been eliminated by complete loss. Intermediate stages of degeneration can be discerned in such molecules as bergenin, which has always been found in plants in association with ellagitannins.

Attention is drawn to the very high concentrations of tannins that may be present in the leaves of these plants. In several instances more than 30% of the dry weight of the leaf is present as tannin, either as ellagitannin or leucoanthocyanin, or both. Regarded as a character in classification, this is associated both with the habit of the plant and with its systematic affinities. This is particularly evident in perennial herbaceous plants in the Rosaceae, Saxifragaceae and Geraniaceae.

An ideal classification is one which divides the objects to be classified into classes themselves divisible into subclasses, and so on, such that the ultimate divisions give rise to no ambiguity between one and another.

This can hardly be said to be true at the present time of the systems in use for the classification of the angiosperms. The primary division into monocots and dicots is *almost* unambiguous, but neither of these classes has been further divided in a way which is universally acceptable. In neither case is there an agreed primary cleavage, such that the affinities of subsequent classes are clear and unambiguous. It is possible that there is in fact no natural justification for such treatment; the affinities may not be ramose but reticulate, the natural system not in fact dichotomous.

While this seems likely to be true of the monocots, the situation in the dicots possesses possibilities of a successful dichotomy being achieved. The most widely used system, that of Engler, is in fact a dichotomous one in which orders with (mainly) choripetalous flowers are separated from those with (mainly) gamopetalous ones—the Archichlamydeae and Metachlamydeae respectively, but the former is really a lumping together of two classes previously distinguished by Bentham and Hooker, their Monochlamydeae and Polypetalae. Skottsberg redistributed the orders in these two groups and made them into one larger group, Polycarpicae, so that his system was essentially dichotomous. So also was Hutchinson's division into Lignosae and Herbaceae, but this was achieved by separating at the outset such closely related orders as the Magnoliales and the Ranales.

More recent authors divide the dicots as a class into a number of subclasses of equal rank: Cronquist, 6; Takhtajan, 7. The former has 5 branches arising out of the Magnoliidae, but the latter recognises a dichotomy into 2 stocks, one arising from the Magnoliidae and the other from the Hamamelidae and the Dilleniidae taken together (here referred to as H-D). This dichotomy derives substantial support from certain chemical features which appear to be distinctive for the separate stocks.

Nobel 25 (1973) Chemistry in botanical classification

Of these, the most important is the presence in the H-D of ellagitannins. These are esters (with glucose in every presently known instance) of hexahydroxydiphenic (HHDP) acid which, on hydrolysis, yield its dilactone, ellagic acid:

(*1*) Hexahydroxydiphenyl ion; (*2*) Ellagic acid;

It seems that these esters are absent from the Magnoliidae (although they are present in certain orders placed by Takhtajan in the Magnoliidae which can justifiably be regarded as doubtfully belonging to that stock). They are, however, also absent from many of the H-D orders, some, perhaps, because these orders are wrongly placed in that stock, but more often because they have been shed in the process of evolutionary advancement. It is the mechanism of this shedding process, the evidence for it, and the reasons for it which will be the main objectives of the present paper. In this respect it is an expansion of a short note published in Nature in 1972 [1].

Chemistry of the Ellagitannins

Because of the importance these substances will assume in the future development of chemical phylogeny it is worth while providing a brief resume of their history. The name "acide ellagique" is itself a play on words, "ellag" being simply the reverse of "galle", so named by Braconnet after its isolation from oak galls in 1818 by Chevreuil. It occurs partly free in woody fruits such as myrobalans (*Terminalia chebula,* Combretaceae) and heartwoods, partly in ester form, often together with gallic acid and breakdown products, such as chebulic acid, in complex combination. Our knowledge of the chemistry of these compounds is largely due to the work of O. T. Schmidt and his colleagues. It is these combined forms which have the property of tanning leather, earning for them the name "ellagitannins". In healthy leaves and other fresh tissues there seems to be little if any uncombined ellagic acid, but this is readily produced from the HHDP esters by acid hydro-

lysis. Almost everything that is known about the systematic distribution of HHDP in plants derives from the detection of ellagic acid on paper chromatograms of the hydrolysates of leaves. This in itself tells us nothing about the precursor of the ellagic acid present in the hydrolysate. There is, however, a simple means of determining whether the precursor is, in fact, an ester of HHDP. When these are present, a methanolic extract of the leaf, treated with nitrous acid, gives a brown colour followed, when the ester is present in quantity, by an intense prussian blue. The brown colour is due to the production of a red pigment which rapidly changes to deep blue, then more slowly to deep yellow. Ellagic acid itself (and also gallic acid, caffeic acid, and many other phenolic acids) only forms a yellow reaction product. The sequence of colour changes described appears to be specific for esters of HHDP. It can, in fact, be regarded as a specific reaction for ellagitannins.

Systematic Distribution of Ellagitannins

The systematic distribution of ellagic acid in hydrolysates of leaves was included in a survey of the phenolic constituents of plants [2]. Most of the species positive for ellagic acid in that survey had also given a positive reaction for ellagitannins, but the space allowed did not permit these results to be reported at that time. As described above, the test is not sensitive to low concentrations of ellagitannin because the green colour of chlorophyll masks both the red reaction and the later blue one, but a refinement of the procedure now allows not only the detection of ellagitannin in low concentration, but also its quantitative determination [3]. The leaves of many species of dicots have now been examined by the new procedure, and in every case in which ellagic acid was present in the hydrolysate, the reaction for HHDP in ester form was positive, and the extract also reacted positively for tannin by precipitating protein.

Ellagic acid has, however, been reported from green algae and liverworts. *Spirogyra* spp. have tannins said to resemble tannic acid [4] yielding glucose, gallic acid and *m*-digallic acid on hydrolysis. We found ellagic acid in *S. arcta,* but the extract gave no blue reaction with nitrous acid and, therefore, presumably contained no HHDP ester. The liverworts *Lophocolea bidentata* and *Plagiochila asplenioides* have recently been reported [5] to contain ellagic acid. Chromatography of a hydrolysate of the former did not confirm the presence of ellagic acid, nor did an extract give any blue colour with nitrous acid. The present evidence is therefore all in favour of the ellagitannins being confined to the dicots.

Systematic Distribution of Ellagic Acid in the Dicots

The distribution of ellagic acid in the light of the Englerian system, which was adopted in the original survey [2], gave no clear indication of systematic relationship. Families positive for ellagic acid occurred in 14 of the 30 orders of the Archichlamydeae and 3 of the 10 orders of the Metachlamydeae. Some indications of orderliness in its distribution emerged in 1965 [6]. The families in which it is most consistently present are those in the Englerian orders 20 to 29, while it is most consistently absent from those in the orders 15 to 19. These clusters of families I designated for my own purposes the "rosalian" and "ranalian", respectively, from their most prominent and, possibly, most central members.

It was with some excitement therefore that when Takhtajan's system appeared in its English version (1969), it became evident that this author recognised two stocks into which my two clusters could be fitted very closely. The result, published in 1972 [1], is reproduced (by permission of the proprietors of Nature) in figs 1 and 2.

For the dendrograms to be at all intelligible, the 250 or more families had to be represented at the ordinal level, and their value obviously depends on the validity of the orders as Takhtajan sees them. They are nearly all heterogeneous as regards the ellagitannin status of the families; in fact the only one which is homogeneous in this respect is the Myrtales (excluding Hippuridaceae). Several orders have only one family in which ellagic acid is present; these are represented as positive. Similarly many families have only a few species which contain ellagic acid, and these again are regarded as

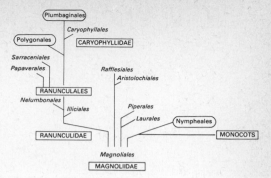

Fig. 1. Orders ringed are those containing ellagitannins.

positive. Particular cases are discussed later; it is necessary first to consider the phylogenetic implications of the distribution in its broader aspects.

In the Magnoliidae (fig. 1) there are only three orders in which species containing ellagic acid have been reported, Polygonales, Plumbaginales and Nympheales. In each case, in only one or two species has ellagic acid been reported, but in the first two orders hydrolysable, i.e. gallo-tannins are frequently present, contrary to the situation in the remainder of the Magnoliidae. In *Nuphar* (Nymphaeaceae) ellagitannin was conclusively demonstrated [7]. It would not be unjustified to suggest that these orders may be wrongly assigned to the Magnoliidae. My "ranalian" cluster would then correspond with the Magnoliidae with these orders removed from it.

The absence of ellagic acid from many of the orders of the Hamamelidae-Dilleniidae (fig. 2) does not, however, of itself give rise to similar doubts about its corresponding with my "rosalian" cluster. Many of the orders positive for ellagic acid in the H-D contain families which are negative, many of the families contain only a few genera which are positive. So long as any one member of a family contains typical ellagitannin, it is assumed that its absence from other members (so long as there are adequate grounds for their inclusion) is to be regarded as evolutionary loss.

It is assumed that the possession of ellagitannins is a primitive condition of the H-D stock. This is, in fact, the essence of the present thesis: that the acquisition of these tannins conferred such an advantage on the plants pos-

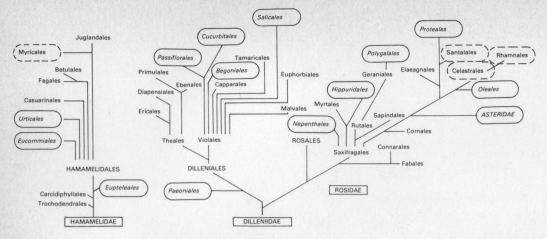

Fig. 2. Orders ringed are those *not* containing ellagitannins; broken rings ellagitannins doubtfully present.

sessing them that they became a dominating element in the world flora. Along with ellagitannins, the phenolic pattern retained the other *vic*-trihydroxy constituents, the flavonoids myricetin and leucodelphinidin, already present in ferns and gymnosperms. In contrast, the Magnoliidae, with the exception of the Nymphaeales, are extremely poor in these constituents (ref. [2] table 3).

Myricetin and leucodelphinidin often persist in H-D species from which ellagic acid is absent, so that the "rosalian" cluster can be broadened to include these. Impoverishment of these and other primitive constituents with evolutionary advancement is evident in the development of the dendrogram in fig. 2. It is the orders at the fringes of the branches that are lacking in ellagic acid, and these tend to be lacking not only in myricetin and leucodelphinidin but in leucocyanidin also. These developments are associated with increasing tendency to herbaceous habit.

Impoverishment can be gradual or abrupt. In the case of ellagic acid, the first stage seems to be the blocking of hydroxyl groups, reversibly by glycosylation, irreversibly by methylation. Indications of the former were first observed [8] in *Geranium phaeum* and *G. reflexum*, in which a red reaction product with HNO_2, λ_{max} 530 nm was produced in varying amount along with the blue reaction product, λ_{max} 600 nm, although ellagic acid only was formed

on hydrolysis. The presence of *O*-methylated and methylenedioxylated HHDP was reported by Lowry [9] in the wood and bark of many species in the Myrtales, and similar compounds have now been found in the leaves of several of these species, and also in those of the Cornaceae [10]. Here again, when the monomethyl ethers of HHDP esters are present, the red reaction product is formed. The protein-precipitating action of glycosylated or methylated HHDP esters is much less than that of the unsubstituted esters.

A constituent which may be formed as the result of interference with ellagitannin synthesis is bergenin. Hegnauer [11] pointed out that this can be regarded as derived from 4-*O*-methylgalloyl glucose by ring closure in position 2:

(3) 2-Galloylglucose; (4) Bergenin

It is always associated in its occurrence with ellagic acid-containing species [6].

Finally, it was suggested [2] that sinapic acid in unusual abundance indicated "the retention of the tendency to tri-substitution of the phenolic constituents in their more advanced condition". Later [12] this was more specifically related to interference with the biosynthesis of ellagic acid through the hypothetical stage of tri-hydroxycinnamic acid, which itself has so far not been found in any plant, but is replaced, in situations where (by extrapolation from the series *p*-coumaric acid-caffeic acid) it might be expected, by ellagic acid itself. More usually, sinapic acid occurs as a minor constituent as an extrapolation of the series *p*-coumaric acid-ferulic acid, just as syringyl groups occur along with *p*-hydroxyphenyl and vanillyl groups as a constituent of dicot lignin. The mere presence of sinapic acid in a leaf hydrolysate does not therefore imply any systematic relationship with species containing ellagitannins; it is its presence in unusual quantities that is significant. Outstanding examples are *Viscum* (Loranthaceae: ellagic acid abundantly present in *Nuytsia*); *Buxus* (Sapindales Engler; Euphorbiales, Takhtajan); *Salacia* (Hippocrateaceae, Sapindales, Engler; Celastrales, Takhtajan), and *Elaeagnus* (Myrtiflorae, Engler; Elaeagnales, Takhtajan). It is interesting to note that the systematic position of nearly all of these is undecided.

These evidences of impoverishment are relatively rare; most often ellagic acid is lost without trace. This can happen in just a few species in a family, as in the Dipterocarpaceae and Geraniaceae, or in the great majority of species, as in the Rosaceae, Saxifragaceae and Ericaceae. Clearly there must be families from which ellagic acid has been completely eliminated. This is assumed to have happened to some at least of the ellagic acid-negative families in the Hamamelidae-Dilleniidae, but there may be some which do not properly belong in this group. Some help is afforded in distinguishing between these two possibilities by taking into consideration the other *vic*-trihydroxy constituents, myricetin and leucodelphinidin. Their occurrence is listed under the heading *a b* in table 3 of Bate-Smith [2]. Except for those families in which ellagic acid is present (Plumbaginaceae, Polygonaceae, Nymphaeaceae) they rarely occur in Takhtajan's Magnoliidae, but are present in many of the ellagic acid-negative families in his H-D. They are present, for instance, in the Salicaceae and Ulmaceae, which might otherwise be considered out of place in this group.

As regards myricetin specifically, Kubitzki

[13] is inclined to regard its loss as one aspect of the progressive loss of hydroxylation through the sequence myricetin-quercetin-kaempferol, not as the abrupt loss of trihydroxylation per se. There are, however, so many cases of "all or none" presence or absence of myricetin and leucodelphinidin that we must regard this as a special case, distinct from the progressive decrease in hydroxylation of the other two flavonols, often accompanied by *O*-methylation, and their eventual replacement by flavones, flavanones, and other reduced or otherwise degraded flavonoid classes. The corresponding derivatives of myricetin either do not exist, or (as in the cases of ampelopsin and tricetin) occur so rarely as to be regarded as metabolic abnormalities.

Kubitzki [13] observed that in the Dilleniaceae the impoverishment of flavonoid hydroxylation was correlated with morphological advancement and geographical dispersal. The same is true also of the Geraniaceae [14]. In this case a "flavonoid score" was awarded on the basis of marks given for trihydroxylation and for preponderance of quercetin over kaempferol, and deducted for methylation of hydroxyl groups and presence of flavones, etc., following Kubitzki's original argument [13]. In *Geranium* the species with highest scores were those with ranges in central Eurasia and S. Africa, those with lowest scores at the extreme limits of the range of the genus, in Macaronesia, eastern USA and Hawaii. A similar treatment of Kubitzki's data for *Dillenia* shows that the species with high scores are those in S.E. Asia east of the Wallace line, decreasing progressively in species with northerly and westerly distribution into and across the Asiatic mainland.

In *Ulmus* [15] there is a similar northerly and westerly trend from a hypothetical S.E. Asian centre of origin, but the effect is less marked, perhaps because the species of *Ulmus* vary so little in their arboreal habit. It was in *U. macrocarpa* and *U. minor* that infraspecific variation in myricetin and leucodelphinidin was observed (cf above) without any discoverable associated morphological differences. As remarked earlier, ellagic acid is absent from the Urticales, but the presence of these *vic* tri-hydroxyflavonoids in the Ulmaceae and Urticaceae supports their inclusion within the Hamamelidae-Dilleniidae.

Further Indications of Chemical Differences between the Magnoliidae and the Hamamelidae-Dilleniidae

Outside the Magnoliidae there are few woody families which do not contain ellagic acid and/ or other *vic*-trihydroxy constituents, but in both divisions there are numerous mainly or wholly herbaceous families from which these are completely absent. The question is whether there are any other chemical indications which can be used as evidence of the correct assignment of such families to one or other division.

One such chemical character has been identified in the shape of the so called iridoid compounds, the prototype of which in the botanical context is aucubin. These substances have not been reported in any member of the Magnoliidae, but they are present not only in some of the more or less woody families of the H-D, but also in quite a number of the herbaceous families, especially those of the Gentianales, Scrophulariales and Lamiales.

Swain has indicated the importance attaching to the acquisition of the isoquinoline alkaloids in the evolution of the angiosperms, and Kubitzki [13] has pointed out their restriction in the dicots mainly to families in the Polycarpicae (Ranales, *sensu lato*). Mabry [16] has further considered the distribution of sesquiterpene lactones, which are conspicuously represented in the Compositae and otherwise present in 11 families, 9 of which are in the Magnoliidae. Finally, there is the pattern of substitution found in the "degenerate" molecules produced when the primitive phenolic compounds are eliminated. Dr Swain has mentioned one such pattern, the methylation or methylenedioxylation of adjacent hydroxyl groups in many phenolic compounds, especially frequent again in the families of the Magnoliidae, but also frequent, for example, in the Moraceae, Rutaceae and Compositae. With regard to the distinctive patterns revealed in adventitious hydroxylations in the 6- and 8 positions, C-glycosylation, chalcone, flavonone, stilbene, isoflavone and rotenoid compounds and the like, Dr Harborne has enlightened us (p. 103). It will be surprising if they do not fall into place in the suggested dichotomy, and help in the more accurate positioning of particular families in one or the other division.

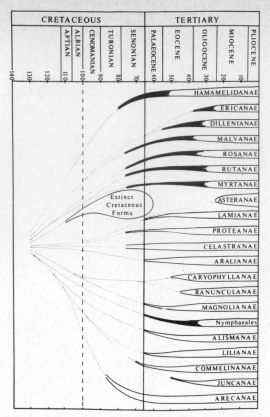

Fig. 3. Diagram, supplied by Dr P. D. W. Barnard, based on the superorders of flowering plants given in the appendix to A. Takhtajan's "Flowering plants, origin and dispersal" (Edinb. ed. 1969). For each superorder a first occurrence has been recalculated from "The fossil record" (ed Harland) together with other publications notably J. Muller. Superorders containing ellagitannins have been blocked in, in part, by the present author. The Celastranae have one member only, *Nuytsia* (Loranthaceae), containing ellagitannin; its inclusion in this superorder is questionable (see text).

Which came first?

Ellagitannins appear to be exclusively present in the dicots. The other *vic*-trihydroxy derivatives, myricetin and leucodelphinidin, are present in the ferns, gymnosperms, monocots and H-D, but rarely in the Magnoliidae. Yet the Magnoliidae are commonly regarded as antecedent both to the monocots and all the other dicots.

An idea of the time-scale involved in the evolution of the angiosperms can be got from fig. 3. This diagram is one given by P. D. W. Barnard to undergraduate students at Reading

University to stimulate discussion. Barnard writes:

"The point of this diagram to me as a palaeobotanist is to emphasise certain features of the fossil record. The earliest fossils which we can with confidence ascribe to the angiosperms are found towards the end of the Lower Cretaceous, Barremian and early Aptian about 112 million years ago. Although it is common practice of palaeobotanists to assign fossil remains of Turanian and older Angiosperms to extant genera this is in my opinion not justified. I feel that in the present inadequate state of our knowledge of the majority of Cretaceous Angiosperm remains they can only be placed in some taxonomic limbo of Extinct Cretaceous Forms. The rest of the diagram therefore attempts to show how far back in time we can safely, in my opinion, trace the main groups of Angiosperms. I have chosen the major groups, the superorders of Takhtajan, as the most convenient groups to handle in preparing a chart of this kind."

"I think the relatively short records of the groups Asteranae, Ranunculanae and Caryophyllanae may be genuine reflections on the more recent occurrence of these herbaceous forms. They may also indicate a differential effect of the fossil record. Trees having more chance of their parts becoming fossilised in comparison to herbs.

I have a personal conviction that the Angiosperms are monophyletic and only came into existence in early Cretacean times. If therefore more research was devoted to tackling the problems of classifying fossil Cretaceous Angiosperms remains then we might get an entirely different and more reliable picture of Angiosperm phylogeny."

In only one respect have I made any change in Barnard's diagram. This is by blocking in the lower ends of those superorders (and that of the Nymphaeales) which contain ellagitannins. This indicates in a rather spectacular way the distribution of these tannins in a particular area of the dicots, and the anomalous position, in this respect of the Nymphaeales, separated from the Ranunculanae by Barnard because of their *supposed* linking of the dicots with the monocots.

On this showing, the earliest dicots are the

Hamamelidanae, but the monocotyledonous Arecanae (palms) are coeval with them. The Magnolianiae (on Barnard's diagram) occur later in the record. Except for ellagitannins, the monocots contain all the kinds of phenolic constituents we have been considering. It becomes increasingly probable that the Magnoliidae are an independent offshoot of the dicotyledonous stock, but so distinct in their chemical features that the dichotomy suggested here is a real one; but where exactly that dichotomy should be made remains an equally real question. Meeuse, who comes to similar conclusions with regard to an early major dichotomy of the dicots [19] is inclined to associate the Sapindales, Rutales, Araliales, Boraginales, Polemoniales, Campanulales and Asterales with the Magnoliidae rather than with the Hamamelidae-Dilleniidae. The present chemical evidence could not exclude at least some of these taxa (so long as they are appropriately delimited) from such a cleavage.

The Importance of Tannins in the Evolution of the Angiosperms

The importance of tannins in the defence of higher plants against all kinds of predators is only just beginning to be appreciated. Their effectiveness against insects is dealt with especially by Feeny [20]. They appear also to act as a "broad spectrum" defensive mechanism against herbivores and pathogens ([21]; Swain, p. 81). Their effectiveness against fungi is especially relevant to the emergence of the angiosperms in the Cretaceous tropical environment. While all tannins, due to their protein-binding capacity, are effective in greater or lesser degree, Hart & Hillis [22] have shown that ellagitannins are solely responsible for the inhibition of wood-rotting fungi in the heartwood of *Quercus alba*.

The protein-binding capacity, or astringency, is readily determined by precipitation of blood protein [23]. Individual tannins vary greatly in astringency. If tannic acid, which appears to consist mainly of heptagalloylglucose, is taken as a unit standard of reference, the relative astringency (RA) of several typical ellagitannins was between 0.5 and 0.76, that of procyanidin dimers only 0.10. The hydrolysable tannins are therefore much more effective as protein precipitants than the condensed tannins, and this could well have given the plants which acquired them a critical advantage in the highly competitive Cretaceous environment. It can be no coincidence that the earliest identified dicot genus in Muller's table, *Ctenolophon,* is one which (at the present day) contains ellagitannin.

I would like to thank Dr P. D. W. Barnard for allowing me to reproduce fig. 3 and his explanation of it; Professors R. Hegnauer, A. D. J. Meeuse and K. Kubitzki for their helpful letters; Dr K. R. Sporne for his guidance in palaeobotanical matters; and Dr T. Swain for many helpful discussions.

References

1. Bate-Smith, E C, Nature 1972, 236, 353.
2. — J Linn soc (Botany) 1962, 58, 95.
3. — Phytochemistry 1972, 11, 1153.
4. Nakabashi, T et al., in Hegnauer, R, Chemotaxonomie der Pflanzen, vol. 1. Birkhäuser, Basel and Stuttgart, 1962.
5. Mues, R & Zinsmeister, H, Österr bot Z 1973, 121, 151.
6. Bate-Smith, E C, Bull soc bot Franç mem 1965.
7. — Phytochemistry 1968, 7, 459.
8. — Ibid 1972, 11, 1755.
9. Lowry, J B, Phytochemistry 1968, 7, 1803.
10. Swain, T et al. In preparation.
11. Hegnauer, R, Chemotaxonomie der Pflanzen, vol. 4. Birkhäuser, Basel and Stuttgart, 1966.
12. Bate-Smith, E C & Swain, T, Lloydia 1965, 28, 313.
13 a. Kubitzki, K, Ber deut bot Ges 1967, 80, 757.
13 b. — Ibid 1968, 81, 238.
14. Bate-Smith, E C, Bot j Linn soc. In press.
15. Bate-Smith, E C & Richens, R H, Biochemical systematics 1973, 1.
16. Mabry, T, Brittonia. In press.
17. Harland, W B et al. (ed) The fossil record, Geological Society, London, 1970.
18. Muller, J, Biol rev 1970, 45, 417.
19. Meeuse, A D J, Acta bot néerl 1970, 19, 90, 133.
20. Feeny, P P, Ecology 1970, 51, 565.
21. Whittaker, R H & Feeny, P P, Science 1971, 171, 757.
22. Hart, J H & Hillis, W E, Phytopathol 1972, 62, 620.
23. Bate-Smith, E C, Phytochemistry 1973, 12, 907, 1809.

Discussion

Bu'Lock: In considering questions of "homology" and so forth in the gallic/ellagic acid series of products we are faced with the particular necessity of resolving a well-established biosynthetic ambiguity. There are two clearly separate pathways to gallic acid. One of these, by way of the aromatic amino acid/ammonia lyase/cinnamic acid routes, is capable of adaptation to your presentation, for example of the manner in which some plant groups set about the task of divesting themselves of gallotannins. However, the other pathway, direct from the C_6C_1 stages of aromatic biosynthesis, cannot be so related. So far as I know, both pathways exist in higher plants, and we know that in some fungi they can actually coexist, but I do not know how widely each is really distributed. Have you any actual biosynthetic information on the real incidence of these two disparate pathways which might somewhat clarify your discussion of the apparent situation?

Bate-Smith: Homology can be inferred in a number of different ways. The way open to me is to establish the identity of the form of combination in which ellagic acid is present by the reaction with nitrous acid producing the blue, λ_{max} 600 nm, reaction product given by esters of hexahydroxydiphenic acid. The biosynthetic pathway considered most likely is through trihydroxycinnamic acid, which is itself not found in plants, but which might be expected to be present as an extrapolation of the series p-coumaric acid-caffeic acid. Ellagic acid is always found in hydrolyzates where trihydroxycinnamic acid might have been expected to be present.

Cronquist: The distribution of ellagitannins will surely prove to be of great taxonomic interest, but I must confess to a certain confusion with regard to your paper. My confusion stems from trying to integrate your charts with your words. You speak of a Hamamelid-Dilleniid line, in contrast to another line for all other angiosperms, but half of the Rosidae are on one side and half on the other. I shall have to read your paper at my leisure, in hope of resolving my confusion.

Also I want to point out that some groups can be carried considerably further back into the fossil record than your charts show. Dr Leo Hichey tells me that the platanoid line can be traced back to the Albian with considerable confidence. Furthermore, the earliest angiosperm pollen is all manosulcate. Manosulcate pollen in modern angiosperms is confined to the magnoliidae and the monocotyledons. Monocot pollen has its own characteristic ornamentation, which does not begin to show in the fossil record until some time after the appearance of magnoliid type pollen, although the divergence did begin relatively early. Furthermore, the earliest angiosperms leaf fossils have a characteristic, relatively disorganized type of venation which today is seen chiefly in certain families of the Magnoliidae, although it is not wholly confined to this group. The association of leaf and pollen fossils strongly suggests that the earliest angiosperms were of general Magnoliid type.

Bate-Smith: The Hamamelidae-Dilleniidae include the whole of the Rosidae, but not all of the orders in the Rosidae contain ellagitannins —it is presumed that they, and especially those which are mainly herbaceous—will have lost them in the process of evolutionary advancement.

The views on the antiquity of the orders are those of the palaeobotanist who supplied the diagram which I showed and are, of course, his own personal interpretation of the data from the sources quoted.

Turner: In listening to your presentation it was not clear to me whether or not the "fundamental cleavage" which you note in the angiosperms might not be looked upon as a kind of cleavage proposed by the late John Hutchinson. That is, can you comment upon the likelihood that your data reflect a distribution based upon absence *primarily* in herbaceous groups and presence *primarily* in woody groups?

Bate-Smith: Hutchinson's dichotomy starts off with an unacceptable cleavage between Magnoliales and Ranunculales. The cleavage I have suggested takes into account the presence of herbaceous and woody taxa, but is not dependent on these characters, but only upon the presence and absence of ellagitannins in the first place, and provides reasons for their absence from forms which might, on other grounds, have been expected to contain them.

Heywood: In considering the distribution of compound against phylogenetic systems, it is highly desirable that as many diverse systems as possible should be taken into account, not just those of Takhtajan or Cronquist, however excellent these may be. It is possible that the distribution of the ellagitannins might show interesting and suggestive correlations with the systems of Emberger, Meeuse and others.

Bate-Smith: Distributions are always considered against the numerous and diverse systems currently advocated. Against the Engler-Prantl system, for instance, the distribution of ellagic acid is little more than random. The "fit" with Takhtajan's system is remarkably informative and almost equally so is that with Cronquist's. No others provide so much of a fit. These systems may not be "final" (as Professor Cronquist admitted of his own) but they are getting much closer to a naturally final system than any other extant system.

Hegnauer: On which arguments is your assumption based, that the first angiosperms contained both condensed tannins and ellagitannins?

Homology: Your concept is based on structural similarity (identity), e.g. ellagic acid is present as such or is derived from esters of hexahydroxydiphenic acid, or proanthocyanidin, may be present in the form of the A- or B-group.

Remarks: (a) Basing homologies on biosynthetic pathways is a sounder concept than basing it on structural similarities.

(b) For a taxonomic interpretation of the distributional patterns of different types of tannins, one should never forget that very often seeds, fruits, leaves, barks and woods of one and the same species contain different types of compounds (e.g. the A-group of proanthocyanidins seems to be more characteristic for fruits and seeds than for leaves and barks for which latter the B-group seems to be more common).

Bate-Smith: Tannins, whether hydrolysable (i.e. gallo- and ellagitannins) or condensed (i.e. leucoanthocyanins) are present in all classes of vascular plants and must be presumed to have been present also, therefore, in the earliest angiosperms. The hydrolysable tannins are ex-

ceptionally efficient as tannins, and may therefore have given the angiosperms in which they
are present (i.e. the dicots) an advantage. The
present-day representatives of the earliest dicot
forms in the palynological record possess ellagitannins.

The A-group of procyanidins, which are more
oxidised than the B-group, may represent a taxonomically distinct class. They yield much more
anthocyanidin when treated with acid than the
B-group, and appear to be more efficient tannins than these—almost as good as the hydrolysable tannins.

(N.B. The "A-group" and "B-group" refer
to Haslam's classes of procyanidins; not to the
benzene rings of the flavanoid molecule.)

Swain: (to Bu'Lock's remarks). We now have
good evidence that the biosynthesis or cinnamic
acids and benzoic acids are under separate control in plants. I therefore believe that there is
only one biosynthesic route to ellagitannins:
that is directly from shikimic acid. The production of hydroxybenzoic acids via phenylalanine
and cinnamic acids is not a biosynthetic route,
but a degradative one. The production of gallic
acid by the latter route has to be found.

Flavonoids as Systematic Markers in the Angiosperms

J. B. Harborne

Phytochemical Unit, Plant Science Laboratory, University of Reading, Reading, UK

Summary

Flavonoids are probably the most useful class of secondary plant constituent for systematic purposes, because of their widespread distribution, stability and ease of detection. One well established correlation between flavonoids and angiosperm evolution is in the marked difference in leaf flavonoids between woody and herbaceous plants. More recently, the distribution patterns of the more uncommon leaf flavonoids, particularly those with extra hydroxyl and methyl substituents, have been explored and the possible significance of the distinctive patterns thus revealed is discussed. Of particular interest is the finding of 8-hydroxyflavonols, especially the yellow pigment gossypetin, in the Ranales, Dilleniidae and Rosidae and of 6-hydroxyflavones in the Asteridae. The distribution of gossypetin within the different tribes of the Ericaceae is also mentioned.

While all classes of flavonoid known in the dicotyledons also occur in the monocotyledons, their evolutionary significance may well be different. This is probably true of the presence of proanthocyanidins in the monocotyledons. We have extensively explored the flavonoid patterns in leaf and inflorescence of families related to the Gramineae. The chemical results generally correlate with present views of family relationships in this area of Monocotyledoneae. One major difference, however, is that the flavonoid data suggest a close relationship between the Gramineae, Cyperaceae and Palmae. All three families have glycoflavones, tricin and luteolin glycosides in the leaves and flavonols are relatively uncommon. A special link between the Palmae and genus *Saccharum* of the Gramineae is suggested by the presence in both groups of flavonoids occurring as potassium bisulphate salts.

Flavonoid potassium bisulphates also occur in the dicotyledons, in the Compositae and Umbelliferae and in several families of the Parietales. The distribution of these and other chemical markers in this order is discussed in relation to taxonomy.

Flavonoid pigments are recognised today as being among the most valuable of all classes of secondary constituent as systematic markers in plants. They have the advantages of universal distribution and great ease of detection. Structural variation is immense, since well over a thousand different flavonoids have been described [1]. Furthermore, they can be detected in small fragments of leaf from herbarium specimens, so that it is possible to extend surveys which would otherwise be restricted to the rather limited range of living plant collections. This means that representative sampling of plant groups can readily be achieved and that complete ascertainment of flavonoid characters is often possible.

For flavonoids to be accepted by taxonomists as being useful in systematic studies, they must obviously show correlations in their distribution patterns with morphological or other biological characters. Many such correlations have, indeed, been observed. One of the first to be noted was in the genus *Pinus,* where the heartwood flavonoids vary according to whether the species have a single or a double vascular bundle of needle-leaves [2]. Other good examples are in the Plumbaginaceae, where flavonoid patterns are correlated with pollen morphology and tribal divisions [3] and in the Gesneriaceae, where they are related to geography and anisocotyly in the seedling [4, 5]. Very recently, a number of interesting correlations with phytogeography have been observed [6, 7]. In *Briza,* for example, the diploid Eurasian species have a different set of flavonoids in the leaves from the tetraploid South American species. In this case, there is also correlation with cytology, since within the one Eurasian species which has both diploid and (auto)tetraploid forms, namely *B. media* L., it is possible to distinguish diploid from tetraploid by the flavonoid chromatograms [8].

Much can be learnt from a knowledge of the general distribution pattern of flavonoids within the angiosperms as a whole. Flavonols (and especially myricetin) and proanthocyanidins characteristically occur in primitive woody plants, and they gradually disappear from the more advanced herbaceous families [9]. These have, instead, some flavonol (but only kaempferol and/or quercetin) but mainly flavone [10]. There is also a tendency in the more highly specialized families for the flavones to be *O*-methylated. Substitution of an extra hydroxyl group in the A-ring of flavonoids seems to follow the same pattern, woody plants having 8-hydroxyflavonols (e.g. gossypetin) and herbaceous ones having 6-hydroxyflavones (e.g. 6-hydroxyluteolin) [11].

Variations in flavonoids of the petal are more frequently controlled by natural selection for those colours which are preferred by the pollinating animals that occur within a particular ecological niche. In tropical climates, there is evolution towards bright red colours, favoured by humming birds, and pelargonidin or apigeninidin are formed as a response in plants such as the Gesneriaceae [4, 5]. In temperate climates, blue is a colour favoured by bees, so that there is natural selection for delphinidin types here. Yellow is another flower colour preferred by bees, and while this is often due to carotenoid, it can be given by yellow flavonols, by chalcones or aurones [12]. In *Primula,* one has the interesting situation where yellow flavonol (gossypetin) is selected for in one section of the genus, whereas carotenoid is selected for in another [13]. In more highly evolved plants, such as the Compositae, there are a number of yellow-flowering plants (e.g. *Coreopsis*) which have both carotenoid and yellow flavonoid pigments in their petals [12].

While flavonoids are clearly important in flower petals of plants for providing colour, their purpose in leaves is more obscure, although there is increasing evidence available that they are concerned in animal, and particularly in insect, feeding preferences [e.g. 14, 15]. The differences in function between leaf and flower must always be borne in mind in systematic studies. While the flavonoids of leaf and flower are often related in structure, there are also many differences, particularly in the class of flavonoid found in the two types of organ. Comparisons of flavonoid patterns should necessarily refer to distribution in leaf *or* flower (*or* other tissue) whenever possible.

Flavonoids are valuable chemical markers to study at all levels of plant classification. Thus, they can be useful at the species level, e.g. for identifying hybrids in populations of the Appalachian *Asplenium* complex [16]. Equally, they can be important at the ordinal level, e.g. for deciding what families should be included or excluded from the order Centrospermae [17, 18]. I propose here to concentrate on the utility of flavonoids at the family level, and to the use that can be made of them for detecting family groups within orders. It is true that experimental taxonomists are not often concerned with relationships at this level of classification. Nevertheless, for this very reason, existing classifications are often based on rather arbitrary decisions about morphological or anatomical similarities so that chemical data, when they are available, may be particularly welcome.

I would stress the importance of taking into consideration with flavonoid data, similar data from other chemicals and I am a firm believer that making judgements about relationships, based on the presence/absence of a single chemical or single class of chemical is a relatively useless exercise. In all the cases referred to here, more than one type of chemical marker are considered before making any chemotaxonomic pronouncements.

Flavonoid Patterns in the Parietales

The Parietales (or Violales) are defined by Cronquist [17] as an assemblage of some 21 families with more than 5200 species. Most taxa have a unilocular, compound ovary with parietal placentation. It is an order which is taxonomically difficult in the sense that it represents a heterogenous group of families, which however appear to be related in having a common ancestry from the Theales. The classification is problematical because as Cronquist puts it "any consistent attempt to divide the order into smaller, more homogenous ones leads into a morass in which perhaps as many as a dozen orders might have to be recognized". Many other arrangements of these families, besides that of Cronquist, have been proposed; Hutchinson, for example, does distribute these families among a number of orders [19].

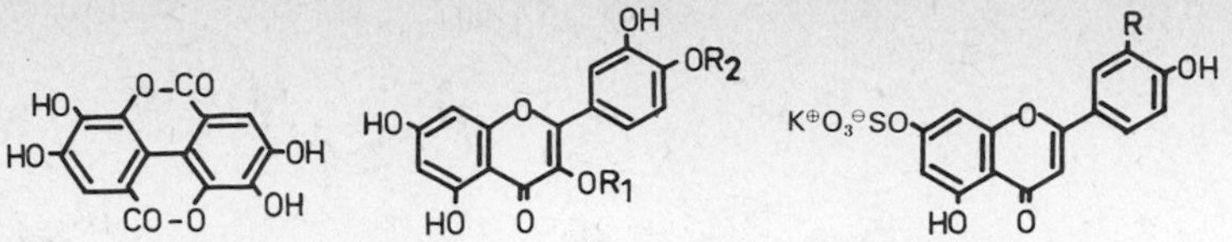

(1) Ellagic acid

(2) (R_1=H, R_2=CH₃) Tamarixetin
(3) (R_1=CH₃, R_2=H) Quercetin 3-methyl ether
(4) (R_1=SO₃⁻ K⁺;R_2=CH₃) Tamarixetin-
    3-bisulphate

(5) (R=H) Apigenin 7-bisulphate
(6) (R=OH) Luteolin 7-bisulphate

Fig. 1. Flavonoids of the Parietales.

Flavonoid patterns were first studied in the order by Bate-Smith [9], who recorded proanthocyanidins and flavonols in some families (e.g. Tamaricaceae, Cistaceae) but found that many families (e.g. Passifloraceae) were either partly or completely lacking in recognisable constituents. Ellagic acid (*1*), a phenol related in structure to the flavonoids, was recorded by him in the Frankeniaceae, Tamaricaceae, Bixaceae and Cochlospermaceae.

A more extensive survey of the Parietales s.l. was later carried out by Lebreton & Bouchez [20], who discovered that most families consistently contained proanthocyanidins (mainly procyanidin) and the flavonols kaempferol and quercetin. Flavones were found infrequently, mainly in the Violaceae and Flacourtiaceae. A significant finding was of methylated flavonols, particularly 3'-, 4'- and 7-methyl ethers of kaempferol and quercetin, in Frankeniaceae, Tamaricaceae, Violaceae and Cistaceae. Tamarixetin, or quercetin 4'-methyl ether (*2*), had earlier been isolated from *Tamarix* [21], but its wider distribution was not then suspected. Yet a further quercetin methyl derivative, the 3-methyl ether (*3*), has very recently been detected as a regular component of leaves of *Cistus* (Cistaceae) [22] and it is interesting that the same compound was earlier reported in *Begonia* [23], since the Begoniaceae is sometimes included in the Parietales, s.l. [see 20].

The present author was first led to study the flavonoids of this group of plants, by an observation in an earlier survey [9] that *Bixa orellana* L. might contain a 6-hydroxylated flavonoid. However, a careful re-examination of the leaves of this plant failed to reveal any unusual flavonoid; acid hydrolysis gave the two common flavones, luteolin and apigenin. A study of the combined forms in which these flavones occurred in the leaf proved more rewarding,

since negatively charged compounds proved to be present. Indeed, further analysis showed that four compounds were present: the 7-glucosides and 7-potassium bisulphates of apigenin and luteolin (*5, 6*).

Flavonoid bisulphates have rarely been recorded in the dicotyledons up to now and these two flavone conjugates represent new compounds. Among the earlier reports of bisulphates, however, is one of a flavonol derivative (*4*) of tamarixetin, in *Tamarix laxa* (Tamaricaceae) [24]. A more thorough survey of the Parietales for these charged compounds therefore became a matter of some interest and the results, to date, are collected in table 1, together with other flavonoid data on the families that have been examined so far. The survey was based mainly on herbarium and live material available in Reading, and on herbarium material, kindly provided by the director of the Botanic garden, Rio de Janeiro, Brazil.

The survey for bisulphates indicate that they are uniformly present in Bixaceae, Tamaricaceae and Frankeniaceae, occasionally present in Cistaceae and Guttiferae, and otherwise absent. Even more interesting, however, is the fact that these compounds are closely correlated with the distribution of ellagic acid in these families (table 1). Indeed, the only exception is Cochlospermaceae, which has ellagic acid in three *Cochlospermum* species (*C. religiosum* (L.) Alston [9], *C. vetifolium* (Willd.) Spreng. and *C. orinocense* (HBK.) Stend. [this survey]) but no charged flavonoids. The other chemical data also generally fit in with a grouping of the families so far examined into three: those with flavonoid bisulphates regularly present, those with occasional occurrences, and those lacking them. Clearly, more extensive sampling is needed to confirm these groupings.

In general, it is satisfying that the chemical results fit in quite well with Cronquist's arrangement of the Parietales [17]. They fit in rather

Table 1. *Flavonoid profiles in families of the Parietales*

Family	Genera/ species surveyed	Charged flavon- oids	Ellagic acid	Proantho- cyanidins	Flavonols	Flavonol methy- lation	Flavones
	Presence/absence[a] of						
Bixaceae[b]	(1/1)	+	+	+	−	−	+
Tamaricaceae	(2/8)	+	+	−	+	+	−
Frankeniaceae	(1/6)	+	+	(+)	+	+	(+)
Cochlospermaceae	(1/2)	−	+	−	+	−	−
Guttiferae[c]	(1/13)	(+)	(+)	+	+	−	−
Cistaceae[d]	(5/22)	(+)	+	+	+	+	−
Violaceae	(2/15)	−	−	+	+	+	(+)
Turneraceae	(3/4)	−	−	−	−	−	−
Flacourtiaceae	(4/4)	−	−	+	−	−	(+)

[a] +, present in all (or majority of) species surveyed; (+), present in one or a few species; −, absent.
[b] Bixaceae is generally regarded as containing only one genus and one species, *B. orellana*. In Index Kewensis, at least six *Bixa* species are recorded. Examination of *B. arborea* Hub., supplied by Rio de Janeiro Bot. Garden, showed that charged flavonoids were present, but there was a difference in combined form from *B. orellana*.
[c] Three of 13 *Hypericum* species had charged flavonoids: *H. coadnatum* Chr.Sm., *H. grandiflorum* Choisy and *H. elodes* L. Ellagic acid is recorded in three of 24 *Hypericum* species by Lebreton & Bouchez [20].
[c] The one Cistaceae with charged flavonoids is *Helianthemum squamatum*.

less well with Takhtajan's system [25], in which Tamaricaceae and Frankeniaceae are placed in a separate order, the Tamaricales (table 2), they certainly do not fit in at all with Hutchinson's scheme [19].

The position in angiosperm systems of the Bixaceae, which contains the single species *Bixa orellana* L., has been a matter of much discussion. The present data agree with its separation from the Flacourtiaceae and also confirm that, chemically, it shows affinities with the Frankeniaceae and Tamaricaceae. Its supposed close relationship with Cochlospermaceae is only part-

ly borne out by the present data, but further sampling of the latter family is necessary before a conclusive judgement can be made.

Flavonoids in the Gramineae and Related Families

A study of the flavonoid pigments in the Gramineae and related families has been in progress in this laboratory for the last seven years. A first paper, on anthocyanins in the grasses, appeared in 1967 [26] and the most recent of the series on flavonoids in the Palmae has been pub-

Table 2. *Different arrangements of some families of the Parietales*

Cronquist [17][a]		Takhtajan [25]	
Violales	1 Flacourtiaceae	Violales	Flacourtiaceae
	4 Violaceae	(Cistales)	Violaceae
	5 Turneraceae		Bixaceae
	6 Passifloraceae		Cochlospermaceae
	8 Bixaceae		Cistaceae
	9 Cistaceae	Tamaricales	Tamaricaceae
	10 Tamaricaceae		Frankeniaceae
	11 Frankeniaceae	Theales	Guttiferae
Theales	Guttiferae		

[a] Cronquist includes Cochlospermaceae in the Bixaceae.

Table 3. *Flavonoid patterns in the Gramineae and related families*

Classification key[b]	Family	Leaf flavonoid pattern
2.6.1 2.6.2 2.3.1 2.3.2	Sparganiaceae Typhaceae[a] Flagellariaceae Anarthraceae	Quercetin and kaempferol (myricetin in *Sparganium* only)
2.2.1 2.3.4	Ericaulaceae Restionaceae[a]	6-Hydroxyflavonols (Eriocaulaceae) or 8-hydroxyflavonols (Restionaceae)
2.1.4 3.4.1 3.4.2	Commelinaceae Araceae[a] Lemnaceae	Flavonols and glycosylflavones
2.5.2 2.5.1 3.1.1 2.4.1	Gramineae Cyperaceae[a] Palmae[a] Juncaceae	Luteolin and glycosylflavones in all four. Bisulphate salts in Palmae (50%), Gramineae (17%) and Juncaceae (25%). Tricin common in all except the Juncaceae. Flavone 5-glucosides in all four

[a] Majority of species in these families are proanthocyanidin positive. Proanthocyanidins are found in the Gramineae, but quite rarely.
[b] From Cronquist [17], 2 subclass Commelinidae; 3, subclass Arecidae. Missing numbers refer to families, for which little or no flavonoid data are available. Pandanaceae (3.3.1) has been examined, but apparently completely lacks any of the common flavonoid constituents.

lished very recently [27]. The programme has been carried out in conjunction with H. T. Clifford, who has been responsible for taxonomic aspects and who has applied numerical computer analyses to the new data.

The results suggest it is possible to define a flavonoid pattern at the family level, based on the flavonoids present in the majority of species since, within the monocotyledons, families are relatively consistent in their flavonoid attributes. Many of the families under consideration are very large, especially the Gramineae, so that there is a considerable problem of sampling. As far as possible, wide surveys at the generic level have been combined with indepth sampling of particular species (i.e. populational surveys) and particular genera. Most work has been done on leaf flavonoids, although some data on inflorescence pigments are also available.

A summary of the results to date is given in table 3, along with Cronquist's arrangement of the families studied. The flavonoid data suggest four groups of families in an evolutionary sequence with the first group (including Typhaceae), with flavonols alone, being the most primitive and the fourth group (including Gramineae), with tricin, being the most advanced [28]. Perhaps the most startling result is

the suggestion that the Palmae are related to the grasses, since the palms have generally been regarded as a more primitive group and also they differ considerably in gross morphology. The chemical data, however, are very convincing in the sense that five leaf flavonoid characters (presence of glycosylflavones, luteolin, tricin, bisulphate complex and 5-glucoside) link the two families. Furthermore, a numerical analysis carried out on morphological and anatomical characters, using a widely representative sample of genera, confirmed this close relationship between the two groups [29].

It is interesting, too, that the Gramineae and Cyperaceae, which are usually placed very near each other in most systems of classification, have basically the same flavonoid pattern. There are, however, minor differences: bisulphates, present in a few grasses, have not yet been detected in the sedges, in spite of a survey for them among forty species [30]. Inflorescence pigments also differ in the two groups: grasses have cyanidin-based pigments whereas sedges instead have the aurone, aureusidin [31].

The position of the Juncaceae vis-à-vis the grasses is also being actively explored at the moment [30]. The Juncaceae show many similarities in leaf flavonoids, although one of the

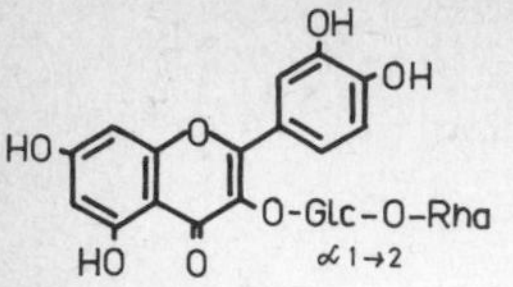

(7) Quercetin 3-neohesperidoside

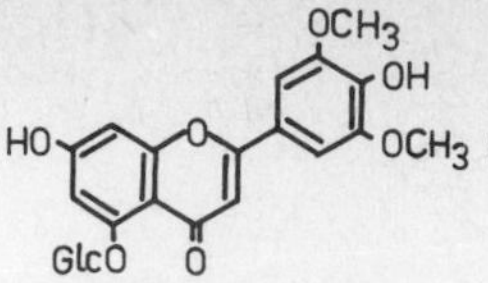

(8) Tricin 5-glucoside

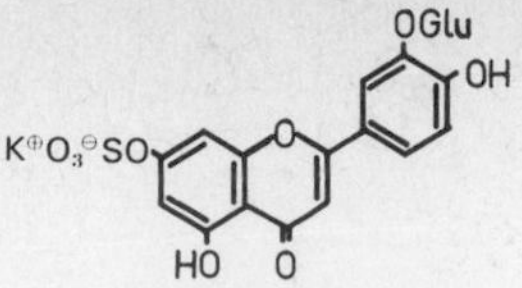

(9) Luteolin 7-bisulphate-3'-glucoside

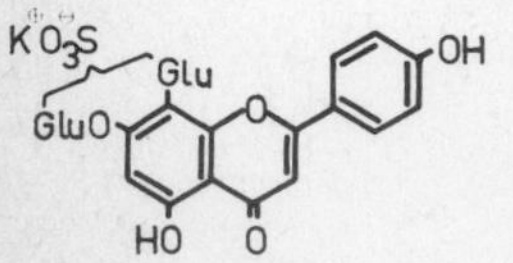

(10) Vitexin 7-glucoside bisulphate

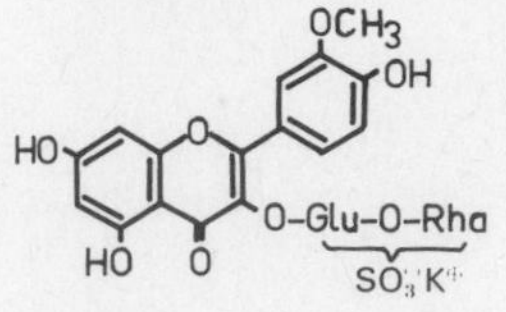

(11) Isorhamnetin 3-rutinosidebisulphate

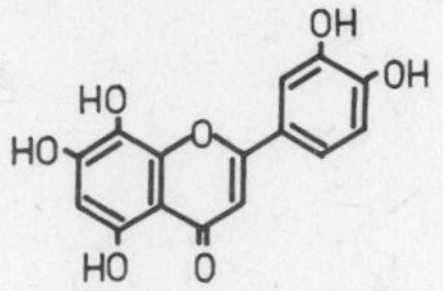

(12) Hypolaetin

Fig. 2. Flavonoids of Gramineae and related families.

key markers in the Gramineae, namely tricin, seems to be absent. There is another difference in the flavonoids of the inflorescences, since desoxyanthocyanins have just recently been detected in five species of *Juncus* and *Luzula* (p. 121).

One by-product of this intensive systematic survey of monocotyledonous families is the discovery of new flavonoids and new flavonoid types. The structures of some of the new compounds are illustrated in fig. 2. Among flavonol glycosides one might mention the discovery of quercetin 3-neohesperidoside (7) in the reedmace, *Typha latifolia* L.; this is the first definitive report of this apparently rare isomer of probably the commonest of all flavonol glycosides, namely rutin, the 3-rutinoside [28]. The flavones occur in these plants as a bewildering array of O- and C-glycosides. Tricin and luteolin very frequently occur in the Palmae, Cyperaceae and Gramineae as the 5-glycosides (e.g. [8]). These derivatives are normally only present in trace amount, but because of their intense blue fluorescence, they are readily detected on two-dimensional chromatograms.

A novel series of flavone potassium bisulphate salts have also been detected, especially in the Palmae. These salts range from compounds such as luteolin 7-bisulphate-3'-glucoside (9) of *Mascarena verschaffeltii* (Wendl.) Bailey to vitexin 7-glucoside bisulphate (10) in *Washingtonia robusta* Wendl. Flavonols are also found in the Palmae, conjugated with bisulphate, a typical example being isorhamnetin 3-rutinoside bisulphate (11) from *Arecastrum romanzoffianum* (Cham.) Becc. [27].

Surprisingly few new flavonoid aglycones have been detected so far. One of these is illustrated in fig. 2, namely hypolaetin (12) or 5,7,8,-3',4'-pentahydroxyflavone, discovered for the first time in *Hypolaena fastigiata* (Restionaceae) [32]. Another erioflavonol, yet to be fully characterised, occurs along with quercetagetin and patuletin in *Eriocaulon* (Eriocaulaceae) [33].

Gossypetin as a Taxonomic Marker in Plants

One way of relating flavonoid patterns in plants to systematics is to study intensively their distribution within a small group of closely related families; two examples of this type of approach have already been given. An alternative is to study the distribution throughout the angiosperms of a particular flavonoid marker, which is of interest because of some unusual structural feature. It should preferably be a substance which is readily distinguished during chromatographic surveys from the more common flavonoid types. This is true of flavone and flavonol 5-methyl ethers and 5-glycosides, because they exhibit intense fluorescence when examined as chromatographic spots in UV light and are readily observed, even at very low concentrations [34, 35]. It is also true of herbacetin (13) and gossypetin (14), since these pigments appear in UV light as dark black spots on paper, the colour being unaffected by ammonia, and they are also unusual in appearing as visible yellow spots on chromatograms viewed in daylight. Furthermore, they can be distinguished from the structurally closely related querce-

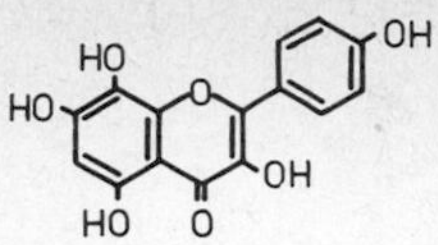

(13) Herbacetin

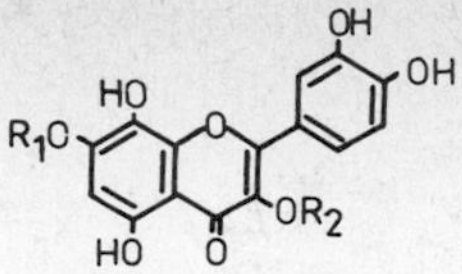

(14) (R_1 = R_2 = H) Gossypetin
(15) (R_1 = H, R_2 = Gal.) Gossypetin 3-galactoside
(16) (R_1 = H, R_2 = Glur.) Gossypetin 3-glucuronide
(17) (R_1 = Glu., R_2 = H) Gossypetin 7-glucoside

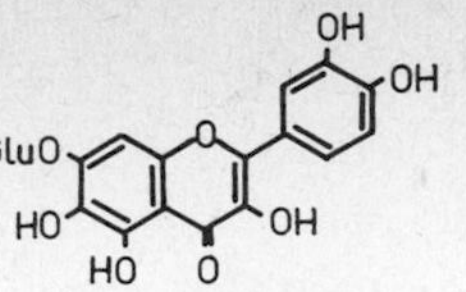

(18) Quercetagetin 7-glucoside

Fig. 3. Gossypetin derivatives of plants.

tagetin pigments by a simple colour test on paper with sodium acetate [36].

Gossypetin was first discovered in plants as the yellow flower pigment of cotton flowers, *Gossypium,* and it is now known as a flower pigment both in the Malvaceae and in a number of other families. Its distribution has been studied in the angiosperms, principally in the author's laboratory [36], and the results are collected together in table 4.

It will be noted from this table that, besides a number of miscellaneous occurrences, gossypetin occurs as a yellow pigment in *Ranunculus* (Ranunculaceae), *Primula* (Primulaceae) and *Rhododendron* (Ericaceae). For some reason, which is not yet clear, it also occurs in the leaves of plants related to those where it is present as a corolla pigment. This is particularly true in *Rhododendron* where it occurs, as the 3-galactoside (*15*), in the flowers of a handful of species, perhaps a dozen, but is present in the leaves of many more, in fact in 76% of 206 species surveyed [37].

Gossypetin also occurs in leaves, but apparently not flowers, of other genera related to *Rhododendron,* indeed in over 14 genera of the Ericaceae. It, incidentally, is present consistently in all known taxa of the Empetraceae, a family closely allied morphologically to the Ericaceae [6]. In fact, gossypetin represents a systematic marker of some significance in the Ericaceae; in the subfamily Rhododendroideae, for example, it links the two tribes Rhodoreae and Phyllodoceae. This and other chemical evidence (e.g. presence of flavonol 5-methyl ethers) suggests that these two tribes need to be placed closer together than most earlier taxonomic treatments allow [38].

A survey of other families in the order Ericales, in addition to the Empetraceae, showed

Table 4. *Distribution of gossypetin in the angiosperms*

Family	Genera (and species)	Frequency and Form[a]
Ranunculaceae	*Ranunculus* spp.	in flowers, as a glycoside
Papaveraceae	*Meconopsis* and *Papaver* spp.	in flowers, as glycosides in leaf
Leguminosae	*Acacia constricta*	
	Lotus corniculatus	in flowers, with the 7- and 8-methylethers
Crassulaceae	*Sedum album, S. acre*	in flowers, herbacetin 8-methylether
Malvaceae	*Abutilon, Gossypium, Hibiscus* spp.	frequent in flowers, also in leaves
Sterculiaceae	*Chiranthodendron pentadactylon,*	in leaves
	Fremontia californica	in flowers and leaves
Rutaceae	*Xanthoxylum acanthopodium*	in seeds, as 8,4′-dimethyl ether
Ericaceae	*Rhodendron* and 15 other genera	occasionally in flowers, often in leaves
Empetraceae	*Empetrum, Corema,* and *Ceratiola* spp.	in leaves of all taxa
Primulaceae	*Primula, Dionysia* and *Douglasia*	as flower pigment, sometimes with herbacetin
Compositae	*Chrysanthemum segetum*	as flower pigment
Restionaceae	*Calorophus lateriflorus, Restio* spp.	in stems and inflorescence

[a] For references, see [36].

that gossypetin was otherwise absent. There was one single exception, the monotypic genus *Galax aphylla,* which is usually referred to the Diapensiaceae. What is exciting here is that, for quite other reasons, Hutchinson [39] felt it was misplaced in this family; the chemical data not only agree with this conclusion but also indicate that, with gossypetin in its leaves, there might be a home for *Galax* among the Ericaceae.

Our most recent discovery is of gossypetin in the Sterculiaceae. This was quite accidental, while searching the remarkable Hand-flower tree, *Chiranthodendron pentadactylon* Larreat., for the reputed presence of desoxyanthocyanins. In fact, the pigment of the sepal and calyx turned out to be the most common anthocyanin of all, cyanidin 3-glucoside [40]. However, the leaves of the plant were examined at the same time and they yielded considerable amounts of gossypetin, as the 3-glucuronide (*16*). This is taxonomically not so surprising, since the Sterculiaceae have always been closely linked in angiosperm classification with Malvaceae, the original source of gossypetin. There are also other chemical links between the two families, notably in the cyclopropene fatty acid, sterculic acid, which is present in the seed fat of both groups [41]. A search of other Sterculiaceae for gossypetin is in progress; so far, it has been detected in one further species, in leaves and flowers of *Fremontia californica.*

It is clear, from its present known distribution, that the ability to synthesize gossypetin must have arisen several times during plant evolution. How else can one explain its presence in a number of quite unrelated plant groups? The fact that gossypetin and herbacetin have been identified in very early land plants, in *Equisetum* species [42], suggests that gossypetin synthesis is a primitive trait. It will also be noted that practically all its occurrences are in woody or otherwise less specialized plant groups. It is, for example, conspicuous in *Rhododendron,* a very ancient genus of plants, and here it is interesting that it appears to have disappeared from the most highly evolved species, i.e. those which have travelled geographically furthest from the centre of origin of the genus.

The only exception to the above thesis is the presence of gossypetin 7-glucoside (*17*) [43] together with quercetagetin 7-glucoside (*18*) [44] in the flowers of the supposedly highly specialized *Chrysanthemum segetum,* Compositae. However, it is the exceptions which are often most interesting and the evolutionary significance of gossypetin versus quercetagetin will be discussed further in the last section of this paper.

6-Hydroxyflavonoids as Systematic Markers

6-Hydroxyflavonoids represent another class of unusual flavonoid structures in plants, which I suggest are of systematic interest [45]. The best known flavonol in the series is quercetagetin or 6-hydroxyquercetin (*19*), first isolated from *Tagetes erecta* (Compositae) as a flower pigment. It co-occurs with the 6-methyl ether, patuletin (*20*) in *T. patula.* It is, in fact, quite rare as a flower pigment, only occurring elsewhere as such in plants of the *Chrysanthemum* complex (also Compositae). More frequently, however, it occurs in partly methylated form in leaves of plants. For example, the 3,6,7,3'-tetramethyl ether (*21*) is present as a lipid soluble leaf component in *Matricaria chamomilla*; a considerable range of other partial methyl ethers are also known. Rather fewer of the corresponding derivatives of 6-hydroxy-kaempferol and 6-hydroxymyricetin have been described and neither of these compounds has yet been isolated from plants except in O-methylated form.

In the case of the 6-hydroxyflavone series, the first to be isolated was scutellarein (*22*), discovered over 70 years ago in the leaves of several *Scutellaria* species (Labiatae). 6-Hydroxyluteolin (*23*) was a much more recent discovery (1967), occurring as it does in leaves of *Catalpa bignonioides* (Bignoniaceae) [4, 5]. As with quercetagetin, a range of partial methyl ethers are known. Particularly frequently found are the 6-methyl ethers and 6,4'-dimethyl ethers of these two flavones. The 6,4'-dimethyl ether of scutellarein has the trivial name pectolinarigenin (*24*).

The known distribution pattern of 6-hydroxyflavonoids is summarized in table 5. From this, it is apparent that they are only abundant in families of the Tubiflorae (e.g. Globulariaceae)

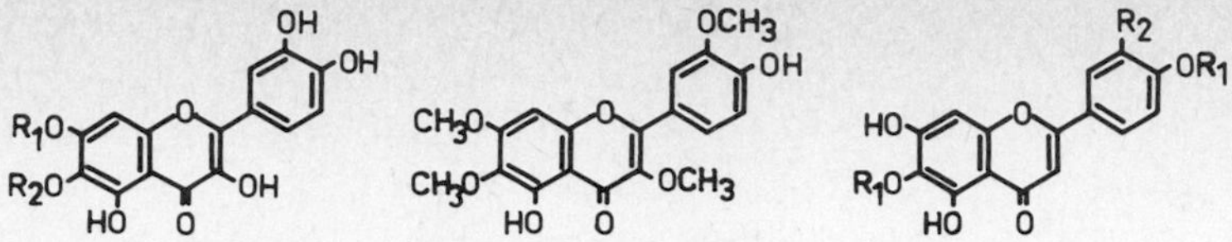

(19) ($R_1=R_2=H$) Quercetagetin
(20) ($R_1=H, R_2=CH_3$) Patuletin

(21) Quercetagetin 3,6,7,3'-tetramethyl ether

(22) ($R_1=R_2=H$) Scutellarein
(23) ($R_1=H, R_2=OH$) 6-Hydroxyluteolin
(24) ($R_1=CH_3, R_2=H$) Pectolinarigenin

Fig. 4. 6-Hydroxyflavonoids of plants.

and in the Compositae, i.e. in Cronquist's subclass Asteridae. Because of the large number of reports of these substances in the latter family, it has been impossible to list more than a fraction in table 5. They are certainly widespread and have been detected in most tribes of this family. In one recent report, Carman et al. [46] noted 6-methoxyapigenin (also known as hispidulin) in 34 species from the genera *Ambrosia, Gaillardia, Helenium, Hymenoxys, Plummera* and *Ratibida*. In our own laboratory, during a detailed examination of the tribe Anthemideae, we have found quercetagetin as a flower pigment in *Anthemis, Chrysanthemum* and *Dendranthema* [47], and methylated derivatives in leaves of the majority of the taxa in the group [48].

The two facts that (1) 6-hydroxyflavones are much more common than 6-hydroxyflavonols and (2) 6-hydroxyflavonoids occur predominantly in the Tubiflorae and Compositae suggest that during evolution of flavonoids, a switchover occurred in the specificity of the enzyme oxidizing the A-ring from the 8- to the 6-position. How else can one explain the point that the position of extra A-ring hydroxylation is so well correlated with the evolutionary trend from flavonols in woody plants to flavones in herbaceous taxa?

The switchover may be due to the fact that 6-hydroxy compounds are more effective than the 8-hydroxy compounds as toxins to predating animals. There are at least indications that methylated 6-hydroxyflavones are toxic to animals; tangeritin from the orange for example can cause neonatal lethality in rats [49]. From the purely systematic point of view, this switchover is of interest, mainly because of the exceptional presence of 6-hydroxyflavonoids in plants not expected to have them and because of their absence from plants which might, in contrast, be expected to have them. This can

be illustrated by discussing briefly, in turn, their occurrences in the four subclasses, as indicated in table 5.

Their occurrence in the subclass Magnoliidae is surprising, if we are to regard them as evolutionarily advanced characters, since the family Betulaceae where they have been found is a relatively primitive one. However, one could explain this as a specialized occurrence linked to function, in that they are produced, not in leaves or flowers, but in the excretion from the woody bud of plants of *Alnus* and *Betula* [50, 51]. They presumably have a special purpose here in providing a protection from microbial infection.

The presence of 6-hydroxyflavonoids in two families of the Centrospermae and in the Polygonaceae is again surprising, but these may represent isolated occurrences rather than part of a general distribution pattern. Certainly, it would be valuable to survey other families of the Centrospermae, and especially the Caryophyllaceae, to see whether 6-hydroxyflavonoids occur anywhere else in the group.

6-Hydroxyflavonoids in the subclass Rosidae are much less remarkable and indeed almost expectable since many of these families, and especially the Leguminosae, show many signs of evolutionary advancement in morphological features. Also, the Rosidae are often regarded as the precursor group of the Asteridae (see discussion in ref. [17]). Hegnauer [52] has shown that there are many chemical features linking the Rosidae and the Asteridae.

Finally, with regard to the Asteridae itself, the main source of plants with 6-hydroxyflavonoids, it is worth comparing those families which have them with those which apparently lack them. Unfortunately, negative phytochemical evidence is often not very strong; apparent absence of a character may just mean that no one has looked hard for it in a particular plant group. In such a case, I can only speak from

Table 5. *Distribution of 6-hydroxyflavonols and 6-hydroxyflavones in the Angiosperms*

Family	Genera and species	Frequency and form[a]
Subclass Magnoliidae		
Betulaceae	*Alnus glutinosa & A. sieboldiana*	bud excretion as methylated derivatives
Subclass Caryophyllidae		
Amarantaceae	*Alternanthera phylloxeroides*	6-methoxyluteolin
Chenopodiaceae	*Spinacea oleracea*	patuletin and quercetagetin 6,3'-DME
Polygonaceae	*Polygonum* and *Atraphaxis* spp.	scutellarein and 5-methyl-6-hydroxykaempferol
Subclass Rosidae		
Saxifragaceae	*Chrysosplenium* spp.	methylated flavonols
Rosaceae	*Kerria japonica & Sorbaria* spp.	scutellarein, 6-methoxyluteolin
Leguminosae	*Distemonanthus, Prosopis* and *Tephrosia*	patuletin and other methylated derivatives
Subclass Asteridae		
Polemoniaceae	*Ipomopsis* and 8 other genera.	patuletin, patuletin 7-methyl ether and 6-hydroxykaempferol 6,7-dimethyl ether
Verbenaceae	*Callicarpa, Cyanostegia Lippia* and *Vitex*	methylated flavonols and 5,6,7,-tri-methoxyflavone
Labiatae	many genera (e.g. *Scutellaria*)	6-hydroxyluteolin, scutellarein and derivatives
Plantaginaceae	*Plantago*	6-hydroxyluteolin in 10/26 spp.
Buddleiaceae	*Buddleia* and *Chilianthus*	6-hydroxyluteolin
Scrophulariaceae	*Digitalis, Linaria Scrophularia, Veronica* spp.	6-hydroxyluteolin and methylated derivatives
Myoporaceae	*Eremophila fraseri*	6-hydroxymyricetin 3,6,7,4'-tetramethyl ether
Globulariaceae	*Globularia, Lytanthus* spp.	6-hydroxyluteolin in all spp. examined
Bignoniaceae	*Catalpa, Jacaranda, Millingtonia, Stereospermum* and *Tecoma* spp.	scutellarein and 6-hydroxyluteolin frequent
Pedaliaceae	*Pedalium* and *Sesamum* spp.	6-methoxyluteolin
Rubiaceae	*Gardenia lucida*	highly methylated 6-hydroxyflavones
Valerianaceae	*Valerianella*	6-hydroxyluteolin in 6 of 40 spp
Compositae	*Chrysanthemum, Coreopsis, Cirsium, Lasthenia* and many others	6-hydroxyflavonols and 6-hydroxy-flavones as such, or methylated, frequent

[a] Taxonomic arrangement follows Cronquist [17]. Most references can be found in those given in text, namely [44–48, 50, 51, 53, 56, 57]. For other examples, see papers and phytochemical reports in Phytochemistry 1971–1973.

my own experience. For example, in connection with other work I have studied the flavonoid patterns in the Solanaceae, Nolanaceae and Convolvulaceae and there is also much published work on these plants. In no case is there any mention of 6-hydroxyflavonoids in these plants. And yet, these three families are included by Cronquist in the same order as Polemoniaceae, which has recently been shown to have patuletin and two related compounds in half the 18 genera which constitute this family [53]. Thus,

chemically, one would prefer to see the Polemoniaceae separated from the Solanaceae and placed in the orders Lamiales or Scrophulariales of Cronquist. To put it another way, the traditional treatment of Polemoniaceae places it in the Tubiflorae [54] and in this case, chemistry would seem to support tradition.

Let us now consider an example of the opposite situation from the above, which exists in the Scrophulariales of Cronquist. The great majority of the twelve families in this order are united in having 6-hydroxyflavonoids (table 5). By contrast, the Oleaceae, which I have extensively surveyed for flavonoids for quite other purposes, shows no signs of having 6-hydroxyflavonoids; quercetin is very common in this family and luteolin and apigenin are also present. If we now turn to the earlier treatments of the Oleaceae, we find that it is not usually placed near the Scrophulariaceae, Gesneriaceae or Globulariaceae. Thus, there is some negative chemical evidence (absence of 6-hydroxyflavonoid) for keeping it isolated in its own order Oleales, as favoured by Engler [54]. A colleague, who knows the taxonomy of the Oleaceae well, supports this view [55]. Cronquist, himself, obviously has reservations about placing the Oleaceae in the Scrophulariales [17], so that this might be one instance where the chemical data might be said to carry some weight.

Having discussed the utility of the 6-hydroxyflavonoid character at the family level, I would like to conclude by pointing out that it can also be useful at lower levels of classification. Greger and Ernet, for example, have found that 6-hydroxyluteolin occurs in the genus *Valerianella* in the more advanced North American and Mediterranean species, rather than in the Asiatic ones. Indeed, the chemical results correlate well with cytology and morphology in suggesting a new treatment of the genus [56]. The same compound, 6-hydroxyluteolin, has been useful at the species level during a comparison of flavonoid patterns in Northern and Southern hemisphere populations of *Plantago maritima* [57]. The only major amphitropical discontinuity in flavonoids was the 7-glucoside of 6-hydroxyluteolin, which characterised the Northern hemisphere plants.

I am grateful to the curators of various herbaria for generous supplies of leaf fragments for flavonoid surveys and I thank Miss Christine Williams for allowing me to quote from her unpublished work on the Juncaceae.

References

1. Harborne, J B, Mabry, T J & Mabry, H, The flavonoids. Chapman and Hall, London. In press.
2. Erdtman, H, Organic chemistry and conifer taxonomy, in Perspectives in organic chemistry (ed A Todd) p. 453. Interscience, New York, 1956.
3. Harborne, J B, Phytochem 1967, 6, 1415.
4. — Ibid 1966, 5, 587.
5. — Ibid 1967, 6, 1643.
6. Moore, D M, Harborne, J B & Williams, C A, Bot j Linn soc 1970, 63, 277.
7. Mabry, T J, Chemistry of geographical races, in Chemistry in evolution and systematics (ed T Swain) p. 377. Butterworth's, London, 1973.
8. Williams, C A & Murray, B G, Phytochem 1972, 11, 2507.
9. Bate-Smith, E C, Bot j Linn soc 1962, 58, 39.
10. Harborne, J B, Recent adv in phytochem 1972, 4, 107.
11. Harborne, J B & Williams, C A, Phytochem 1971, 10, 367.
12. Harborne, J B, Flavonoids: distribution and contribution to plant colour, in Chemistry and biochemistry of plant pigments (ed T W Goodwin) p. 247. Academic Press, London, 1965.
13. Harborne, J B, Phytochem 1968, 7, 1215.
14. Zielske, A G, Simons, J N & Silverstein, R M, Phytochem 1972, 11, 393.
15. Doskotch, R W, Mikhail, A A & Chatterji, S K, Phytochem 1973, 12, 1153.
16. Smith, D M & Harborne, J B, Phytochem 1971, 10, 2117.
17. Cronquist, A, The evolution and classification of flowering plants. Thomas Nelson, London, 1968.
18. Mabry, T J, Betacyanins and betaxanthins, in Comparative phytochemistry (ed T Swain) p. 231. Academic Press, London, 1966.
19. Hutchinson, J, Evolution and phylogeny of flowering plants. Academic Press, London, 1969.
20. Lebreton, P & Bouchez, M-P, Phytochem 1967, 6, 1601.
21. Gupta, S R & Seshadri, T R, J chem soc 1954, 3063.
22. Poetsch, I & Reznik, H, Ber deut bot Ges 1972, 85, 209.
23. Harborne, J B & Hall, E, Phytochem 1964, 3, 453.
24. Utkin, L M, Chem Abs 1966, 65, 15309.
25. Takhtajan, A, Flowering plants origin and dispersal. Oliver & Boyd, Edinburgh, 1969.
26. Clifford, H T & Harborne, J B, Proc Linn soc Lond 1967, 178, 125.
27. Williams, C A, Harborne, J B & Clifford, H T, Phytochem 1973, 12, 2417.
28. — Ibid 1971, 10, 1059.
29. Clifford, H T, Bot j Linn soc 1970, 63, suppl. 1, 25.
30. Williams, C A. Unpublished data.
31. Clifford, H T & Harborne, J B, Phytochem 1969, 8, 123.
32. Harborne, J B & Clifford, H T, Phytochem 1969, 8, 2071.
33. Bate-Smith, E C & Harborne, J B, Phytochem 1969, 8, 1035.
34. Harborne, J B, Phytochem 1969, 8, 419.
35. Glennie, C W & Harborne, J B, Phytochem 1971, 10, 1325.
36. Harborne, J B, Phytochem 1969, 8, 177.
37. Harborne, J B & Williams, C A, Phytochem 1971, 10, 2727.
38. — Bot j Linn soc 1973, 66, 37.

39. Hutchinson, J B, Genera of flowering plants, vol. 3. Clarendon Press, Oxford. In press.
40. Harborne, J B & Smith, D M, Z Naturforsch 1972, 276, 210.
41. Shorland, F B, Distribution of fatty acids in plant lipids, in Chemical plant taxonomy (ed T Swain) p. 253. Academic Press, London, 1963.
42. Saleh, N A M, Majak, W & Towers, G H N, Phytochem 1972, 11, 1095.
43. Geissman, T A & Steelink, C, J org chem 1957, 22, 946.
44. Harborne, J B, Heywood, V H & Saleh, N A M, Phytochem 1970, 9, 2011.
45. Harborne, J B & Williams, C A, Phytochem 1971, 10, 367.
46. Carman, N J, Watson, T, Bierner, M W, Averett, J, Sanderson, S, Seaman, F C & Mabry, T J, Phytochem 1972, 11, 3271.
47. Harborne, J B, Heywood V H & Saleh, N A M, Phytochem 1970, 9, 2011.
48. Harborne, J B, Heywood, V H & Glennie, W. Unpublished work.
49. Stout, M G, Reich, H & Huffman, M N, Cancer chemother rep 1964, 36, 23.
50. Wollenweber, E & Egger, K, Z Pflanzenphysiol 1971, 65, 427.
51. Wollenweber, E, Bouillant, M L, Lebreton, P & Egger, K, Z Naturforsch 1971, 266, 1188.
52. Hegnauer, R, Chemotaxonomie der Pflanzen III Dicotyledoneae: Acanthaceae-Cyrillaceae. Birkhäuser, Basel, 1964.
53. Smith, D M, Glennie, C W & Harborne, J B, Phytochem 1971, 10, 3115.
54. Melchior, H (ed), A Enger's Syllabus der Pflanzenfamilien, 12th edn. vol. 2. Gebrüder Borntraeger, Berlin, 1964.
55. Green, P S. Personal communication.
56. Greger, H & Ernet, D, Phytochem 1973, 12, 1693.
57. Moore, D M, Williams, C A & Yates, B, Bot notiser 1972, 125, 261.

Discussion

Hegnauer: 6-Hydroxyflavonoids: Why did you not include 6-hydroxyapigenin? If you include it the distribution of 6-hydroxycompounds would be wider than indicated by your slide.

Large tropical families (e.g. Flacourtiaceae)? The taxa so far investigated represent a comparatively small fraction of the family. We are far from having a true picture of the flavonoid pattern in the Flacourtiaceae. The same is true for all other types of compounds and for many large families.

Harborne: 6-Hydroxyapigenin (or scutellarein) was in fact included in the survey. Shortage of time prevented the mention of this fact.

With regard to the Flacourtiaceae, it is obviously true that the present survey should be extended. However, it is possible that in spite of the morphological heterogeneity in the family, the flavonoid pattern may be simple, i.e. morphological variation is not necessarily correlated with chemical variation. Lebreton & Bouchez examined a larger sample of Flacourtiaceae for flavonoids than I did and they recorded common substances in all species.

Grant: You stated that the phenols were identified from herbarium specimens. In the genus *Lotus* (Leguminosae), I have compared chromatograms from leaf material dating back to the 1890's with those from fresh leaves. In general, the pattern was the same for the herbarium material as that from the fresh leaves; however, there was a greater number of compounds present in the chromatograms from the fresh leaves. Certain compounds presumably disintegrated with age and I assume this would happen in the material you examined from herbarium specimens. Is it not possible that this would have an effect on the comparative results which you have reported if herbarium material of different ages was examined?

Harborne: One explanation for finding fewer compounds in old herbarium specimens would be the hydrolysis of several glycosides giving the same aglycone and this process undoubtedly occurs during drying of plant material and subsequent storage.

What is remarkable about herbarium leaf material is that it is frequently possible to detect what are labile and easily oxidized flavonoids such as myricetin and gossypetin. I have been able to find these compounds consistently in such leaf specimens, some of which were collected up to 100 years ago.

Wagner: If it is accepted, that the flavonoids have also any function in plants, it would be of interest, whether e.g. the tasteless flavanone-7-O-rutinosides and the bitter tasting flavanone-7-O-neohesperidosides occur together in the same citrus species, and secondly whether the presence or absence of one of these two glycosidic types has anything to do with the attraction for or protection against insects or other animals, respectively.

Harborne: The bitter tasting flavanone glycosides could well play a role in plant-animal interactions in citrus, but as far as I am aware, there is no experimental evidence available on this point.

With regard to the function of different glycosides of flavonoids, it is interesting that in the case of the silk moth-mulberry interaction, quercetin 3-glucoside has been shown to be a biting attractant whereas the 3-rhamnoside, fed artificially to the silk worm, is a repellent.

Ourisson: I hope some of our botanists in the audience will bring us chemists up-to-date on our evolutionary thinking.

To me, many of the remarks made to explain "why" this or that compound has appeared in a group sound neo-Lamarckian and therefore (with my old-fashioned training), are an anathema.

I accept that compound A may play a role, biological or ecological, in species A', but I am not sure this must always be of importance, especially when species B', in the same niche, does not contain A. This is why I hope I can be helped to revise my thinking, if needed.

Furthermore, I am not ready to accept Bate-Smith's view that "bitterness is bitterness", and therefore repellent whatever the species. Certainly many insects live on plants bitter to us!

Flavonoids in Bryophytes[1]

E. Nilsson and G. Bendz

Chemical Institute, University of Uppsala, S-751 21 Uppsala, Sweden

Summary

Several species of bryophytes have been investigated as to their content of flavonoids. The flavones and anthocyanidins so far isolated and fully identified are all of the 3-desoxy type. From the moss *Dicranum scoparium* (L.) Hedw. we have also isolated a new natural biflavone identified as 5′,8″-biluteolin.

In 1962 Hegnauer in his book "Chemotaxonomie der Pflanzen" pointed out that bryophytes, that is liverworts and mosses, may be considered as "new territory" for chemists, and even today these members of the plant kingdom are relatively unknown chemically. The bryophytes include some 24 000 species.

For a long time the liverworts and mosses have been considered unable to synthesize flavonoids. Consequently, half a century elapsed between Molisch's first tentative identification of the flavone-C-glycoside saponarin in the liverwort *Porella platyphylla* (L.) Lindb. [1] and the first fully established structures viz. those of the anthocyanins luteolinidin-5-glucoside and luteolinidin-5-diglucoside occurring in the moss *Bryum cryophilum* Mårt. [2]. Recently there has been a significant increase in interest in the flavonoid chemistry of bryophytes and several investigations have now been published [3].

Anthocyanins

Reddish liverworts and mosses have been tested with respect to the nature of their red pigments and only one type of anthocyanins, i.e. glycosides of luteolinidin, has hitherto been found. In the genus *Bryum* there are several reddish species but luteolinidin derivatives have only been detected in 3 species (*B. cryophilum*, *B. weigelii* Spreng. *and B. rutilans* Brid.) belong-

ing to the subgenus Leucodontium Amann [4]. It seems that cell-sap-soluble anthocyanins are present in appreciable amounts only in species of this subgenus, other reddish *Bryum* species probably having their pigments bound to the cell wall.

Splachnaceae is the only other family of the Bryopsida investigated so far that contains derivatives of luteolinidin which have been isolated from *Splachnum rubrum* Hedw. and *S. vasculosum* Hedw. Among other reddish bryophytes we have investigated some species belonging to the genera *Drepanocladus, Calliergon* and the bluish pigmented rhizoids of *Ricciocarpus natans* (L.) Cda. (Hepaticae). From the latter we never succeeded in obtaining even a coloured extract. It is characteristic of the other two that the pigments can only be extracted to a certain degree and that they exhibit some of the features of a polymer. It seems probable that a transient pigment that is extractable is formed during the early stages of the development of the plants and is later incorporated into the cell wall material. Some evidence for this hypothesis has been obtained through a preliminary study of the seta of *Ceratodon purpureus* (Hedw.) Brid. A violet pigment could be extracted when the seta were freshly red, while this was not the case when they later acquired the more reddish-brown appearance [5].

Some *Sphagnum* species are reddish pigmented and of these we have investigated *S. magellanicum* Brid. och *S. nemoreum* Scop. The two violet pigments isolated from these species have been proved by spectroscopic and polarographic studies to possess many properties in common with 3-desoxyanthocyanidins, although they are neither known glycosides or aglycones [3]. These pigments seem to be strongly bound

---

[1] Read by G. Bendz.

Table 1. *Flavonoid tests on bryophytes*

Genus	Number of species tested	Result
A Bryopsida		
Bryum	several	*O*-glycosides and anthocyanins
Dicranum	several	*O*-glycosides and biflavones
Mnium	several	*O*- and *C*-glycosides
Hedwigia	one	*O*- and *C*-glycosides
Pohlia	several	positive
Sphlachnum	two	anthocyanins
Sphagnum	several	negative
Drepanocladus	three	negative
Calliergon	several	negative
Aulacomnium	two	negative
Grimmia	three	negative
B Hepaticopsida		
Porella	several	*C*-glycosides
Jungermannia	one	positive
Marchantia	one [16]	*O*- and *C*-glycosides
Hymenophytum	one [15]	*C*-glycosides
Monoclea	one [17]	polysaccharide
Plagiochila	one	negative
Symphyogyna	one [15]	negative

to the cell wall material and the poor extractibility they exhibit implies that they may to some extent be built into the cell walls.

Since only the rare 3-desoxyanthocyanidin luteolinidin has so far been found in mosses, we have tested different species for the occurrence of potential precursors (proanthocyanidins) to the common anthocyanidins which all contain a hydroxyl group in the 3 position. However all the species investigated responded negatively to the proanthocyanidin test [6]. The absence of this type of compounds in bryophytes has also been verified by other authors [7, 8].

Flavone derivatives

In 1967 McClure & Miller detected flavonoids or flavonoid-like compounds in 34 of 70 examined species of bryophytes. Their tests were performed on an acidified methanolic extract by means of colour reactions, R_F-values and response to spray reagents [9]. During our studies we have performed similar tests on neutral extracts on several species of both liverworts and mosses. Table 1 gives only a few examples and includes references to works of some other authors.

From the genera *Bryum* and *Dicranum* we have hitherto isolated only flavonoid-*O*-glycosides. The *Bryum* species mentioned above, in addition to anthocyanins, contain a complex mixture of flavone derivatives. One of these has been identified as the 7-glucoside of scutellarein (*1*), a relatively rare flavone which is

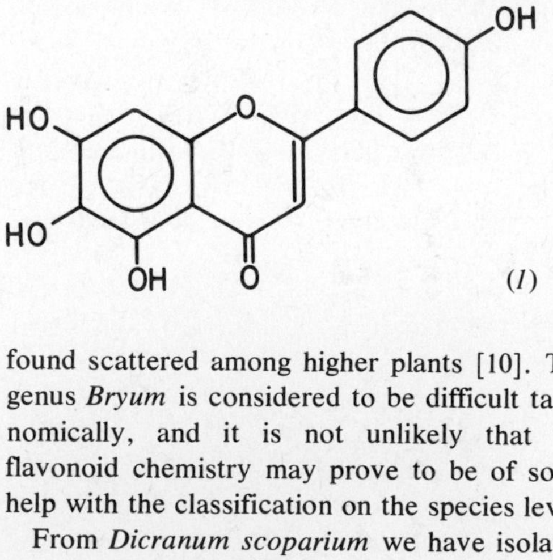

(*1*)

found scattered among higher plants [10]. The genus *Bryum* is considered to be difficult taxonomically, and it is not unlikely that the flavonoid chemistry may prove to be of some help with the classification on the species level.

From *Dicranum scoparium* we have isolated two flavone glycosides. One of these is a branched apigenin 7-glycoside identified as apigenin-7-*O*-[2,4-di-*O*-(α-ʟ-rhamnopyranosyl)]-β-ᴅ-glucopyranoside (*2*) [11]. The sugar sequence α—L—Rhap $\xrightarrow{1\quad 4}$ (β—D—Glup)—O

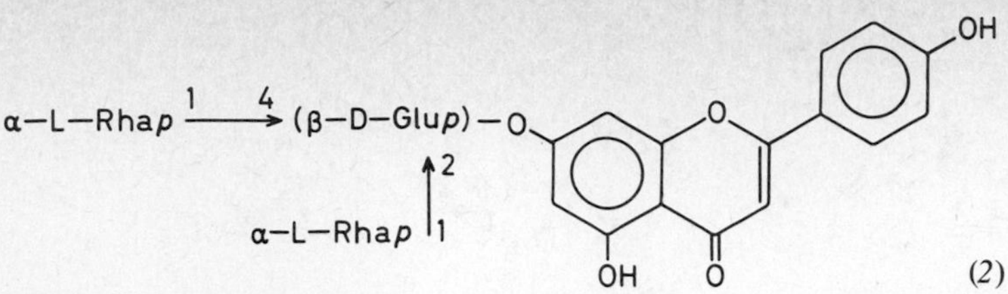

α—L—Rhap |1

(2)

analysis was achieved by a method known from carbohydrate chemistry. We used the GC-MS analysis of the alditol-acetates which were obtained by exhaustive methylation, hydrolysis, reduction of the methylated sugars to the alditols and acetylation of the alditols. By this method we could prove that the rhamnose units are linked to glucose at the 2- and 4-oxygens. The configuration at the glucosidic linkages was proved by partial hydrolysis and identification of L-rhamnose and by enzymatic hydrolysis of the remaining apigenin-7-monoglucoside indicating this to be a β-D-glucoside. The second flavone has been identified as luteolin-7-rhamnoglucoside [11]. From this species we have also isolated a new natural biflavone. It has been identified as 5′,8″-biluteolin (3) and

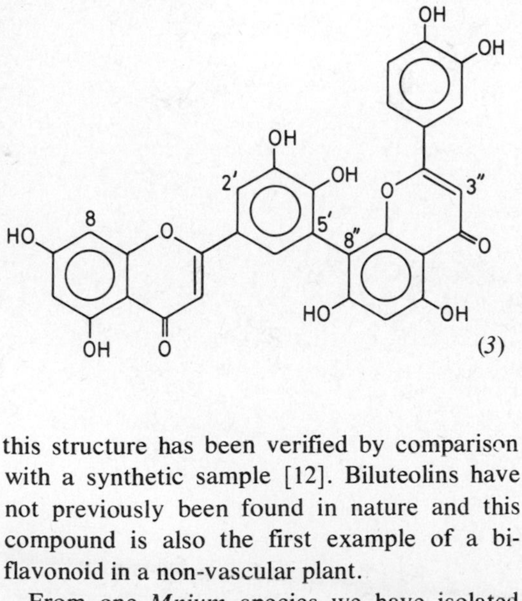

(3)

this structure has been verified by comparison with a synthetic sample [12]. Biluteolins have not previously been found in nature and this compound is also the first example of a biflavonoid in a non-vascular plant.

From one *Mnium* species we have isolated both *C*- and *O*-glycosides of apigenin and from other species preliminary results also indicate the presence of these two types of glycosides.

In Sweden the genus *Hedwigia* is represented by one single species *H. ciliata* (Hedw.) P.

Beauv. This species has been found to contain glycoflavones of both vicenin- and lucenin-type. In addition to these we have isolated three *O*-glycosides, one of which is a pentaglycoside of luteolin. All these compounds are under investigation.

In addition to flavones, species of the genus *Pohlia* were also shown to contain a diglucoside ester of caffeic acid, and it seems that cinnamic acid derivatives have a widespread occurrence in the mosses [3].

According to McClure & Miller the genus *Sphagnum* is probably richer in flavonoids than their results suggested. However, when screening some 20 species for common flavones we obtained negative results which may indicate the absence in this genus of easily extractable flavone derivatives.

Flavone tests on species of *Drepanocladus* and *Calliergon* have given negative results. Since also McClure & Miller could not detect any flavones in the pleurocarpous mosses they investigated, this indicates that these mosses possess a very limited capacity or perhaps lack the ability to synthesize these pigments.

At our laboratory we have investigated only a few liverworts with respect to their flavonoid content. The early suggestion of the presence of the flavone-*C*-glycoside saponarin in *Porella platyphylla* has been verified by us and by other authors [13, 14]. However, Markham and co-workers have investigated several liverworts and detected both *O*- and *C*-glycosides of flavones [15–18]. Of special interest is the flavone polysaccharide they were able to isolate from a *Monoclea* species, the aglycone of which was suggested to be the rare flavone 8-methoxyluteolin [17]. If their suggestion is right, it is of great interest since the presence of oxygenation at the 6- and 8-position of flavonoids as well as *O*-methylation have been considered phylogenetically advanced characters in higher plants. The fact that flavones are bound to polysaccharides in bryophytes is also of some importance since it indicates the possibility that they may be tied to the cell wall of these plants.

All fully identified flavonoids so far isolated from bryophytes are of the 3-desoxy type. Preliminary reports of flavonols from the liverworts *Corsinia coriandrina* (Spreng.) Lindb. [19] and the moss *Mnium arizonicum* [20] have, unfortunately, not been followed up by structure determination. The substantiation of these results is essential, since all other investigations hitherto made suggest that mosses and liverworts are unable to synthesize 3-hydroxy flavonoids which are very common constituents of higher plant groups.

We are very indebted to the Swedish Natural Science Research Council which has, for many years, financially supported this investigation.

We also wish to thank our collaborators, O. Mårtensson, G. Lindberg and B.-G. Österdahl, especially Dr Mårtensson who has collected a large part of the bryophytes.

References

1. Molisch, H, Ber deut botan Ges 1911, 29, 487.
2. Bendz, G, Mårtensson, O & Terenius, L, Acta chem Scand 1962, 16, 1183.
3. Nilsson, E, Studies of anthocyanidins and bryophyte flavonoids. Abstracts of Uppsala Dissertations from the Faculty of Science 1973, 239.
4. Bendz, G & Mårtensson, O, Acta chem Scand 1963, 17, 266.
5. Mårtensson, O & Nilsson, E, Lindbergia. In press.
6. Bendz, G, Mårtensson, O & Nilsson, E, Acta chem Scand 1966, 20, 277.
7. Cambie, R C, Cain, B F & LaRoche, S, New Zealand j sci 1961, 4, 731.
8. Bate-Smith, E C & Lerner, N H, Biochem j 1954, 58, 126.
9. McClure, J W & Miller, H A, Nova Hedwigia 1967, 14, 111.
10. Nilsson, E, Arkiv kemi 1969, 31, 475.
11. Nilsson, E, Lindberg, G & Österdahl, B-G, Chem scripta 1973, 4, 66.
12. Lindberg, G, Österdahl, B-G & Nilsson, E, Chem scripta. In press.
13. Nilsson, E, Acta chem Scand 1969, 23, 2910.
14. Tjukavkina, N A, Benesova, V & Herout, V, Coll Czech chem comm 1970, 35, 1306.
15. Markham, K R, Porter, L J & Brehn, B G, Phytochemistry 1969, 8, 2193.
16. Markham, K R & Porter, L J, Phytochemistry 1973, 12, 2007.
17. Markham, K R, Phytochemistry 1972, 11, 2047.
18. Markham, K R, Mabry, T J & Averett, J E, Phytochemistry 1972, 11, 2875.
19. Reznik, H & Wiermann, R, Naturwiss 1966, 53, 230.
20. Melchert, T E & Alston, R E, Science 1965, 150, 1170.

Discussion

Reichstein: I was interested that Dr Bendz found the sugar linkage in a bryophyte glycoside to be α-L-rhamnoside and β-D-glucoside. This corresponds exactly to Klyne's rule for glycosides in higher plants, having always the same absolute configuration at the anomeric center (α-L identical to β-D). This does not hold for glycosides from microorganisms.

Lavie: I wonder what do we know about the role of the flavones in the plants. In the previous lecture, it was described how they could be used for some classification purpose and in the present we hear that they are missing in several species. I would like to report about two cases in which the role of a flavone has been determined in our laboratory. One in which a flavone was found to protect a *Citrus*, the mandarine tree from a fungal disease, the so called "Malsecco" which is induced by the growth of a fungus in the vascular system, whereas the lemon tree which is very sensitive to the disease, does not contain this specific flavone, although other flavones were present. The second case refers to a flavone which induces the biting of a beetle and oviposition in the bark of apricot and plum trees. This would initiate the infestation of these trees by the attacking insect. Two cases in which one is protection and the other a kind of self destruction. In the latter case certain concentrations of the flavone are required for the initiation of the infestation.

Red Pigments in the Juncaceae family[1]

A. Fredga, Gerd Bendz and A. Apell

Chemical Institute, University of Uppsala, Uppsala, Sweden

Summary

Juncus effusus, Juncus articulatus, Juncus bulbosus var. *fluitans, Luzula pilosa* and *Luzula multiflora* have been investigated. All species contain glycosides of luteolinidin, a 3-desoxyanthocyanidin. No ordinary anthocyanins have been found.

If the results are representative of the family, the Juncaceae seem to hold a rather unique position among the angiosperms. However, very little is known about the colouring matters of the related family Cyperaceae.

The Juncaceae is a rather small family of 8 genera and some 320 species. Of these, at least 70% belong to the genus *Juncus* and about 25% to *Luzula*. These two genera are cosmopolitan. The remaining 6 genera, each containing one or a few species, are restricted to the Southern Hemisphere. The flowers are inconspicuous but the leaves and the stem may show red pigmentation. Very little is known about the colouring matters but in a Swedish dissertation from 1905 it is said that certain *Juncus* and *Luzula* species may contain an anthocyanin of unusual type [1].

We are investigating the red pigments in 3 species of *Juncus* and two of *Luzula,* all easily accessible in Sweden. In *Juncus effusus* L. the red pigment is located in the basal sheaths. In *Juncus articulatus* L. the whole plant may show red pigmentation, especially if the habitat is dried up during the summer. *Juncus bulbosus* L. var. *fluitans* (Lam.) Fr. is submersed and often sterile; the whole plant may be more or less red.

Juncus articulatus and some other species may be infested by larvae of a small insect, *Livia juncorum*. In that case the inflorescences are transformed into conspicuous clusters of small leaves, generally with strong red pigmentation.

No red pigments have so far been found in *Juncus filiformis* L. and *Juncus gerardi* Lois. In *Luzula pilosa* (L.) Willd. the leaves often turn red in the course of the summer, sometimes only on the upper side. In *Luzula multiflora*

(Retz.) Lej. single leaves are often red and at the end of the summer the whole plant may show a bright red colour.

All the red pigments were found to be glycosides of luteolinidin (*1*), a 3-desoxyanthocyanidin.

In *Juncus effusus* we have found at least two anthocyanins. The main component has two or three molecules of sugar, located in the 5-position. After hydrolysis, both glucose and galactose could be identified by means of paper chromatograms. The second component contained less sugar and may be an artefact, formed by partial hydrolysis of the main component during the operations. The other species seem to contain the same pigments. The sugars have not yet been definitely identified but the nature of the aglycone is well established.

3-Desoxyanthocyanins are found in mosses [2] and ferns but only in few cases in angiosperms [3]. Their occurrence is often regarded as a phylogenetically "primitive" character. In angiosperms they are generally found together with ordinary anthocyanins, but in our work only luteolinidin glycosides were found. The pigment pattern seems quite uniform and the species investigated represent different sections of the two dominating genera of the Juncaceae. Of course it can be questioned if they are representative of the whole family, but if they are, the Juncaceae seem to hold a unique position among the angiosperms. It is our intention to investigate more species, if possible also some representatives of the 6 genera growing in the Southern hemisphere.

There is also another problem. The Juncaceae are regarded as related to the Cyperaceae and the Gramineae but also to the Liliaceae. I think there

[1] Read by A. Fredga.

are also chemical reasons for that. In the Gramineae ordinary anthocyanins are often found; cyanidin-3-glucoside seems to be most frequent. In the Liliaceae ordinary anthocyanins are also common but to our knowledge very little is known about the colouring matters of the Cyperaceae. We have therefore started a search for extractable red pigments in this family and noted that certain *Eleocharis* species have red basal sheaths, very similar to those of *Juncus effusus*. An investigation of *Eleocharis palustris* (L.) R. & S. is in progress and we have good hope to establish the presence of luteolinidin in this plant.

This may indicate a special connection between the Juncaceae and the Cyperaceae. What we have presented here may be useful as a small piece in a big puzzle, but I think we should leave to the botanists to draw the conclusions.

References

1. Gertz, O, Studier öfver anthocyan, p. 45. Dissertation, Lund 1905.
2. Bendz, G, Mårtensson, O & Terenius, L, Acta chem Scand 1962, 16, 1183.
3. Harborne, J B, Phytochemistry 1966, 5, 589.

Discussion

Harborne: I would comment that the discovery of desoxyanthocyanins, and of luteolinidin in particular, in the Juncaceae is very satisfying from the point of view of the biosynthetic relationship between anthocyanins and flavones since the other major constituent of Juncaceae is the closely related structure luteolin. Remarkably, this occurs free in large quantitites in the inflorescences of *Juncus*; small amounts of luteolin glycosides are also present and their structures are under investigation in our laboratory. While desoxyanthocyanins are primitive pigments in the sense that they are the main anthocyanin types of mosses and ferns, their presence in the Juncaceae must, I am sure, be regarded as an advanced character in this highly specialized monocotyledonous family. Advancement has presumably preceded by loss mutation, i.e. loss of a specific flavanone 3-hydroxylating enzyme. The co-occurrence with luteolin, and *not* the corresponding flavonol, emphasizes this point. Also, an exactly parallel situation occurs in the dicotyledons where the highly specialized sympetalous family, the Gesneriaceae, accumulate desoxyanthocyanins in the flower; in this instance it is in response to natural selection for a bright red or orange flower colour in a tropical habitat.

Bate-Smith: Luteolinidin and apigenidin occur in *Sorghum vulgare* hypocotyls (H. Stafford). The tribe Andrapogoneae of the Graminae contain leucoluteolinidin, which forms a complex luteolinidin in the seed coat colouring it red or even black. This is a flavan-4-ol (catechins being flavan-3-ols).

Fredga: In *Sorghum vulgare* (as in other angiosperms) the 3-desoxyanthocyanins occur *together with* ordinary anthocyanins. In this respect there is a clear difference between *Sorghum* and the Juncaceae species.

It has been pointed out in the discussion that certain *Carex* species, e.g. *Carex caespitosa*, have red or reddish sheaths. To my knowledge these pigments are not extractable and probably cell-wall bound. In ref. [1] of my paper there are some statements to this effect.

Ourisson: Have the cell-wall bound pigments been studied, e.g. by reflection UV spectroscopy or reflection fluorescence?

Fredga: Not to my knowledge. Certainly not by us.

Merxmüller: There is almost no disagreement amongst botanists concerning the close relationship of the Juncaceae and Cyperaceae. Especially some cytological characters (like the diffuse centromeres and chromosome fragmentation) are so unusual that there cannot be much doubt. Special red pigments in these both groups also underline this somewhat isolated problem.

Birch: I would like to underline the point made by Dr Harborne. One would like biosynthetic pathways to pass a pointer into direction of evolution. Unfortunately, they do not. Since loss mutations are more common than gain mutations, if there is any good reason for a plant to use a product obtained with less trouble it may do. With this proviso a more "advanced" plant might produce a more "primitive" compound. The only safe ground seems to me to lie in consideration of a number of pathways and all of the other evidence.

Volatile Constituents, Especially Terpenes, and their Utility and Potential as Taxonomic Characters in Populational Studies [1]

R. H. Flake and B. L. Turner

Department of Engineering and Department of Botany, University of Texas at Austin, Austin, Tex. 78712, USA

Summary

Volatile constituents are found in nearly all of the major plant groups, and because of the ease and rapidity of identification and quantification using combined GLC-mass spectroscopy, they make ideal characters for systematic purposes, especially at the generic level and below. Terpenes, in particular, have been widely studied and it is clear that their qualitative and quantitative expression has a genetic base. Environmental factors can be expected to cause variable, mainly quantitative, expressions of such characters in nature, but using populations as the evolutionary units for chemical analysis and treatment, and with statistical and appropriate computer programs, such variability can be treated in a meaningful way. Thus considerable insight can be obtained about speciation or adaptational processes occurring within a given taxon, especially at the specific level or lower. In working with populations from which systematic samples have been made over a large area, algorithms have been developed which present single-character or multiple-character sets in topograpic or three-dimensional form. Such figures are drawn by the computer and make possible highly sophisticated interpretations of subspecific variations, including single and multiple character clines, especially as these relate to climatic and topographic features or historical events. They also permit relatively objective analysis of regional variation, i.e., whether this might be due to either gradual accommodation to an environmental gradient or through the more turbulent form of gene accommodation expected under hypotheses for introgressive hybridization. Examples of such studies are presented using recently obtained chemical data from populations of *Juniperus virginiana* collected over a 500 000 square mile area.

Volatile constituents, especially terpenes, have been used as taxonomic characters since the time of Aristotle and presumably earlier, for even primitive societies made extensive use of their noses in classifying appropriate medicinal plants. In fact, it is because of their widespread distribution, and presumably independent origin, among many orders that has largely precluded their use as effective phylogenetic characters, especially at the familial level or higher. But hardly any experienced plant systematist would deny their utility, in combination, as diagnostic markers. Thus the mint family, Lamiaceae, is readily distinguished (most of the time!) from the Scrophulariaceae by the high concentration of terpenes in the former and their absence in the latter; the tribes Anthemidae and Tagetinae of the family Asteraceae are likewise relatively easily distinguished from neighboring tribes by their strong aromaticity; etc. Except as distinguishing characters, terpenes have not been especially useful as taxonomic characters at the generic level or higher. The same might be said of flavonoids, with notable exceptions stressed by Harborne [1, 2], Mabry [3], and especially Bate-Smith [4] and Kubitzki [5].

There have been, however, several exceptional and exceedingly interesting systematic studies using terpenoids at the generic level or lower. These include those of Irving [6] in which monoterpenes were used to distinguish a relatively simple situation in which three sympatric taxa of *Hedeoma* were coexisting without evidence of hybridization (as judged by chemical data!), as contrasted with the more classical morphological and biosystematic data which would have at least two of these taxa hybridizing so as to form the third [7]. Other workers however, have used terpenoids to document natural hybrids [8, 9]; indeed, many studies are now available to suggest strongly that most chemical components, including their quantitative expression, are under rigorous genetic control

[1] Read by B. L. Turner.

[8]. In addition to the documentation of hybridization, terpenoids have been quite useful in working out likely migration routes and centers of origin for selected populations, usually within a species [7, 10, 11, 12].

Like flavonoids, then, terpenoids have been used extensively in the study of systematic problems. But while flavonoids are generally more numerous and widespread than are terpenoids, they are nonetheless more difficult to work with at the technical level, at least as contrasted with the relatively few monoterpenes which are readily identified by the novice using combined gas liquid chromatography (GLC) and mass spectrometry. Further, while volatile compounds are not so numerous and widespread among plants generally as are flavonoids, they are usually much more numerous as to kinds within a given species or plant. For example, using GLC, the passion fruit (*Passiflora edulis* Sims) was found to contain over 250 discrete peaks [13]. Such a wealth of systematic data, if fully identified and quantitated, would impress any systematist, for an equal bevy of morphological characters from any one species would prove an enormously laborious undertaking using the current techniques of eyeballs, caliper, and ruler. In fact, we conclude that, because of the ease of quantitation using GLC, terpenes are better characters than flavonoids for systematic studies at the species level or lower, simply because the existence of quantitative variation in a polygenically controlled chemical character within a population or group of populations permits the use of rather powerful statistical methods in the treatment of data derived from such populations. This contrasts sharply with most flavonoid data, the compounds themselves being usually under relatively simple genetic control, being present or absent within a given plant, so that, unlike isoenzymes (which can be distinguished in the heterozygous condition such that gene frequencies can be calculated and from this a "genetic distance" established), flavonoids have rarely been used within a species to resolve effectively purely population problems in which hybridization has not been a factor [7].

As indicated, it is relatively well established that the quantitative expression of most terpenes is under some form of genetic control. But what of the environmental influences upon these characters? No doubt environmental factors do affect the quantitative expression of terpenoids, as has been shown by a number of workers [8, 13]. Indeed, it is clear that terpenoids are under a nearly constant state of metabolic flux in the living plant [14]. But such environmentally induced variation need not preclude their use in systematic studies of the type reported here if populations (as contrasted with individuals) are used as the unit of taxonomic study since, with appropriate sampling and statistical treatment, the environmentally induced perturbations of a given terpene within a population can be "sounded out" and leveled to a meaningful statistic or else, in the case of excessively variable compounds, these can be "singled out" and eliminated from one's study as being significantly "too much affected by environmental factors" so as to be used effectively as systematic characters.

It is the purpose of this paper, then, to present some of our research bearing on the problem of infraspecific variation in *Juniperus virginiana* L. (Eastern red cedar), a common, widespread, weedy tree of the eastern United States. We believe that in this way we can best exemplify the utility of such characters in systematic studies and at the same time point out the potential of their use for future studies. Much of the preliminary work on this has been reported upon in previous publications, and it is not our intent to repeat here these results [15, 16]. Rather we wish to present a brief, preliminary report of newly acquired, unpublished terpenoid data which bear upon the problem of clinal or regional variation in this highly variable species. At the same time, we would like to introduce new approaches to the presentation of these data, approaches which we feel are superior to those used in our earlier presentations.

The present authors, in collaboration with others, first began their studies of *J. virginiana* in 1968 [15, 16]. This was undertaken to ascertain if a meaningful clustering of discreet populations across a 1500 mile transect might be obtained using newly developed numerical treatments [17] with which to evaluate the 9 populations and 90 plants which were examined for their 37 or more volatile constituents. The results (fig. 1) proved so striking that the same sites were subsequently examined in 1970, but this time the number of plants per population was doubled and a new peripheral population was added to the transect. This work has been recently reported upon in some detail [16] and we need only note here that while the second year's results largely confirmed those of the first, addition of the new population made apparent the existence of presumably regional structuring of the populations in Texas. This stimulated us to extend the investigations into yet another session of sampling, this time re-sampling yet a third time the same populations sampled during the first and second years of our investigation, in addition to others as will be reported below.

The data obtained from populations sampled during the years 1968 and 1970 were originally portrayed in a relatively simple two-dimensional diagram such that levels of significance and "terpene-topography" were not easily compre-hended by the interested, or at least casual, reader. Other workers have also portrayed che-mical data from plant populations of *Juniperus* in this fashion [18, 19], but they used clustering algorithms and computer procedures not em-ployed in the present study.

More recently, powerful algorithms and com-puter techniques have been developed which

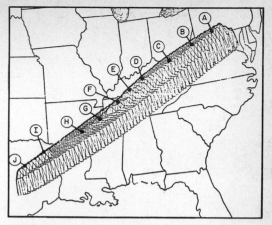

Fig. 2. Cluster analysis of 10 populations of *Juniperus virginiana* collected in 1972. Data from 20 plants at each population; 37 volatile constituents taken in combination were used to construct the profile.

portray multiple-character affinity states, as obtained over a broad region, in three-dimen-sional, topographic form [20]. We have applied these techniques (using an algorithm developed by the junior author, unpublished) to both the first and second year's data referred to above. The results for the first year are shown in fig. 1. It can be seen that the topographic form of the multi-character cline across the approx. 1500 miles concerned is one of relative homogeneity of populations with respect to the chemical characters used, especially in the northeastern United States, the more southwestern popula-tions becoming decreasingly associated with those centered in the Northeast; i.e., they cluster at successively lower levels of similarity. How-ever, as already indicated, when an additional population was added to the extreme end of the cline in central Texas, the entire aspect of that end of the cline changed (fig. 2). In fact, the new population clustered so strongly and at such a high level of similarity with its adjacent popu-lation, that a portion of the new topographic cline became statistically disjointed at its south-western most extreme. Clearly, the two Texas populations seemed to be part of a regionally isolated, much larger group of populations with a presumably different history or origin, at least as contrasted with those populations found near to or northeast of the Mississippi River.

Fig. 1. Cluster analysis of populations of *Juniperus virginiana* collected in 1968 (from eastern Texas to Washington, D. C.). Data from 10 plants at each site; 37 volatile constituents taken in combination were used to construct the profile. Only the statistically significant contours are shown.

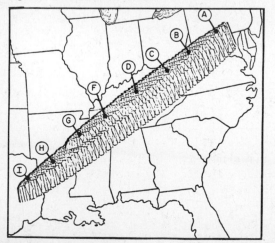

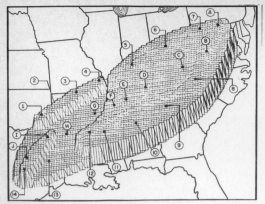

Fig. 3. Cluster analysis of 24 populations of *Juniperus virginiana* collected in 1973. Data from 503 plants (20 or more for each site); 37 volatile constituents taken in combination were used to construct the profile. Only the statistically significant contours are shown.

The unexpected recognition of statistically significant, regionally clustering, populations in Texas led us to extend our studies into a third year (1972), this time not only sampling again the original populations, but extending the sample to include populations both north and south of the transect so that a large regional sample (about 500 × 2000 km) over the central range of the species might be had. Chemical data from these 24 populations were subjected to a clustering algorithm similar to that used in previous studies, and the preliminary results are shown in fig. 3. The data have not as yet been subjected to rigorous scrutiny by techniques that we feel are capable of being developed, but such an analysis with explanatory rationale is in progress. But even with preliminary results, what does seem evident in fig. 3 is that the species is relatively homogeneous, populationally speaking, throughout much of its range, except along the southwestern periphery where there appears to be at least two disjointed regional populations; one centered in eastern Texas and another centered in the highland regions (Ozark Mountains) of Arkansas and adjacent Oklahoma.

It should be reemphasized that fig. 3 reflects data from 503 plants distributed over 24 populations, each separated from the other by approx. 150 miles so as to form a grid system across the central range of the species. Further, the "topographic affinity structure" of the dia-

gram itself is based upon a continuous spectrum of data from each of the 37 terpenes for which data were available. It would be possible to construct such a diagram for each of the terpenoid characters examined, and while such diagrams might prove instructive for ascertaining the relative quantitative distribution of a given terpene, and perhaps instructive from the standpoint of relating this to possible predator ranges, climatic factors, etc., these would have little meaning as systematic guides to multicharacter, albeit weighted, clustering which is what taxonomy is all about [17].

Fig. 3, then, reflects a weighted (using a computor-derived figure based upon intra- and interpopulational variance of the characters concerned) clustering based on the entire set of 37 chemical characters. As such, this is a "terpenoid affinity profile" of the species, and one might well ask how this might relate to a "morphological affinity profile" of the species.

A topographic model of the type shown in fig. 3 is not, of course, available for the taxon in the region covered by our sampling. However, Hall [21] has been able to recognize a number of regional races in *Juniperus virginiana* as indicated in fig. 4. Hall's treatment of the species was largely derived from habital features of the plant and it is interesting to compare fig. 3 and 4 as to the number and size of the regional races recognized. It will be noted that Hall's treatment recognizes a southwestern race which our results indicate is divided into an Ozark race and a Texas race. Both treatments recog-

Fig. 4. Schematic diagram of the races recognized by Hall [21] in the southern and central portions of the range of *Juniperus virginiana*.

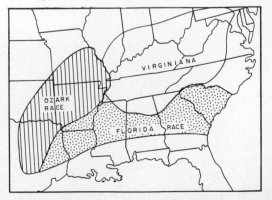

nize a much larger, presumably more homogeneous northeastern race in the present results. We have not, however, detected a southern or southeastern race, as proposed by Hall. Further, we would ascribe the geographical patterning as being due to factors other than introgression from the distantly related allopatric species, *J. ashei* Bucholtz, although it is possible that introgression from the closely related, allopatric species *J. scopulorum* has affected the peripheral populations of *J. virginiana,* as proposed by Van Haverbeck [22]. Alternatively, the well structured peripheral populations may merely reflect long-time regional isolation, climatic adaptation, and presumably habital and terpenoid differentiation of the three groups of populations or geographic races. Chemical data bearing on the likely origin of such populational structuring have now been assembled and are currently under investigation by the present authors.

This research has been supported by an NSG Grant GB-31048.

References

1. Harborne, J B, Comparative biochemistry of the flavonoids. Academic Press, London and New York, 1967.
2. Harborne, J B, Evolution and function of flavonoids in plants, in Recent advances in phytochemistry (ed C V Runeckles & J C Watkins) p. 108. Appleton-Century-Crofts, New York, 1972.
3. Mabry, T J, Brittonia (1973). In press.
4. Bate-Smith, E C, Flavonoid patterns in the monocotyledons, in Perspectives in phytochemistry (ed J B Harborne & T Swain) p. 167. Academic Press, London and New York, 1969.
5. Kubitzki, K, Taxon 1969, 18, 360.
6. Irving, R S, Doctoral dissertation. University of Texas, Austin, Tex., 1968.
7. Turner, B L, Taxon 1969, 18, 134.
8. Irving, R & Adams, R, Genetic and biosynthetic relationships of monoterpenes, in Terpenoids: Structure, biogenesis, and distribution (ed V C Runeckles & T J Mabry) p. 187. Academic Press, London and New York, 1973.
9. Adams, R, Southwestern Naturalist 1973. In press.
10. Turner, B L, Molecular approaches to populational problems at the infraspecific level, in Phytochemical phylogeny (ed J B Harborne) p. 187. Academic Press, London and New York, 1970.
11. Von Rudloff, E, Can j bot 1972, 50, 1025.
12. Zvarin E & Snajberk, K, Phytochemistry 1972, 11, 1407.
13. Murray, K E, Shipton, J & Whitfield, F B, Aust j chem 1972, 25, 1921.
14. Loomis, D W & Croteau, R, Biochemistry and physiology of lower terpenoids, in Terpenoids: Structure, biogenesis, and distribution (ed C V Runeckles & T J Mabry) p. 147. Academic Press, London and New York, 1973.
15. Flake, R H, von Rudloff, E & Turner, B L, Proc natl acad sci US 1969, 64, 487.
16. Flake, R H, von Rudloff, E & Turner, B L, Confirmation of a clinical pattern of chemical differentiation in *Juniperus virginiana* from terpenoid data obtained in successive years, in Terpenoids: Structure, biogenesis, and distribution (ed C V Runeckles & T J Mabry) p. 215. Academic Press, London and New York, 1973.
17. Flake, R H & Turner, B L, J theor biol 1968, 20, 260.
18. Adams, R & Turner, B L, Taxon 1970, 19, 728.
19. Adams, R, Systematic zool 1970, 19, 385.
20. Pauken, R J & Metter, D E, Systematic zool 1971, 20, 434.
21. Hall, M T, Ann Missouri bot garden 1952, 39, 1.
22. Van Haverbeck, D F, Univ Nebraska studies 1968, 38, 1.

Discussion

Lavie: Would you please explain more explicitly what one was actually seeing as the results of the computer data produced by the analysis of the characteristics discussed, how to understand the drawings given in the slides?

Turner: The contours in figs 1–3 represent affinity values for the populations concerned as determined from the analyses of 37 terpenes in combination. Thus, populations located at or near the top of a given "topographic" region show affinity values of 90% or better; those clustering at lower "topographic" levels on the same "massif" will have lower affinity values, etc. Affinity values were calculated for all of the populational clusters, but only the statistically significant contours are shown in the computer drawn figures.

Herout: Have you studied the variation of essential oil composition during different seasons of a year, on a single individual?

Turner: Such variation on a plant to plant basis has been investigated fairly extensively by my exstudent, Dr Robert Adams of Colorado State University. He has shown that there are statistically significant variations in the volatile components of *Juniperus* from month to month in natural populations, especially in the spring, summer and fall. But during the winter months, when metabolic activity is low, such variation was found to be minimal. Our samples, consequently, were taken during a several-day period in mid-winter so as to minimize environmental influences.

Some Aspects of the Distribution of Diterpenes in Plants*

G. Ourisson

Institut de Chimie, Université Louis Pasteur, Strasbourg, France

Summary

A brief review is given of some of the close correlations often exhibited by families of higher plants as regards their diterpene constituents. Order and chaos appear to prevail, in that any apparent order finds its rapid counterpart in "exceptions", or looks meaningless. It is then recalled that some diterpenes are obligatory constituents of higher plants (geranyl-geraniol and *ent*-kaurene) but that they are normally present in minute amounts. It is postulated that the abundant diterpenes are all products of deviations of the normal metabolism leading to the normal, minor diterpenes ("target compounds" in Birch's terminology), and that they normally serve no physiological role. They are considered as appendages (like the horns of deer; for another comparison, see Conclusion), dispensible with from a physiological point of view, but useful in the plant eco-system, in ways that are not always evident and that may have lost their significance.

The variety of diterpenes is large, but not vast. In Scott & Devon's Index (1972) about 500 individual structures of naturally occurring diterpenes are listed, and each year a few dozen more become known. We shall try to demonstrate that the distribution of these substances can only be explained by the interplay of necessity and randomness. The diterpenes are of course not the only group of substances illustrating what we consider to be general principles of chemical evolution, but this paper will deal with diterpenes.

Diterpenes as Chemosystematic Markers

Many of the 18 structural families of diterpenes are found to be chemosystematic markers: there is often a strict correlation between a given plant family or sub-family, and a given diterpene family. Let us start with a few examples, chosen among many others.

Conifers are general sources of diterpenes of the "normal" series (10β), in various structural families derived biosynthetically from a common bicyclic precursor. Some representative structures are those of communic acid (*1*) and manoöl (*2*) (bicyclic diterpenes), pimaric acid (*3*) and abietic acid (*4*) (tricyclic resin acids). Variants include strobic acid (*5*).

A striking observation is that kaurene occurs both in the "normal" series (*6*) (e.g. in one *Podocarpus*) and in the "*ent-*" (10α) series (*7*), (e.g. in other *Podocarpus,* or in *Agathis* species containing also some bicyclic diterpenes of the "normal" series).

Caesalpinioideae often contain diterpenes in their resins (copals of the sub-family Amherstiae). Representative examples are daniellic acid (*8*) (the enantiomer of which is known in a *Pinus*), *ent*-kaurenoic acid (*9*), hardwickic acid (*10*) and hydroxytrachylobanic acid (*11*).

---

* (1) The purpose of this review is not to provide the reader with the kind of information normally found in a secondary reference, but to raise a few critical issues. No references are given in the text, where they would serve no real, scholarly purpose. The Bibliography at the end gives a list of key references to exhaustive or critical reviews where facts and primary references will be found.

(2) The main ideas of this review were developed progressively for terpenoids, alkaloids and other natural products in graduate lectures given at the University of Strasbourg: they were presented more systematically in lectures on terpenoids given at Cambridge during the winter term of 1967–1968. They converge, with a slightly different emphasis, with the ideas expressed by A. J. Birch in his lecture on "Biosynthetic pathways in chemical phylogeny", presented in New Delhi in February, 1972.

(3) Our own contributions to the field are limited to the study of one diterpene from Euphorbiaceae and about a dozen from Caesalpinioideae.

(4) This paper was not orally presented.

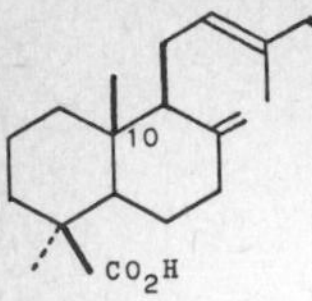

Some diterpenes of conifers

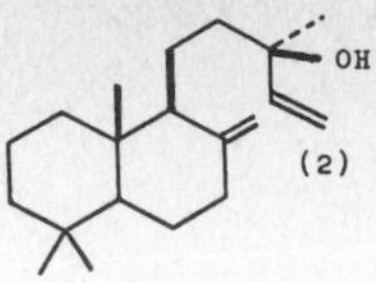

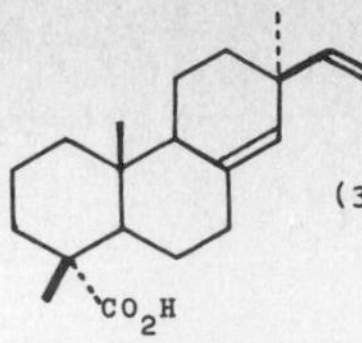

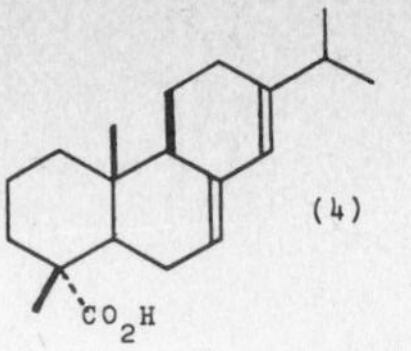

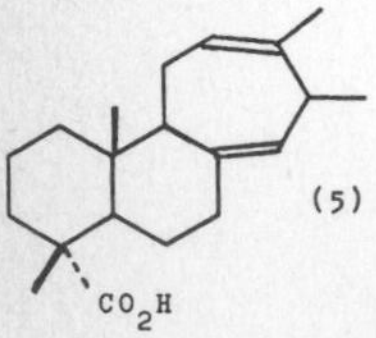

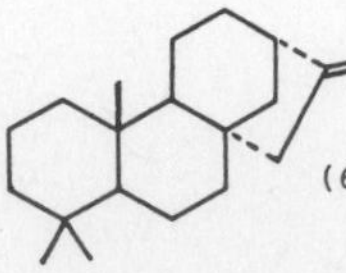

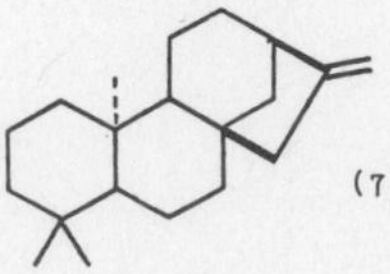

Labiatae contain diterpenes of the two enan-tiomeric series, and all of the di- and tricyclic ones reported so far are of the "normal", 10β series, while all the tetra- and pentacyclic examples known are of the "ent-" series. Representative examples of the first group are sclareol (12) and salviol (13) from *Salvia*, lagochilin (14) from *Lagochilus*, marrubiin (15) from *Marrubium*, nepetaefolin (16) from *Leonotis*.

Sideritis contain many ent-diterpenes, such as sideridiol (17) and trachylobane (18). *Isodon* contains also a wide range of ent-diterpenes, all of them extensively hydroxylated or oxidised. Isodonol (19) is one of the simplest ones.

It remains to be seen whether this criterion has infra-familial taxonomic value.

Euphorbiaceae (and Thymeleaceae): *Ent*-diterpenes are again found, both in the tri- and in the tetracyclic series, in Euphorbiaceae, as shown by structures (20) (*Macaranga*), (21) and (22) (*Cleistanthus*), (23) (*Beyeria*) and (24) (*Ricinocarpus*). However, there are other, original diterpenes in other Euphorbiaceae; these are derived from the macrocyclic hydrocarbons cembrene (25) (not found yet in Euphorbiaceae) and mostly casbene (26) (found in germinating *Ricinus* seeds). Representative ex-

amples of this growing family are bertyadionol (27) (*Bertya*), jatrophone (28) (*Jatropha*), phorbol (29) (*Croton*), ingenol (30) (*Euphorbia*) and huratoxin (31) (*Hura*).

The last one of these irritant, toxic, cytostatic or cocarcinogenic, substances is an ortho-ester entirely analogous to mezerein (32), the ortho-benzoate toxin from *Daphne*. The implication is evident in this case: Thymeleaceae like *Daphne* had already been considered to be related morphologically to Euphorbiaceae, and the co-occurrence of complicated, highly specific, structures like (31) and (32) is a strong chemical argument in favour of this contention.

Ericaceae: Many species of Ericaceae (*Rhododendron, Kalmia, Leucothoe,* etc.) contain very characteristic, toxic, tetracyclic diterpenes, with the rearranged skeleton of andromedol (33). This is obviously derived from ent-kaurene.

Ranunculaceae: The same derivation from ent-kaurene is found in the many, varied, diterpene alkaloids isolated from the poisonous plants *Aconitum* and *Delphinium*. Structures (34) (napelline) and (35) (atisine) are easily understood; heteratisine (36) requires more insight, or more previous knowledge, to be recognised as a derivative of ent-kaurene.

We have thus mentioned in passing about

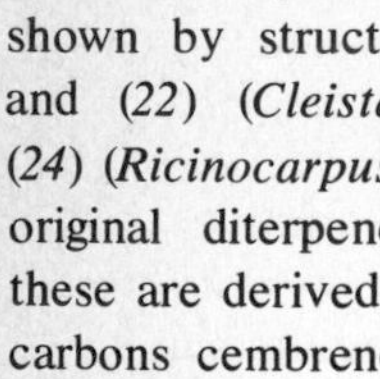

Some diterpenes of Caesalpinioideae (Amherstiae)

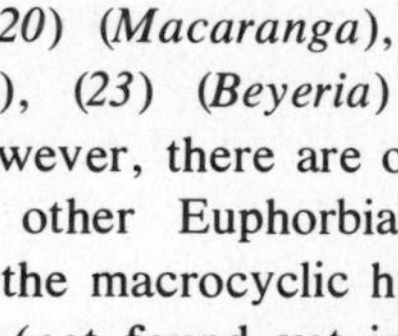

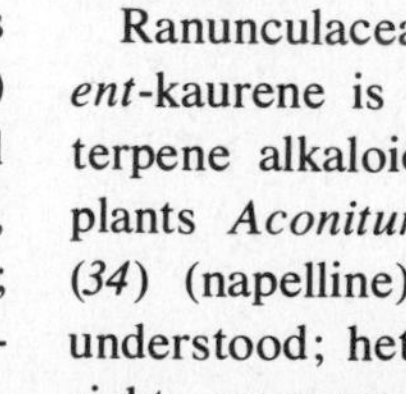

Nobel 25 (1973) Chemistry in botanical classification

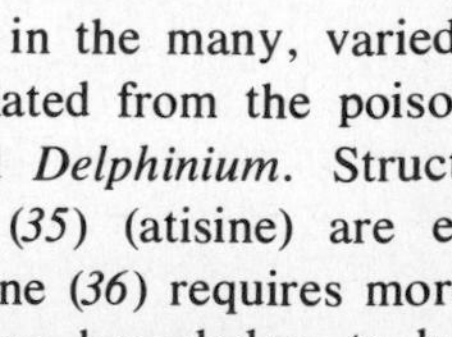

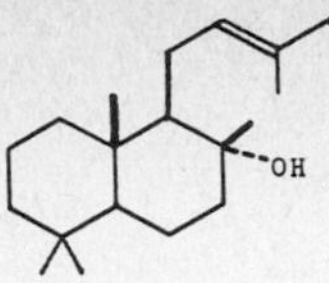

Diterpenes from Labiatae (bicyclic, 10 β)

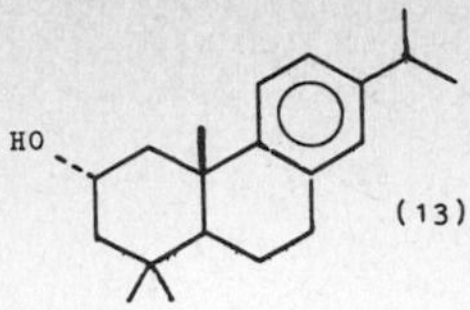

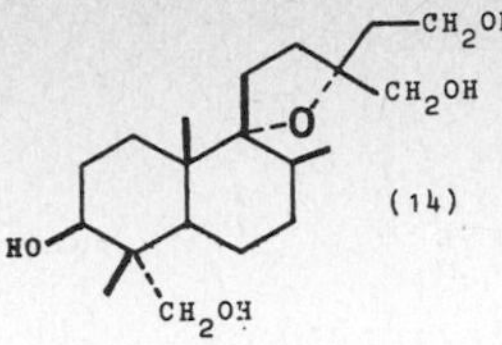

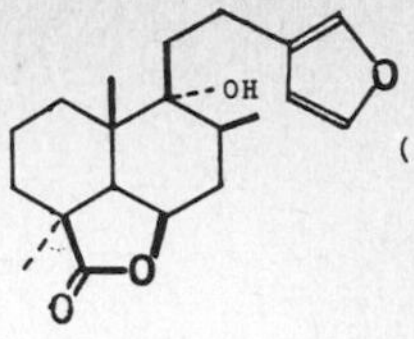

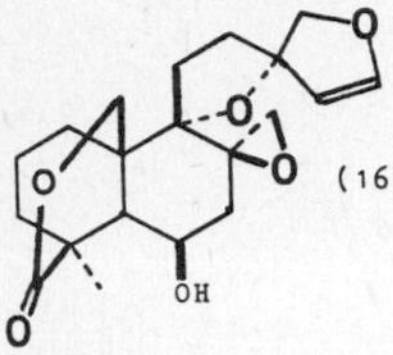

8% of the known diterpenes; it is already obvious that their distribution is of two types: while derivatives of the "normal" series appear to be restricted to a few families, derivatives of *ent*-kaurene appear in various guises and forms in many unrelated groups. In fact, they have also been isolated in lichens, ferns, Compositae, Annonaceae, Rubiaceae, Rutaceae, Convolvulaceae, etc. (I am of course aware of the danger of "obvious correlations" in the field of chemical plant systematics, based on the incomplete inventory obtained from structural work. "The chemist finds only what he looks for" is not a really wrong statement (it has in fact been implied by Prof. Cronquist in his remarks on anthochlors) and one who isolates an O_9 toxic diterpene will surely not spend much time looking at the hexane extract! We can therefore only rely on gross comparisons, gaining weight statistically as more and more substances happen to be isolated.)

"Special" groups (like the toxic substances *27–36*) are of course more varied than we have indicated: taxane derivatives in Taxaceae, gingkolides in *Gingko*, ryanodine, eremolactone, etc., represent other cases; they are all restricted to one family, or one genus, or one species.

The wide occurrence of one of the tetracyclic diterpene series thus stands in marked contrast with the chaos surrounding the distribution of the other types; before we discuss this point, we shall first review the biochemical role of diterpenes in plants.

The Biochemical Role of Diterpenes in Plants

Diterpenes are all transformation products of their acyclic precursor, geranyl-geraniol (*37*). This is a universal constituent of all green plants, because it is also an obligatory precursor of the carotenoids, and chlorophyll can only function in plants when it is protected from photodegradation by carotenoids: this is shown by experiments with green carotenoid-free mutants of *Rhodopseudomonas,* or with diphenylamine-treated, carotenoid-deprived plants. This implies that every green plant contains the necessary diterpene precursor; diterpene elaboration can be a branching-off of a general biosynthetic pathway: the polycondensation of mevalonic acid to the ubiquitous polyprenols.

Another diterpene is also probably universally present in higher plants, and this is *ent*-kaurene. It is indeed known that this is an obligatory precursor of the plant hormones,

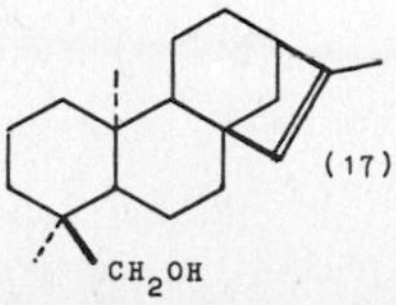

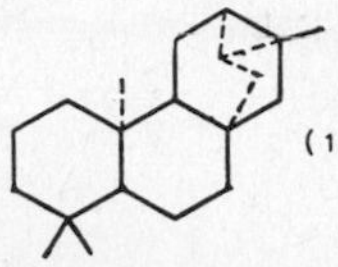

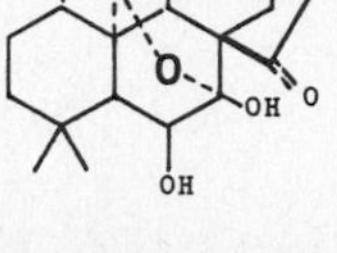

Diterpenes from Labiatae (tetracyclic, *ent*)

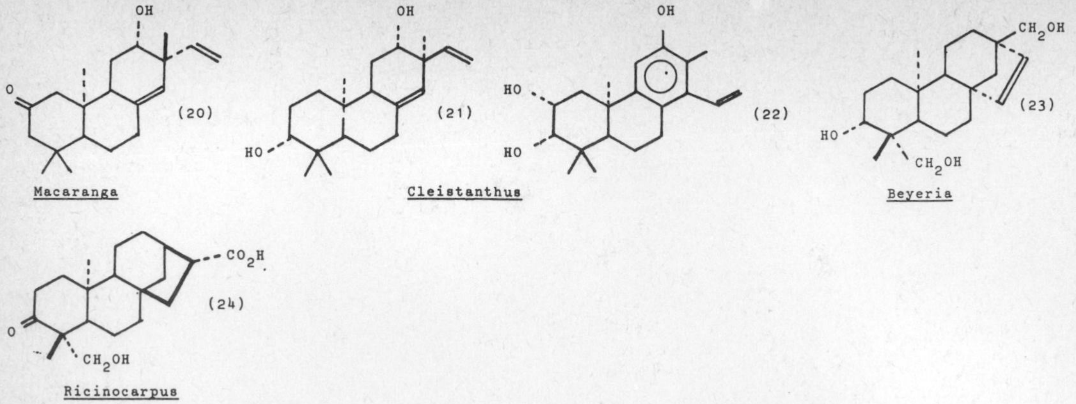

Tri- and tetracyclic *ent*-diterpenes from Euphorbiaceae

the gibberellins, such a gibberellin A_{12} (*38*) or gibberellic acid (*39*), themselves found so far in every plant species where they have been looked for!

We arrive therefore at a most important observation, which we shall formulate forgetting all the necessary qualifications, as follows: Every green plant contains the required precursor for both carotenoid and gibberellin biosynthesis, geranyl-geraniol, and the enzymatic armamentorium required for its cyclisation to *ent*-kaurene, itself a universal plant diterpene, and for the oxidation of diterpene substrates.

We must, however, immediately stress that the steady-state concentrations of geranyl-geraniol and of *ent*-kaurene in any plant are minute, probably not detectable in most cases except by very elaborate techniques, whereas the other diterpenes discussed here are important constituents of the plants. These diterpenes are therefore "abnormal" both in that they represent "abnormal"derivatives of geranyl-geraniol, and because they are present in

"abnormally" large amounts. We like to think these are related observations: a "normal" plant, biochemically balanced, would, in our view, contain no more than the small steady-state amounts of geranyl-geraniol and *ent*-kaurene. Any unbalance would lead to increased production of these substances, and could trigger excretion into resins, or alternate modes of disposal: alternate cyclisations of geranyl-geraniol, oxidations, degradations, etc., by using existing enzymatic systems or by inducing them. The only absolute requirement, at the cellular level and for the organism, is that such an unbalanced overproduction accompanied by disposal of the excess must not be lethal.

In fact, it is quite probable that the pathway geranyl-geraniol→carotenoids, which must be present in the simplest photosynthetic organisms, is more primitive, and may have later developed into the more complex pathway leading to *ent*-kaurene and the gibberellins (this is one of the ideas put forward, and docu-

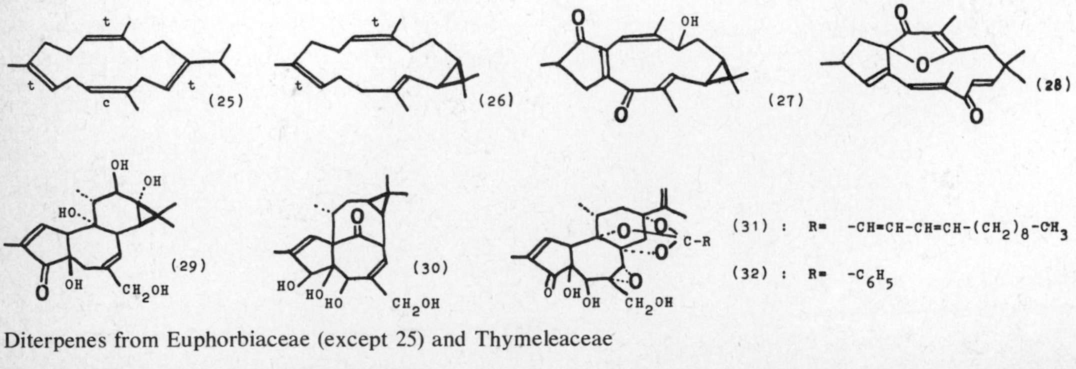

Diterpenes from Euphorbiaceae (except 25) and Thymeleaceae

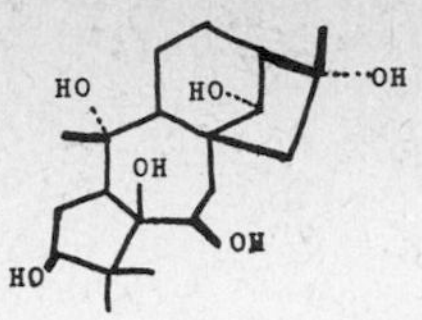

Tetracyclic *ent*-diterpene from Ericeae

mented, in a forthcoming chapter by T. S. Swain on chemical evolution, in Florkin, Comprehensive biochemistry); it may be that a randomly happened cyclisation followed by a randomly happened oxidation sequence, has led to a physiologically useful product, a gibberellin. This might have been a sufficient bonus to fix the process. From this point of view, it would be interesting to study the distribution and the possible regulatory role of the active substances isolated from *Podocarpus,* such as inumakilactone A-15-glucoside (*40*), a close relative of antibiotics isolated from a fungus, *Acrostalagmus,* such as (*41*); these lactones are potent inhibitors of the expansion and division of plant cells.

The Ecological role of Diterpenes

A biochemically indifferent, or even slightly unfavourable, mutation can have positive value from the point of view of evolution if it provides the organism with better survival or spreading chances in its milieu. Unfortunately, the evaluation of such characters with positive selective value is a most difficult and perilous task: it entails comparing an observed macro-system with its postulated variants, usually with no possible recourse to experiments.

For example, is the monstrously increased production of geranyl-geraniol in Conifers, which leads to the excretion of large quantities of resin acids in the pine oleoresin, of positive selection value because it leads to faster, cleaner,

healing of accidental wounds; or because it provides a defense against insects; or because it provides symbiotic soil micro-organisms with nutrients? Is it really of positive value? Or is it a neutral—or even an unfavourable—side-effect of some other process, itself of positive selective value? I doubt whether we can yet undertake scientifically the analysis of such problems; but I believe the weight of evidence is overwhelming, that whatever role resin acids play, it is not directly related to specific structures, as there can be a large degree of variability within the "type" resin acid, in intimately related species.

In other cases however, a striking biological activity of one substance apparently signals a possible ecological boost. For instance, the bitter, highly poisonous diterpene alkaloids of *Delphinium* and *Aconitum* are obviously, in our chosen mode of presentation, catabolic products derived from excess *ent*-kaurene; they may well provide the plants containing them with a protection from phytophagous predators (even though my garden *Aconitum* are, to-day, dying from an uninhibited attack of immune Aphids!).

A similar protection might be provided to *Marrubium* by the bitter marrubiin (*15*), or to *Daphne* by the highly irritant mezerein (*32*). But how could we know without proper experimentation whether the sweet stevioside (a diterpene glycoside) tastes like candy or gall to snails or to caterpillars? And all the "toxic" plants mentioned have their many immune specialists predators.

Anyway, we have so far no indication that diterpenes play any widespread role in the important interaction insects/plants, like the sesquiterpenes surely do, except in the trail scent of some termites, which is a cembrene derivative.

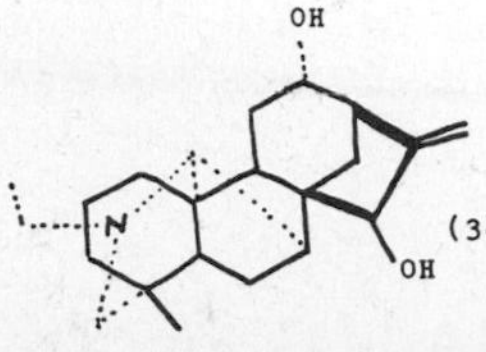

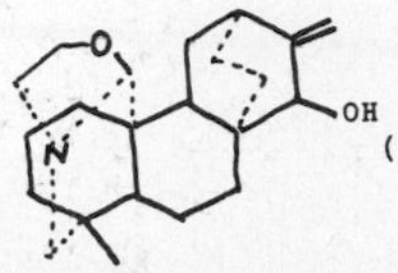

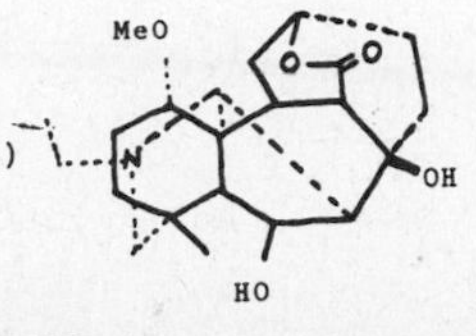

Tetracyclic *ent*-diterpene alkaloids from Ranunculaceae

Gibberellins

Conclusion

I come to the term of this very superficial analysis. With diterpenes, just like with any other class of "secondary metabolites", we are confronted with contradictory facts: The apparent permanence of types, linked with the limited variability of the gross biosynthetic pathways, and leading to relations between the families of substances and the families of plants; and the apparent inexhaustibility of the fancy of Nature, producing not one, not two, not a few variants of the basic type, but a flurry.

I believe we have, not scientifically, maybe not convincingly, and with surely the humble skepticism of one who understands the danger of posing as a "Natural phylosopher", illustrated a system of interpretation converging with that of Birch. A basic, universal, biochemically necessary pathway can, given unbalance, confront the organism with the necessity or with the opportunity of coping with excess material, and leads it to use other general molecular processes to effect physical or chemical disposal: secretion, or functional modifications. Closely related plants, in taxa of good taxonomic status, make often use of closely related processes: e.g. Ericaceae, at least of the sub-family Rhododendroideae, Euphorbiaceae exclusive of Phyllanthoideae (which are "chemically"not Euphorbiaceae), etc.

Finally, we have taken refuge in the contention that, at best, diterpenes may play a role in the ecological selection of organisms, but that this rôle is not obvious. I must conclude by evoking the possibility, anathemous to some, that the variability of diterpenes be sometimes of *no* detectable selective significance whatsoever, and that they be purely neutral, indifferent, characters, happening by necessity, but dispensible with. After all, what is the present evolutionary value of hair in man? (For another comparison, just as useful but less personal, see Summary.) Straight, or kinked; fair, or dark; red or auburn; overabundant or sparse; is hair of negative selection value (as some hairy students complain), of positive selection value (as blondes have taught us to believe), or perfectly neutral (as balding quinquagenarians like to hope)?

References

The structures indicated are all referred to in
1. Hanson, J R, The tetracyclic diterpenes. Pergamon Press, Oxford, 1968.
2. — Recent advances in the chemistry of the tetracyclic diterpenes, in Progress in phytochemistry (ed L Reinhold & Y Liwschitz) p. 165, vol. 1. Interscience, New York, 1968.
3. The chemical society, Specialist periodical reports, Terpenoids and steroids, vol. 1, 1971; vol. 2, 1972.
4. Devon, T K & Scott, A I, Handbook of naturally occurring compounds, vol. 2, Terpenes. Academic Press, New York and London, 1972.
5. Goodwin, T W, Carotenoids, in Phytochemistry (ed L P Miller) vol. 1, p. 112. Van Nostrand and Reinhold, New York, 1973.

Other aspects concerning the distribution of diterpenes have been discussed in
6. Ponsinet, G, Ourisson, G & Oehlschlager, A C, Systematic aspects of the distribution of di- and triterpenes, in Recent advances in phytochemistry (ed T J Mabry, R E Alston & V C Runeckles) p. 271, vol. 1. Appleton-Century-Crofts, New York, 1968.
7. Birch, A J, Biosynthetic pathways in chemical phylogeny. Pure and appl chem 1973, 17, 33.
(and this volume, p. 261).

Biologically active diterpene lactones

Chemical Constitution and Botanical Affinity in *Artemisia*[1]

T. A. Geissman and M. A. Irwin

Department of Chemistry, University of California, Los Angeles, Calif. 90024, USA

Summary

The sesquiterpenoid components of 22 taxa within the Tridentatae group of *Artemisia* have been studied with a view to the possibility of using new chemical characters in assessing botanical affinities. The data provide grounds for suggesting several broad revised groupings within the Section.

The genus *Artemisia* (Family Asteraceae, tribe Anthemideae) comprises about 400 species which occur largely in temperate regions of the northern hemisphere. The genus is ordinarily divided into four sections: Abrotanum, Absinthium, Dracunculus and Seriphidium, but whether these groupings are natural remains uncertain. Certain sub-groups within these Sections can be recognized as belonging to complexes within which certain affinities are readily apparent. Two examples that can be mentioned are the group allied to *A. vulgaris* within the Section Abrotanum, and the Tridentatae (Rydberg) within the Section Seriphidium.

The Section Tridentatae

The Tridentatae comprise somewhat over a dozen species and subspecies characterized by the North American sagebrushes, and make up a plant population of vast scope and range, principally in the Western United States. The selection of this group for examination of their chemical constitution was guided by their recognizable affinities, variety, accessibility, and the fact that most of them are rich in sesquiterpene lactones which include several of the most prevalent structural classes. Seriphidium as a whole and Tridentatae as a class have been studied in detail by several taxonomists [1–4] with general agreement; but the taxonomic difficulties within the Section are considerable. It was

hoped the discovery of new characters describing constitution would provide additional information that would be useful in assessing the taxonomy of the group.

A total of 22 taxa in 54 collections were examined, all of them by chromatographic methods, and 23 collections were subjected to extraction and isolation procedures for the isolation of pure compounds [5]. Thirty-five sesquiterpenes, nearly all lactones, were isolated, most of them new natural products. Included were 7 germacranolides, 10 eudesmanolides (and 3 non-lactonic eudesmane derivatives), 14 guaianolides (and 1 non-lactonic guaiane derivative). Some of these compounds are of little chemotaxonomic significance for they are simple derivatives (acetates, dihydro compounds) of others, but the majority are not related in so direct a manner.

Some of the taxonomic difficulties involved in the study of this Section became apparent when it was discovered that in a few cases collections regarded as the same species were chemically dissimilar, and in one or two others, plants identified as different species were found to be chemically identical by chromatographic criteria. That these observations can be accounted for by infraspecific variation rather than by problems of identification is a possibility, for when one considers the morphological variations within defined taxa, variations in chemical constitution would not be surprising.

Although the results of this study provide only a glimpse of the possibilities of the application of chemical characters to the taxonomic and phylogenetic problems remaining, they do suggest that continued investigation along these lines will be rewarding. While few of the taxa examined were found to contain chemotaxonomic characters that could be regarded as

---

[1] Read by T. A. Geissman.

uniquely definitive, it will be seen that the species studied were found to arrange themselves into several broad groupings whose chemistry was characteristic. The principal aberrations from a clear cut chemotaxonomic picture were found in certain subspecies whose chemical divergence from the species *typica* was often extreme. We do not, however, regard even this comparative wealth of new data as an adequate basis for taxonomic revisions in the Section, but feel that with the accretion of additional information of this kind several such reassessments might be made. It is, of course, recognized that the evaluation of the importance of individual characters in defining the ranks assigned to individual groups is a relative matter, subject to opinion. The chemical constituents of a group, however, are usually new characters and thus offer surer ground for decisions as to rank than does the reevaluation of old characters.

The present studies have shown, first of all, what is to be expected, namely, that there are many points of similarity within the Section taken as a whole, and that its identity as a supraspecific group is, with the possible exception of *A. pygmaea,* consistent with its chemistry. Secondly, the results suggest affinities, on chemical grounds, that differ somewhat from those proposed on grounds of ecological behavior and morphology. Thirdly, the findings support—as do the results of chemical studies on other plant groups—a "splitting" rather than a "lumping" within the group, but, as in most such matters, do not require it.

Holbo & Mozingo's [6] study of the Tridentatae by chromatographic examination of (unidentified) constituents revealed by ultraviolet visualization provides broad support for the generally accepted classification of the group, but led to the suggestion that 3 of the species, *A. pygmaea, A. rigida* and *A. bigelovii* Gray, be excluded. Our examination of *A. pygmaea* supports this conclusion, for the single specimen available to us was entirely lacking in sesquiterpene lactones. It did contain, however, 2 sesquiterpene alcohols, cryptomeridiol and pygmol, and thus is sharply distinguished from all of the other Tridentatae investigated by us.

All of the other Tridentatae examined in this study contain sesquiterpene lactones, none of which, however, is ubiquitous in the Section, but several of which occur in more than one species.

Individual Taxa

A. bigelovii was earlier found [7] to contain arbiglovin, a lactone bearing a close similarity to the matricarins that are characteristic of several species of the Tridentatae. It is worthy of note that *A. bigelovii* is placed in the Seriphidium by Beetle and by Hall & Clements, but placed in Abrotanum by Ward. The presence of matricarin and its deacetyl derivative (as well as a stereoisomer of matricarin) in *A. tilesii* [8] is also of special interest, for this species, placed in Abrotanum, bears synonyms which relate it to the *A. vulgaris* group. Our own studies of *A. vulgaris* and several related species, which will not be described here, has disclosed the presence of no lactones closely related to the guaianolides of the matricarin group. Thus, while *A. pygmaea* can be regarded as (chemically) disjunct from the other members of the Section, the position of *A. bigelovii* is not so clear, and *A. tilesii* is quite distinct from the *vulgaris* group. The inclusion of the latter in Tridentatae would appear to be more justifiable than that of *A. pygmaea*, although it might be argued that the absence of lactones from *A. pygmaea* may be of no more importance than, for instance, the absence of betacyanins from the Caryophyllaceae.

Two species, *A. arbuscula* (ssp *arbuscula* and *thermopola*) and *A. rothrockii* show a clear chemical relationship by their content of the compounds shown in fig. 1, as well as by the absence of guaianolides. In contrast to our observations, however, Bhadane [9] has found arbusculins-A, -B and -C in a number of collections of *A. tridentata* ssp *vaseyana* of western Montana, and the absence of arbusculins in two Montana *A. arbuscula* ssp *arbuscula*. Although, as will be noted below, we found no arbusculins in *A. tridentata* ssp *vaseyana,* our collections of this plant were also clearly different from other *tridentata* subspecies, suggesting that this form deserves specific recognition.

A. tripartita ssp *rupicola* bears little chemical resemblance to *A. tripartita* ssp *tri-*

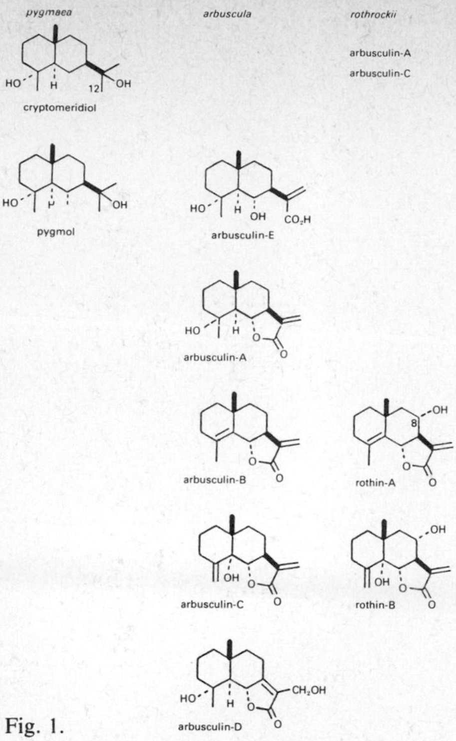

Fig. 1.

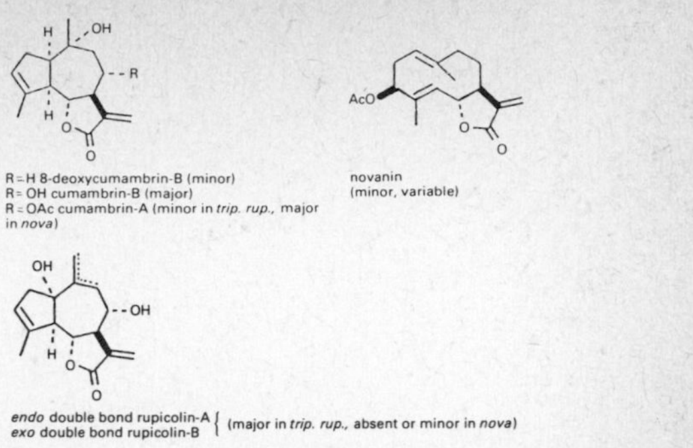

pygmaea — cryptomeridiol

arbuscula

rothrockii — arbusculin-A, arbusculin-C

pygmol

arbusculin-E

arbusculin-A

arbusculin-B

rothin-A

arbusculin-C

rothin-B

arbusculin-D

Fig. 2.

R = H 8-deoxycumambrin-B (minor)
R = OH cumambrin-B (major)
R = OAc cumambrin-A (minor in *trip. rup.*, major in *nova*)

novanin (minor, variable)

endo double bond rupicolin-A
exo double bond rupicolin-B } (major in *trip. rup.*, absent or minor in *nova*)

terized by the presence of one or more of matricarin, deacetoxymatricarin and deacetyl-matricarin. These compounds, of which only 1 or 2, and occasionally all 3 could be found in a given specimen, are not found throughout the Section, and are distinctive of the taxa mentioned. Matricarin, it should be noted, also occurs in *Matricaria* [10] and, as mentioned earlier, in *A. tilesii*.

Chemistry and Botanical Affinity

There can be no doubt that the large and structurally diverse group of lactones occurring in the plants of this Section are biosynthetically related. Indeed, one of the more interesting consequences of these studies has been the recognition that the compounds can be placed into rational sequences of structural elaboration by biosynthetic processes of which epoxidation appears to play a prominent role [11]. It is therefore not surprising to find certain compounds, which appear to occupy intermediate

partita, but shows a close similarity to *A. nova* (fig. 2). Both taxa, and no other Tridentatae examined in this study, contain cumambrins-A and -B, 8-deoxycumambrin-B, and novanin. The fact that *A. tripartita* ssp *rupicola* alone was found to contain 6 additional lactones (cumambrin-B oxide, rupin-A and -B, ridentin, ridentin-B and artecalin) might be attributed to the much greater effort spent on the species. It should be noted also that thin-layer chromatograms of total extracts of this species showed only a few of this large number of constituents, indicating the shortcomings of t.l.c. as a basis for taxonomic characterization. One collection of *A. tripartita* ssp *rupicola* regarded as an ecotype (a "shade form") was found to be chemically identical with 4 *A. arbuscula* ssp *arbuscula* specimens, an indication either of the gross morphological similarities existing between some of these taxa, of environmental effects, or of difficulties in identification.

A. tripartita ssp *tripartita* showed a close chemical similarity to *A. tridentata* (two subspecies), and *A. cana* (two subspecies). These plants, along with *A. argilosa*, were charac-

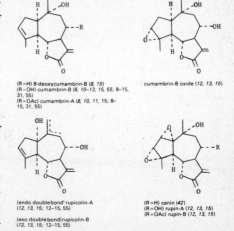

(R = H) 8-deoxycumambrin-B (8, 15)
(R = OH) cumambrin-B (8, 10–13, 15, 55; 8–15, 31, 55)
(R = OAc) cumambrin-A (8, 10, 11, 15; 8–15, 31, 55)

cumambrin-B oxide (12, 13, 15)

(endo double bond) rupicolin-A (12, 13, 15; 12–15, 55)

(exo double bond) rupicolin-B (12, 13, 15; 12–15, 55)

(R = H) canin (42)
(R = OH) rupin-A (12, 13, 15)
(R = OAc) rupin-B (12, 13, 15)

Compounds (From plant numbers given; *isolated or identified*):

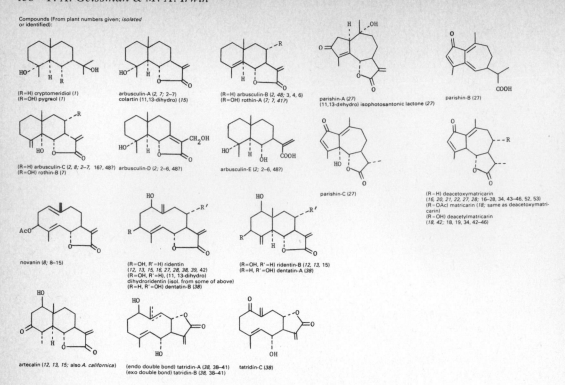

(R=H) cryptomeridiol *(1)*
(R=OH) pygmaol *(1)*

arbusculin-A *(2, 7; 2–7)*
colartin (11,13-dihydro) *(15)*

(R=H) arbusculin-B *(2, 48; 3, 4, 6)*
(R=OH) rothin-A *(7; 7, 41?)*

parishin-A *(27)*
(11,13-dehydro) isophotosantonic lactone *(27)*

parishin-B *(27)*

(R=H) arbusculin-C *(2, 8; 2–7, 16?, 48?)*
(R=OH) rothin-B *(7)*

arbusculin-D *(2; 2–6, 48?)*

arbusculin-E *(2; 2–6, 48?)*

parishin-C *(27)*

(R=H) deacetoxymatricarin
(16, 20, 21, 22, 27, 28; 16–28, 34, 43–46, 52, 53)
(R=OAc) matricarin *(18;* same as deacetoxymatricarin)
(R=OH) deacetylmatricarin
(18, 42; 18, 19, 34, 42–46)

novanin *(8; 8–15)*

(R=OH, R'=H) ridentin
(12, 13, 15, 16, 27, 28, 38, 39, 42)
(R=OH, R'=H), (11, 13-dihydro)
dihydroridentin (isol. from some of above)
(R=H, R'=OH) dentatin-B *(38)*

(R=OH, R'=H) ridentin-B *(12, 13, 15)*
(R=H, R'=OH) dentatin-A *(38)*

artecalin *(12, 13, 15;* also *A. californica)*

(endo double bond) tatridin-A *(38, 38–41)*
(exo double bond) tatridin-B *(38, 38–41)*

tatridin-C *(38)*

positions in these presumptive synthetic processes, occurring more or less randomly throughout the Section. The compound ridentin, for instance, was found in collections of *A. tripartita* ssp *rupicola, A. tripartita* ssp *tripartita,* in several of the *A. tridentata* subspecies, and in *A. cana* ssp *cana.* Ridentin can be regarded as a product formed by simple steps proceeding from novanin, and only *A. tripartita* ssp *rupicola* and *A. nova* contain the latter. Thus, *A. tridentata, A. tripartita* ssp *tripartita* and *A. cana* appear to possess additional oxidative capacity in common, for the eventual formation of the matricarins can be most satisfactorily regarded as the result of successive stages of dehydration and epoxidation starting at cumambrin, the most characteristic constituent of *A. tripartita* ssp *rupicola* and *A. nova,* but absent in the matricarin-containing plants and in *A. arbuscula* and *A. nova.* (Because the cumambrins (like the rupins and the matricarins) are compounds with 8-H, 8-OH or 8-OAc, reference to them will occasionally be made by the use only of the parent name.)

Occasionally, unique and distinctive compounds are found in single specimens of taxa whose other constituents are the expected ones. One collection of *A. tridentata* ssp *tridentata* and three collections of *A. tridentata* ssp *vaseyana* f. *spiciformis* contained the new lactones tatridin-A and -B, which were observed in no other plants of those studied. The fact that this specimen of *A. tridentata* ssp *tridentata* was also devoid of matricarins raises the question of its identity, and the near identity of chromatograms of these four specimens makes it likely that they are the same, i.e., ssp *vaseyana* f. *spiciformis.*

For practical reasons it was not feasible to examine all of so large a number of collections in complete detail, with the isolation of all lactonic constituents, and so negative results obtained by t.l.c.—the lack in one sample of a compound found in another—should not be given too much weight. As an example, only one collection of *A. tridentata* ssp *tridentata* f. *parishii* yielded a group of novel lactones, one of which was the known (by earlier synthesis) isophotosantonic lactone and the others modifications of the matricarins (parishins-A, -B and -C).

The finding of isophotosantonic lactone in a

member of the Seriphidium raises a point of interest. The best and longest known of the sesquiterpene lactones is santonin, a constituent of an Old World group of *Artemisia* species belonging to the Section Seriphidium, and typified by *A. maritima*. The presence of isophotosantonic lactone, a well known photo-isomerization product of santonin, in a member of the Section might lead to the supposition that it is an artefact, having been derived in vivo from santonin by photochemical change in the plant. This conjecture can be rejected, for in none of the North American members of the Section has any trace of santonin been discovered. From our experience with authentic santonin-containing species its presence would surely have been detected if it had been present in the specimens used in our studies. Indeed, the biosynthetic formation of isophotosantonic lactone can be readily formulated as arising by way of a simple acid-catalyzed rearrangement of a 4,5-epoxy guaianolide. It is significant to note that accompanying isophotosantonic lactone in our specimen was the corresponding 11,13-dehydro lactone. The related 11,13-dehydrosantonin is not known in nature.

Conclusions

The summary of our observations is that in the species of Tridentatae included in our study there appear to be 3 groups characterized by the structures of their principal lactonic constituents:

(1) *A. arbuscula* and *A. rothrockii* contain only simple eudesmanolides with 11-methylene groups (arbusculins, rothins).

(2) At the other extreme are found *A. tridentata, A. tripartita* ssp *tripartita, A. cana,* and *A. argilosa,* which contain the highly oxidized matricarins containing 11-methyl groups, as well as less oxidized guaianolides and eudesmanolides.

(3) *A. tripartita* ssp *rupicola* and *A. nova* occupy an intermediate position. The principal constituents, often present in surprisingly large amount, are the cumambrins. Although no matricarins were found in these taxa, they are rich in a variety of lactones, including germacranolides, eudesmanolides, and guaianolides, and thus appear to span the intermediate range

from the relatively simple arbusculins to the structurally more complex matricarins.

As would be anticipated in a group of plants so clearly allied by accepted taxonomic criteria, a number of the lactones were found to be distributed over several of the species. Two collections, one of *A. tripartita* ssp *tripartita* and one of *A. cana* ssp *viscidula,* were found to contain, along with their characteristic matricarins, small amounts of arbusculins (by t.l.c.). Arbusculin-B was isolated from the latter.

The only species regarded as a member of the Tridentatae that was entirely distinct from the others was *A. pygmaea* which was found to contain no lactones. This is in accord with its early assignment to a separate Section [12], and with the conclusion of Holbo & Mozingo that it be excluded from the Tridentatae.

Beetle's [1] extensive study of the Tridentatae led him to suggest that there are 2 well-defined sub-sections. In one, *A. cana* and *A. tripartita* are closely related, but *A. rigida* is disjunct. In the other, *A. tridentata, A. nova, A. bigelovii,* and *A. longiloba* are closely related but *A. pygmaea* is disjunct. We were unable to devote sufficient attention to *A. longiloba* and *A. rigida* to establish their chemical relationships in the Section, but chromatographic study of extracts of these plants disclosed no similarities to the other taxa examined. It is clear from these chemical studies that the groupings suggested by Beetle bear little or no relationship to their lactonic constituents, and that the chemical findings support quite another grouping. Although it would be rash to suggest that chemical characters be given such weight as to use them as a basis for regrouping these taxa, it is recognized that natural systems should reflect evolutionary relationships [13] and it can be suggested that chemical characters are perhaps the most direct reflection of enzymatic (thus, genetic) systems. The difficulties arise when it is recalled that chemical characters in single "species" can differ in diploid and polyploid plants [14], in new growth and mature growth [15], and can vary as a result of factors that can be described as "ethological" [16].

Finally, it should be recalled that certain chemical constituents are found in quite unallied genera, tribes, or even families. Cumambrin-B, for example, is found in *Ambrosia* and *Artemisia*. The lactone psilotropin is found in *Psilostrophe* (Helenieae) and its stereoisomer (epimeric at the 8-position) in *Geigera* (Inuleae) [17]. If it were to be argued that this stereoisomerism is an important distinction, it will be recalled that isomers differing only in this way have also been found in the same plant (aromatin and aromaticin). Matricarin is found in *Matricaria*, *Achillea* and *Artemisia*. Costunolide is found in Asteraceae and Magnoliaceae. Numerous other examples of this kind are known, particularly among the alkaloids.

Species studied: (Identification or authentication by Drs A. A. Beetle and R. O. Asplund (University of Wyoming), G. H. Ward (Knox College).)

1	*A. pygmaea* Gray
2–3	*A. arbuscula* Nutt. ssp *arbuscula*
4–5	*A. arbuscula* Nutt. ssp *thermopola* Beetle
6	*A. tripartita* Rydb. ssp *rupicola* Beetle ("shade form")
7	*A. rothrockii* Gray
8–11	*A. nova* Nels
12–15	*A. tripartita* Rydb. ssp *rupicola* Beetle
16–17	*A. tripartita* Rydb. ssp *tripartita*
18–24	*A. tridentata* Nutt. ssp *tridentata*
25–28	*A. tridentata* Nutt. ssp *tridentata* f. *parishii*
29–30	*A. arbuscula* Nutt. ssp *arbuscula* (?)
31–33	*A. tridentata* Nutt. ssp *wyomingensis* Beetle and Young
34–37	*A. tridentata* Nutt. ssp *vaseyana* (Rydb.) Beetle
38	*A. tridentata* Nutt. ssp *tridentata*
39–41	*A. tridentata* Nutt. ssp *tridentata* f. *spiciformis* (Ost.) Beetle
42–45	*A. cana* Pursh ssp *cana*
46–48	*A. cana* Pursh ssp *viscidula* (Ost.) Beetle
49	*A. cana* Pursh ssp *bolanderi* (Gray) Ward
50	*A. longiloba* (Ost.) Beetle
51	*A. rigida* Nutt. Gray
52–53	*A. argilosa* Beetle
54	*A. palmeri* Gray
55	*A. tripartita* Rydb. ssp *rupicola* Beetle

Species	Common form and habit
A. bigelovii	dwarf; 2–3 dm high
A. pygmaea	dwarf, cushion
A. nova	low shrub; 1–3 dm high
A. arbuscula arbuscula	dwarf shrub; to 5 dm high
A. arbuscula thermopola	dwarf shrub; to 3 dm high
A. tridentata tridentata	erect shrub; 1–2 m high
A. tridentata tridentata parishii	erect shrub; occas. to 3 m high
A. tridentata vaseyana	shrub; to 1 m high
A. tridentata vaseyana spiciformis	large leaves
A. longiloba	dwarf; to 3 dm high
A. rothrockii	low shrub; 2–8 dm high
A. cana cana	erect; to 1.5 m high
A. cana viscidula	erect; under m high
A. cana bolanderi	low rounded shrub; 3–6 dm high
A. rigida	low shrub; 1–3 dm high
A. tripartita tripartita	erect; to 2 m high
A. tripartita rupicola	dwarf shrub; to 1.5 dm high
A. argilosa	erect shrub; 5–8 dm high

References

1. Beetle, A A, A study of sagebrush. Univ Wyoming agr exp stn bull 1960, 368.
2. Beetle, A A, Rhodora 1959, 61, 82.
3. Ward, G H, Contr Dudley herbarium 1953, 4, 155.
4. Hall, H M & Clements, F E, Carnegie inst Wash publ 1923, 326, 1.
5. Irwin, M A, PhD dissertation. Univ of California, Los Angeles, 1971.
6. Holbo, H R & Mozingo, H N, Amer j bot 1965, 52, 970.
7. Herz, W & Santhanam, P S, J org chem 1965, 30, 4340.
8. Herz, W & Ueda, K, J Am chem soc 1961, 83, 1139.
9. Bhadane, S. Priv communication.
10. Cekan, Z, Prochazka, V, Herour, V & Šorm, F, Coll Czech chem comm 1959, 24, 1994.
11. Geissman, T A, Proc XI ann symp phytochem soc North America, Monterey, Mexico, October, 1971. Academic Press, New York 1973.
12. Rydberg, P A, N Amer Flora 1916, 34, 244.
13. Cronquist, A, The evolution and classification of flowering plants. Houghton Mifflin Co, Boston, 1968.
14. Geissman, T A & Matsueda, S, Phytochemistry 1968, 7, 1613.
15. Lee, K H, Matsueda, S & Geissman, T A, Phytochemistry 1971, 10, 405.
16. Payne, W W, Geissman, T A, Lucas, A J & Saitoh, T, Biochem systematics 1973, 1, 21.
17. Anderson L A P, deKock, W T, Pachler, K G & Brink, C V, Tetrahedron 1967, 23, 4153.

Discussion

Mabry: We have observed that for many Compositae genera the chemically less complex sesquiterpene lectones and flavonoids as well as fewer numbers of compounds altogether appear in what are considered on classical morphological grounds to be the most advanced members of the genus under investigation. If this phenomenon also operates in *Artemisia* then *A. pygmea* might be an advanced taxon relative to the other members of the genus. I would be interested in yours and Professor Cronquist's views with regard to this matter.

Geissman: In my experience, the simplest structures are more likely to be widely distributed across a number of families, genera or species. It would appear that the synthetic capabilities of plants at the simple structural level are less likely to reflect individual genetic individuality, and that the development of such individuality is accompanied by increasing complexity in structural elaboration.

Lavie: Would you see the different compounds described in your presentation and being present in large number in certain species as being the different steps of oxidative transformation or degradation sequences. In each case then the different combinations of compounds observed should represent the accumulation of certain specific products due to different enzymatic combinations and systems operating in this oxidative sequence.

Geissman: I agree. This is the subject of a different lecture. I should add that although I have selected for this Symposium to discuss the chemical-botanical relationships, our most interesting and to us significant findings are those that reveal probable biosynthetic relationships between the lactones present in related taxa of separate sections or subsections. We have in fact discussed this in a paper published about a year ago.

Herout: After all observations it seems to me that the first step in producing sesquiterpenoid γ-lactones in Compositae is the oxidizing of isopropyl (head) end of the farnesol-pyrophosphate precursor just to the γ-lactone entity. Thus, *Artemisia pygmaea* may miss the enzyme system necessary to such an oxidation. As a remark, only plants of Senecioneae tribes seems to use another biosynthetic pathway in the synthesis of eremophilanolides: the intermediate step is very probably a furano-derivative (autooxidation of furanoeremophilane derivatives to eremophilanolides was demonstrated).

As a summary of Professor Geissman's lecture I can state that all his findings fit well

with the theory that *Artemisia* species (as a member of Anthemidae tribus) produces lactones of germacrane→eudesmane or guaiane type; not a single pseudoguaianolide derivative was observed yet.

The last remark: the first "exception" from such a presumption was recently described. Professor Stefanovic (Beograd) isolated a cadinanolide from *A. annua*. This is the first lactone belonging to the carbon skeleton found in the Compositae family.

Geissman: There are several hundred species of *Artemisia,* and we have examined less than 20% of these. So far we have not observed pseudoguaianolides, cadinanolides, xanthonolides and modified lactones related to these.

Cronquist: It is not surprising that the chemical data in this group or *Artemisia* present a complex pattern that is not easily correlated with the existing classification. As a student of the Asteraceae and of western American flora I have had to pay some attention to the *Artemisia tridentata* group. This group has defied the attentions of taxonomists for many years. In 1923 Hall & Clements reduced the plethora of then existing names to a small number of accepted taxa. Their treatment was an improvement over the past, but it may have lumped too much. In the early 1950's George Ward did a little more splitting, and in the 1960's Alan Beetle carried the splitting still further. I am not fully satisfied with any of these treatments, and I am not sure that the group is susceptible to a satisfactory treatment. The pattern of variation is highly complex. In some cases one can see a population, and know that it is different from others; yet it is most difficult to put down a set of features that will enable even one's self to identify other specimens in the absence of reliable data on habits and geographic origin. The situation is complicated by polyploidy. At some places one can move in one step from a population of one taxon to a population of another taxon, and have no trouble drawing the distinction. In at least some such cases the adjacent populations are at different ploidy levels. In other places two things occur together and intergrade so freely that any distinction is very hard to draw. Therefore I am not surprised that the identification of some of your specimens was changed after the chemical analysis was made.

Within the next two or three years I will have to look at this group again, because I must prepare a treatment for the Intermountain Flora. This time I will have the chemical as well as other data to consider, but I am not confident that I will be able to offer any improvement on existing treatments.

One thing that does show clearly in the chemical data is the separation of *Artemisia pygmaea* from the *A. tridentata* complex. This is not surprising, because in my opinion *A. pygmaea* does not belong to the *A. tridentata* group. Unfortunately I cannot offer any phylogenetic interpretation in response to Dr Mabry's question. *A. pygmaea* is taxonomically rather isolated; I don't know what to put it with, within the genus. The members of the *A. tridentata* group are so closely related among themselves that I hesitate to suggest what might be more primitive and more advanced.

Geissman: I hope that further study of these chemical data will help Dr Cronquist in his reevaluation of this complex genus. We have additional information on other sections (e.g. the *vulgaris* complex) in which he will find further interest.

Weimarck: Is the variation of chromatographically separated spots a variation between sister plants of a population or between geographically distant populations?

Could chromatographic technique obscure or contribute to variation? It would be most interesting from a botanical point of view to continue that part of the investigation.

Geissman: We do not care to rely upon chromatographic patterns in studies of this kind. Different compounds can occupy the same region of a chromatogram, and many minor components are often not observed at all.

Herz: (in reply to Dr Weimarck). I would like to second Professor Geissman's remarks on the difficulty of identifying plants of this group only on the basis of thin-layer chromatograms of their sesquiterpene lactones. We are currently studying the genus *Liatris*, a fairly large genus of North America. Several well-defined species

exhibit very similar, if not identical sesquiterpene lactone profiles on t.l.c. However, when the components of the t.l.c. are isolated, they are found to differ chemically.

Cronquist: If your material of *A. rothrockii* really came from Wyoming, the name probably does not apply. The type of *A. rothrockii* came from the Sierra Nevada, and according to George Ward's interpretation which I think is probably correct, this taxon is largely confined to the Sierra Nevada. Of course you have to take the names as given to you by Dr Beetle, but I mention this item as a further example of the taxonomic difficulty of the group.

Geissman: We have relied upon the authority of the botanist who provided our material, and could provide no independent assessment of our own.

Hegnauer: (1) Does sesquiterpene lactone chemistry not change during the development of plants during one season?
(2) *Artemisia tridentata* is one of the Californian plants reported in literature (Muller) to inhibit the development of annual plants in its surroundings (allelopathy). Do you know if there is a relationship between the lactones present in a given taxon of the aggregate species *A. tridentata* and its allelopathic potency, in other words whether some of the C_{15}-lactones are engaged in allelopathy?

Geissman: In answer to the first question, indeed it does. In several cases we have observed a dramatic qualitative difference in the chemical constitution of seedlings and mature plants of the same species (and in one case, of young and mature growth from the same rootstock). As to the second question, I have discussed this with Dr Muller but so far no systematic studies of the allelopathic properties of these compounds has been carried out.

Cronquist: According to present interpretation, several of the taxa in the group contain both diploid and tetraploid forms. When the taxa occur together at the diploid level, they can probably be distinguished fairly readily. When they occur together at the tetraploid level it is likely to be much more difficult. When diploids and tetraploids occur together the distinction may be very clear, but it may not always conform to the taxonomic pattern.

Geissman: We are aware of the possible complications of unrecognized polyploidy. However, the chemist is often required to rely upon the botanist who provides or authenticates his material. A further complication is almost insuperable: it is that proper chemical examination usually requires a considerable quantity of plant material. Consequently, collection of a large number of plants from an extensive population may well include individuals that differ in detail due to ploidy or other factors. For the separate study of individuals of a population, reliance would have to be placed on chromatographic methods, and these have grave shortcomings when complex mixtures of compounds, some of them novel, are present.

Chemotypen von *Chrysanthemum vulgare* (L.) Bernh.[1]

K. Forsén & M. v. Schantz

Abteilung für Pharmakognosie der Universität Helsinki, Helsinki, Finland

Summary

In careful investigations of about 600 wild growing tansy plants from different localities in Finland the following chemotypes, named according to their main component, were detected: champhor type, thujone type, borneol type, chrysanthenyl acetate type, chrysanthemum-epoxide type, umbellulone type, artemisiaketone type, cineol type, isopinocamphone type, monoterpene hydrocarbon type and sesquiterpene type.

The chemotypes are not only characterized by their specific main components, but also by typical accompanying components.

The distribution of the oil components in the progeny obtained by crossing between thujone and champhor types, were investigated. Every step in the biosynthesis is genetically controlled. A schematic model for gene control and interactions in the biosynthesis of thujone, sabinene and camphor was presented. Crossings between thujone type and camphor type resulted primarily in offspring of the thujone type. Some new chemotypes were even detected among the progeny. The camphor type is, however, the most common type in Finland.

Der Rainfarn, *Chrysanthemum vulgare* (L.) Bernh. (*Tanacetum v.* L.), ist sowohl in Europa als auch in Amerika weit verbreitet und kommt häufig an sandigen Standorten vor. Die Blütenkörbchen enthalten ein ätherisches Öl, das in der Pharmazie als Wurmmittel gebraucht wird. In letzter Zeit ist das Interesse an der Zusammensetzung des ätherischen Öles gross geworden, da gaschromatographische und spektrometrische Methoden eine genaue Analyse von Rainfarnölen einzelner Individuen ermöglicht haben. Es wurde nachgewiesen, dass der morphologisch sehr einheitliche Rainfarn auf Grund chemischer Merkmale in eine Anzahl chemischer Taxa eingeteilt werden kann. Da die Nomenklaturfragen und die taxonomische Bedeutung dieser Taxa von Tetenyi [1] schon eingehend behandelt worden sind, beschränke

ich mich auf die nähere Charakterisierung sowie auf die Ergebnisse von Kreuzungen der verschiedenen chemischen Taxa von *Chrysanthemum vulgare* (L.) Bernh. Auch wenn diese Taxa genetisch einheitlich sind, wird die systematische Einheit (forma, varietas, Rasse etc.) im folgenden nicht näher spezifiziert. Der mehr neutrale Ausdruck Chemotyp wird gebraucht, und die Chemotypen werden nach dem Hauptbestandteil des ätherischen Öles genannt.

Die Entdeckung verschiedener Chemotypen von *Chrysanthemum vulgare* (L.) Bernh. hat mit den Arbeiten von Rudloff [2], Stahl & Schmitt [3], Järvi [4], v. Schantz & Järvi [5] begonnen. Diese Arbeiten weisen uns darauf hin, dass in der Natur einzelne Individuen mit ätherischem Öl von verschiedenartiger Zusammensetzung in den Blütenkörbchen vorkommt. Ohne weiteres ist es aber nicht klar, dass ein Individuum mit abweichender chemischer Zusammensetzung einen gewissen Chemotyp repräsentiert. Es ist notwendig zu beurteilen, ob die Zusammensetzung des Öles einen konstanten Charakter für ein gewisser Chemotyp darstellt und nicht nur zufällig variiert. Von Interesse war erstens zu untersuchen, ob diese Zusammensetzung auch bei Standortwechsel konstant bleibt, ob verschiedene Düngung sie beeinflusst und ob ontogenetische Faktoren eine Änderung dieser Zusammensetzung hervorrufen. Um diese Fragen zu beantworten, wurde Material von etwa 600 verschiedenen Standorten Finnlands eingesammelt und auf einem Versuchsfeld in der Nähe von Helsinki ausgepflanzt. Auf Grund dieses Materials wurden später Kreuzungen zwischen verschiedenen Chemotypen gemacht und die Zusammensetzung des ätherischen Öles aus den Blütenkörbchen der Nachkommen eingehend studiert, um das Entstehen von verschiedenen Chemotypen besser kennenzulernen.

---

[1] Read by M. v. Schantz.

Die Zusammensetzung des ätherischen Öles verschiedener ausgepflanzter Chemotypen wurde jetzt 7 Jahre lang verfolgt, und weder qualitative noch signifikante quantitative Änderungen konnten festgestellt werden. Der Ölgehalt aber kann unter besonderen äusseren Bedingungen verschieden sein. Bei Düngungsversuchen konnte stärkeres Wachstum beobachtet werden, und Schwankungen des Ölgehalts traten öfters auf, aber nie signifikante Änderungen in der Zusammensetzung des Öles.

Während der Ontogenese wurde beobachtet (vgl. [6]), dass in kleinen, noch grünen Blütenkörbchen die quantitative Zusammensetzung des ätherischen Öles der Blütenkörbchen nicht der der vollentwickelten entspricht. Besonders deutlich ist das in Bezug auf den Hauptbestandteil zu erkennen. Doch muss gesagt werden, dass die endgültige Zusammensetzung des Öles schon vor dem Öffnen der Blütenkörbchen erreicht wird und bis über die Vollblüte hinaus konstant bleibt. Von Rudloff & Underhill [7] haben nicht bedeutende Schwankungen in Gehalt an Hauptkomponente des ätherischen Öles beobachtet, aber wahrscheinlich nicht gleich alte Blütenkörbchen studiert. Einige der unter 10% Anteil vorkommenden Komponenten wiesen doch erhebliche Variation auf.

Verbreitung der bisher bekannten Chemotypen von *Chrysanthemum vulgare* (L.) Bernh.

In kommerziellem Rainfarnöl findet man (+)-iso-Thujon als Hauptbestandteil. Nach v. Rudloff [2] und v. Rudloff & Underhill [7] wurden in Rainfarn amerikanischer Herkunft nur (+)-iso-Thujon und (–)-Thujon als Hauptbestandteile des Öles gefunden. Auch in Europa kommt (+)-iso-Thujon öfters als Hauptbestandteil des Rainfarnöles vor. Ausser dem Thujontyp ((+)-iso-Thujon und (–)-Thujon) haben Stahl & Schmitt [3] in mitteleuropäischem Material den Campher-Typ mit (–)-Campher und Camphen als Hauptbestandteile sowie einen Terpen-Typ mit β-Pinen als Hauptbestandteil gefunden. Weiter wurden ein Sesquiterpentyp und ein Estertyp erwähnt. Später sind ein Artemisiaketontyp, ein Umbellulontyp und ein Chrysanthemumepoxidtyp gefunden worden [8–11].

Das Studium des Vorkommens der verschiedenen Chemotypen in Finnland zeigte, dass der Campher-Typ am weitesten verbreitet ist. Dieser Chemotyp kommt von der Südküste bis Lappland vor. Auch im Schärenhof ist er der häufigste. Der in Amerika und Europa am weitesten verbreitete Thujon-Typ kommt in Finnland nur im südlichsten Teil des Landes ziemlich häufig vor. Ausser dem Camphertyp wurde der Chrysanthenylacetattyp (von Scheu [11] als Monoterpenester-Typ erwähnt) zwar ziemlich selten, aber auch bis Lappland verbreitet gefunden. Insgesamt machten diese drei Typen etwa 75% des untersuchten Materials aus. Weiter wurde ein Borneoltyp mit entweder Borneol oder Bornylacetat als Hauptbestandteil, ein 1,8-Cineoltyp, ein Chrysanthemumepoxidtyp, ein Artemisiaketontyp, ein Umbellulontyp, ein Isopinocamphontyp, ein Monoterpenkohlenwasserstofftyp mit entweder α-Pinen oder γ-Terpinen, nicht aber mit Camphen als Hauptbestandteilen, und ein Sesquiterpentyp mit einigen noch unbekannten Sesquiterpenkohlenwasserstoffen als Hauptbestandteilen und ein noch nicht entgültig untersuchten Chemotyp mit dem Sesquiterpenderivat $C_{15}H_{24}O_2$ als Hauptbestandteil (vgl. auch [12, 13]) gefunden. Etwa 15% des ganzen Materials stellten sog. Mischtypen dar, die keine Klassifizierung in Chemotypen ermöglichten.

Die Zusammensetzung des ätherischen Öles einiger Chemotypen

Bei der weiteren Untersuchung der Chemotypen wurde beobachtet, dass ein Chemotyp nicht nur durch den Hauptbestandteil charakterisiert ist, sondern jeder Chemotyp enthält auch besondere, für ihn charakteristische Begleitstoffe, die immer zusammen mit dem Hauptbestandteil auftreten. Ausser diesen Bestandteilen wurde in jedem Chemotyp ubiquitäre Stoffe gefunden, die mehr oder weniger häufig in allen Chemotypen vorkommen. Besonders bei der Analyse der Nachkommen von Kreuzungen wurde die genaue Kenntnis der Nebenbestandteile von grosser Bedeutung. Die übliche Methode, die Peaks der Gaschromatogramme durch Retentionszeiten und Zumischen von Vergleichstoffen zu identifizieren, gab keine eindeutigen Ergebnisse. Deshalb wurde jeder Bestandteil noch massenspektrometrisch untersucht. Es wurde gefunden, dass öfters im selben Peak viele Verbindungen zu finden waren. Die Zu-

Tabelle 1. *Zusammensetzung der ätherischen Öle*

Peak	Substanz	Campher-Typ	Thujon-Typ	Sabinen-Typ	Chrysanthenyl-acetat	Chrysanthemum-epoxid
1	Tricyclen	0,2	–	0,1	–	–
1 a	α-Pinen	0,6	0,7	1,0	0,1	–
2	Camphen	3,4	0,3	1,1	–	14,0
3	β-Pinen	0,6	0,3	0,2	–	0,6
3 a	Sabinen	0,8	5,4	36,5	0,1	–
6	α-Terpinen	0,1	0,3	2,1	–	–
7	1,8-Cineol	3,2	1,9	3,6	–	5,4
7 a	Chrysanthemum-epoxid	–	–	–	–	48,9
8	γ-Terpinen	0,4	0,7	4,3	0,1	–
9	p-Cymol	0,6	0,2	0,9	0,1	–
9 a	Terpinolen	–	0,1	0,8	–	–
10 a	Unbekannt	–	–	–	–	3,8
11 a	(–)-Thujon	0,2	0,5	–	–	–
11 b	(+)-iso-Thujon	0,1	81,0	21,7	–	–
12	Chrysanthenol	–	–	–	1,5	–
13	Unbekannt	Spuren	–	–	–	–
14	Campher	64,0	0,7	11,5	0,3	11,2
14 a	Chrysanthenylacetat	–	–	–	81,6	–
15 a	Pinocarvon	0,4	–	–	–	–
16	Bornylacetat	3,5	–	–	–	–
16 a	Chrysanthenylpropionat	–	–	–	1,7	–
16 b	Chrysanthenylisobutyrat	–	–	–	1,0	–
17	Terpinen-4-ol	0,9	1,2	6,6	–	0,6
17 b	β-Caryophyllen	0,5	–	0,7	–	–
17 c	$C_{11}H_{18}O_2$	0,5	–	0,7	–	–
18	$C_{15}H_{24}$	0,4	–	–	–	Spuren
18 a	$C_{15}H_{24}$	0,4	–	–	–	–
19	$C_{10}H_{14}O$	0,4	–	–	–	–
19 a	$C_{12}H_{18}O_2$	0,3	–	–	–	–
20	Thujylalkohol	–	0,4	–	–	–
20 a	Chrysanthenylvalerat	–	–	–	0,6	–
21	Borneol	1,8	–	–	–	Spuren
22	Cadinen	5,3	2,4	1,9	2,8	Spuren
23	Carvon	1,3	–	–	–	–
24	Monoterpenalkohol $C_{10}H_{16}O$	7,2	–	0,9	–	–

sammensetzung des ätherischen Öles jedes Chemotypes wurde analysiert, Massenspektren von jedem Peak genommen, und in unsicheren Fällen wurden die Stoffe präparativ isoliert und davon noch Massenspektren, IR-, UV- und NMR-Spektren aufgenommen.

Im Öl des Camphertyps wurden 25 Bestandteile identifiziert [14]. Der Camphergehalt des ätherischen Öles des Campher-Typs variiert in den Ölen der verschiedenen untersuchten Individuen im allgemeinen zwischen 27 und 76% [15]. Für diesen Chemotyp sind als Begleitstoffe besonders Camphen und Tricyklen zu erwähnen.

Weiter gehören Borneol und Bornylacetat zu den Begleitstoffen dieses Typs; sie sind zumeist in relativ grossen Mengen vorhanden (vgl. Tabelle 1). Ein Monoterpenalkohol, wahrscheinlich Camphenol, dessen Struktur noch nicht endgültig aufgeklärt ist, kommt in Mengen bis 10% im Öl dieses Chemotyps vor.

Im Öl des Thujontyps wurden 15 Bestandteile identifiziert. Der Thujongehalt im ätherischen Öl der verschiedenen untersuchten Individuen des Thujontyps variiert zwischen 37 und 96%. Für diesen Chemotyp ist der hohe Gehalt an Hauptbestandteil charakteristisch. Damit wird

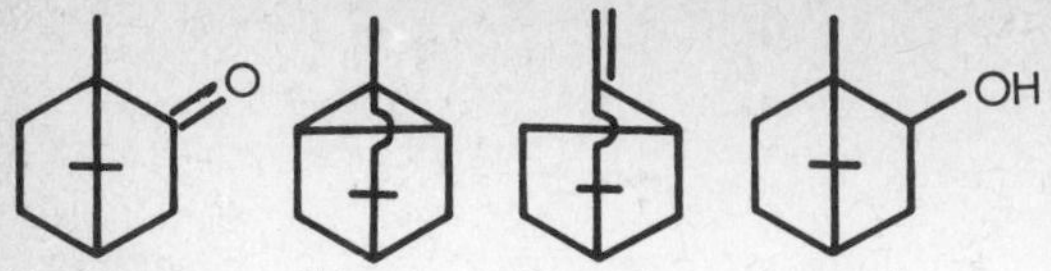

				Chrysanthenol	R=H
				Chrysanthenylacetat	R=CH₃CO
				Chrysanthenylpropionat	R=C₂H₅CO
				Chrysanthenylbutyrat	R=C₃H₇CO
Campher	Tricyclen	Camphen	Borneol	Chrysanthenylvalerat	R=C₄H₉CO

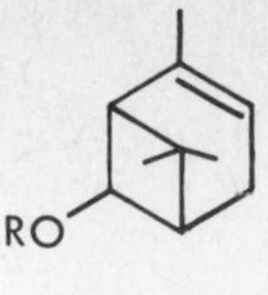

der prozentuelle Anteil der Begleitstoffe klein, und auch bei Benutzung von empfindlichen gaschromatographischen Methoden in Kombination mit Massenspektroskopie wurden viel weniger Begleitstoffe identifiziert als bei den Camphertypen. Besonders typisch für den Thujontyp sind Thujylalkohol und Sabinen. Beide kommen in ziemlich kleinen Mengen vor, aber sind in allen untersuchten Exemplaren des Thujontyps zu finden. Campher und Camphen treten im allgemeinen in sehr geringen Menge auf oder diese Komponenten fehlen ganz.

Im Öl des Chrysanthenylacetattyps wurden insgesamt 11 Bestandteile identifiziert. Der Gehalt an Chrysanthenylacetat im Öl der verschiedenen untersuchten Individuen variiert zwischen 32 und 78%. Dieser Chemotyp ist ausser durch den Hauptbestandteil durch Chrysanthenol und verschiedene Chrysanthenylester als Begleitstoffe charakterisiert.

Im Öl des Chrysanthemumepoxid-Typs wurden insgesamt 10 Bestandteile identifiziert. Der Gehalt an Chrysanthemumepoxid schwankt zwischen 22 und 49%. Als wichtigste Begleitstoffe wurden Campher und Camphen gefunden, ein Umstand, der in einer Beziehung auf eine nähere Verwandschaft zum Campher-Typ hinweist. Man kann sich, wie schon Scheu [11] angenommen hat, die Bildung des Epoxids durch eine Aufspaltung des Campher-Moleküls an zwei Stellen denken.

Kreuzung der Chemotypen

In Zusammenarbeit mit dem Genetischen Institut der Universität Helsinki wurden die Chemotypen weiter durch Kreuzung miteinander studiert. Hier werden nur die Kreuzungen der

Thujon- und der Campher-Typen näher beschrieben (vgl. auch [16]).

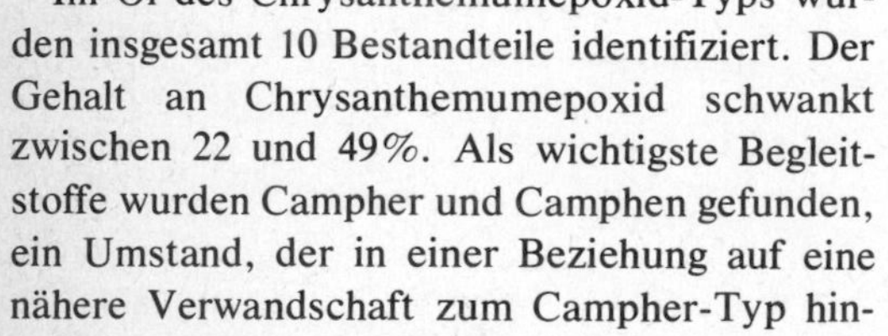

Acht ausgewählte Pflanzen wurden auf 31 verschiedene Weisen gekreuzt und insgesamt 513 Nachkommen auf die Zusammensetzung des ätherischen Öles des Blütenkörbchens gaschromatographisch analysiert.

Die technische Ausführung der Kreuzungen war nicht so einfach wie erwartet, da Selbstbestäubung bei Rainfarn möglich ist. Darum wurde zuerst diese eingehend untersucht. Es ergab sich jedoch, dass bei der Selbstbestäubung die Samenproduktion gestört wird. Nach Fremdbestäubung beträgt die Samenproduktion im Mittel 48% derjenigen der normalen Insektenbestäubung nach Selbstbestäubung aber im Mittel weniger als 1% davon. Darum kann auf die Auswirkung einer Selbstbestäubung neben der Fremdbestäubung bei der Auswertung der Nachkommen verzichtet werden.

Auch die unterschiedliche Blütezeit der nordfinnischen und südfinnischen Rainfarne bereitete gewisse Schwierigkeiten, und ferner mussten alle Elternpflanzen gegen pollenübertragende Insekten geschützt werden, damit nicht Nachkommen durch Befruchtung mit Pollen von anderen Chemotypen entstehen. Zu diesem Zweck wurden besondere Terylentüten benutzt, die eine gute Durchlüftung gestatten, damit die normale Entwicklung der Fruchtknoten bis zur Samenreife nicht gestört wurde. Im Herbst wurden die Samen in den Originaltüten geerntet, einige Monate in Kühlschrank aufbewahrt und erst dann in einem Kasten im Gewächshaus ausgesät. Anfang Juni wurden 20 Keimlinge jeder Kreuzung ausgewählt und auf das Versuchsfeld ausgepflanzt. Auf diese Weise wurden schon im ersten Sommer blühende Pflanzen erhalten. Die Proben für die gaschromatographische Analyse wurden Anfang August bei Vollblüte eingesammelt und tiefgefroren bis Analyse aufbewahrt.

Unter den Nachkommen wurden folgende Chemotypen gefunden: Campher-Typ (27–76% Campher), Sabinen-Typ (22–70% Sabinen),

Thujon	Thujylalkohol	Sabinen

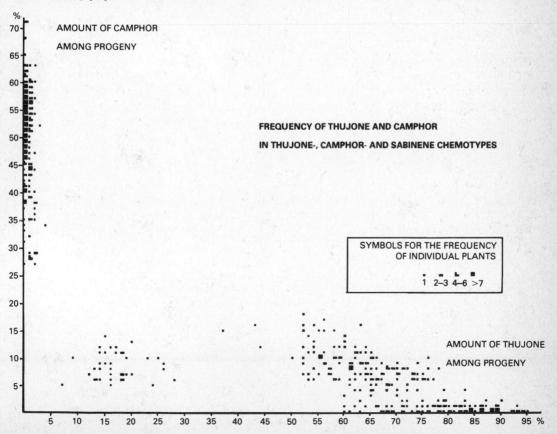

Chrysanthemum-epoxid   Campher   Chrysanthemum-epoxid

Thujon-Typ A (37–78% Thujon) und Thujon-Typ B (60–90% Thujon). Es ist also festzustellen, dass, auch wenn man aus der Natur Chemotypen mit möglichst hohem Gehalt an Hauptbestandteil auswählt, reine Homozygoten nicht oder nur mit äusserst geringer Wahrscheinlichkeit zu erreichen sind. Die nahe Verwandtschaft der Chemotypen lassen erkennen, dass auch in der Natur Kreuzung zwischen den verschiedenen Chemotypen nach gewissen Regeln stattfindet und mehr oder weniger deutliche Mischtypen auftreten können.

Eine Zusammenstellung des Campher-, Thujon- und Sabinengehalts der Nachkommen ist aus Abb. 1 abzulesen. Die Campher-Typen bil-

den eine Gruppe, die nahe an der Ordinate liegt und immer einen sehr kleinen Thujongehalt aufweist, während die Thujontypen an der Abscisse eine Gruppe mit sehr niedrigem Campher-Gehalt bilden. Die Pflanzen, die den Sabinentyp repräsentieren, liegen dazwischen mit mehr Thujon als in den Camphertypen. Mischtypen wurden unter den Nachkommen nicht gefunden.

Aus Tabelle 2 sind die nach Kreuzung erhaltenen Chemotypen zu ersehen. Sogar bei Selbstbestäubung erhält man nicht nur Nachkommen desselben Chemotyps. Ausgehend vom Camphertyp z. B. wurden von 58 Nachkommen 45 vom Camphertyp, 4 vom Thujontyp, 4 vom Chrysanthenylacetattyp und 5 von anderen Chemotypen erhalten. Bei Kreuzung von zwei Camphertypen miteinander waren von 114 Nachkommen 106 vom Camphertyp, 3

Abb. 1. Frequenz von Thujon und Campher in den Nachkommen nach Kreuzung vom Camphertyp mit dem Thujontyp [16].

Tabelle 2. *Kreuzungen von Chemotypen von Chrysanthemum vulgare (L.) Bernh.*

	Campher-Typ	Thujon-Typ	Sabinen-Typ	Chrysanthenyl-acetat-Typ	Andere Chemotypen	Anzahl Kreuzungen
Campher selbstpoll	45	4	–	4	5	58
Campher × Campher	106	3	–	1	4	114
Thujon selbstpoll	–	5	–	–	–	5
Thujon × Thujon	4	94	–	–	4	102
Campher × Thujon	27	72	27	3	4	133
Thujon × Campher	6	72	10	1	12	101
						513

Tabelle 3. *Anzahl verschiedene Chemotypen in den Nachkommen*

	Anzahl der untersuchten Nachkommen	Campher	Thujon	Sabinen	Andere Chemotype
Campher × Campher					
$K_1 \times K_1$ Selbstbestäubung	21	16	–	–	5
$K_1 \times K_2$	20	19	–	–	1
$K_1 \times K_3$	19	17	1	–	1
$K_2 \times K_1$	19	18	–	–	1
$K_2 \times K_2$ Selbstbestäubung	19	16	2	–	1
$K_2 \times K_3$	18	18	–	–	–
$K_3 \times K_1$	19	19	–	–	–
$K_3 \times K_2$	19	15	2	–	2
$K_3 \times K_3$ Selbstbestäubung	18	13	2	–	3
	172	151	7	–	14
Thujon × Thujon					
$T_1 \times T_2$	9	–	9	–	–
$T_2 \times T_1$	2	–	2	–	–
$T_2 \times T_4$	19	1	18	–	–
$T_3 \times T_1$	16	1	13	–	2
$T_3 \times T_2$	19	–	18	–	1
$T_3 \times T_4$	19	2	17	–	–
$T_5 \times T_4$	18	–	17	–	1
$T_5 \times T_5$ Selbstbestäubung	5	–	5	–	–
	107	4	99	–	4
Campher × Thujon					
$K_1 \times T_1$	19	–	9	10	–
$K_1 \times T_2$	19	6	13	–	–
$K_2 \times T_1$	19	1	11	7	–
$K_2 \times T_2$	19	7	12	–	–
$K_3 \times T_1$	20	4	4	10	2
$K_3 \times T_2$	19	6	10	–	3
$K_3 \times T_4$	18	3	13	–	2
	133	27	72	27	7
Thujon × Campher					
$T_1 \times K_1$	8	1	5	1	1
$T_1 \times K_2$	19	1	9	9	–
$T_2 \times K_1$	1	–	–	–	1
$T_2 \times K_3$	18	–	18	–	–
$T_3 \times K_1$	18	–	12	–	6
$T_3 \times K_3$	18	4	9	–	5
$T_4 \times K_1$	19	–	19	–	–
	101	6	72	10	13

vom Thujontyp, einer vom Chrysanthenyl-acetattyp und 4 von anderen Chemotypen. Leider war die Samenproduktion nach Selbstbestäubung der Thujontypen so schlecht, dass nur 5 Nachkommen erhalten wurden, die aber alle vom Thujontyp waren. Das Material war zu klein, um bezüglich der Streuung der Nachkommen Schlüsse ziehen zu können. Bei der Kreuzung von 2 Thujontypen miteinander wurden von 102 Nachkommen 94 vom Thujon-typ erhalten, 4 vom Campher-Typ und 4 von anderen Chemotypen. Auch die zur Kreuzung gebrauchten Thujontypen waren also keine Homozygoten. Bei der Kreuzung von ♀ Camphertyp × ♂ Thujontyp wurden von 133 Nachkommen 72 vom Thujontyp und nur 27 vom Camphertyp gefunden, und bei der Kreuzung von ♀ Thujontyp × ♂ Camphertyp wurden von 101 Nachkommen 72 vom Thujon-typ und nur 6 vom Camphertyp erhalten. In beiden Fällen trat ein neuer noch nicht in der Natur gefundener Chemotyp, der Sabinentyp auf, und zwar bei etwa 15% der Nachkommen. Weiterhin würden vier Chrysanthenylacetat-typen und 16 andere Chemotypen erhalten.

Bei näherem Studium der zur Kreuzung ver-wendeten Eltern (Tabelle 3) wird ersichtlich, dass die Anlagen zum Sabinen-Typ eigentlich nur bei einem einzigen Exemplar der Eltern zu erkennen sind, und zwar bei dem Thujontyp der mit T_1 bezeichnet ist. Es handelt sich um ein Individuum mit hohem Thujongehalt und nur wenig Sabinen (7,5%), das also zu dem Thujontyp gerechnet wird, aber doch eine ziem-lich starke Neigung zur Sabinenbildung hat.

Auf Grund dieser Ergebnisse wurde an-schliessend an frühere Theorien [17–21], die sich auf die Annahme hypothetischer Carbo-nium-ionen nach Ruzicka [22] stützen, ein Modell aufgestellt, das die genetischen Bezie-hungen der Bildung von Thujon, Sabinen und Campher erklären kann (Abb. 2). Wenigstens fünf Loci sind im Genom von Rainfarn vor-handen, die an dem Process teilnehmen, der entweder das Elektron der Carboniumionen in die Stellung bringen kann, die die Biosynthese des Thujons und des Sabinens ermöglicht oder in eine Stellung, welche zur Campher-Synthese führt. Vier von diesen Loci funktionieren als Multiple (A_1, A_2, B_1 und B_2), die von dem fünften Locus reguliert werden.

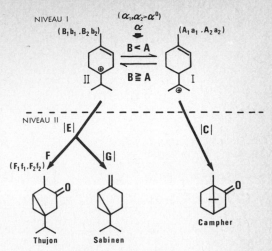

Abb. 2. Modelle für die genetische Kontrolle der Biosynthese von Thujon, Sabinen und Campher in *Chrysanthemum vulgare* (L.) Bernh.

Es ist offenbar, dass in diesem Locus drei Allele vorkommen, unter denen die Dominanz-ordnung folgende ist: $\alpha_1 > \alpha_2 > \alpha^0$. Wenn in dem Genom die Anzahl der dominanten A-Gene die der dominanten B-Gene überschreitet, wird nur Campher produziert, und wenn umgekehrt die Zahl der dominanten B-Gene die der dominan-ten A-Gene überschreitet oder beide in gleicher Anzahl vorhanden sind, wird die Thujon-Linie bevorzugt. Die Allele von Locus α-kontrol-lieren die Biosynthese so, dass α_1 die Funktion von A_1 hemmt, während α_2 die Funk-tion von beiden A-Loci hemmen kann und da-mit der Weg zum Campher verschlossen wird.

Bei der Biosynthese in Richtung Thujon ist der erste Schritt durch eines oder mehrere mo-nomorphe Gene E kontrolliert, welche die Bio-synthese entweder in Richtung Thujon oder in Richtung Sabinen weiterführen.

Die Biosynthese von Sabinen wird durch das Gen (die Gene) G kontrolliert. Die Funktion von G wird wieder durch das Gen F (oder Gene F_1, F_2) gehemmt, das für die die Thu-jonbiosynthese kontrollierenden Enzyme verant-wortlich ist.

Auf diese Weise können die genetischen Be-dingungen bei der Biosynthese der obengenann-ten Terpene erklärt werden.

Literaturverzeichnis

1. Tetenyi, P, Infraspecific chemical taxa of medi-cal plants. Académia Kiadó, Budapest, 1970.

2. v Rudloff, E, Can j chem 1963, 41, 1, 1737.
3. Stahl, E & Schmitt, G, Arch Pharm 1964, 297, 385.
4. Järvi, M, Farm aikak 1965, 74, 11.
5. v Schantz, M & Järvi, M, Proc 25th congr pharm sci Prague 24–27 August, 1965, p. 255, 1966.
6. v Schantz, M, Järvi, M & Kaartinen, R, Planta med 1966, 14, 421.
7. v Rudloff, E & Underhill E, W, Phytochemistry 1965, 4, 11.
8. Stahl, E & Scheu, D, Naturwiss 1965, 52, 394.
9. — Arch Pharm 1967, 300, 456.
10. Stahl, E. Pharm weekblad 1971, 106, 237.
11. Scheu, D, Dissertation. Saarbrücken, 1966.
12. v Schantz, M, Planta med 1968, 16, 395.
13. Forsén, K & v Schantz, M, Arch Pharm 1971, 304, 944.
14. v Schantz, M & Forsén, M, Farm aikak 1971, 80, 122.
15. Forsén, K & Lokki, J, Farm aikak 1972, 81, 67.
16. Lokki, J, Sorsa, M, Forsén, K & v Schantz, M, Hereditas 1973, 74, 225.
17. v Schantz, M, Farm aikak 1967, 76, 265.
18. Banthorpe, D V & Wirz-Justice, A, J chem soc (C) 1969, 571.
19. Banthorpe, D V & Baxendale, D, Ibid 1970, 2694.
20. Banthorpe, D V, Mann, J & Turnbull, K W, Ibid 1970, 2689.
21. Zavarin, E, Phytochemistry 1970, 9, 1049.
22. Ruzicka, L, Experientia 1953, 9, 357.

Discussion

Swain: Is there any evidence that different insects act as pollinators for the different chemical races?

von Schantz: No there is no evidence.

Weimarck: Professor Swain hat gefragt, ob es einen möglichen Zusammenhang zwischen den verschiedenen Geruchsstoffen und der Pollination durch Insekten gibt. Professor von Schantz hat nun gesagt, dass in der Natur bei *Chrysanthemum vulgare* Fremdbestäubung am meisten vorkommt. Wenn diese Stoffe einen spezifischen Anlockungsmekanismus darstellen, sollte man erwarten, dass eine gewisse reproduktive Isolation zwischen den Phänotypen erkennbar wäre, weil Insekten den einen oder anderen Typus wählen. Es wäre dann zu erwarten, dass dieses auch zu einer morphologischen Diskontinuität geführt hätte. Solch eine Korrelation mit morphologischer Variation ist aber nicht vorhanden.

Wir haben vorher die Möglichkeit diskutiert, dass gewisse Stoffe eine spezifische, sowohl öko-logische wie physiologische Bedeutung haben. Hier haben wir vielleicht einen Fall, wo eine solche Bedeutung, zumindest für die Blütenökologie, nicht gefunden werden kann, obwohl man es erwarten könnte.

Hegnauer: All available evidence indicates that essential oil chemotypes are often the result of selection by abiotic factors (i.e. climate, soil, not insects) of biotopes. The distribution of the camphor and thujon chemotypes of *Tanacetum* in Finland is suggestive of a similar situation, just as are recent investigations performed in France with *Thymus vulgaris*.

Pseudoguaianolides in Compositae

W. Herz

Department of Chemistry, The Florida State University,
Tallahassee, Fla 32306, USA

Summary

Pseudoguaianolides and modified pseudoguaianolides found in Compositae can be divided into two groups which differ in stereochemistry at C-10. Tables showing their distribution in the various species and genera of Compositae are presented. Representatives of one group, the ambrosanolides with H-10 *alpha*, are characteristic secondary metabolites of Ambrosiinae and *Parthenium* (Melampodiinae). Pseudoguaianolides with H-10 *beta*, the helenanolides, are typical of certain genera in Heleniinae although occasionally encountered elsewhere. Supporting evidence for the hypothesis that the two classes owe their origin to cyclization of different *trans, trans* and *cis, trans-* germacradiene precursors is given. The difference may be of use in illuminating certain relationships within the family.

Sesquiterpene lactones seem to be characteristic constituents of Compositae, although instances of their occurrence in other plant families are coming to light [1 (ref. 1–8), 2–11]. In the present contribution, recent developments concerned with the distribution and possible origin of one important class of sesquiterpene lactones, the pseudoguaianolides, which was last reviewed in 1966 [12, 13], will be discussed.

In scheme I, taken from a review by Herout [14], various types of known sesquiterpene lactones are arranged in order of what might be called "increasing biogenetic complexity". Such a scheme is useful for illuminating relationships among plants, as has been done at the tribal level of Compositae [14, 15], as well as for systematizing the approx. 600 sesquiterpene lactones which are now known. The germacranolides might be said to represent the most primitive stage or stage I, elemanolides, eudesmanolides and guaianolides stage II. Stage III is represented by examples showing further modifications such as migrations of methyl

groups which lead to eremophilanolides and pseudoguaianolides. Ring cleavage of stage II compounds gives rise to stage III compounds such as xanthanolides and other types which are not included in the scheme. Further modification of stage III compounds leads to stage IV, a few representatives of which are depicted in the last column.

It is not widely appreciated that two different classes of pseudoguaianolides and modified pseudoguaianolides exist. These may be exemplified by ambrosin (*1*a), on the one hand, and aromatin (*2*) on the other. In compounds related to ambrosin, the so-called ambrosanolides [17], the methyl group attached to C-10 is β-oriented and the lactone ring is generally, but not invariably, closed to C-6. Ambrosanolides are characteristic constituents of the genus *Parthenium* and certain genera in Ambrosiinae (see table 1).

In compounds of the second group, for which I would like to propose the name helenanolides, the C-10 methyl group is α-oriented and the lactone ring is invariably closed to C-8. These substances are characteristic constituents of certain genera in the tribe Heleniae (see table 2).

Historically, the existence of the pseudoguaianolide skeleton was first recognized [17, 18] in ambrosin (*1*a), a lactone originally discovered in *Ambrosia maritima* [19, 20], and parthenin (*1*b), the main sesquiterpene lactone constituent of *Parthenium hysterophorus* [21, 22]. Chemical studies [17, 18, 22–24] followed by X-ray crystallographic analysis of 3-bromoambrosin [25] led unequivocally to the relative and absolute stereochemistry depicted in *1*a and *1*b.

Almost simultaneously, it was realized [26,

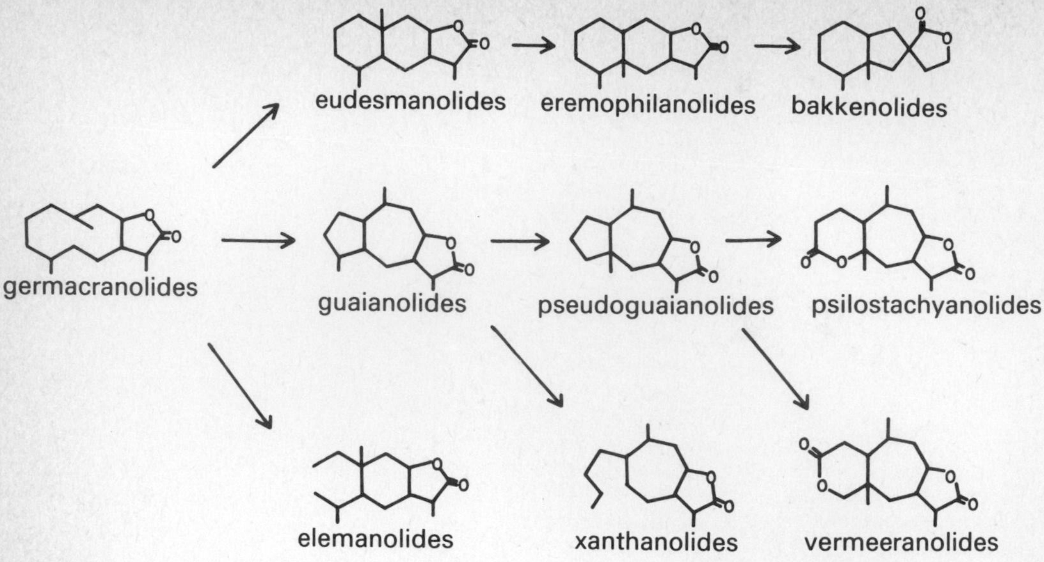

eudesmanolides   eremophilanolides   bakkenolides

germacranolides   guaianolides   pseudoguaianolides   psilostachyanolides

elemanolides   xanthanolides   vermeeranolides

Scheme I

27] that the properties of the sesquiterpene lactones tenulin [28–31] from *Helenium amarum* [32, 33] and helenalin [34, 35], from *Helenium autumnale* [36, 37] and *Balduina angustifolia* [38], were accommodated by formulas *3* and *4*. Subsequent work [39–41] established the relative and absolute stereochemistry of these substances.

To understand how these two classes of pseudoguaianolides and their derivatives might be formed in the plant, it is necessary to consider briefly current hypotheses concerning sesquiterpene biogenesis [42]. Scheme II shows the central role attributed to the 10-membered ring germacradiene system—itself formed by cyclization of farnesol pyrophosphate—in the genesis of certain types of widely-distributed sesquiterpenes and the sesquiterpene lactones of scheme I characteristically found in Compositae. It has been stated or implied that this involves the *trans*-1,10-*trans*-4,5-germacradiene *5* de-

picted in the center of scheme II which has its origin in a *trans, trans*-farnesene precursor, but as will be shown subsequently this can probably no longer be accepted as the exclusive precursor postulated previously. For the purpose of the present discussion, we are less interested in the class of elemanes or the corresponding lactones, the elemanolides, which are formed from germacradienes by a Cope rearrangement (arrows) than in the eudesmanes, guaianes or the corresponding lactones which are their origin to a second cyclization of a germacradiene.

Stereospecific enzyme-mediated Markownikow-oriented cyclization of a chair-folded germacradiene, whether analogous to the reaction illustrated at the top of scheme III (where X is some electrophile) or whether acid-catalyzed as shown at the bottom, could lead to the hypothetical precursors *6* or *7* or the eudesmanolide series. The absolute stereochemistry

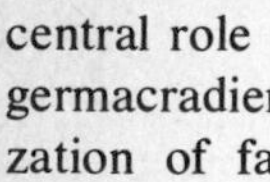

1 a R=H
  b R=OH

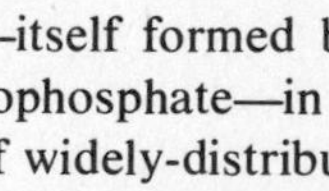

2

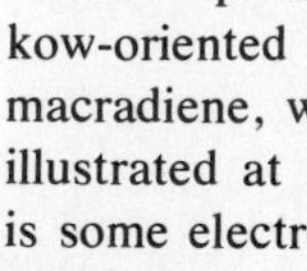

3

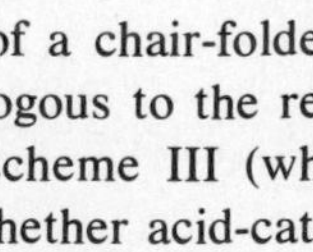

4

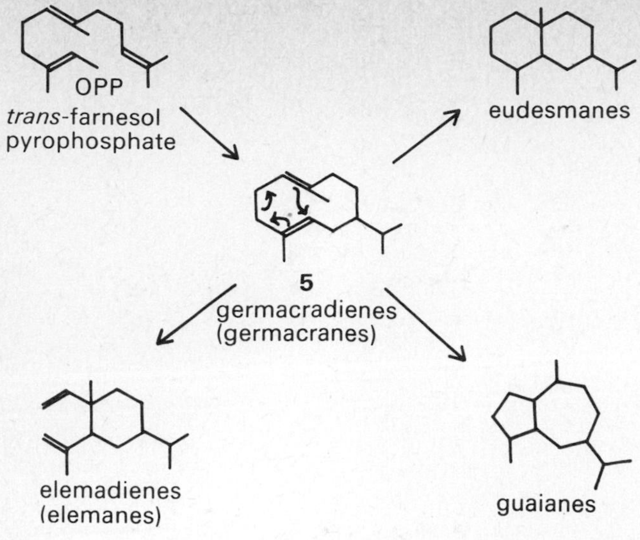

Scheme II

of naturally-occurring eudesmanolides (H-5 α, C-10 methyl β) is in accordance with this hypothesis. Many laboratory analogies for such a cyclization exist; the existence of chair-folded germacradienes with the "crossed" double bonds necessary for cyclization to bicyclic compounds has been inferred repeatedly from NMR spectroscopy and verified by X-ray crystallography.

The second large group of stage II lactones is that of the guaianolides. At a time when the stereochemistry of naturally-occurring guaianes and guaianolides was not known, Hendrickson [43] suggested that anti-Markownikow-orientated cyclization of a chair-folded *trans,trans*-germacradienyl cation might be responsible for the formation of the guaianes. As illustrated in scheme IV this would lead to *cis*-fused guaianes and guaianolides. In fact, with very few exceptions which will be considered subsequently, the stereochemistry required by this scheme (H-1, H-5 α) has turned out to be the stereochemistry of all guaianolides whose structure has been elucidated in toto. However, the single laboratory analogy reported so far [44] for the postulated conversion of a germacranolide to a guaianolide remains ambiguous because of lack of information about stereochemistry of starting material and product.

The *cis*-fused cation 8 resulting from this cyclization process or the lactone analog 9 can be invoked as a biogenetic precursor of the ambrosanolides. The series of migrations indi-

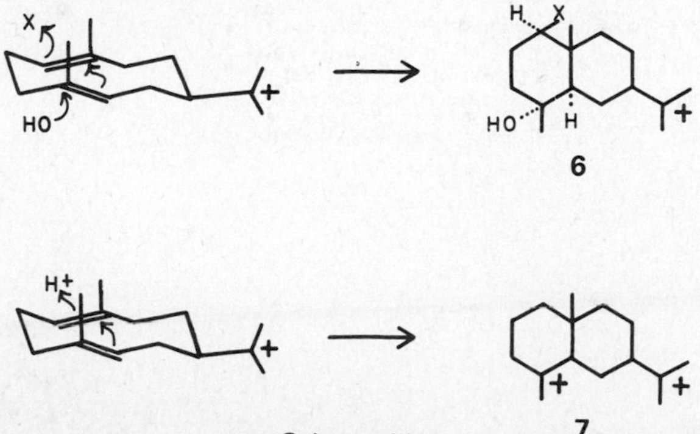

Scheme III

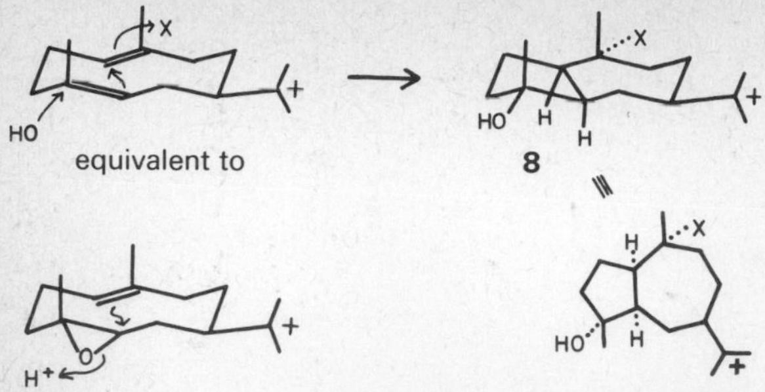

Scheme IV

cated by the arrows would lead to lactones whose stereochemistry (H-1 α, C-5 and C-10 methyl groups β) is in accord with that actually found in the ambrosanolides. (A difficulty with the biogenetic scheme presented in scheme V is that both protons at the ring junction are *cis*. Hence the series of migrations envisioned in the scheme is not likely to be concerted. Perhaps this explains the wide distribution of *cis*-fused guaianolides in Compositae, the capacity for their conversion to ambrosanolides being highly specialized.) Subsequent biological oxidation of the 5-membered ring can be envisioned as leading to modified ambrosanolides (psilostachyinolides) such as psilostachyin *(10)* or psilostachyin C *(11)* whose stereochemistry is in accord with this hypothesis.

Naturally-occurring ambrosanolides and psilostachyinolides have so far been found only in the genus *Parthenium* (Heliantheae, Melampodiinae) which invariably produces such compounds and in three of the 5 genera (*Ambrosia, Hymenoclea, Iva, Dicoria, Xanthium*) ordinarily included in subtribe Ambrosiinae of Heliantheae.

Table 1 enumerates the species containing ambrosanolides and psilostachyinolides and cites these substances as well as all other sesquiterpene lactones isolated from the listed species, without regard to occasionally quite vari-

able results from specific collections as for example, in members of the large and diversified ambrosioid complex which includes *Ambrosia artemisifolia* [45–48], *A. cumanensis* [49–51], *A. peruviana* [49, 52, 53] and *A. psilostachya* [47, 50, 53–64] and is probably related through *A. confertiflora,* also a chemically variable species [49, 65–70], to the more primitive franseroid [71] core of the genus. This heterogeneity however appears to be under genetic control; the distribution of such chemical races is quite possibly linked to evolutionary trends within species and subgenera [48]. In any event, those germacranolides which are found in Ambrosiinae possess; with one exception (*vide infra*), the *trans*-$\triangle^{4,10}$-double bond referred to earlier cyclization of which should result in the formation of *cis*-fused guaianolides derived from 9 as required by the biogenetic hypothesis. Guaianolides with this stereochemistry have been found in certain collections of *A. acanthicarpa* [63] and *A. cumanensis* [51] and are widely distributed in species outside Ambrosiinae. Species which yielded germacranolides or eudesmanolides, but no ambrosanolides include *A. camphorata* (Greene) Payne [70], *A. chamissonis* (Less.) [72], *A. ilicifolia* (Gray) Payne [73] and *A. "castanesis"* [70]. Species which yielded no lactones are *A. bidentata* Michx. [48], *A. cheiranthifolia* Gray [74], *A. eriocentra* (Gray) Payne [70, 75], *A. grayi*

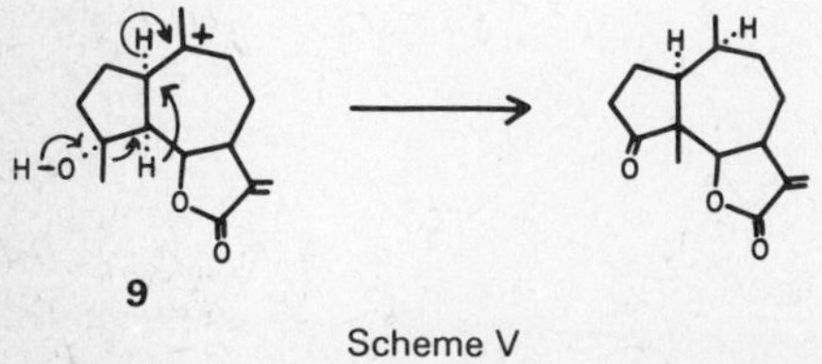

Scheme V

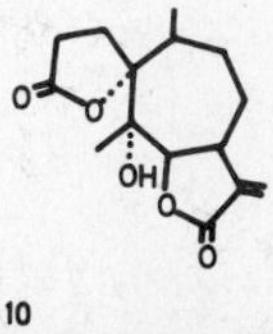

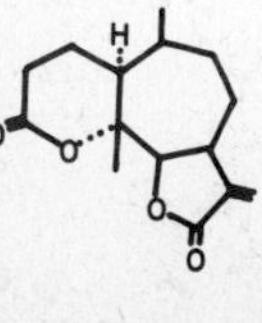

Table 1. *Ambrosanolides in Compositae*

Species	Compound (formula) pp 154, 156, 165–166	Ref.
1 Heliantheae Cass.		
A Ambrosiinae Less.		
1 *Ambrosia acanthicarpa* (Hook.) Cav.	confertiflorin (*39* b)	[69a]
	desacetylconfertiflorin (*39* a)	[69a]
	psilostachyin C (*11*)	[69a]
	germacranolides	[62, 69a]
	guaianolides	[79]
A. ambrosioides (Cav.) Payne	damsin (*28* a)	[66, 70, 133]
	franserin (*47*)	[66]
	damsinic acid (*32*)[a]	[133]
A. artemisiifolia L.	psilostachyin (*10*)	[45]
	peruvin (*48* b)	[46]
	germacranolides	[45, 47]
var. *elatior*	coronopilin (*28* b)	[48]
A. canescens (Gray) Payne	canambrin (*52*)	[134]
A. chenopodiifolia (Benth.) Payne	damsin (*28* a)	[13, 70]
A. confertiflora DC.	confertiflorin (*39* b)	[65, 70]
	desacetylconfertiflorin (*39* a)	[65, 66]
	confertin (*48* a)	[66, 69]
	peruvin (*48* b)	[69]
	psilostachyin (*10*)	[49, 69]
	psilostachyin B (*51*)	[69]
	psilostachyin C (*11*)	[49, 169]
	germacranolides and eudesmanolides	[67, 68, 135]
A. cordifolia (Gray) Payne	psilostachyin C (*11*)	[77]
	cordilin (*53*)	[77]
	psilostachyin B (*51*)	[70]
A. cumanensis H.B.K.	psilostachyin (*10*)	[50]
	psilostachyin B (*51*)	[50]
	psilostachyin C (*11*)	[49, 50]
	cumanin (*50* a)	[50]
	ambrosin (*1* a)	[49]
	damsin (*28* a)	[49]
	guaianolides	[51]
A. deltoidea (Torr.) Payne	damsin (*28* a)	[56]
	psilostachyin C (*11*)	[56, 70]
A. dumosa (Gray) Payne	coronopilin (*28* b)	[54, 135]
	psilostachyin (*10*)	[135]
	psilostachyin C (*11*)	[135]
	ambrosiol (*35*)	[135]
	burrodin (*38*)	[135]
	apoludin (*36* a)	[135]
	germacranolides	[70, 135]
A. hispida Pursh.	ambrosin (*1* a)	[136]
	damsin (*28* a)	[136]
"*A. jamaicensis*"[b]	damsin (*28* a)	[70]
	ambrosin (*1* a)	[70]
	psilostachyin (*10*)	[70]
A. maritima L.	ambrosin (*1* a)	[17–20, 22–25]
	damsin (*28* a)	[17–20, 22–25]
A. peruviana Willd.	peruvin (*48* b)	[52]
	peruvinin (*49*)	[53]
	tetrahydroambrosin (*27*)	[49]
	psilostachyin (*10*)	[49]

Table 1. (cont).

Species	Compound (formula) pp 154, 156, 165–166	Ref.
A. psilostachya DC.	psilostachyin (10)	[50, 55, 57]
	psilostachyin B (51)	[50, 58]
	psilostachyin C (11)	[50, 56]
	coronopilin (28 b)	[47, 50, 54, 59, 63, 64]
	parthenin (1 b)	[50, 55, 59, 63]
	ambrosiol (35)	[50, 55]
	cumanin (50 a)	[50, 63, 64]
	3-acetylcumanin (50 b)	[63]
	2,3-diacetylcumanin (50 c)	[63]
	ambrosin C (1 a)	[50]
	3-hydroxydamsin (29)	[50, 59]
	germacranolides	[60–62]
A. pumila (Nutt.) Gray	psilostachyin (10)	[49]
	psilostachyin C (11)	[49]
	desacetylconfertiflorin (39 a)	[70]
A. tenuifolia Spreng.	psilostachyin (10)	[49]
2 Hymenoclea monogyra T.+G.	ambrosin (1 a)	[137]
	anhydrocoronopilin (31)	[137]
	psilostachyin (10)	[137]
	psilostachyin C (11)	[137]
	germacranolide	[137]
H. salsola T.+G.	ambrosin (1 a)	[138]
	anhydrocoronopilin (31)	[138]
	coronopilin (28 b)	[139]
	dihydrocoronopilin (30)	[139]
	salsolin (36 b)	[139]
	hymenin (34)	[139]
	hymenolin (33)	[139]
3 Iva acerosa (Nutt.) Jackson	coronopilin (28 b)	[140]
I. nevadensis M. E. Jones	coronopilin (28 b)	[140]
	parthenin (1 b)	[140]
	nevadivalin[c]	[140]
I. xanthifolia Nutt.	coronopilin (28 b)	[141, 142]
	ambrosin (1 a)	[142]
	ivoxanthin (37)	[142]
	anhydrocoronopilin (31)	[142]
B Melampodiinae Less.		
1 Parthenium alpinum (Nutt.) T.+G.	tetraneurin A (44 a)	[143]
var. tetraneuris (Barneby) Rollins	tetraneurin B (43 b)	[144]
	tetraneurin C (45 b)	[144]
	tetraneurin D (45 a)	[144]
P. bipinnatifidum Ortega	ambrosin (1 a)	[145]
	hysterin (42)	[145]
P. confertum Gray	tetraneurin A (44 a)	[146]
var. lyratum (Gray) Rollins	tetraneurin E (46 a)	[146]
	tetraneurin F (46 b)	[146]
P. confertum Gray (?)	conchosin A (40)	[147]
	conchosin B (41)	[147]
	hymenin (34)	[147]
P. fruticosum Less.	tetraneurin A (44 a)	[148]

Nobel 25 (1973) Chemistry in botanical classification

Table 1. (cont.)

Species	Compound (formula) pp 154, 159, 165–166	Ref.
var. *fruticosum*	chiapin A (*44*b)	[148]
	chiapin B (*44*c)	[148]
var. *trilobatum* Rollins	tetraneurin B (*43*b)	[144]
	tetraneurin C (*45*b)	[144]
	tetraneurin D (*45*a)	[144]
	fruticosin[c]	[144]
P. *hysterophorus* L.	parthenin (*1*b)	[17, 18, 21, 22]
P. *incanum* H.B.K.	coronopilin (*28*b)	[48, 149]
	ambrosin (*1*a)	[18, 149]
	anhydrocoronopolin (*31*)	[149]
	ligulatin B (*43*a)	[149]
P. *integrifolium* L.	tetraneurin C (*45*b)	[146]
	tetraneurin E (*46*a)	[146]
P. *lozianum* Bartlett.	tetraneurin B (*43*b)	[144]
	tetraneurin C (*45*b)	[144]
	tetraneurin D (*45*a)	[144]
P. *schottii* Greenm. ex. Millspaugh and Chase	tetraneurin B (*43*b)	[148]
	tetraneurin D (*45*a)	[148]
	ligulatin B (*43*a)	[148]
	coronopilin (*28*b)	[148]
	confertin (*48*a)	[148]
II Eupatorieae Cass.		
A Eupatoriinae Dumort.		
1 *Stevia rhombifolia* H.B.K.	stevin (*14*)	[98]

[a] Not a pseudoguaianolide, but biogenetically related.
[b] Apparently a new species.
[c] Unknown structure.

Nels. Shinners [70, 76], *A. linearis* (Rydb.) Payne [70], *A. tomentosa* Nutt [70, 76] and *A. trifida* L. [48, 70].

Of the 4 other genera generally included in Ambrosiinae, *Hymenoclea* is rich in ambrosanolides and psilostachyinolides (table 1), but *Dicoria* yielded no lactones [77]. Only 3 species in *Iva*, all in section Cyclochaena [78] yielded ambrosanolides (table 1), 2 other species in the same section—*I. ambrosiaefolia* [79] and *I. dealbata* Gray [80, 81]—elaborate xanthanolides such as ivalbin (*12*) typical of the lactones generally found in *Xanthium* (Ambrosiinae) and also in the monotypic genus *Parthenice* (Melampodiinae). On the other hand, all members of section *Iva* and section

Linearbractea elaborate eudesmanolides or guaianolides [82–89] with one noteworthy exception. This is *I. frutescens*, from which the *cis*-$\triangle^{1,10}$-*trans*-$\triangle^{4,5}$-germacradienolide frutescin (*13*) was isolated last year [90]. Other *cis*-$\triangle^{1,10}$-*trans*-$\triangle^{4,5}$-germacranolides have so far been found only in two *Polymnia* species (Melampodiinae) [91, 92], in *Melampodium leucanthemum* T. & G. [93, 94 a] and in *Enhydra fluctuans* Lour. [95–97].

The implications of these and other findings for the suggestion that Melampodiinae may be linked to the Ambrosiinae through *Parthenium* and *Parthenice* remain to be explored. On the other hand the discovery [98] of an ambrosanolide *14* in *Stevia rhombifolia* (Eupatorieae),

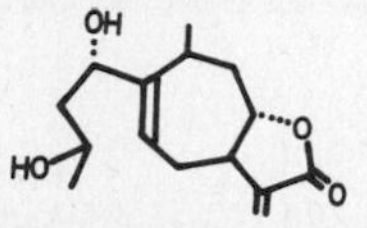

12 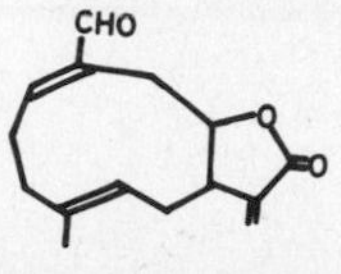   13 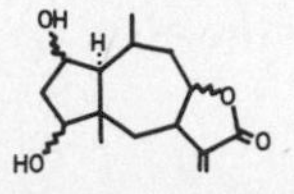   14 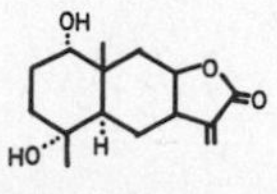   15 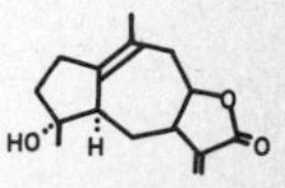   16

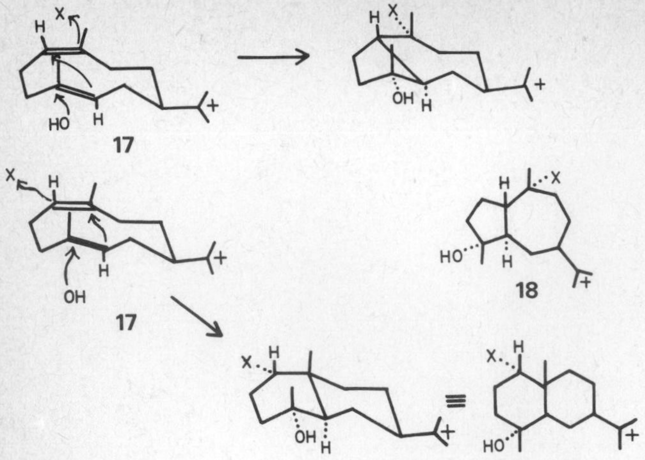

Scheme VI

has so far remained unique outside Heliantheae and cannot, at least at this time, be ascribed any taxonomic significance.

Before I deal with the second class of pseudoguaianolides, the helenanolides, mention must be made of the discovery, in my laboratory, of a chemical race of *Iva microcephala* Nott. the only instance at this time of writing [210] of a plant which simultaneously elaborates a eudesmanolide, microcephalin (*15*) [84] and a guaianolide, pseudoivalin (*16*) [85, 97]. These two compounds are clearly derived from a common 10-membered ring precursor. Now the stereochemistry of microcephalin at C-1 is opposite to that expected from the *trans*-anti-parallel cyclization of a chair-folded germacradiene illustrated in scheme III and the formation of pseudoivalin in terms of the Hendrickson scheme (scheme IV) would require the non-concerted loss of HX from the intermediate *8*. To rationalize this finding, Parker, Roberts & Ramage [42] suggested the intermediacy of the *cis-trans*-germacradienyl cation *17* in scheme VI. As shown in the scheme, Markownikow-type *trans* antiparallel cyclization of *17* would lead directly to a compound with the structure

and stereochemistry of microcephalin, while anti-Markownikow-orientated cyclization of 17 would lead to a *trans*-fused guaiane *18* with a stereochemistry appropriate for concerted loss of HX to pseudoivalin.

The recent discovery of just such *cis-$\triangle^{1,10}$-trans-$\triangle^{4,5}$*-germacranolides required by the Parker-Roberts-Ramage hypothesis—the so-called melampolides [94], in *Iva frutescens* and certain Melampodiinae has already been touched upon. It is noteworthy that all melampolides which have been isolated so far are unlikely candidates for further acid-catalyzed—and presumably also enzyme-mediated—cyclization because the carbonyl group at C-14 (see *13*) serves to deactivate the $\triangle^{1,10}$-double bond; as a corollary one may speculate that melampolides with a non-deactivated $\triangle^{1,10}$-double bond have so far escaped detection because they are only transitory intermediates on the path to *trans*-fused guaianolides of type *18*.

Now, in contrast to the *cis*-fused guaiane *8* of scheme IV, the *trans*-fused guaiane *18* of scheme VI can undergo further *concerted* rearrangement (scheme VII), the overall result being a lactone with an α-oriented methyl group at C-10, an α-oriented hydrogen at C-1 and a β-oriented methyl group at C-5.

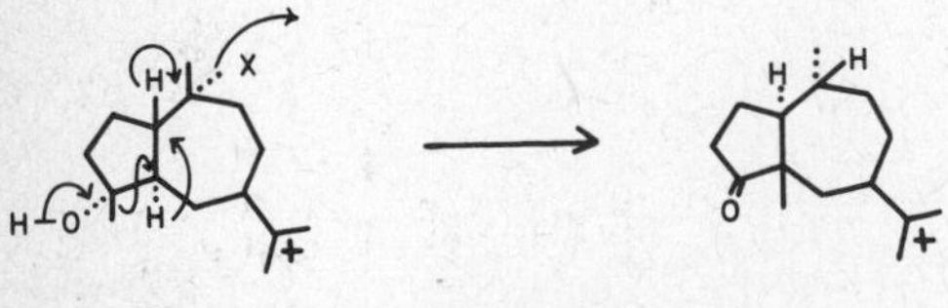

18   Scheme VII   19 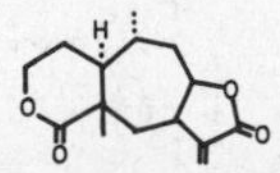   20

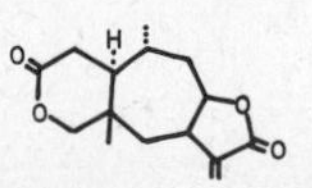

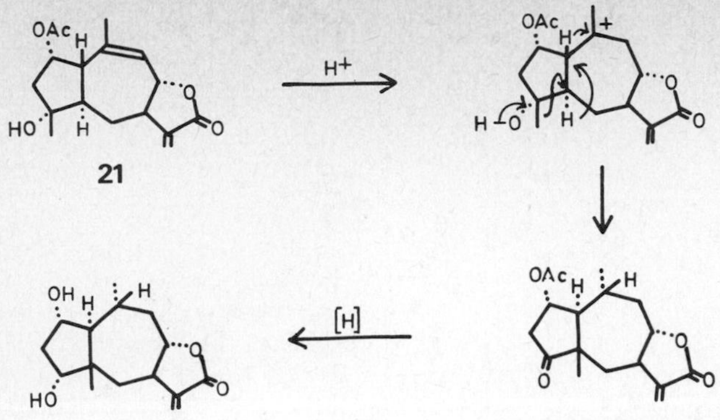

21

22

Scheme VIII

This is exactly the stereochemistry of the pseudo-guaianolides found in *Helenium* and related species, and the stereochemistry of the modified pseudoguaianolides such as vermeerin *(19)* and greenein *(20)*, derived from them by biological oxidation of the 5-membered ring.

Although *trans*-fused guaianolides of type *18* are quite rare in nature, those that have been found occur either in conjunction with helenanolides or in chemical races of species which normally elaborate helenanolides, an observation which reinforces the hypothesis that *cis*-$\Delta^{1,10}$-*trans*-$\Delta^{4,5}$-germacranolides, *trans*-fused guaianolides of type *18* and helenanolides lie on the same biogenetic pathway. Thus the *trans*-fused guaianolide gaillardin *(21)*, found by Kupchan et al. [100–102] in a chemically and cytologically distinct race of *Gaillardia pulchella*, can be transformed on paper into pul-chellin *(22)*, the main sesquiterpene lactone of the coastal race of *G. pulchella* [102–107)] by the usual series of hydride and methyl shifts adumbrated in scheme VIII followed by a reduction step. (A Western race of *G. pulchella* elaborates not helenanolides, but a series of eudesmanolides [108–110] which has also been found in a Wyoming race of *G. aristata* [111].)

Similarly, while the morphologically [112] and chemically quite variable *Helenium*

autumnale customarily elaborates pseudoguai-anolides of the helenalin type (see table 2) a Japanese cultivar yielded a guaianolide *23* of unspecified configuration [113] which I presume to possess the stereochemistry depicted in the formula and a chemical race from North Carolina furnished only the *trans*-fused guaiano-lides *24* a and *24* b [114 a]. The features common to *21, 23* and *24* (α-oriented hydroxyl at C-4, *trans*-ring fusion and 9,10-double bond) indicate that all are directly derived from a postulated intermediate of type *18*.

Table 2 enumerates species containing helena-nolides and modified helenanolides as well as other lactones isolated from the listed species, again without regard to variations in sesqui-terpene lactone content of different collections of the same species. Elaboration of this group of lactones is characteristic of all species belonging to *Helenium* and *Gaillardia*, two very closely related genera in Heleniae, subtribe Heleniinae [115] (Gaillardianae of Rydberg [116]) except for *Helenium pinnatifidum* (Nutt.) Rydb. which produces a eudesmanolide [117], *H. virginicum* Blake which produces a guaianolide [118] and *H. hoopesii*, recently returned to *Dugaldia* [116], (private communication from Dr M. W. Bierner) which yields no lactones [119]. *Dugaldia* and *Amblyolepis setigera* DC., which likewise contains no sesquiterpene lactones [120], have been placed in subtribe Gaillardianae [118], but appear to be less closely

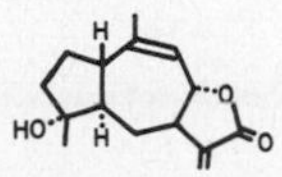

23

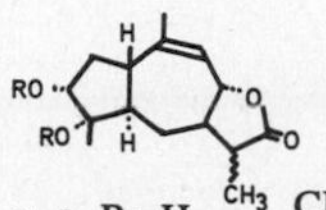

24 a, R=H
b, R=CO—C=CH

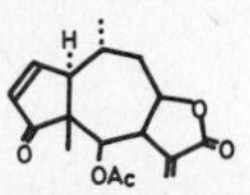

25

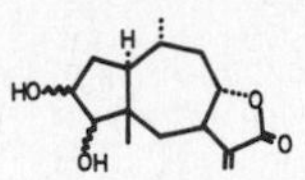

26

Table 2. *Helenanolides in Compositae*

Species	Compound (formula) pp 154, 161, 166–167	Ref.
1. Heleniae Benth. and Hook.		
A Heleniinae Less.		
1 *Helenium alternifolium* (Spreng.) Cabrera	tenulin (*3*)	[150]
	linifolin A (*62* b)	[150]
	brevilin A (*64* b)	[150]
	alternilin (*66* c)	[150]
H. amarum (Raf.) Rock	tenulin (*3*)	[26, 28–33, 39, 40, 152]
	aromaticin (*68*)	[152]
	amarilin (*70*)	[152]
H. arizonicum Blake	isotenulin (*63* b)	[153]
H. aromaticum (Hook.) Bailey	helenalin (*4*)	[154]
	mexicanin I (*62* a)	[154]
	aromatin (*2*)	[154]
	aromaticin (*68*)	[154]
H. autumnale L.	helenalin (*4*)	[27, 34–37, 39, 41, 151, 155]
	2-acetylflexuosin A (*66* b)	[151]
	autumnolide (*81*)	[151, 156]
	dihydromexicanin E (*72*)[a]	[155]
	tenulin (*3*)	[157]
	linifolin A (*62* b)	[157]
	mexicanin I (*62* a)	[157]
	autumnalin (?)	[157]
	helenium lactone (*23*)[b]	[113]
	carolenin (*24* b)[b]	[114 a]
	carolenalin (*24* a)[b]	[114 a]
H. badium (Gray) Greene	tenulin (*3*)	[33, 158]
H. bigelovii Gray	bigelovin (*61*)	[159, 160]
	tenulin (*3*)	[160]
	isotenulin (*63* b)	[160]
	desacetylisotenulin (*63* a)	[160]
H. brevifolium (Nutt.) A. Wood	brevilin A (*64* b)	[150, 161]
	brevilin B[c]	[161]
	brevilin C (mexicanin I, *62* a)	[161, 162]
	brevilin D[c]	[161]
	gaillardipinnatin (*83* b)	[162]
H. campestre Small	helenalin (*4*)	[163]
H. chihuahensis Bierner[d]	tenulin (*3*)	[153]
H. elegans DC		
var. *elegans*	tenulin (*3*)	[32, 158]
var. *amphibolum* (Gray) Bierner	tenulin (*3*)	[164]
	mexicanin I (*62* a)	[164]
H. flexuosum Raf.	flexuosin A (*66* a)	[151, 165]
	flexuosin B (*57*)	[151, 165]
H. laciniatum Gray	helenalin (*4*)	[153]
H. linifolium Rydb.	linifolin A (*62* b)	[150, 153]
	linifolin B (*60*)	[153]
H. mexicanum H.B.K.	helenalin (*4*)	[27]
	neohenalin (mexicanin D, *56*)[e]	[27, 166]
	mexicanin A (*59*)	[27, 39, 166]
	mexicanin B[c]	[166]

Table 2 (cont.)

Species	Compound (formula) pp 154, 161, 166–167	Ref.
	mexicanin D (*64*a)	[39, 166]
	mexicanin E (*71*)[a]	[166–170]
	mexicanin F[c]	[166]
	mexicanin G[c]	[166]
	mexicanin H (*73*)	[166, 171]
	mexicanin I (*62*a)	[157, 172]
H. microcephalum DC		
var. *microcephalum*	helenalin (*4*)	[34, 158]
var. *ooclinium* (Gray) Bierner	neohelenalin (*56*)	[153]
	mexicanin E (*71*)	[153]
H. montanum Nutt.	tenulin (*3*)	[33]
H. plantagineum (DC) MacBr.	linifolin A (*62*b)	[173]
	mexicanin I (*62*a)	[173]
H. quadridentatum Labill.	helenalin (*4*)	[33, 158, 174]
H. scorzoneraefolium (DC) Gray	helenalin isomer[f]	[153]
	helenalin (*4*)	[164]
	linifolin A (*62*b)	[164]
H. thurberi Gray	tenulin (*3*)	[160]
	thurberilin (*65*)	[160, 175]
H. vernale Walt.	helenalin (*4*)	[161]
2 *Gaillardia amblyodon* Gay	gaillardipinnatin (*83*b)	[162, 176]
	desacetylgaillardipinnatin (*83*a)	[162]
	amblyodin (*84*)	[162, 176]
	amblyodiol	[162]
G. aristata Pursh.	spathulin (*88*)	[177, 178, 179]
	aristalin[c]	[177]
	eudesmanolide	[178]
	eudesmanolide	[111]
G. arizonica Gray	gaillardilin (*67*)	[180]
G. fastigiata Greene	fastigilin A (*85*a)	[181]
	fastigilin B (*85*b)	[181]
	fastigilin C (*86*a)	[181, 182]
G. grandiflora van Houtte	spathulin (*88*)	[177]
	aristalin[c]	[177]
G. megapotamica (Spreng.) Baker	helenalin (*4*)	[183]
G. mexicana Gray	spathulin (*88*)	[177]
	unknown lactone	[177]
	neoleonin (*90*)[g]	[183a]
G. multiceps Greene	helenalin (*4*)	[183]
G. parryi Greene	flexuosin A (*66*a)	[177]
G. pinnatifida Torr.	helenalin (*4*)	[180]
	bigelovin (*61*)	[180]
	mexicanin I (*62*a)	[180]
	gaillardilin (*67*)	[180]
	gaillardipinnatin (*83*b)	[176, 180, 181]
G. pulchella Foug. coastal race	pulchellin (*22*)	[103–107]
	neopulchellin (*75*)	[107, 184–186]
	pulchellidine (*74*)	[107, 184–186]
	neopulchellidine (*76*)	[107, 184–186]
Rio Grande race	spathulin (*88*)	[177]

Table 2 (cont.)

Species	Compound (formula) pp 154, 160, 161, 166–167	Ref.
Texas race	gaillardin (21)[b]	[100–102]
	isogaillardin[b]	[101]
G. spathulata Gray	spathulin (88)	[177]
B Tetraneurinae Rydb.		
1 Hymenoxys acaulis (Pursh.) K.F. Parker	fastigilin C (86 a)	[187]
H. anthemoides (Juss.) Cass.	vermeerin (19)	[188]
	psilotropin (floribundin, 91)	[188]
	anthemoidin (93)	[188]
	themoidin (92)	[188]
H. grandiflora (Pursh.) K. F. Parker <	paucin (69)	[189]
	hymenograndin (77)	[189]
	florigrandin (79)	[189]
	hymenoflorin (78)	[189]
H. greenei (Cockll.) Rydb.	psilotropin (floribundin, 91)	[188]
	greenein (20)	[188]
H. linearifolia Hook.	linearifolin A (86 b)	[182]
	linearifolin B (87)	[182]
H. linearis T.+G.	mexicanin I (62 a)	[164]
	neohelenalin (56)	[164]
H. odorata DC.	paucin (69)	[188]
	hymenoratin (80)	[164, 190]
	hymenoxynin (94)	[188]
	hymenolide (95)[h]	[188]
H. richardsonii (Hook.) Cockll. var. floribunda	{ vermeerin (19) }	[188]
	{ psilotropin (floribundin, 91) }	[188]
H. rusbyi (Gray) Cockll.	paucin (69)	[182]
	psilotorpin (floribundin, 91)	[182]
H. subintegra Cockll.	psilotropin (floribundin, 91)	[182]
C Riddelliinae Hoffm.		
1 Baileya multiradiata Harv. and Gray	fastigilin C (86 a)	[123]
	baileyolin[i]	[123]
	germacranolides	[123]
B. pauciradiata Harv. and Gray	paucin (69)	[123]
B. pleniradiata Harv. and Gray	paucin (69)	[123]
	plenolin (82)	[123]
	baileyolin[i]	[123]
	fastigilin (86 a)	[123]
	pleniradin (96)	[124]
	radiatin (85 c)	[124]
	germacranolides	[123]
2 Psilostrophe cooperi (Gray) Greene	psilotropin (floribundin, 91)	[125]
II Heliantheae Cass.		
A Galinsoginae Benth. and Hook.		
1 Balduina angustifolia (Pursh.) Robins	helenalin (4)	[38]
B. atropurpurea Harper	atropurpurin[c]	[191]
B. uniflora Nutt.	balduilin (25)	[27, 39, 192]
III Inuleae Cass.		
A Buphthalminae Less.		
1 Geigeria africana Gries	vermeerin (19)	[193]
	guaianolides	[194]
	xanthanolides	[194–197]

Table 2. (cont.)

Species	Compound (formula) pp 160, 161, 166–167	Ref.
G. aspera Harv.	vermeerin (19)	[198]
	geigerinin (26)	[132, 199]
	guaianolides	[198, 200–203]
B Inulinae Dumort.		
1 *Inula britannica* Bieb.	britannin (89?)	[204, 205]
IV Senecioneae Cass.		
A Senecioninae Dumort.		
1 *Arnica foliosa* Nutt.	arnifolin (58)	[206, 207]
	xanthanolides	[208]
A. montana L.	arnifolin (58)	[206]
	dihydrohelenalin (54 a)	[209]
	tetrahydrohelenalin (55)	[209]
	arnicolide A (54 b)	[209]
	arnicolide B (54 c)	[209]
	arnicolide C (54 d)	[209]
	arnicolide D (54 e)	[209]
	arnicolide E^c	[209]

[a] A norpseudoguaianolide of abnormal sterochemistry.
[b] Guaianolide precursors of pseudoguaianolides.
[c] Unknown structure.
[d] Presumably this recently described species is identical with the unpublished *H. bloomquistii* Rock [153].
[e] Rearrangement product of helenalin.
[f] Unknown stereochemistry.
[g] Partial stereochemistry assumed.
[h] Possibly an artifact.
[i] An isomer of fastigilin A (85 a).

related to Helenium and Gaillardia than *Hymenoxys* (subtribe Tetraneuraneae of Rydberg [118]), all taxa of which produce either helenanolides or variants in which the five-membered ring has undergone further oxidation. A study of three Palafoxianae, *Polypteris integrifolia, Palafoxia linearis* (Cav.) Lag. and *Palafoxia feayi* Gray, gave no lactones (unpublished experiments)).

Other genera of the tribe Heleniae whose naturalness has been questioned [121, 122] have not been investigated thoroughly, although further work is clearly desirable and may contribute to the realignment now under discussion on other grounds. So far, only *Baileya* [123, 124] and *Philostrophe* species [125] in subtribe Riddel-

lianeae have furnished helenanolides, while germacranolides or guaianolides have been isolated from scattered representatives of subtribes Jaumanae [126], Bahianae [127–129], Chaenactidanae [126] and Eriophyllanae [126, 130, 131].

Table 2 reveals that helenanolides and modified helenanolides have been isolated from a few taxa belonging to other tribes. Least surprising perhaps is the occurrence of helenalin (4) or balduilin (25) in *Balduina,* generally placed in Heliantheae but considered to be very closely related to *Helenium.* More interesting is the appearance of vermeerin (19) and geigerinin (26, unspecified configuration; the partial stereochemistry suggested is partially based on the reported [132] NMR spectrum) in some

27 28 a R=H b R=OH 29 30 31 32

Geigeria species of Inuleae, a tribe which characteristically produces eudesmanolides, and the discovery of helenanolides closely related to helenalin in *Arnica foliosa* and *Arnica montana* which are generally included in Senecioneae, a tribe well known for its elaboration of eremophilanes. Possible implications of these findings have been discussed elsewhere [14].

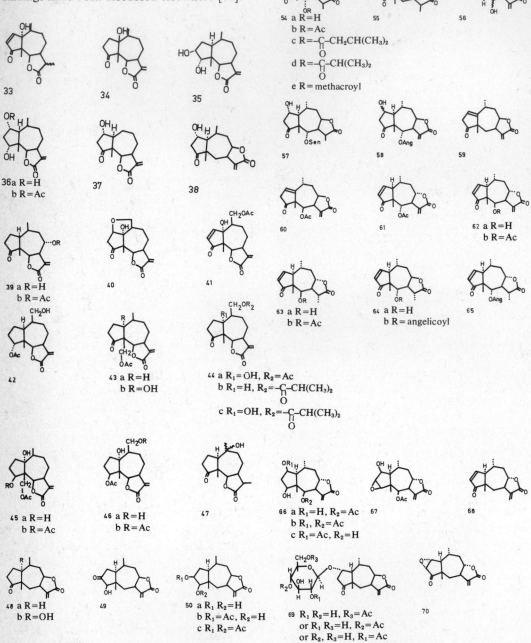

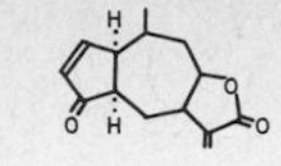

71 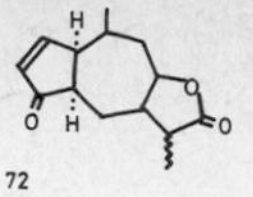 72 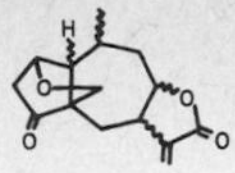 73

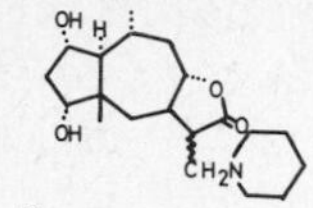

74 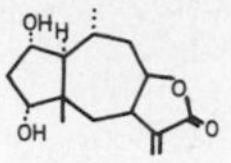 75 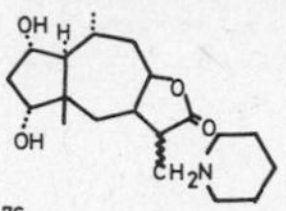 76

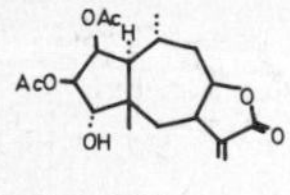

OAc
AcO
77 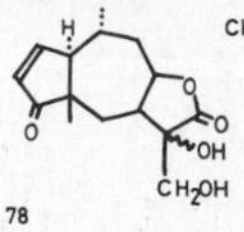 78 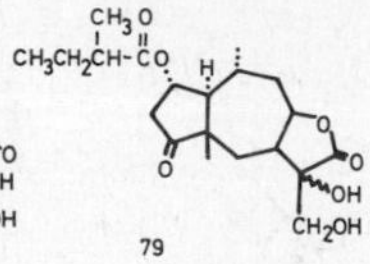 79

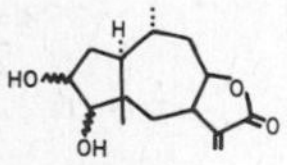

HO
80 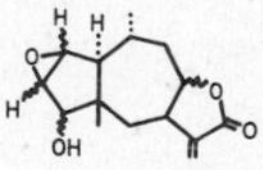 81 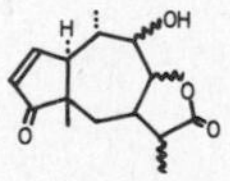 82

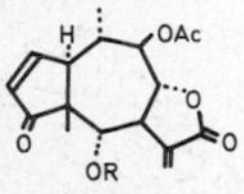

OAc   OAc   OH
83 a R=H   84 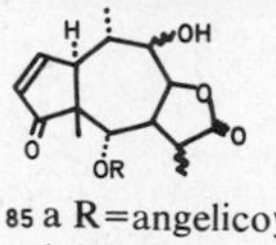   85 a R=angelicoyl
   b R=Ac   b R=senecioyl
                    c R=methacroyl

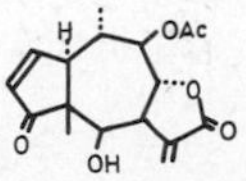

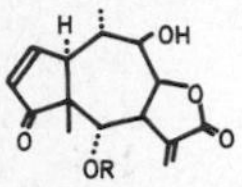

OH   OH   OAc
86 a R=senecioyl 87   88
   b R=tigloyl

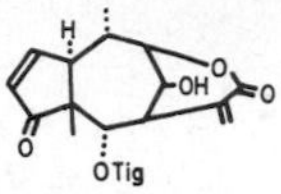

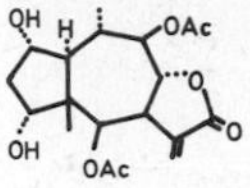

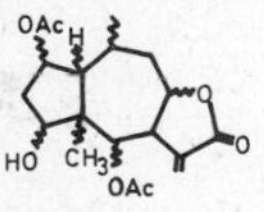

OAc   OH OAc
89 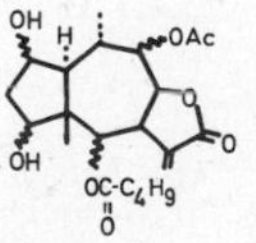   90 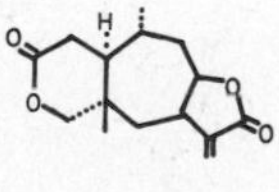   91

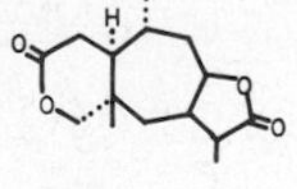

92 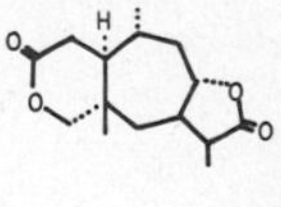 93 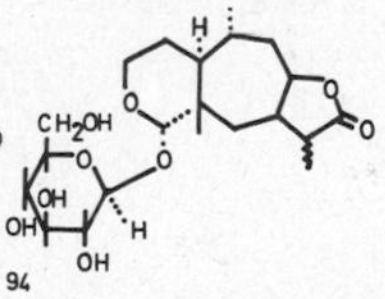 94

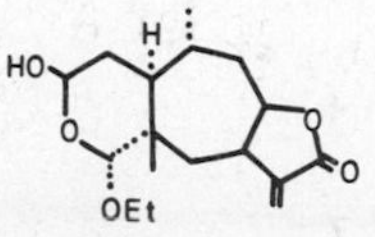

HO   OH
OEt   HO
95 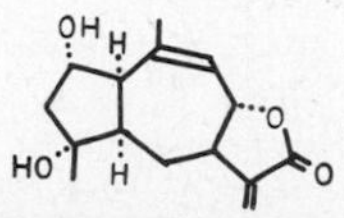   96

Work in the author's laboratory has been supported by grants from the USPHS (CA-13121) through the NCI, the NSF (GP-12582) and Hoffman-La Roche Inc.

References

1. Herz, W, Sesquiterpene lactones in Compositae, in Pharmacognosy and phytochemistry (ed H Wagner & L Hörhammer) p. 64. Springer, Berlin, Heidelberg, New York, 1971.
2. Talapatra, S K, Patra, A & Talapatra, B, Chem comm 1970, 1534.
3. Tada, H & Takeda, K, Chem comm 1971, 1390.
4. Serkherov, S V, Khim prir soedin 1971, 7, 590, 667; 1972, 8, 63.
5. Kiryalov, N P, Bukreeva, T V & Gindin, V A, Khim prir soedin 1972, 8, 446.
6. Mincione, E & Iavarone, C, Chem ind (Milan) 1972, 54, 525.
7. Doskotch, R W, Keely, S L & Hufford, C D, Chem comm 1972, 1137.
8. Holub, M, Motl, O, Samek, Z & Herout, V, Coll czechoslov chem commun 1972, 37, 1186.
9. Holub, M, Samek, Z & Herout, V, Phytochemistry 1972, 11, 2053.
10. Perold, G W, Muller, G-C & Ourisson, G, Tetrahedron 1972, 28 5797.
11. Connolly, J D & Thornton, I M S, Phytochemistry 1973, 12, 631.
12. Romo, J & Romo de Vivar, A, The pseudoguaianolides, in Fortschritte der Chemie organischer Naturstoffe (ed L Zechmeister) vol. 25, p. 90. Springer, Wien, New York, 1967.
13. Herz, W, Pseudoguaianolides in Compositae, in Recent advances in phytochemistry (ed T J Mabry, R E Alston & V C Runeckles) vol. 1, p. 229. Appleton-Century-Crofts, New York, 1968.
14. Herout, V, Chemotaxonomy of the family Compositae (Asteraceae), in Pharmacognosy and phytochemistry (ed H Wagner & L Hörhammer) p. 64. Springer, Berlin, Heidelberg, New York, 1971.
15. Herout, V & Šorm, F, Chemotaxonomy of the sesquiterpenoids of the Compositae, in Perspectives in phytochemistry (ed J B Harborne & T Swain) p. 139. Academic Press, New York and London, 1969.
16. Šorm, F & Dolejs, L, Guaianolides and germacranolides. Herrmann, Paris, 1965.
17. Herz, W, Miyazaki, M & Kishida, Y, Tetrahedron lett 1961, 82.
18. Herz, W, Watanabe, H, Miyazaki, M & Kishida, Y, J Am chem soc 1962, 84, 2601.
19. Abu-Shady, H & Soine, T D, J Am pharm assoc 1958, 42, 387.
20. — Ibid 1954, 43, 365.
21 a. Arny, H V, J pharm 1890, 121.
21 b.— Ibid 1897, 169.
22. Herz, W, Watanabe, H & Miyazaki, M, J Am chem soc 1959, 81, 6088.
23. Bernardi, L & Büchi, G, Experientia 1957, 13, 466.
24. Šorm, F, Suchy, M & Herout, V, Coll czechoslov chem commun 1959, 24, 1548.
25. Emerson, M T, Caughlan, C N & Herz, W, Tetrahedron lett 1966, 3151.

26. Herz, W, Rohde, W A, Rabindran, K, Jayaraman, P & Viswanathan, N, J Am chem soc 1962, 84, 3857.
27. Herz, W, Romo de Vivar, A, Romo, J & Viswanathan, N, J Am chem soc 1963, 85, 19.
28. Ungnade, H E & Hendley, E C, J Am chem soc 1948, 70, 3921.
29. Ungnade, H E, Hendley, E C & Dunkel, W, Ibid 1950, 72, 3818.
30. Barton, D H R & de Mayo, P, J chem soc 1956, 143.
31. Brown, B H, Herz, W & Rabindran, K, J Am chem soc 1956, 78, 4423.
32. Clark, E P, J Am chem soc 1939, 61, 1836.
33. — Ibid 1940, 62, 597.
34. Adams, R & Herz, W, J Am chem soc 1949, 71, 2546, 2551, 2554.
35. Büchi, G & Rosenthal, D, J Am chem soc 1956, 78, 3860.
36. Lamson, P D, J pharmacol exptl therap 1913, 4, 471.
37. Clark, E P, J Am chem soc 1936, 58, 1982.
38. Herz, W & Mitra, R B, J Am chem soc 1958, 80, 4878.
39. Herz, W, Romo de Vivar, A, Romo, J & Viswanathan, N, Tetrahedron 1963, 19, 1359.
40. Rogers, D & Mazhar-ul-Haque, Proc chem soc 1963, 92.
41. Emerson, M T, Caughlan, C N & Herz, W, Tetrahedron lett 1964, 621.
42. Parker, W, Roberts J S & Ramage, R, Quart rev 1967, 21, 331.
43. Hendrickson, J B, Tetrahedron 1959, 7, 82.
44. Govindachari, T R, Joshi, B S & Khamat, V N, Tetrahedron 1965, 21, 1509.
45. Bianchi, F, Culvenor, C C J & Loder, J W, Aust j chem, 1968, 21, 1108.
46. Augustine, R L. Private communication.
47. Porter, T H, Mabry, T J, Yoshioka, H & Fischer, N H, Phytochemistry 1970, 9, 199.
48. Herz, W & Högenauer, G, J org chem 1961, 26, 5011.
49. Herz, W, Anderson, G, Gibaja, S & Raulais, D, Phytochemistry 1969, 8, 877.
50. Miller, H E, Mabry, T J, Turner, B L & Payne, W W, Am j bot 1968, 55, 316.
51. Romo, J, Romo de Vivar, A, & Diaz, E, Tetrahedron 1968, 24, 5625.
52. Joseph-Nathan, P & Romo, J, Tetrahedron 1966, 22, 1723.
53. Romo, J, Joseph-Nathan, P, Romo de Vivar, A & Alvarez, C, Tetrahedron 1967, 23, 529.
54. Geissman, T A & Turley, R J, J org chem 1964, 29, 2553.
55. Mabry, T J, Renold, W, Miller, H E & Kagan, H B, J org chem 1966, 31, 681.
56. Kagan, H B, Miller, H E, Renold, W, Lakshmikantham, M V, Tether, L R, Herz, W & Mabry, T J, J org chem 1966, 31, 1629.
57. Mabry, T J, Miller, H E, Kagan, H B & Renold, W, Tetrahedron 1966, 22, 1139.
58. Mabry, T J, Kagan, H B & Miller, H E, Tetrahedron 1966, 22, 1943.
59. Miller, H E & Mabry, T J, J org chem 1967, 32, 2929.
60. Yoshioka, H, Mabry, T J & Miller, H E, Chem comm 1968, 1679.
61. Yoshioka, H & Mabry, T J, Tetrahedron 1969, 25, 4767.
62. Tori, K, Horibe, I, Yoshioka, H & Mabry, T J, J chem soc (B) 1971, 1084.
63. Geissman, T A, Griffin, S, Waddell, T G & Chen, H H, Phytochemistry 1969, 8, 145.
65. Romo, J, Joseph-Nathan, P & Siade, G, Tetrahedron 1966, 22, 1499; see also Romo, J, Romo de Vivar, A & Diaz, E, Tetrahedron 1968, 24, 5625.
65. Fischer, N H & Mabry, T J, Tetrahedron 1967, 23, 2527.
66. Romo, J, Romo de Vivar, A, Velez, A & Urbino, E, Can j chem 1968, 46, 1535.
67. Fischer, N H, Mabry T J & Kagan, H B, Tetrahedron 1968, 24, 4091.
68. Renold, W, Yoshioka, H & Mabry T J, J org chem 1970, 35, 4214.
69. Yoshioka, H, Renold, W, Fischer, N H, Higo, A & Mabry, T J, Phytochemistry 1970, 9, 823.
70. Higo, A, Hamman, Z, Timmermann, B N, Yoshioka, H, Lee, J, Mabry, T J & Payne, W W, Phytochemistry 1971, 10, 2241.
71. Payne, W W, J Arnold arbor 1964, 45, 401.
72. Geissman, T A, Turley, R J & Murayama, S, J org chem 1966, 31, 2269.
73. Herz, W, Chikamatsu, H & Tether, L R, J org chem 1966, 31, 1632.
74. Mabry, T J, Phytochemical phylogeny (ed J B Harborne) p. 269. Academic Press, London and New York, 1970.
75. Herz, W, Fitzhenry, B & Anderson, G D, Phytochemistry 1973, 12, 1181.
76. Herz, W & Anderson, G D. Unpublished.
77. Herz, W, Raulais, D & Anderson, G D, Phytochemistry 1972, 12, 1415.
78. Jackson, R C, Univ Kans sci bull 1960, 41, 793.
79. Yoshioka, H, Higo, A, Mabry, T J, Herz, W & Anderson, G D, Phytochemistry 1971, 10, 401.
80. Herz, W, Chikamatsu, H, Viswanathan, N & Sudarsanam, V, J org chem 1967, 32, 682.
81. Chikamatsu, H & Herz, W, J org chem 1973, 38, 585.
82. Herz, W & Högenauer, G, J org chem 1962, 27, 905.
83. Herz, W & Viswanathan, N, J org chem 1964, 29, 1022.
84. Herz, W, Högenauer, G & Romo de Vivar, A, J org chem 1964, 29, 1700.
85. Herz, W, Romo de Vivar, A & Lakshmikantham, M V, J org chem 1965, 30, 118.
86. Herz, W, Sudarsanam, V & Schmid, J J, J org chem 1966, 31, 3232.
87. Herz, W & Sudarsanam, V, Phytochemistry 1970, 9, 895.
88. Herz, W, Chikamatsu, H, Viswanathan, N & Sudarsanam, V, J org chem 1967, 32, 682.
89. Anderson, G, McEwen, R & Herz, W, Tetrahedron lett 1972, 4423.

90. Herz, W, Bhat, S V & Sudarsanam, V, Phytochemistry 1972, 11, 1829.
91. Herz, W & Bhat, S V, J org chem 1970, 35, 2605.
92. — Phytochemistry 1973, 12, 1737.
93. Fischer, N H, Wiley, R & Wander, J D, Chem comm 1972, 137.
94. Neidle, S & Rogers, D, Chem comm 1972, 140.
94 a. Bhacca, N S, Wiley, R A, Fisher, N H & Wehrli, F W, Chem comm 1973, 614.
95. Joshi, B S, Kamat, V N & Fuhrer, H, Tetrahedron lett 1971, 26, 2373.
96. Kartha, G, Go, K T & Joshi, B S, Chem comm 1972, 1327.
97. Ali, A, Dastidar, P P G, Pakrashi, S C, Durham, L J & Duffield, A M, Tetrahedron 1972, 28, 2285.
98. Rios, T, Romo de Vivar, A & Romo, J, Tetrahedron 1967, 23, 4265.
99. Anderson, G D, Gitany, R, McEwen, R S & Herz, W, Tetrahedron lett 1973, 2409.
100. Kupchan, S M, Cassady, J M, Bailey, J & Knox, J R, J pharm sci 1965, 54, 1703.
101. Kupchan, S M, Cassady, J M, Kelsey, J E, Schnoes, H K, Smith, D H & Burlingame, A L, J Am chem soc 1966, 88, 5292.
102. Dullforce, T A, Sim, G A, White, D N J, Kelsey, J E & Kupchan, S M, Tetrahedron lett 1969, 973.
103. Herz, W, Ueda, K & Inayama, S, Tetrahedron 1963, 19, 483.
104. Aota, K, Caughlan, C N, Emerson, M T, Herz, W, Inayama, S & Mazhar-ul-Haque, J org chem 1970, 35, 1448.
105. Yanagita, M, Inayama, S & Kawamata, T, Tetrahedron lett 1970, 131.
106. Sekita, T, Inayama, S & Iitake, Y, Tetrahedron lett 1970, 135.
107. Yanagita, M, Inayama, S, Kawamata, T, Okura, T & Herz, W, Tetrahedron lett 1969, 207, 4170.
108. Herz, W & Inayama, S, Tetrahedron 1964, 20, 341.
109. Herz, W & Roy, S K, Phytochemistry 1969, 6, 661.
110. Yoshioka, H, Mabry, T J, Dennis, N & Herz, W, J org chem 1970, 35, 627.
111. Mitchell, D R & Asplund, R O, Phytochemistry 1973, 12, 2541.
112. Bierner, M W, Brittonia 1972, 24, 331.
113. Hikino, H, Kuwano, D & Takemoto, T, Chem pharm bull 1968, 16, 1601.
114. Furukawa, H, Lee, K H, Shingu, T, Meck, R & Piantadosi, C, J org chem 1973, 38, 1722.
115. Solbrig, O, Taxon 1963, 12, 229.
116. Rydberg, P A, North American flora, vol 34, part 2, p. 119, 1915.
117. Herz, W, Mitra, R B, Rabindran, K & Viswanathan, N, J org chem 1962, 27, 4041.
118. Herz, W & Santhanam, P S, J org chem 1967, 32, 507.
119. Wagner, H, Iyengar, M A, Michahelles, E & Herz, W, Phytochemistry 1971, 10, 2547.
120. Herz, W & Bhat, S V, Phytochemistry 1970, 9, 817.
121. Cronquist, A, Am midl nat 1955, 53, 478.
122. Solbrig, O, J Arnold arboretum 1963, 44, 436.
123. Waddell, T G & Geissman, T A, Phytochemistry 1969, 9, 2371.
124. Yoshitake, A & Geissman, T A, Phytochemistry 1969, 8, 1753.
125. de Silva, L B & Geissman, T A, Phytochemistry 1970, 9, 59.
126. Geissman, T A & Atala, S A, Phytochemistry 1971, 10, 1075.
127. Romo de Vivar, A & Ortega, A, Can j chem 1969, 47, 2849.
128. Herz, W, Bhat, S V, Crawford, H, Wagner, H, Maurer, G & Farkas, L, Phytochemistry 1972, 11, 371.
129. Herz, W & Bhat, S V, J org chem 1972, 37, 906.
130. Torrance, S J, Geissman, T A & Chedekel, M R, Phytochemistry 1969, 8, 2381.
131. Saitoh, T, Geissman, T A, Herz, W & Bhat, S V, Rev latinoam quim 1971, 2, 69.
132. de Villiers, J P & Pachler, K, J chem soc 1963, 4989.
133. Doskotch, R W & Hufford, C D, J org chem 1970, 35, 486.
134. Romo, J & Rodriguez-Hahn, L, Phytochemistry 1970, 9, 1610.
135. Geissman, T & Matsueda, S, Phytochemistry 1968, 7, 1613.
136. Herz, W & Sumi, Y, J org chem 1964, 29, 3438.
137. Toribio, F P & Geissman, T A, Phytochemistry 1969, 8, 313.
138. Geissman, T A & Toribio, F P, Phytochemistry 1967, 6, 1563.
139. Toribio, F P & Geissman, T A, Phytochemistry 1968, 7, 1623.
140. Farkas, L, Nogradi, M, Sudarsanam, V & Herz, W, J org chem 1966, 31, 3228.
141. Novikov, V I, Rybalko, K S & Koreshchuk, K E, J obshch khim 1964, 34, 4120.
142 a. Novikov, V I, Forostyan, Y N, Popa, D P, Khim prir soedin 1968, 4, 193, 386.
142 b. — Ibid 1969, 5, 375, 487.
142 c. — Ibid 1970, 6, 29.
142 d. Samek, Z, Holub, M, Novikov, V I, Forostyan, Y N & Popa, D P, Coll czech chem comm 1970, 35, 3818.
143. Rüesch, H & Mabry, T J, Tetrahedron 1969, 25, 805.
144. Yoshioka, H, Rüesch, H, Rodriguez, E, Higo, A, Mears, J A, Mabry, T J, Calzado, J G & Dominguez, X A, Tetrahedron 1970, 26, 2167.
145. Romo de Vivar, A, Bratoeff, E A & Rios, T, J org chem 1966, 31, 673.
146. Yoshioka, H, Rodriguez, E & Mabry, T J, J org chem 1970, 35, 2888.
147. Romo de Vivar, A, Aguilar, M, Yoshioka, H, Rodriguez, E, Mears, J A & Mabry, T J, Tetrahedron 1970, 26, 2775.
148. Rodriguez, E, Yoshioka, H & Mabry, T J, Rev latinoam quim 1972, 2, 184.
149. Romo de Vivar, A, Guerrero, C & Wittgreen, G, Rev latinoam quim 1970, 1, 39.

150. Herz, W, Gast, C M & Subramaniam, P S, J org chem 1968, 33, 2780.
151. Herz, W, Subramaniam, P S & Dennis, N, J org chem 1969, 34, 2915.
152. Lucas, R A, Rovinski, S, Kiesel, R J, Dorfman, L & McPhillamy, H B, J org chem 1964, 29, 1549.
153. Herz, W, J org chem 1962, 27, 4043.
154. Romo, J, Joseph-Nathan, P & Diaz, F, Tetrahedron 1964, 20, 79.
155. Lucas, R A, Smith, R G & Dorfman, L, J org chem 1964, 29, 2101.
156. Pettit, G R. Private communication.
157. Herz, W & Subramaniam, P S, Phytochemistry 1972, 11, 1101.
158. Herz, W & Barry, Jr. J L. Unpublished.
159. Parker, B A & Geissman, T A, J org chem 1962, 27, 4127.
160. Herz, W & Lakshmikantham, M V, Tetrahedron 1965, 21, 1711.
161. Herz, W, Mitra, R M, Rabindran, K & Rohde, W A, J Am chem soc 1959, 81, 1481.
162. Herz, W & Srinivasan, A. Unpublished.
163. Herz, W, Jayaraman, P & Watanabe, H, J Am chem soc 1960, 82, 2276.
164. Romo, J, Romo de Vivar, A & Aguilar, M, Bol inst quim univ nacl auton mex 1969, 21, 66.
165. Herz, W, Kishida, Y & Lakshmikantham, M V, Tetrahedron 1964, 20, 979.
166. Romo de Vivar, A & Romo, J, Ciencia (Mex) 1961, 21, 33.
167. — J Am chem soc 1961, 83, 2326.
168. Romo, J, Romo de Vivar, A & Herz, W, Tetrahedron 1963, 19, 2317.
169. Herz, W, Lakshmikantham, M V & Mirrington, R N, Tetrahedron 1966, 22, 1709.
170. Caughlan, C N, Mazhar-ul-Haque & Emerson, M T, Chem comm 1966, 151.
171. Romo, J, Romo de Vivar, A & Joseph-Nathan, P, Tetrahedron lett 1966, 1029.
172. Dominguez, E & Romo, J, Tetrahedron 1963, 19, 1415.
173. Silva, M, J pharm sci 1967, 56, 922.
174. Giral, F & Ladabaum, S, Ciencia (Mex) 1961, 21, 35.
175. Romo de Vivar, A, Rodriguez, L, Romo, J, Lakshmikantham, M V, Mirrington, R N, Kagan, J & Herz, W, Tetrahedron 1966, 22, 3279.
176. Herz, W & Srinivasan, A, Phytochemistry 1972, 11, 2093.
177. Herz, W, Rajappa, S, Lakshmikantham, M V, Raulais, P & Schmid, J J, J org chem 1967, 32, 1042.
178. Herz, W, Subramaniam, P S & Geissman, T A, J org chem 1968, 33, 3743.
179. Herz, W & Srinivasan, Phytochemistry. In press.
180. Herz, W, Rajappa, S, Lakshmikantham, M V & Schmid, J J, Tetrahedron 1966, 22, 693.
181. Herz, W, Rajappa, S, Roy, S K, Schmid, J J & Mirrington, R J, Tetrahedron 1966, 22, 1907.
182. Herz, W, Aota, K & Hall, A L, J org chem 1970, 35, 4117.
183. Herz, W, & Inayama, S, Tetrahedron 1964, 20, 341.

183 a. Dominguez, X A, Butruille, D & Aubad, A Y, Rev latinoam quim 1970, 1, 136.
184. Yanagita, M, Inoyama, S & Kawamata, T, Tetrahedron lett 1970, 131.
185. Yanagita, M, Inayama, S & Kawamata, T, Tetrahedron lett 1907, 3007.
186. Inayama, S, Kawamata, T & Yanagita, M, Phytochemistry 1973, 12, 1941.
187. Herz, W, Aota, K & Hall, A L, J org chem 1970, 35, 4117.
188. Herz, W, Aota, K, Holub, M & Samek, Z, J org chem 1970, 35, 2611.
189. Herz, W, Aota, K, Hall, A L & Srinivasan, A. Unpublished.
190. Ortega, A, Romo de Vivar, A & Romo, J, Can j chem 1968, 46, 1538.
191. Herz, W & Kannan, R. Unpublished.
192. Herz, W, Mitra, R B & Jayaraman, P, J Am chem soc 1959, 81, 6061.
193. Anderson, L A P, de Kock, W T, Pachler, K G R & Brink, C v d M, Tetrahedron 1967, 23, 4153.
194. de Villiers, J P, S Afric ind chem 1959, 13, 194.
195. — J chem soc 1961, 2048.
196. Anderson, L A P, de Kock, W T, Nel, W & Pachler, K G R, Tetrahedron 1968, 24, 1687.
197. de Kock, W T & Pachler, K G R, Tetrahedron 24, 1701.
198. Rimington, C & Roets, G C S, Onderstepoort j vet sci 1936, 7, 485.
199. de Villiers, J P, J chem soc 1959, 2412.
200. Perold, G W, J chem soc 1957, 47.
201. Barton, D H R & Levisalles, J E D, J chem soc 1958, 4518.
202. Hamilton, J A, McPhail, A T & Sim, G A, Proc chem soc 1960, 278.
203. Barton, D H R, Pinhey T J & Wells, R J, J chem soc 1964, 2518.
204. Rybalko, K S, Sheichenko, V I, Maslova, G A, Kiseleva, E Y & Gubanov, I A, Khim prir soedin 1968, 4, 251.
205. Chugunov, P V, Sheichenko, V I, Bankovskii, A F & Rybalko, K S, Khim prir soedin 1971, 7, 276.
206. Evstratova, R I, Sheichenko, V I, Rybalko, K S & Bankovskii, A K, Khim farm zh 1969, 3, 39.
207. Evstratova, R I, Bankovskii, A I, Shevchenko, V I & Rybalko, K S, Khim prir soedin 1971, 7, 270.
208. Holub, M, Samek, Z & Thoman, J, Phytochemistry 1972, 11, 2627.
209. Poplawski, J, Holub, M, Samek, Z & Herout, V, Coll czech chem comm 1971, 36, 2189.

Added in proof: 114 a. A recent X-ray analysis has led to the conclusion that *24 a* and *24 b* are *cis*-fused guaianolides and possess a *cis*-fused lactone ring as well; McPhail, A T, Luhan, P A, Lee, K H, Furukawa, H, Meck, R, Piantadosi, C & Shingu, T, Tetrahedron lett 1973, 4087. Hence these compounds are probably not guaianolide precursors of pseudoguaianolides. The isolation of a eudesmanolide and a guaianolide from *Laserpitium siler* L. ssp *siler* (Umbelliferae) has been reported recently; Holub, M, Popa, O P, Herout, V & Šorm, F, Coll czechoslov chem commun 1970, 35, 3226; Holub, M, Samek, Z, Popa, D P & Herout, V, Ibid 1973, 38, 1804.

Discussion

Turner: You mentioned that *Helenium hoopesii* was now treated as part of *Dugaldia* by Bierner; the latter genus was said to not show lactones; does *H. hoopesii* possess these?

Herz: We have found none to date, thus supporting Bierner's treatment.

Geissman: Does a report of "no lactones" mean that none could be isolated or that a lactonic chromophore is absent in the extracts?

Herz: In the case of *"Helenium hoopesii"*, if lactones are present, they must be present in very very small quantity.

Geissman: An alternative to the cyclization process can be illustrated by the compounds douglanine and santamarine in which the l-hydroxyl group is β- in one case, α- in the other. The compound artemorin is an α-hydroxygermacronolide which contains an exocyclic methylene group at which cyclization can occur to lead to the l-α-OH compound. Cyclization of the 1/10 epoxide derived from a *trans-trans* germacrodienolide leads to the l-β-OH compound.

Lavie: In the biogenetic sequences you did not refer to the possibility of epoxides being involved in the initiation of rearrangements or cyclization reactions. Such approaches are well accepted now. And secondly, how widespread are the epoxides containing compounds in this series which to my belief should be intermediates in the biogenetic pathway.

Herz: I am of course fully aware that epoxides are likely to be involved in cyclization reactions. Scheme IV shows very clearly that I consider the "old-fashioned" idea of postulating attack by "OH$^+$" equivalent to acid-induced cyclization of an epoxide; if I occasionally use the former representation it is because I am interested in conveying to our botanical colleagues the idea that the *cis* or *trans* nature of the double bonds is important in deciding the stereo-

chemistry of the products. To answer the second part of your question, epoxygermacrenolides which might be intermediates in the biogenetic pathway are found fairly frequently in genera of Composites that do not yield pseudoguaianolides, and have occasionally been found in those species yielding ambrosanolides or helenanolides which I have described as variable. Such species are enumerated in tables 1 and 2.

Ourisson: One should not be too dogmatic about concertedness for the reactions involved in the rearrangements of methyls and hydrogen atoms. In the 5:7 systems, several such rearrangements are known to occur in vitro with NO concertedness (i.e. with CO groups). I refer in particular to work by Luu Bang with several sesquiterpene epoxides.

Turner: You note the distribution of melampolides among several genera which on morphological grounds would appear to be "primitive" types. Could you tell us whether or not in a biosynthetic sense, you consider this group of compounds to also be "primitive".

Herz: They would fall into the so called biosynthetically "primitive" class, in my scheme.

Bu'Lock: Let me make a serious plea that we should all try to avoid unwarranted applications of the word "primitive" in connection with the basic steps in biosynthetic sequences; it exemplifies the dangers which grow from imperfect communication between two groups of people working on different sides of a topic. A phylogenetic dendrogram necessarily has "primitive" species at its base. A biosynthetic chart leading from a basic reaction through alternative steps to ultimate peripheral diversity looks dangerously like such a dendrogram but it is not the same thing. The basic steps are essential in that the peripheral products cannot be realized if they are absent, but an accumulation of the product of the basic steps, rather than of a peripheral product, is not necessarily "primitive" in the phylogenetic sense and indeed it may be quite the reverse; It is, surely, only the totality of evidence that can legitimately lead to statements of such phylogenetic importance.

Herz: I believe that Professor Turner and I have used the term "primitive" only in the sense that melampolides represent an earlier stage in what I have suggested is a possible biogenetic route to the helenanolides.

Geissman: The presence of $\triangle^{4,5}$ santanolides in some plants suggests the intermediacy of a C_4 carbonium ion, stabilized in some way (by an enzyme?), because laboratory cyclizations of some germacronolides which could lead to such santanolides, lead instead to the $\triangle^{3,4}$ or 4-methylene compounds, the expected products of concerted protonation-cyclization-deprotonation reactions.

Herz: I agree. For example, acid-catalyzed cyclization of costunolide leads to $\triangle^{3,4}$- and $\triangle^{4,5}$- cyclocostunolide only. On the other hand I recall a report that thermal cyclization results in the formation of all three possible isomers, the $\triangle^{3,4}$-, $\triangle^{4,5}$- and $\triangle^{4,5}$-cyclocostunolides.

Distribution of Tetracyclic Triterpenoids of Lanostane Group in Pore Fungi (Basidiomycetes)*

S. Natori,[1] A. Yokoyama[1] (née Kanematsu)
and K. Aoshima[2]

[1]National Institute of Hygienic Sciences, Tokyo, and [2]Government Forest Experimental Station, Tokyo, Japan

Summary

Ether extracts of fruit bodies of Aphyllophorales, Basidiomycetes (mainly of Polyporaceae and related families) were examined for their triterpenoidal and steroidal constituents. Although steroids were found to occur in all fungi, triterpene acids of the lanostane group proved to be characteristic metabolites for certain groups of fungi, causing brown-rot of woods. On the other hand, white-rot fungi were found not to produce the triterpene acids.

Lanosterol (1), first isolated from wool-fat, is now well known as the intermediate for the biosynthesis of sterols from squalene in animals and fungi. In higher plants, cycloartenol (2) plays this role instead [1, 2]. Although sterols occur widely in almost all plants and animals, the presence of the triterpenoids of lanostane and the related groups such as protostane, cycloartane and cucurbitane from limited sources have been reported [3, 4]. From Basidiomycetes, chiefly of Polyporaceae, about 40 compounds of the lanostane group, especially of C_{30} or C_{31} carboxylic acids such as trametenolic acid (3) and eburicoic acid (4), have been isolated and the structures elucidated [5, 6]. The rather wide-ranging occurrence of the lanostane derivatives in Basidiomycetes shows a contrast to the occurrence of fusidane and protostane derivatives such as (5) and (6) in some Ascomycetes [6, 7].

Although the chemistry of the lanostane group of triterpenoids has been firmly established [3], the application of modern methods might promote further development in this field. We therefore initiated the survey on the triterpenoids of wood-rotting fungi chiefly of Polyporaceae and the related families. The results so

far obtained [8–11] have not been very fruitful from the point of view of natural product chemistry, but we have learned that the occurrence of the triterpene acids is exclusively limited to brown-rot fungi, though not to white-rot fungi. The results are presented in this paper.

Fungal Materials and Identification of Triterpenes and Steroids

The materials used in this study are the fruit bodies (in some cases, sclerotia) of about 100 species of wood-rotting fungi, mostly of Polyporaceae (tables 1, 2). Most of them were collected in Japan and their taxonomical situation is shown in table 3.

The air-dried fruit bodies including context and trama (ca 5–20 g) were extracted with ether and the extracts were examined preliminarily for the presence of sterols and triterpenoids by thin-layer (tlc) and gas (glc) chromatography. When they showed the presence of the triterpene carboxylic acids, they were treated with hexane to remove hexane-insoluble acids.

The ether extract or the hexane solution was then applied for preparative tlc and the steroid fraction was collected. The steroids were identified by glc before and after trimethylsilylation. For further confirmation the steroidal mixture was applied for mass spectra as the trimethylsilyl ethers and compared with those of the authentic samples [12].

The hexane-insoluble fraction was methylated with diazomethane and the reaction mixture was applied on glc and tlc [11, 12]. The methyl ester recovered from tlc plates was examined by IR and mass spectra and again by glc. Oc-

---

* Read by S. Natori.

Table 1. *Fungi producing triterpene acids*

Species	Family	Compound	Method of identification
Daedalea dickinsii Yasuda	P	*14*	IR, NMR
		11, 12	UV, IR, NMR
		15	glc, NMR
D. tanakae (Murr.) Aoshima	P	*14*	IR, NMR
Fomitopsis pinicola (Swartz ex Fr.) Karst.	P	*10*	IR, NMR
		(14)	NMR
Gloeophyllum abietinum (Bull. ex Fr.) Karst.	P	*4, 3*	glc, IR
G. odoratum (Wulf. ex Fr.) Imaz.	P	*3*	IR, NMR
G. sepiarium (Wulf. ex Fr.) Karst.	P	*4, 3*	glc, IR
G. striatum (Swartz ex Fr.) Murr.	P	*4, 3*	glc, IR
G. trabeum (Pers. ex Fr.) Murr.	P	*4, 3*	glc, IR
Laetiporus sulphureus (Bull. ex Fr.) Bond. et Sing.	P	*4, 3*	glc, IR, NMR, MS
Lentinus lepideus Fr.	T	*4, 3*	glc, IR, NMR, MS, GC–MS
Melanoporia juniperina Aoshima	P	*14*	IR
		(11)	tlc, IR
M. rosea (Alb. et Schw. ex Fr.) Aoshima	P	*14*	IR, NMR
		11	IR, NMR
		25	IR, NMR, MS
M. nigra (Berk. et Curt.) Murr.	P	*(14)*	tlc, IR
		(11)	tlc, IR
		(25)	tlc
M. cajanderi (Karst.) Aoshima	P	*(14)*	tlc, IR
		(11)	tlc, IR
M. purpureacea Aoshima	P	*?*	tlc
Piptoporus betulinus (Bull. ex Fr.) Karst	P	*14*	IR, NMR
Poria cocos (Fr.) Wolf (sclerotium)	P	*13*	glc, IR
		14	IR, NMR
		17	IR, NMR, MS
Spongiporus appendiculatus (Berk. et Br.) Aoshima	P	*4 (3)*	glc, IR
		11	glc, IR
Veluticeps angularis (Lloyd) Aoshima et Furukawa	V	*3*	IR, NMR

P, Polyporaceae; T, Tricholomataceae; V, Veluticepsacea.

casionally, gas chromatography–mass spectrometry [11] and isolation in pure forms followed by identification by spectral methods (IR, NMR) were also carried out [8–11].

The results thus obtained are summarized in tables 1, 2 and fig. 1.

The fungal sterols have been studied from old times [13] and the major sterol of higher fungi has been assumed to be ergosterol *(7).* However, reliable identification by modern methods has been carried out in only a few cases [5, 6, 14–23]. Our survey showed that all fungi examined contained sterols and that the most abundant of these is not ergosterol *(7)* but ergosta-7,22-dien-3β-ol (5,6-dihydroergosterol) *(8),* as far as the fungi examined in this study are concerned. Of course, ergosterol *(7)* is widely distributed as the minor constituent along with

ergosta-7-en-3β-ol (fungisterol) *(9).* In some fungi such as *Grifola umbellata* and *Poria cocos* (where sclerotium was used), as well as in *Fomitopsis pinicola* and *Lentinus lepideus*, ergosterol *(7)* was found to be the major sterol. Although there are many unidentified spots or peaks in tlc or glc there was no sign of the presence of other sterols, methylsterols, dimethylsterols or trimethylsterols (lanostane derivatives lacking the carboxyl group) in abundance except in the case of certain fungi.

The triterpenoids identified in the course of the study were: trametenolic acid *(3),* eburicioic acid *(4),* 3α-acetoxylanosta-8,24-dien-21-oic acid *(10),* tumulosic acid *(11),* dehydrotumulosic acid *(12),* pachymic acid *(13),* polyporenic acid C *(14),* carboxyacetyl-quercinic acid *(15),* and inotodiol *(16).* In

Table 2. *Fungi not producing triterpene acids*

POLYPORACEAE
Abortiporus biennis (Bull. ex Fr.) Sing.
Antrodia mollis (Sommerf.) Karst.
Bjerkandera adusta (Willd. ex Fr.) Karst.
B. fumosa (Pers. ex Fr.) Karst.
Coriolus brevis (Berk.) Aoshima
C. flabelliformis (Klotz) Aoshima
C. hirsutus (Schum. ex Fr.) Quél.
C. meyenii (Klotz) Aoshima
C. pubescens (Schum. ex Fr.) Quél.
C. subluteus Aoshima
C. unicolor (Bull. ex Fr.) Pat.
C. versicolor (L. ex Fr.) Quél.
Cryptoporus volvatus (Peck) Hubbard
Daedaleopsis confragosa (Bolt. ex Fr.) Schroet.
D. lata (Berk.) Aoshima
D. purpurea (Cooke) Imaz. et Aosh.
D. styracina (P. Henn. et Shirai) Imaz.
D. tricolor (Bull. ex Fr.) Bond. et Sing.,
Echinodontium tsugicola (P. Henn.) Imaz.[a]
E. tinctorium Ells et Ev.[a]
Favolus alveolarius (Bosc. ex Fr.) Quél.
F. arcularius (Batsch ex Fr.) Ames
F. varius (Pers. ex Fr.) Imaz.
Fomes fomentarius (L. ex Fr.) Kickx
Fomitella cytisina (Berk.) Aoshima
F. latissima (Bres.) Aoshima
F. rhodophaea (Lév.) Aoshima
Fuscoporia obliqua (Pers. ex Fr.) Aoshima[a]
Ganoderma applanatum (Pers.) Pat.
G. fornicatum (Fr.) Pat.
Gloeoporus amorphus (Fr.) Clem. et Shear
Grifola umbellata (Pers. ex Fr.) Pilát[c]
Hapalopilus croceus (Pers. ex Fr.) Donk
Heterobasidion insularis (Murr.) Aoshima
H. sensitivum (Yasuda) Aoshima,
Hirschioporus abietinus (Dicks. ex Fr.) Donk
H. durus (Jungh.) Aoshima
H. fusco-violaceus (Fr.) Donk
Ishnoderma resinosum (Schrad. ex Fr.) Karst.
Lenzites betulina (L.) Fr.
Leucofomes ulmarius (Sow. ex Fr.) Pouz.
Oxyporus populinus (Schum. ex Fr.) Donk
Perenniporia ochroleuca (Berk.) Donk
P. tephlopora (Mont.) Aoshima
Phaeolus schweinitzii (Fr.) Pat.[b]

Poria nigrescens Bres.
P. subacida (Peck) Sacc.
Porodisculus pendulus (Schw.) Murr.
Pycnoporellus fibrillosus (Karst.) Murr.[b]
Pycnoporus coccineus (Fr.) Karst.
Roseofomes cesatianus (P. Henn.) Aoshima
Trametes gibbosa Fr.
T. orientalis (Yas.) Imaz.
T. palisoti (Fr.) Imaz.
T. suaveolens (L. ex Fr.) Fr.
Tyromyces albellus (Peck) Bond. et Sing.
T. tangerianus Aoshima

HYDNACEAE
Dentipellis echinospora Furukawa sp. nov.

STEREACEAE
Laurilia sulcatum (Burt) Kolt. et Pouz.
Stereum fasciatum (Schw.) Fr.
S. hirsutum (Willd.) Fr.
S. princeps (Jungh.) Lév.
S. taxodii Lentz et McKay

HYMENOCHAETACEAE
Inonotus hispidus (Bull. ex Fr.) Karst.
I. mikadoi (Lloyd) Bond.
I. sciurinus Imaz.
Phellinus gilvus (Schw. ex Fr.) Pat.
P. hartigii (Allesch. et Schnabl) Imaz.
P. igniarius (L. ex Fr.) Quél.
P. linteus (Berk. et Curt.) Aoshima
P. pachyphloeum Pat.
P. robustus (Karst.) Bourd. et Galz.
Porodaedalea abietis (Karst.) Aoshima
P. pini (Thore ex Fr.) Murr.
P. sanfordii (Lloyd) Aoshima

MERULIACEAE
Merulius tremellosus Schrad. ex Fr.

FISTULINACEAE
Fistulina hepatica Fr.

[a] The fungi produce the triterpenes devoid of the carboxyl group, i.e. echinodol derivatives *(18–24)* [10] and inotodiol *(16)*.
[b] The fungi belong to brown-rot fungi.
[c] Sclerotium.

the course of studies, 3β-hydroxylanosta-7,9,-*(11)*, 24-trien-21-oic acid *(27)* [9], seven echinodol derivatives *(18–24)* [10], 12β-hydroxycarboxyacetylquercinic acid *(25)* [11] were newly isolated and the structures were elucidated.

Distribution of Triterpenoids and Their Relation to the Types of Wood-Rotting Fungi

The chemotaxonomy of Basidiomycetes was discussed using amino acids, pigments, urea, amines and other metabolites as chemical

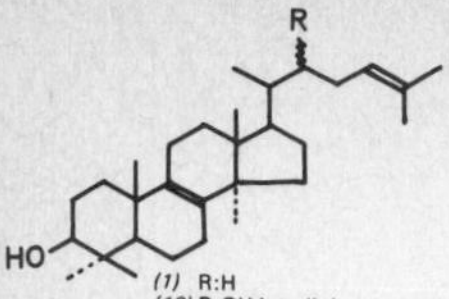

(1) R:H
(16) R:OH intodiol

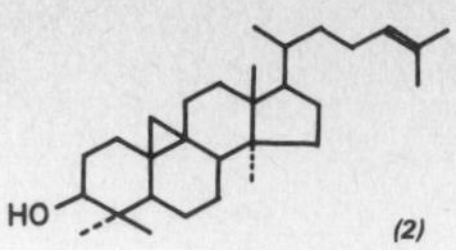

(2)

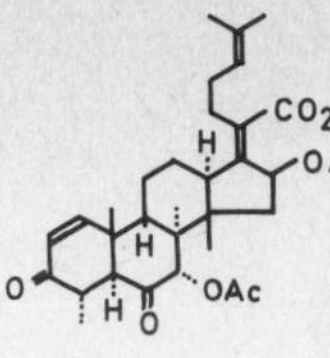

(5)

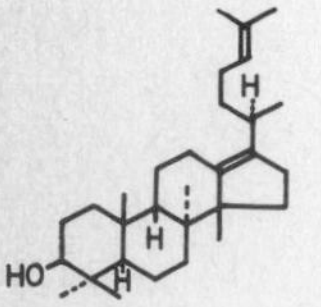

(6)

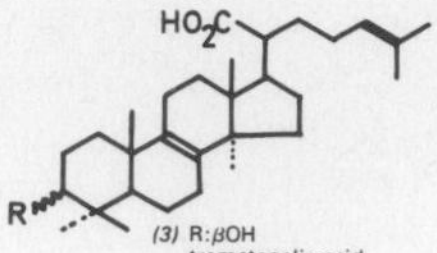

(3) R:βOH
trametenolic acid
(10) R:αOAc

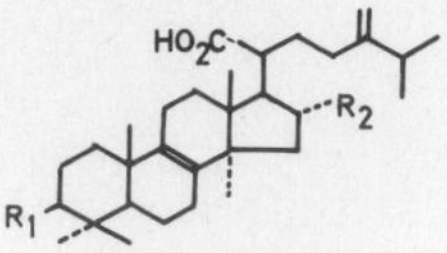

(4) R₁:OH R₂:H eburicoic acid
(11) R₁:R₂:OH tumulosic acid
(13) R₁:OAc R₂:OH pachymic acid

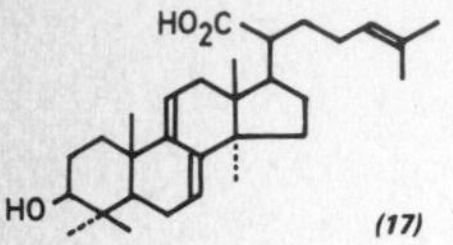

(17)

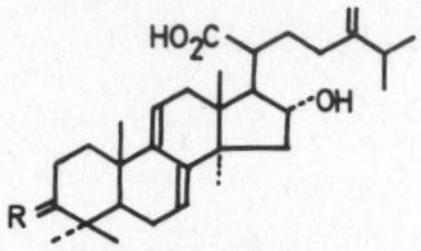

(12) R:βOH dehydrotumulosic acid
(14) R:O polyporenic acid C

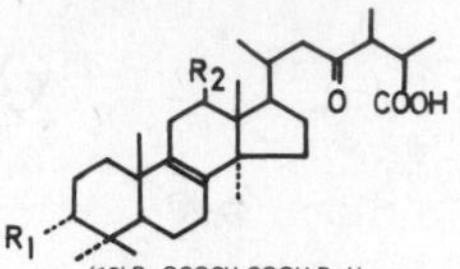

(15) R₁:OCOCH₂COOH R₂:H
carboxyacetylquercinic
acid
(25) R₁:OCOCH₂COOH R₂:OH

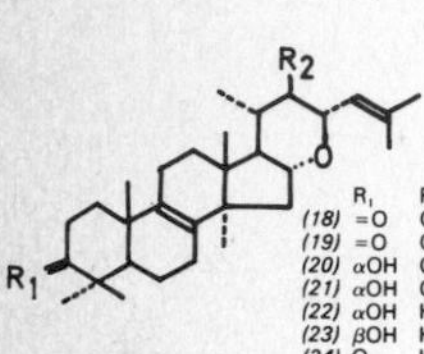

	R₁	R₂
(18)	=O	OAc echinodone
(19)	=O	OH
(20)	αOH	OAc 3-epiechinodol
(21)	αOH	OH
(22)	αOH	H
(23)	βOH	H
(24)	O	H

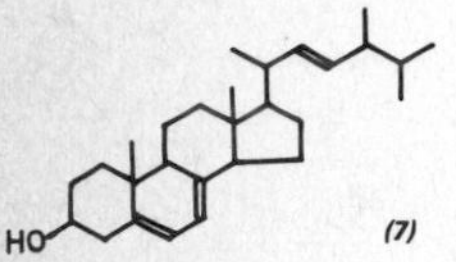

(7)

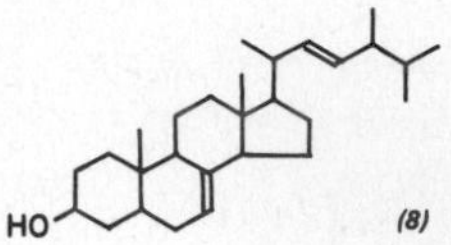

(8)

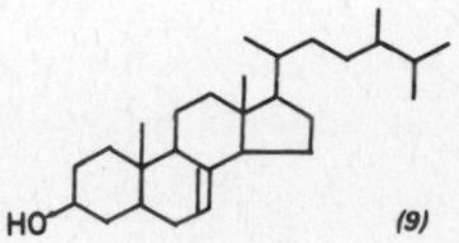

(9)

Table 3. Wood-rotting fungi and classification of Basidiomycetes

	Family	Numbers of fungi examined
Homobasidio-mycetae	Hydnaceae	1
	Polyporaceae	75
Aphyllophorales	Hymenochaetaceae	12
	Stereaceae	5
	Veluticepsacea	1
	Corticiaceae	
	Meruliaceae	1
	Coniophoraceae	
	Fistulinaceae	1
Agaricales	Tricholomataceae	1
	Crepidotaceae	
Heterobasidio-mycetae		
Auriculariales	Auriculariaceae	
Tremellales	Tremellaceae	

markers [24–27]. Fungal phylogeny was also discussed from chemical, metabolic and enzymatic points of view [28].

Since lanostane derivatives were rather widely distributed in higher fungi but not widely in other plants, the triterpenoids are assumed to be usable as chemical markers.

The wood-rotting fungi cover rather wide range of fungi groups, taxonomically, but most of them belong to Basidiomycetes. The main families are shown in table 3, in which the Polyporaceae is predominant. Almost all species in Polyporaceae and related families fall into this category. They are classified into two types: White-rot fungi (lignin decomposer) and brown-rot fungi (cellulose dissolver) and their characteristics are set out in table 4 [29].

At the initial stage of our work one of us (K. A.) pointed out that the fungi producing the

Table 4. *Characteristics of brown-rot and white-rot fungi* [29]

Morphology	Sexual compatibility	Phenol oxidase	Other metabolic system
Brown-rot fungi Mono- or di- mitric hyphae	Bipolar	Tyrosinase (extracellular oxidase lacking)	
Basidiospore mostly smooth surface			
White-rot fungi Di- or tri- mitric hyphae	Tetrapolar	Tyrosinase Laccase oxidase present)	Oxalic acid metabolism
Ornamentation of spore surface			Amyloidity of basidiospore sometimes present

triterpene acids so far reported in literature and studied by us, mostly belong to the brown-rot fungi type. This has been confirmed finally by the fact that 19 species of the brown-rot fungi produce the triterpene carboxylic acids of C_{30} or C_{31} (table 1), with the exception of two only (*Pycnoporellus fibrillosus* and *Phaeolus schweinitzii*). On the other hand, all 75 species belonging to the white-rot fungi failed to show the presence of acids in detectable amounts (table 2).

Three species (*Echinodontium tinctorium, E. tsugicola* and *Fuscoporia obliqua*), which produce triterpenoids lacking the carboxyl group, belong to the white-rot fungi. Of 25 species appearing in the literature as the source of the triterpene acids [5, 6, 30–47], only a few species such as *Phellinus hartigii* (Allescher et Schnabl) Imaz. (*Fomes hartigii* Allescher et Schnabl) [30], *Ishnoderma resinosum* (Schrad. ex Fr.) Karst. (*Polyporus benzoinus* (Wahl.) Fr.) (from cultured mycelia [31], *Pycnoporus sanguineus* (L. ex Fr.) Karst. (*Polyporus sanguineus* L. ex. Fr.) [32] and *Inonotus hispidus* (Bull. ex Fr.) Karst. (*Polyporus hispidus* (Bull. ex Fr.) Fr.) (from cultured mycelia [33]) belong to the white-rot fungi and others belong to the brown-rot fungi.

According to Nobles [48] the production of extracellular oxidase will be assumed to be an important key for the taxonomy of Polyporaceae and the characteristics shown by white-rot fungi indicate a more advanced type than those shown by brown-rot fungi. This relationship is similar to that of woody plants vs. herbaceous plants and may correlate to the fungal phylogeny [29].

At this stage we cannot explain the meaning of the presence or the absence of the triterpene acids in biogenetic terms but our finding adds one more factor in the distinction of metabolism into the two types of wood-rotting fungi of Aphyllophorales.

References

1. Goodwin, T W, in Rodd's chemistry of carbon compounds (ed S Coffey) vol. 2E, p. 54. Elsevier, Amsterdam, 1971.
2. Good, L J & Goodwin, T W, in Progress in phytochemistry (ed L Reinhold & Y Liwschitz) vol. 3, p. 113. Interscience, London, 1972.
3. Ourisson, G, Crabbé, P & Rodig, O R, Tetracyclic triterpenes. Hermann, Paris, 1964.
4. McCrindle, R & Overton, K H, in Rodd's chemistry of carbon compounds (ed S Coffey) vol. 2C, p. 369. Elsevier, Amsterdam, 1969.
5. Shibata, S, Natori, S & Udagawa, S, List of fungal products, p. 119. University of Tokyo Press, Tokyo, 1964.
6. Turner, W B, Fungal metabolites, p. 252. Academic Press, London, 1971.
7. Ponsinet, P, Ourisson, G & Oehlschlager, A C, in Recent advances in phytochemistry (ed T J Mabry, R E Alson & V C Runeckles) vol. 1, p. 271. North-Holland, Amsterdam, 1968.
8. Kanematsu, A & Natori, S, Yakugaku zasshi 1970, 90, 475.
9. – Chem pharm bull (Tokyo) 1970, 18, 779.
10. – Ibid 1972, 20, 1993.
11. Yokoyama (née Kanematsu), A & Natori, S, Chem pharm bull (Tokyo) 1973, 21. In press.
12. Yokoyama, A, Natori, S & Aoshima, K, Phytochem. To be published.
13. Zellner, J, Monatsh chem 1913, 34, 321.

14. Milazzo, F H, Can j bot 1965, 43, 1347.
15. Cambie, R C & LeQuesne, P S, J chem soc (c) 1966, 72.
16. Singh, P & Rangaswami, S, Ind j chem 1965, 3, 575.
17. Munro, H D & Musgrave, O C, J chem soc (c) 1971, 685.
18. Strigina, L I, Elkin, Yu N & Elyakov, G B, Phytochemistry 1971, 10, 2361.
19. Batta, A K & Rangaswami, S, Phytochemistry 1969, 8, 1309.
20. Endo, M, Kajiwara, M & Nakanishi, K, Chem comm 1970, 309.
21. Cambie, R C, Duve, R N & Parnell, J C, New Zealand j sci 1972, 15, 200.
22. Batey, I L, Pinhey, J T, Ralph, B J & Simes, J J H, Aust j chem 1972, 25, 2511.
23. Weete, J D, Phytochemistry 1973, 12, 1843.
24. Hegnauer, R, Chemotaxonomie der Pflanzen, vol. 1, p. 101. Birkhäuser, Basel, 1962.
25. Tyler Jr, V E, in Evolution in the higher basidiomycetes, An international symposium (1968) (ed R H Petersen) p. 29. University of Tennessee Press, Knoxville, 1971.
26. Arpin, N & Fiasson, J-L, in Evolution in the higher basidiomycetes, An international symposium (1968) p. 63. University of Tennessee Press, Knoxville, 1971.
27. Benedict, R G, Ann rev microbiol 1970, 1.
28. Bartnicki-Garcis, S, in Phytochemical phylogeny (ed J B Harborne) p. 81. Academic Press, London, 1970.
29. Aoshima, K, Abstracts of papers presented at the 20th annual meeting of the Japan wood research society, Tokyo, 1970.
30. Canonica, L & Fiecchi A, Gazz chim ital 1959, 89, 818.
31. Birkinshaw, J H, Morgan, E N & Findlay, W P K, Biochem j 1952, 50, 509.
32. Cambie, R C & LeQuesne, P W, J chem soc (c) 1966, 72.
33. Cort, L A, Gascoigne, R M, Holker, J S E, Ralph, B J, Robertson, A & Simes J J H, J chem soc 1954. 3713.
34. Sheth, K, Cataleomo, P & Sciuchett, L A, J pharm sci 1967, 56, 1656.
35. Villaneva, V P, Phytochemistry 1971, 10, 427.
36. Batey, I L, Pinhey, J T, Ralph, B J & Simes, J J H, Aust j chem 1972, 25, 2511.
37. Hikino, H, Kuwano, D & Takemoto, T, Yakugaku zasshi 1969, 89, 1149.
38. Geigert, J, Stermitz, F R & Schroeder, H A, Phytochemistry 1973, 12, 1491.
39. Simes, J J H, Wootton, M, Ralph, B J & Pinhey, J T, Aust j chem 1971, 24, 609.
40. Inouye, H, Tokura, K & Hayashi, T, Yakugaku zasshi 1972, 92, 621, 859.
41. Anderson, C G, Epstein, W W & Van Lear, G, Phytochemistry 1972, 11, 2847.
42. Devys, M & Barbier M, Bull soc chem biol 1969, 51, 925.
43. Munro, H D & Musgrave, O C, J chem soc (c) 1971, 685.
44. De Reinarch-Hirtzbach F & Ourisson, G, Tetrahedron 1972, 28, 2259.
45. Batta, A K & Rangaswami, S, Ind j chem 1969, 7, 1063.
46. Pinhey, J T, Ralph, B J, Simes, J J H & Wootton, M, Aust j chem 1970, 23, 2141.
47. Cambie, R C, Duve, R N & Parnell, J C, New Zealand j sci 1972, 15, 200.
48. Nobles, M K, in Evolution in the higher basidiomycetes. An international symposium (1968), p. 169. University of Tennessee Press, Knoxville, 1971.

Discussion

Farnsworth: Did you take into account the possibility that triterpene glycosides might be present in your samples, since your extraction procedure (ethyl ether) would have excluded this possibility?

Natori: Yes, there are two reasons why we employed the non-polar solvent. (1) If we use a polar solvent, the separation becomes very complicated because of the great amounts of extracts; (2) as far as we are aware, there have been few examples of the occurrence of glycosides in fungal metabolites. As triterpene glycoside, trametoside [40] is the only case.

Bu'Lock: It is very pleasing to me to see attention once more drawn to the distinction between "white-rot" and "brown-rot" fungi. Although initially this is based on what may seem to be a rather minor feature, an accumulation of evidence more and more suggests that it may be a fundamental one—despite the fact that it cuts completely across the major morphological divisions. To the distinctions listed by Natori one might add, for example, that secondary metabolism of C_6C_3 and C_6C_1 compounds seems to be more highly developed in the white rots. So far as the triterpene acids are concerned, we have encountered them rather frequently ourselves, and a spectrophotometric screening which we carried out some years ago showed that the characteristic 242 nm absorption of the doubly-exocyclic diene system is an easily observed and very widespread characteristic; unfortunately we have never analysed these results on a taxonomic basis as you have now encouraged us to do. But may I query whether your work-up procedure would pick out the known cases in which these triterpenoids occur linked to malonic, mevalonic, and possibly other dibasic acids?

Natori: Yes. The identifications of compounds (*15*) and (*25*), in fig. 1, p. 176 are the cases. As for the characteristic 7,9(11)-diene absorption, we have also employed UV for determination of the ratio of 8-enes and the contaminated 7,9, (11)-dienes: On some occasions we learned that the contamination of the diene is rather small in amount.

Ourisson: (1) It is nice to see some order brought to this group, if the subdivision really can be substantiated. (2) Have the *Inonotus* species studied yielded any inotodiol?

Natori: The three species examined did not show any sign of production of the compound though the original source, *I. obliquus* (syn. *Fuscoporia obliqua* in table 2) was found to produce the compound.

Ourisson: It may interest the literary-minded that inotodiol has been reported to be the active constituent of the purportedly anti-tumour drug *tschaga* mentioned by Solzhenitzin in "Cancer Ward". We have work in progress to check this on synthetic material.

Farnsworth: With regard to Ourisson's remarks that the triterpene present in the "tschaga" material of the Russians is probably not the active material: Recently, betulin and oleanolic acid have been shown to have in vivo antitumour activity against the Walker 256 (IM) carcinosarcoma in rats. There are also other triterpenes that have antitumour activity.

Applying Chemistry to Genetics in Certain Solanaceae

D. Lavie

Department of Organic Chemistry, The Weizmann Institute of Science, Rehovot, Israel

Summary

The occurrence of the steroidal lactones of the withanolide type in the different chemotypes of *Withania somnifera* (L.) Dun. (Solanaceae) have been studied. The changes involved in the substitution pattern of these compounds, produced by different genetic combinations of the natural types were investigated through different combinations of cross breeding. These experiments provided an insight into the enzymatic reactions governed by their genetic background, determining thereby the dominant or recessive character of certain chemical groupings in the molecule.

Within the framework of a chemical survey of the Israeli flora, the leaves of *Withania somnifera* (L.) Dun. (Solanaceae) have been extensively investigated for their non alkaloidal constituents. In the early stages of this investigation the structure of withaferin A, a steroidal lactone having structure (*1*), was elucidated [1]. Following the detailed chemical studies, biological bioassays were undertaken, and larger quantities of this material had to be made available. As a result of these needs, quantities of leaves of this plant were collected and, to our dismay, the required compound withaferin A could not be obtained from samples obtained from this plant collected in various areas of Israel. A systematic study of this species was then initiated in order to clarify the possibility of ontogenetic changes, or the chemical variability of the species [2].

Our interest in withaferin A (*1*) and related compounds was raised by some interesting physiological properties ascribed to this compound. In addition to a reported bacteriostatic activity [3], the compound has also been found to display cytotoxic properties on experimental tumors in mice [4].

From the plants obtained from various sources and different geographical areas, a large number of such steroidal lactones have been isolated and identified. The general name of "withanolide" has been given to this group of compounds [5, 6] which is characterised by a basic steroidal skeleton having a nine carbon atoms side chain, being therefore a C-28 system.

Withanolide skeleton

A characteristic six-membered ring lactone is present in the side chain and a 1-keto-Δ^2-system was found to be a rather general feature of the molecule; it has been observed to occur in ring A of the skeleton in almost all cases. From the biogenetic point of view, the withanolide can be considered to have a cholestane type structure with an additional methyl group at C-24, and with various oxygenated groups or double bonds placed at different sites of the skeleton.

These withanolides have been observed to occur as well in certain related genera as *Physalis, Nicandra, Acnistus* and *Jaborosa,* the latter two being known to occur in Central and South America. These species have been studied to various extent by different laboratories.

Withania somnifera is known to have a fairly wide geographical distribution. When 24 populations of this species from different parts of Israel were examined for their leaves content in withanolides, 3 well defined chemotypes were disclosed.

We attribute the name chemotype to groups of plants which morphologically are similar in all respects, however, differ by their chemical constituents, in the present case through their content of withanolides in the leaves. Roots

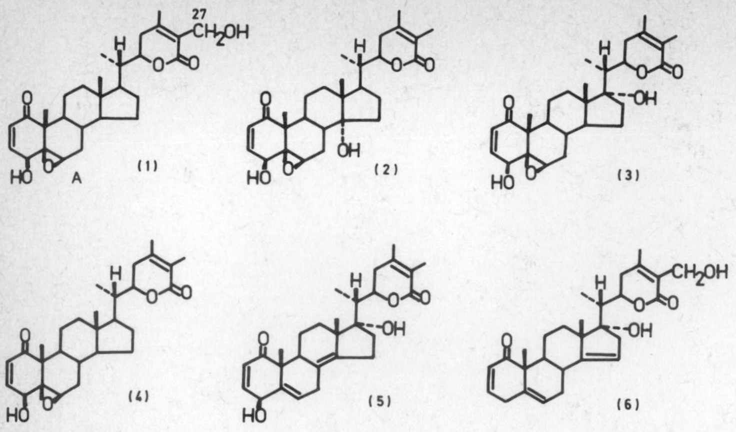

Fig. 1. Chemotype I.

were also analysed, and only in few single species were these compounds detected in that part of the plant.

The 3 chemotypes were identified as I, II and III, each showing a different combination of withanolides with various substitution patterns [2]. During our studies on these chemotypes, exhaustive analyses were performed, attempting very carefully to isolate and identify *all* the steroidal constituents detectable in the leaves. The chemical analysis of these compounds has been the object of several publications. In the present discussion we will present the various compounds found in each of the types refraining from any chemical comments.

The compounds present in chemotype I are shown in fig. 1. The predominant product, the one being accumulated in the leaves is withaferin A (*1*) (4 g from 1 kg of dry leaves), whereas all the other compounds (*2*) to (*6*) were obtained in quantities varying from 15 to 45 mg. By observing these compounds it can be seen that an interesting character inherent to this plant is its ability to introduce OH groups at various sites of the molecule, C-14 or C-17 in (*2*) and (*3*) or C-27 in (*1*). However, an interesting observation in this case is the fact that no OH-group is found at position C-20. Furthermore, the pattern seen in rings A/B, the 4β-OH, 5,6β-epoxy system is present in most cases, and quantity-wise it is a predominant feature, since (*1*) is produced in the largest amount [(*5*) may be considered a precursor of (*3*)].

The compounds identified in chemotype II are shown in fig. 2. In this group the compound

Fig. 2. Chemotype II.

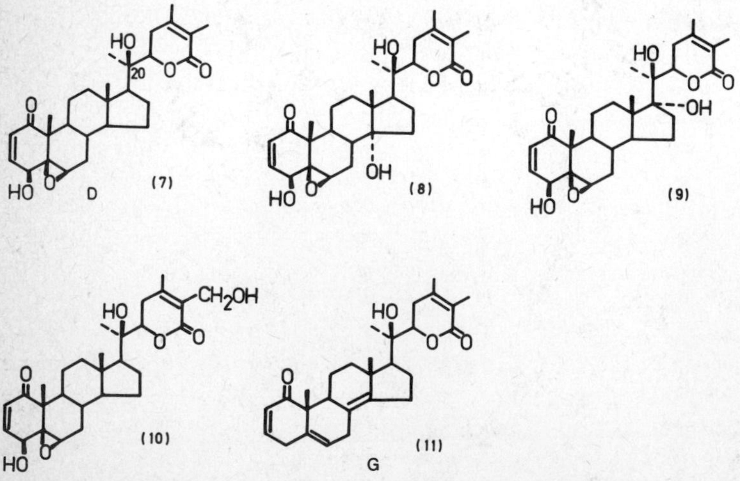

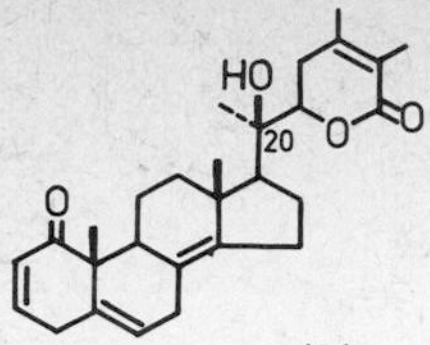

Fig. 3. Chemotype III.

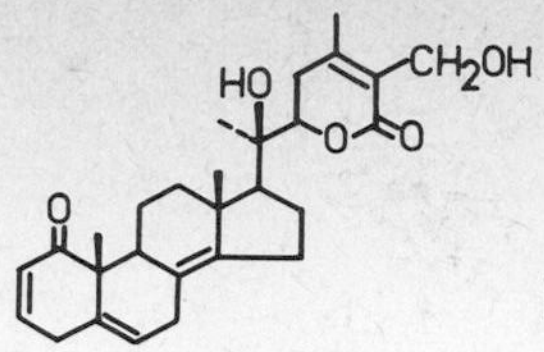

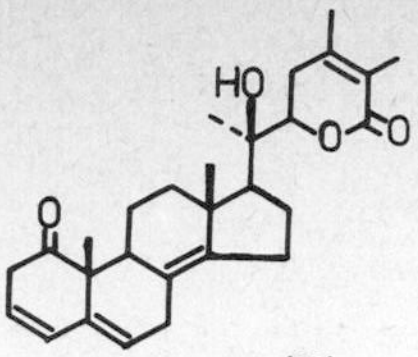

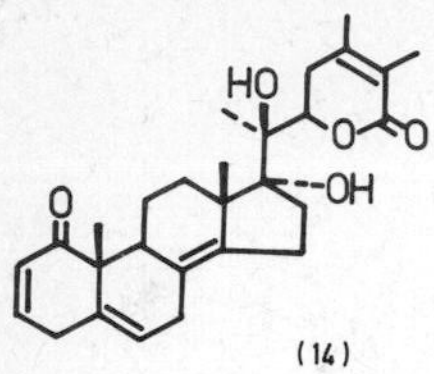

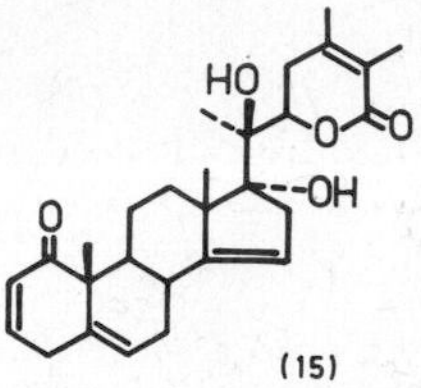

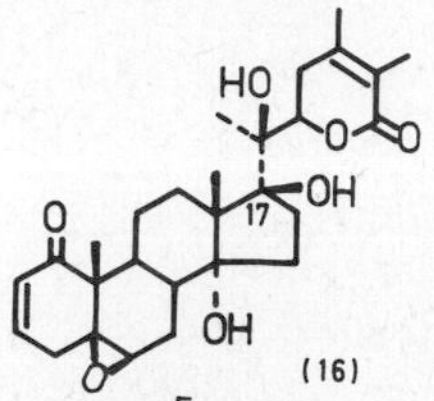

present in the largest quantity, and which accumulates is withanolide D [7] (*7*), (0.53 g/kg dry leaves). All the other components were found in quantities of tens of mg. The outstanding feature in all these compounds is the presence of an OH-group at C-20 in all structures. Here again the 4β-OH, 5,6β-epoxy system is present in all cases, with the exception of withanolide G (*11*) which can be looked upon as a precursor of this group; its presence in small quantities may reflect the detection of some "unreacted material". The ability of the plant to introduce OH-groups in various positions of the skeleton, in addition to the C-20 can be noted again, OH-groups being observed at positions C-14 compound (*8*), C-17 (*9*) and C-27 (*10*). Still, these substances are produced only in small quantities.

In chemotype III 9 steroidal lactones have been isolated from the leaves, they have been labelled withanolide E to M [8]. A selected number of compounds relevant to the present discussion are shown in fig. 3. Two groups of compounds have been identified in this type, those having an unsaturated system: 3 double bonds not conjugated between themselves in rings A, B and C as shown in (*11*) to (*15*), however, all possessing an OH-group at position C-20; and a second group represented by withanolide E (*16*) in which the side-chain is α-oriented. Interestingly this compound has been observed to have a 17β-oriented OH-group and will be referred therefore as 17β-OH; the com-

pound has an *inverted* stereochemistry at C-17 which to our knowledge is the first instance whereby an α-oriented side chain (a complete 9 carbon chain) has been observed to occur in nature. The structure of withanolide E has been determined by X-ray analysis [9]. For a better appreciation of the relation existing between the different compounds in this type, it is worth reporting that from 1 kg of dry leaves the following quantities were obtained: withanolide E (*16*) 5.25 g, G (*11*) 0.22 g and J (*14*) 0.15 g.

In order to ascertain that each of the chemotypes does not consist of mixtures of plant specimens, random collection of single plants in selected populations were examined. The analytical method used primarily thin layer chromatographic techniques, following a careful study of the "finger print" of the spots distribution of each type. In order to determine the reliability of the method, different solvent systems were initially examined. Each population in its proper area was found to belong to one chemotype. These have usually a rather definite geographical distribution area: chemotype I occurring in the southern and central part of Israel, chemotype II being located in the northern part of the country and chemotype III in the southern coastal plain [2].

The possibility of edaphic influences on the formation of the compounds in various types had to be considered. Therefore, samples of each of the types were collected from different places in their natural distribution area and analysed without detecting any alterations in their gross compositions. The constancy of the types were also proven by raising them from seeds in a uniform nursery producing offsprings of the corresponding types showing the same content of withanolides as those obtained from their natural habitat. In parallel, the possible ontogenetic changes in the withanolide content were studied in plants of each of these chemotypes. Plants grown in the nursery were examined in 5 different phenological stages. No qualitative changes of the composition in withanolides was detected during the growing period. Some quantitative changes may be found during the various stages of the development of the plant.

A comparison of the morphological characteristics of the chemotypes was carried out on plants of the same age in the nursery. No differences could be found between the types. Only a difference in flowering time was observed. The only variation is therefore of chemical nature and specifically on the content of the steroidal lactones of the withanolide type.

Since from the above considerations it could be deduced that the differences existing in the formation of the various compounds are of a genetic character, a confirmation was sought through cross-breedings by pollination between the different types. Such crossings would provide an understanding of the mechanisms of the biogenetic pathways, as well as of the transformations taking place in the withanolide skeleton; they would also enable a genetic control leading to new chemotypes. To this end a set of reciprocal crossings was performed.

For example, the cross of chemotypes I×III (or III×I) gave offsprings F_1 which showed in the leaves the accumulation of withanolide D (7). This compound which was never detected in chemotypes I and III, the parties in this cross, happened to be the major constituent of the withanolides present in II. This offspring F_1 was not identical, however, with type II as seen on a chromatoplate, the set of spots observed in each of them not being identical throughout the path of the developed mixture. Compound (1) could not be seen or detected in this offspring, and compound E (16) was present in lesser quantitites than in the parent type. The same was true for substance G (11)

which occurred in a reduced amount. A combination of genes had obviously taken place which induced the observed changes in the oxygenated substitution pattern of the products.

The crossing of I×II produced again primarily withanolide D (7), whereas II×III produced a mixture having compound (7) as well as the 17β-OH compound E (16) as dominating products. These results implied that the enzymatic systems involved in the reactions producing compound (7), namely hydroxylation and epoxidation, were dominant factors in the genetic pool.

These observations could by far be better analysed and understood when the second generation F_2 of I×III was studied. Some 43 single plants of this F_2 generation raised from seeds obtained by self pollination were analysed for their constituents. The identification of the plant groups were done by thin layer chromatography, and whenever required larger quantities of material were isolated from the leaves, and the compounds were identified using nuclear magnetic resonance and mass spectrometric methods.

In order to obtain some statistics, a count was performed on the number of times that a certain substituent would appear at a given position of the skeleton. In this case the compounds involved were not necessarily the same. In this specific experiment, and for this count the contribution of withanolide E was not taken into account, since it was obvious from various considerations that for this system, having an inverted side chain (17β-OH), an enzymatic combination of a rather different nature was involved. Indeed during the various crossings no changes were observed with this molecule. The results have been collected in table 1. It can be seen that the number of times that an OH-group will be present at C-4 is 33, against 10 times when this position will be occupied by an H atom. A similar relationship occurs for a 5,6β-epoxy-group present 33 times against a double bond (Δ^5) observed 10 times only. The 8 and 14 positions will have a double bond 10 times whereas this position will be reduced in 33 cases. The 20-OH will occur in 32 cases whereas the 27-OH will happen only in 11 cases out of the 43 counts.

The relationship between the occurrence and

Table 1. *Count of the repeated occurrence of groupings in the withanolide skeleton in the F_2 generation of the hybridization of chemotypes $I \times III$. Study on 43 single plants*

Position in skeleton	Grouping			
	OH	H	C C O	C=C
C-4	33	10		
C-5,6			33	10
C-8,14		33		10
C-20	32	11		
C-27	11	32		

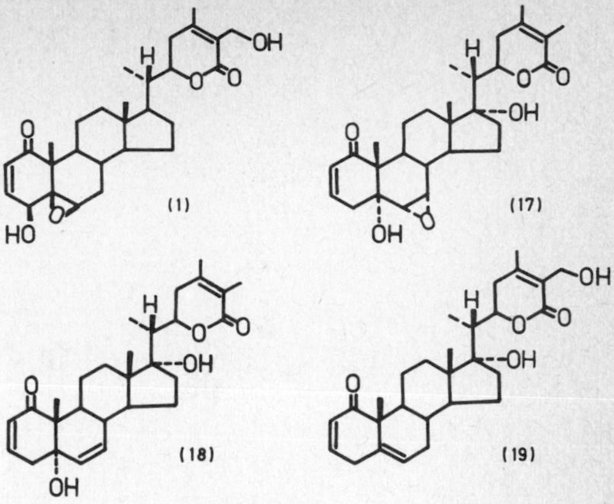

Fig. 4. Chemotype Indian.

non-occurrence of a grouping in a certain position is therefore of about 3:1 or 1:3 depending on the site and the reaction. Since we observe the product of enzymatic reactions which take place in the cell, and are controlled by genetic characters, one can refer to these characters as *dominant* or *recessive* while being observed on the molecular level. For example, the introduction of an OH-group at position C-20 is then a dominant character, whereas the presence of a 27-OH is deliberately a recessive one. The same would be true for the 4β-OH and $5,6\beta$-epoxy groups which are dominant features. It seems, however, that in this case one should separate these two characters from each other since in several cases as in (16), or as with a withanolide called jaborosalactone, (a withanolide from *Jaborosa integrifolia*) [10]), an epoxide may be present at C-5,6 without an OH being attached at C-4. These are characters which have to be studied separately and this is true as one goes around the different sites of the molecule.

In order to widen the scope of these experiments, cross-breedings were also performed with a chemotype of *W. somnifera* of Indian origin. Seeds from this type identified as *Indian I* had been obtained earlier in our laboratory and the plant grown in the nursery. Several samples have been since received from India, and most could be related to this Indian I type. The detailed analyses of this plant disclosed a relatively large number of withanolides of the most varied nature [11]. Some 11 structures were determined out of which the most relevant have been se-

lected for presentation in fig. 4. An outstanding feature is probably the formation of a different substitution pattern observed heretofore, and involving rings A and B, this is the 5α-OH, $6,7\alpha$-epoxy system of compound (17) which accumulates in large quantities, and is therefore a characteristic of this type. A 17α-OH is also present in this structure. Together with this compound, withaferin A (1) is also produced in sizable quantities, so that this plant has the outstanding possibility of producing simultaneously the two systems involving rings A, B, namely the 5α-OH, $6,7\alpha$-epoxide (17) and the 4β-OH, $5,6\beta$-epoxide (1). It is noteworthy, however, that in all the 11 compounds isolated so far from this type, not in a single case is a 20-OH group present in the molecule; obviously this plant does not possess the appropriate enzymatic system required for this specific reaction, and is incapable of introducing an OH-group at this C-20 position.

Following the hybridization of Indian I by chemotype III (fig. 5) an F_1 offspring was obtained in which the compounds which now accumulated were withanolide D (7) having the 20-OH group, and replacing the withaferin A (1), with the 27-OH, and a new withanolide (20) showing the rings A, B features of the Indian type substance, however, bearing also an OH-group at position C-20. A third product obtained from this hybrid is compound (21) having the same substitution pattern in ring A, B but without the 17α-OH, and in which again a 20-OH has been introduced.

Since from our previous crossings the hy-

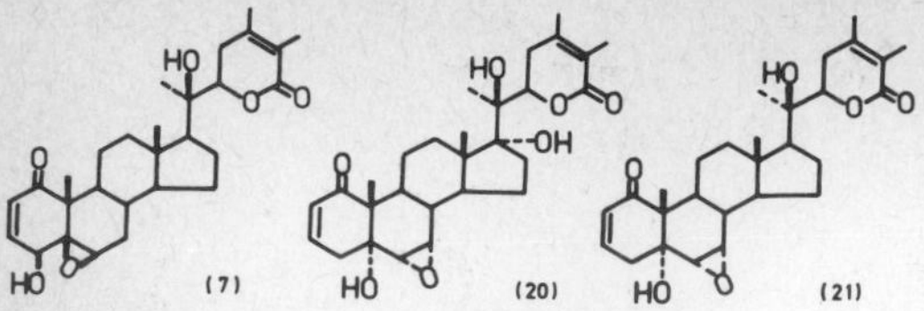

186   D. Lavie

Fig. 5. Cross of Indian I×chemotype III.

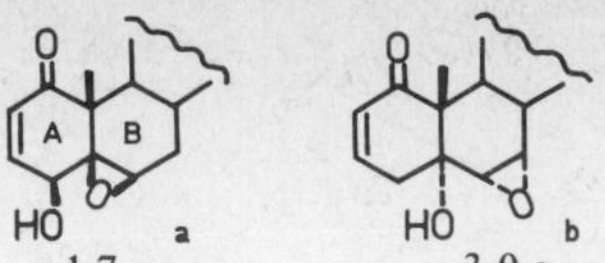

From 1 kg dry leaves	a	b
Indian I type	1.7 g	3.0 g
Indian I×III crossing	2.9 g	1.4 g

Fig. 6. Quantities obtained from different systems.

droxylation at C-20 was found to be a *dominant factor,* the introduction of the 20-OH group in the compound of these offsprings was anticipated, and indeed took place. These results can be considered as a definite planning of a molecule using genetic characters.

Another interesting feature of this hybridization was the fact that withanolide E (*16*) (with the 17β-OH) was produced unchanged in this hybrid, however, in much smaller quantities than in the chemotype III in which it is a predominant component. It is noteworthy that this compound was also formed unchanged in the crossing I×III but again in reduced amount.

So far, only the qualitative aspects of our observations have been succintly described. An attempt was then made to approach the problem in a more quantitative way in order to get a better insight on some of the enzymatic reactions and their limitations. For example, whereas among the substances characteristic for the Indian I type having a 5α-OH, 6,7α-epoxy system, compound (*17*) bearing an OH-group at C-17 was the predominant component (3 g from 1 kg of dry leaves), in the offsprings of the hybridization of Indian I×III, the relation found for the 3 possible substitution patterns shown in (*22*) namely: R_1=OH, R_2=H; R_1=H, R_2=OH; R_1=OH, R_2=OH was roughly of equal amounts (0.40, 0.56 and 0.43 g, respectively). This fact clearly indicated that a definite competition was taking place in the new combination of enzymatic systems produced by the addition of the factor of the 20-OH group,

now coexisting with that operating to introduce the 17α-OH.

A second observation connected with this hybridization deals with the two systems *a* and *b* involving rings A and B (see fig. 6). It has been observed that taking into consideration all the compounds in which these two systems were formed, the quantitative relation between the two is in the Indian I type of 1.7 to 3.0 g. However, following the hybridization with chemotype III (cf fig. 3) this relation was found to be inverted and became 2.9 g for *a* to 1.4 g for *b*. Interestingly, during this crossing, type III did not add into the genetic pool any structural genes enabling the formation of the system described as *a* (the 4β-OH, 5,6β-epoxide). The sizable change in the relation of these two systems has to be attributed therefore to alterations in the appropriate regulating factors (or genes) involved in this hybridization, and accounting for this relative inversion from roughly 33% against 66% in Indian I to 66% against 33% in the hybrid with chemotype III.

It may be too early yet, to draw all the conclusions from these experiments. They do provide, however, a new approach for the chemical understanding of genetic problems, and we propose for such studies the name of "chemogenetics", or the study of genetic problems on the chemical level.

The above description is a succint outlay of a fairly vast series of experiments which will be the subject of appropriate and detailed publications.

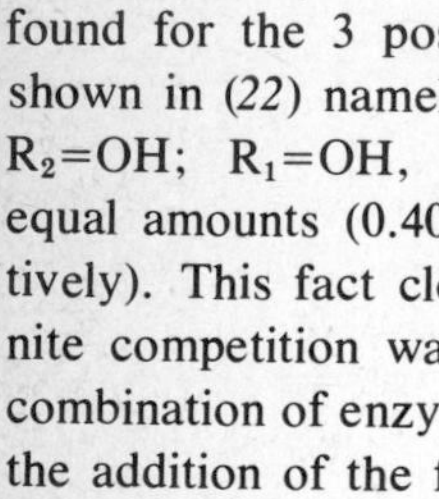

The botanical work has been carried out in collaboration with Mr A. Abraham from the Volcani Centre, Bet Dagan, and is in partial fulfillment of the requirements for a Ph.D. degree to the Hebrew University of Jerusalem, Faculty of Agriculture, Rehovot, Israel.

The chemical work has been performed by Dr I. Kirson and earlier by Dr E. Glotter of our department.

(22)

References

1. Lavie, D, Glotter, E & Shvo, Y, J chem soc 1965, 7517.
2. Abraham, A, Kirson, I, Glotter, E & Lavie, D, Phytochemistry 1968, 7, 957. For the stereochemistry at C-20 which in this publication is erroneous (cf ref. [5]).
3. Ben-Efraim, S & Yarden, A, Antibiot chemotherapy 1962, 12, 576.
4. Shohat, B, Gitter, S & Lavie, D, Int j cancer 1970, 5, 244 (and references cited therein).
5. Kirson, I, Glotter, E, Abraham, A & Lavie, D, Tetrahedron 1970, 26, 2209.
6. Lavie, D, Kirson, I, Glotter, E & Snatzke, G, Tetrahedron, 1970, 26, 2221.
7. Lavie, D, Kirson, I & Glotter, E, Israel j chem 1968, 6, 671 (cf ref. [5]).
8. Glotter, E, Kirson, I, Abraham, A & Lavie, D, Tetrahedron 1973, 29, 1353.
9. Lavie, D, Kirson, I & Glotter, E, Rabinovich, D & Shakked, Z, Chem comm 1972, 877.
10. Tsechsche, R, Schwang, H, Fehlhaber, H W & Snatzke, G, Tetrahedron 1966, 22, 1129.
11. Kirson, I, Glotter, E, Lavie, D & Abraham, A, J chem soc C 1971, 2032.

Discussion

Reichstein: It may be good to point out a special result in Dr Lavie's work which I think is important for practical reasons. When you cross one plant which produces a compound A with a related one producing an other product B (of the same class) it is often found that the hybrid will produce A+B. But it is well known that this is not always so. In some cases the hybrid may produce neither A nor B but a new compound C. Dr Lavie gave an excellent example for such a case and he also suggested the reason. Plant A contains an enzyme introducing a hydroxyl group in a certain position of a basic skeleton giving product A. Plant B contains another enzyme introducing a hydroxyl group in another position of the same skeleton giving B. The hybrid contains both enzymes. If the product of one of them is not attacked by the other a mixture of A+B will be produced. If both enzymes act independently then a compound C (containing both additional hydroxy groups) may result.

Schwarting: If I understood you correctly, you indicated that the Israel chemotypes were in-

distinguishable morphologically. In India the species has been shown to occur in a number of distinct forms. One of these, a form of low growth habit, is cultivated for root production for use as the drug "Ashwaghanda". This form might well represent a form possessing a unique alkaloid content. Have you examined the Israel types for alkaloids and what was the nature of India sample you examined?

Lavie: We have refrained from examining the alkaloidal content of our chemotypes and of the other members of these species since it was outside the scope of our work. I understand that you have been doing that work and we avoided duplication. Concerning the Indian chemotype, it is a *Withania somnifera* from which we had obtained seeds and have grown them in our experimental plots. Since, various samples of *W. somnifera* obtained from India were found to be of the same chemotype as seen on the finger print of their t.l.c. We have also analysed *Withania coagulans* from that country. Concerning "Ashwaghanda", it is a cultivated medicinal plant grown in India since several centuries. Its botanical origin is obscure and unknown to us, it is probably a product of agricultural selection for high yields of roots. There is a possibility that it has not been developed from a local Indian *Withania*, since its chemical composition is quite different. It should be looked upon as a separate species which still requires more work along the lines of our presentation.

Heywood: You mentioned that the distribution of the different chemotypes had a certain geographical basis. Can I ask how extensively you sampled the populations since detailed sampling often reveals that variation forms a clinal pattern.

Lavie: When the material was collected in the field, single plants of each population were carefully tested and identified for their character. The populations of plants were found to be chemically homogenous in each locality. A fairly large number of tests were performed during our survey and for each locality, and no clinal changes could be detected neither on the morphology nor on the chemistry. This plant occurs as patches and not in large continual areas which could enable clinal observations.

Birch: The work is a very good example of the enzymic elasticity often found in connection with secondary metabolites. In primary metabolism, the enzymes appear to be linked very closely to the structure of their normal metabolite. The fact that enzymes of secondary metabolism can often deal with altered metabolites and still transform them, means that alteration at any early stage of biosynthesis can still lead to a series of structures parallel to, but different from, the virginal series. This very general point should be taken into account in considering genetic aspects of modifications in secondary compounds.

Bu'Lock: We have exposed a striking instance of the complications which can arise through genome interaction through our study of the effect of the single-gene mating-type locus in Mucorales. In these fungi, the reactions leading to sex hormone synthesis are manifest only at a repressed level, but the pattern of repression is different for the two mating-type loci. Consequently, single strains carry out, at a low level, different selections from the total reaction sequence and so produce mating-type-distinctive metabolites. In mixed culture—which is of course a much more limited situation than that in hybrids and which in this case does not even involve any plasmogamy—the pathway is fully de-repressed and what is in effect a third type of metabolite (i.e. the sex hormones) is produced. This does not occur simply by the complementation of two incomplete genomes, but because this is the complete expression of the genome in each strain; the mating-type-specific products of the single strains are selective de-repressors for the opposite mating-type.

I mention this concrete instance from lower organisms simply to show just how complex may be the type of interactions we should anticipate in the case of genuine hybridization effects on biosynthetic pathways in higher plants, and so as an amplification of the comments of Reichstein and Birch.

Turner: In connection with this general discussion, I would like to note that the late Dr Alson, with assistance from Dr Mabry and others, about 8 years ago called attention to the existence of "new" flavonoids in hybrid plants, i.e. compounds that were not found in either parent. One explanation of this phenomenon was that genomic interaction caused the expression of such flavonoids in developing leaves which might usually occur in, say, the earliest developmental stages. My question is, did you examine root material for your so-called "hybrid" compound in the plants concerned?

Finally, in connection with Dr Reichstein's remarks, I would like to call attention to a paper published 6 or more years ago by an American worker using isogenic corn strains. He reported the occurrence of "hybrid" protein bonds in crossing experiments among such strains, but it is possible that they also arose through genomic interactions, bringing to the fore (electrophoretically speaking) isozymes not normally active at the later development stages.

Lavie: Concerning the question of the roots content in withanolides, in most cases, they were not found in the roots of *Withania somnifera*. In certain other species, for example *W. coagulans* they may have been detected in that part of the plant.

As to the location where the withanolides are found in relation with the specificity of the types, I would like to refer to an experiment whereby chemotype I was grafted on a different species, *Nicotiana glauca*, which is known not to produce these steroidal lactones in any part of the plant. The result was a nice large branch of leaves of *W. somnifera* of chemotype I, as shown by the leaves analysis, which grew out of the graft. It seems clear therefore that the biosynthetic reactions take place in the leaves.

The Use of Alkaloids in Determining the Taxonomic Position of *Vinca libanotica* (Apocynaceae)*

G. H. Aynilian,[1] *N. R. Farnsworth*[1] *and J. Trojánek*[2]

[1]Department of Pharmacognosy and Pharmacology, College of Pharmacy, University of Illinois at the Medical Center, Chicago, Ill. 60612, USA and [2]Research Institute for Pharmacy and Biochemistry, Prague, Czechoslovakia

Summary

Several classification systems have been proposed for the taxonomic differentiation of *Vinca* species. These systems are reviewed and compared with the various types of indole alkaloids reported from all of the *Vinca* species that have been chemically examined to date. Based on recent results in our laboratories, the following alkaloids were isolated from *Vinca libanotica,* which allowed us to suggest the best taxonomic position for this species: vincamajine, reserpinine, tabersonine, herbamine, herbadine, querbrachidine, rauniticine, vincamidine, vincamidine hydrate, picrinine, venalstonine and vincoline.

On the basis of these findings, it is clear that *Vinca libanotica* should be classified as a distinct species, with the botanical classification system of Novacék & Stary seeming most appropriate for the genus *Vinca* on chemotaxonomic grounds.

Botanical Classifications Involving *Vinca*

The genus *Vinca* was originally established by Linnaeus in 1735 [1]. The main area within which species of *Vinca* are native, extends eastwards from Morocco, Algeria, Portugal, Spain and France over central and southern Europe to southwestern European Russia, including the Crimea and the north Caucasus and northern Iran. Outside this area there is only one species, *V. erecta,* isolated far to the east of it in central Asia (Turkistan and Afghanistan).

Pichon [2, 3], in his reviews on the classification of the Apocynaceae considers *Vinca* as follows:

Family: Apocynaceae
  Subfamily: Plumieroideae
    Tribe: Alstonieae

* Read by N. R. Farnsworth.

Sub-tribe: Catharanthinae
  Genus: *Vinca* L.
    Vinca herbacea Waldst. et Kit.
    (syn. *V. erecta* Regel. et Schmalh.)
      var. *herbacea* M. Pichon
      var. *libanotica* (Zucc.) M. Pichon
      var. *sessilifolia* (DC.) M. Pichon
    Vinca major L.
      var. *major* M. Pichon
      (syn. *V. pubescens* Urv.)
      var. *difformis* (Pourr.) M. Pichon
    Vinca minor L.
      var. *minor*
      var. *nammulariaefolia* Fournier

Pobedimova [4] considered *V. erecta* as a distinct species, independent from *V. herbacea.* He also considered *V. major* and *V. pubescens* completely independent. The classification of the genus *Vinca* according to Pobedimova is as follows:

Family: Apocynaceae
  Subfamily: Plumieroideae
    Genus: *Vinca* L.
      Vinca herbacea Waldst. et Kit.
      Vinca erecta Regel. et Schmalh.
      Vinca major L.
      Vinca pubescens Urv.
      Vinca minor L.

Stearn [5] has an extensive classification of the genus *Vinca,* where he recognizes 16 species belonging to 6 sections. Section Herbacea consists of 7 species, *V. libanotica* being one of them. Thus Stearn considers *V. libanotica* as a distinct species. He also considered *V. major* as comprised of 2 subspecies, *major* and *hirsuta*, and *V. difformis* to be comprised of subspecies *difformis* and *sardoa*. Finally, he recognized a new Section, Balancia, with one species, *V. balancia*. He considered *V. erecta* and *V. minor* as monotypic. The classification of the genus *Vinca* according to Stearn is as follows:

Family: Apocynaceae
  Subfamily: Plumieroideae
    Genus: *Vinca* L.
      Section: Herbacea
        V. herbacea Waldst. et Kit.
        V. pumila E. D. Clarke
        V. libanotica Zucc.
        V. sessilifolia DC.
        V. bottae Jaub. & Spach.
        V. mixta Vel.
        V. hausknechtii Bornm. et Sint.
      Section: Erecta
        V. erecta Regel. et Schmalh.
      Section: Major
        V. major L.
          subspecies: *major*
          subspecies: *hirsuta* (Boiss.) Stearn
          (syn. *V. pubescens* Urv.)
      Section: Difformis
        V. difformis Pourr.
          subspecies: *difformis*
          subspecies: *sardoa* Stearn
      Section: Minor
        V. minor L.
      Section: Balancia
        V. balancia Penzes.

From the above classification it is obvious that the taxonomy of the genus *Vinca* is not well established. Since alkaloids are not ubiquitous in nature, their distribution in the plant kingdom can serve as a very important criterion in chemotaxonomic studies.

Previous Chemotaxonomic Studies

During the last two decades, three important attempts were made to solve the taxonomic uncertainties of the genus *Vinca* from a chemotaxonomic point of view, by studying the distribution of alkaloids in each species.

The first of these studies was conducted by Paris & Moyse [6] in 1957, who carried out paper electrophoresis on several species. Their results were very interesting, in that they found that the alkaloid extracts of whole plants of *V. herbacea* and *V. libanotica* (collected from Lebanon) had basically different electrophoreograms. Analogous differences were also observed in the case of alkaloid extracts obtained from the similarly related taxa *V. major* and *V. difformis*, thus seriously questioning Pichon's classification.

No further studies were conducted until 1965 when Janot and co-workers [7] carried out a detailed chemotaxonomic investigation by studying the distribution of indole alkaloids in *Vinca* species.

By this time it was known, as the result of numerous feeding experiments, that the $C_{9,10}$ components of the indole alkaloids were derived by way of the acetate-mevalonate pathway through an iridoid skeleton, which were incorporated into indole alkaloids through condensation with tryptamine. Since the non-tryptophan aliphatic $C_{9,10}$ units can be incorporated in three basic forms, their incorporation can result in three basic classes of indole alkaloids—A, B and C.

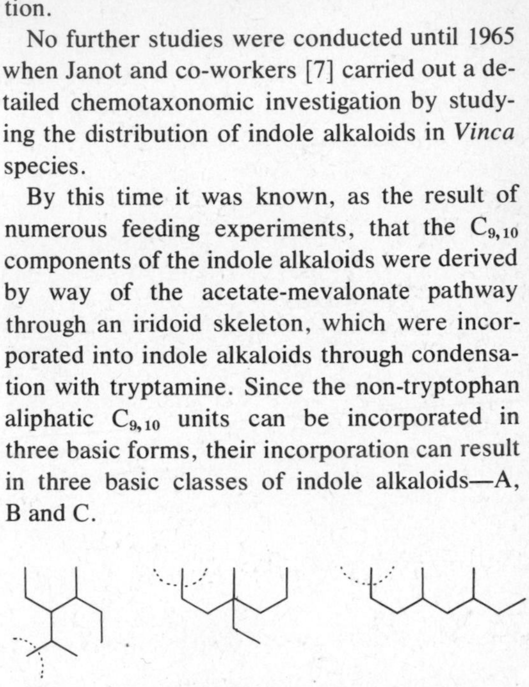

Class A   *Class B*   *Class C*
Yohimbine type Aspidospermine type Iboga type

Janot and co-workers pointed out that *V. minor* contains mainly alkaloids of class B (aspidospermine type), while other *Vinca* species contained alkaloids of class B and class A (yohimbine type), the latter prevailing regularly. Another major conclusion reached by Janot and co-workers was that a great difference of alkaloid composition existed between *V. major* and *V. difformis*, this fact being incompatible with Pichon's classification, in which *V. difformis* was considered as a variety of *V. major*. Finally, Janot indicated that alkaloids of class C (Iboga type) and dimeric bases of the mixed B–C classes, were absent in *Vinca* and seemed to be characteristic of the genus *Catharanthus*.

Table 1. *Distribution of classes A and B of indole alkaloids*

Section	Species	Number of characterized alkaloids	Class A	Class B	Predominant class
Minor	*V. minor*	39	7	32	B
Major	*V. major*	12	9	3	A
	V. difformis	9	6	3	A
	V. pubescens	3	3	0	A
Herbacea	*V. herbacea*	19	15	4	A
	V. libanotica	12	9	3	A
	V. haussknechtii	–	–	–	–
Erecta	*V. erecta*	44	24	20	A

Recently, based on the results of their broad studies of rich herbarium material, and on a critical evaluation of published chorologic data, Trojánek, Novacék & Stary [8] divided the genus *Vinca* into 4 sections, including a total of 8 species. This classification agrees well, not only with the distribution of the basic alkaloid classes A and B, but also with that of corresponding structural subgroups within these classes. The classification of *Vinca* species according to Novacék & Stary is as follows:

Section	Species
1. Minor	1. *V. minor* L.
2. Major	2a. *V. major* L.
	2b. *V. difformis* Pourr.
	2c. *V. pubescens* Urv.
3. Herbacea	3a. *V. herbacea* Waldst. et Kit.
	3b. *V. libanotica* Zucc.
	3c. *V. haussknechtii* Bornm. et Sint.
4. Erecta	4. *V. erecta* Regel. et Schmalh.

Table 1 utilizes the classification of Novacék & Stary and shows the distribution of classes A and B indole alkaloids in the genus *Vinca*.

A careful examination of table 1 shows that there is justification for devoting a separate section to *V. minor* since it is the only species able to synthesize about 80% of its alkaloids as the B type.

Concerning the second section, i.e. Major, *V. major* and *V. difformis* constitute separate species because of the occurrence of structurally different types of alkaloids [7]. This will be made clearer in the phytochemistry section of this review. Although identical alkaloids are present in both species (vincamine, vincamedine), their evaluation as independent species is further supported by differences in the amino acid content [9].

Among the 3 species forming the third section, i.e. Herbacea, *V. herbacea* and *V. libanotica* have been chemically studied. They contain various types of alkaloids, mainly of type A. Four identical alkaloids have been reported to be present in both species, namely: herbamine, herbadine, reserpinine and (–)-tabersonine. The third member of the section, *V. haussknechtii*, has not been studied chemically.

Section Erecta is monotypic and contains only *V. erecta*, which is endemic to a limited area in Russia. This species cannot be subjected to chemotaxonomic studies since it is the only species where almost every type of alkaloid (class A and B) has been reported as isolated. At the present time *V. erecta* should be placed in a separate section because of the fact that it is endemic to a totally different part of the world compared with the other members of the genus.

Just recently a new species was described in the genus *Vinca* [10], named *V. semidesertorum* Ponert, from the badland region of the Azerbeyjian SSR. The population of this species was compared with populations of *V. herbacea* Waldst. et Kit. From the botanical descriptions it can be assumed that this species should be placed tentatively in the section Herbacea until phytochemical works are reported. This species was described after Novacék & Stary proposed their classification of the genus *Vinca*.

The Phytochemistry of *Vinca* Species

In the last two decades, at least 100 different alkaloids were isolated from this genus of plants. Structure elucidation has been accomplished for all but 20 of these bases. In this discussion the classification proposed by Novacék & Stary will be utilized since it is the only one that seems to be chemotaxonomically feasible [8].

The distribution of indole alkaloids of the various *Vinca* species (up to the end of 1972) is presented in table 2.

The alkaloids are divided into two groups, i.e. A (yohimbanoids) and B (aspidosperma-noids), and further divided into subgroups according to their complexity and structural relationship [11]. Table 2 includes the basic structure of each group, including the stereochemistry, the alkaloids, their synonyms if any, their occurrence, and the reference for each species. Whenever the structure or the stereochemistry of an alkaloid is not rigorously established, this is indicated by a note or question mark.

It can be seen from table 2 that the structures of 102 alkaloids have been reported up to the present time, 51 of them belonging to type A and 51 to type B. Of these, the structures of 77 alkaloids have been unequivocally established with their stereochemistry, 38 of which belong to type A and 39 to type B. Of the remaining 25 alkaloids, 14 belonging to type A and 11 to type B, the stereochemistry of 17 have not been worked out, and the structures of the remaining 8 alkaloids are questionable, one of them being suspected to be an artefact, one to be a mixture of 2 isomers and 6 suspected to have incorrect structures.

V. minor has been studied thoroughly by 4 groups in Europe. Mokry, Kompis and their co-workers in Bratislava, Trojánek and co-workers in Prague, Döepke, Meisel and co-workers in Berlin, and finally Plat, Le Men and co-workers in France. These groups have carried out extensive studies on *V. minor,* which includes the isolation and characterization work, together with degradative studies. The stereochemistry of all but 5 of the 39 alkaloids reported from *V. minor* has been established.

V. major has been investigated by two groups, Trojánek and his group and Janot and his co-workers in France. Twelve alkaloids have been reported from this species, most of them by Trojánek's group. The structures with their stereochemistry of all of the alkaloids have been established.

The Russian workers Abdurakhimova & Chknikvadze have investigated *V. pubescens*

and have reported three alkaloids: reserpinine [16], carapanaubine [23], and majoridine [16].

As far as *V. difformis* is concerned, Janot & Le Men have carried out all of the phytochemical investigations. Nine structurally known alkaloids have been reported, all of whose stereochemistry has been established. Their studies have been very important from a chemotaxonomic point of view, since by means of their results, it has been possible to differentiate this species from *V. major.*

The number of structurally known alkaloids isolated from *V. herbacea* is 19. Ognyanov and co-workers in Sofia have carried out all of the investigations on *V. herbacea* grown in Bulgaria. They have reported of the isolation of 10 alkaloids, and elucidated the structures with their stereochemistry of 6 of these. The rest of the alkaloids have been reported by the Russian investigators, Malikov & Yuldashev, on *V. herbacea* grown in Azerbeyjian SSR [49].

Due to the location of *V. erecta,* the isolation, characterization and chemical work of the alkaloids has been a monoply of the Yunusov & Yuldashev group. They have reported the isolation of 44 alkaloids, the highest number reported from any *Vinca* species.

In the future, phytochemical investigations of *V. pubescens* and *V. haussknechtii* will reveal, without doubt, new bases that had not been reported from this genus before.

Very little work has been done on the isolation of non-alkaloid entities from *Vinca* species. No work of this nature has been reported from *V. difformis* [86]. Ursolic acid has been isolated from all other *Vinca* species [87]. *V. minor* has yielded fructose, β-sitosterol, a kaempferol glycoside and a quercetin glycoside [88].

Chemotaxonomic Conclusions

Extensive research in correlating systematic classification based on morphological characters of plants with that of the nature of their chemical constituents has occupied chemotaxonomists for a long time.

Since *V. libanotica* and *V. herbacea* cannot be differentiated easily from each other by morphological characters, this study was undertaken to determine if these two species were indeed the same or distinct from each other,

Table 2. *Distribution of indole alkaloids in Vinca species*

Class A
Group A-1.1

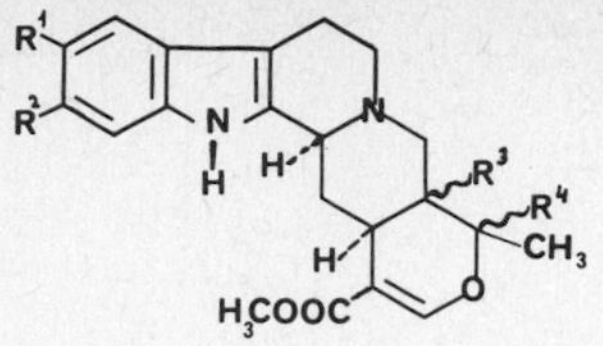

	R¹	R²	R³	R⁴	Occurrence	Ref.
Ervine	H	H	α-H	α-H	V. erecta	[12]
(syn. rauniticine)					V. libanotica	[13]
Ajmalicine	H	H	β-H	β-H	V. erecta	[14]
(syn. vincaine, vinceine)						
Reserpinine	H	OCH₃	α-H	β-H	V. major	[15]
(syn. pubescine)					V. pubescens	[16]
					V. herbacea	[17]
					V. erecta	[14]
					V. libanotica	[13]
Isoreserpiline	OCH₃	OCH₃	α-H	β-H	V. erecta	[18]

Group A-1.2

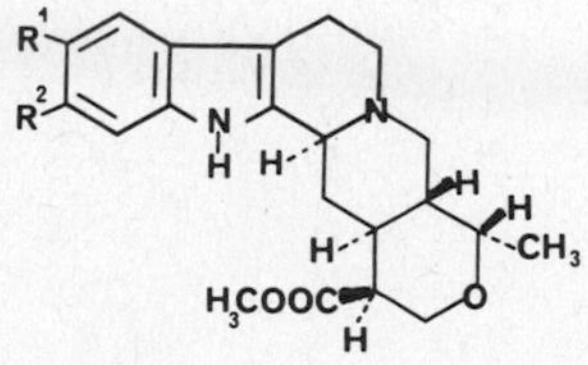

	R¹	R²	Occurrence	Ref.
Herbaceine	OCH₃	OCH₃	V. herbacea	[19]
(syn. vincaherbinine)				
Herbaine (without stereochemistry)	H	OCH₃	V. herbacea	[20]
(syn. vincaherbine)				

Group A-2.1

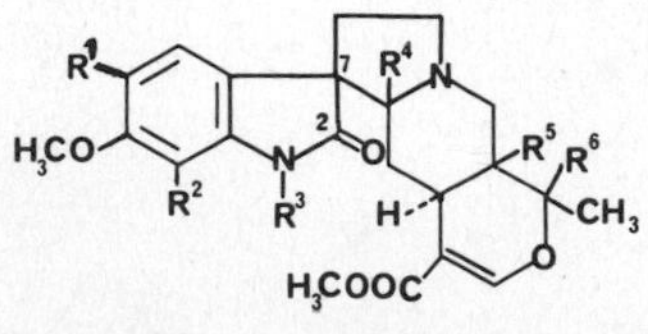

	R¹	R²	R³	R⁴	R⁵	R⁶	C7(2)	Occurrence	Ref.
Ericine	H	H	H	H	H	H	?	V. erecta	[21]
(without stereochemistry)									
Vinerine	H	H	H	H	H	H	?	V. erecta	[22]
(without stereochemistry)									
Vineridine	H	H	H	H	H	H	?	V. erecta	[22]
(without stereochemistry)									
Carapanaubine	OCH₃	H	H	α-H	α-H	β-H	β	V. pubescens	[23]
(syn. vinine, majorixine)									
Majdine	H	OCH₃	H	α-H	α-H	β-H	β	V. major	[15]
(syn. majoroxine)								V. herbacea	[24]
Isomajdine	H	OCH₃	H	α-H	α-H	β-H	α	V. herbacea	[24]
N-Acetylvinerine	H	H	COCH₃	H	H	H	?	V. erecta	[25]
(without stereochemistry)									

Group A-2.2

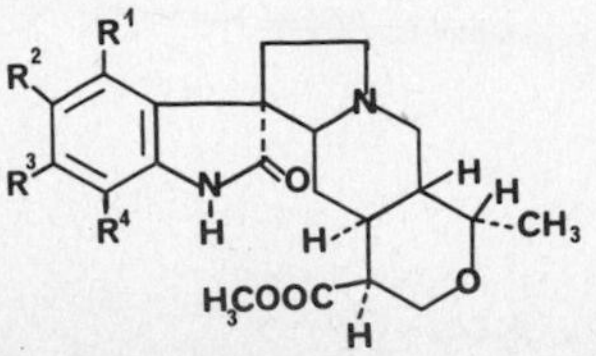

	R^1	R^2	R^3	R^4	Occurrence	Ref.
Herbaline	H	OCH_3	OCH_3	H	*V. herbacea*	[19]
Elegantine	H	H	H	H	*V. major*	[26]
Herbavine (structure is not well established)	CH_3	CH_3	H	H	*V. herbacea*	[27]

Group A-3

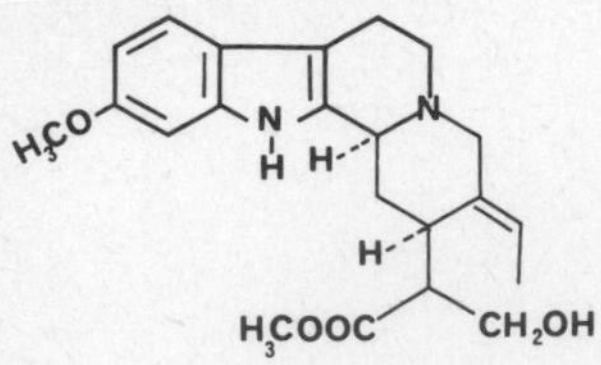

	Occurrence	Ref.
Hervine (syn. 11-methoxyisositsirikine)	*V. herbacea*	[28]

Group A-4.1

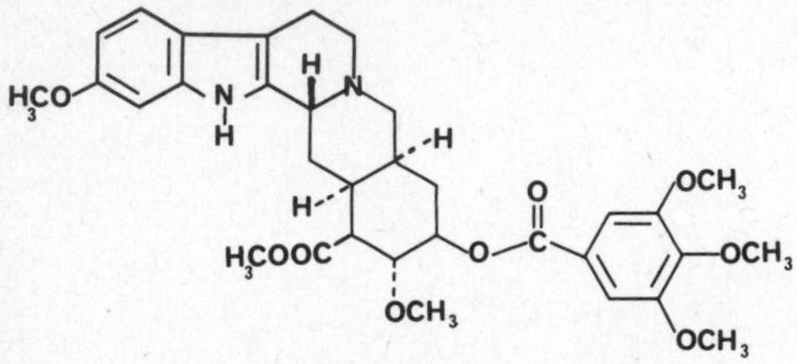

	Occurrence	Ref.
Reserpine	*V. minor*	[29]
	V. herbacea	[20]

Group A-4.2

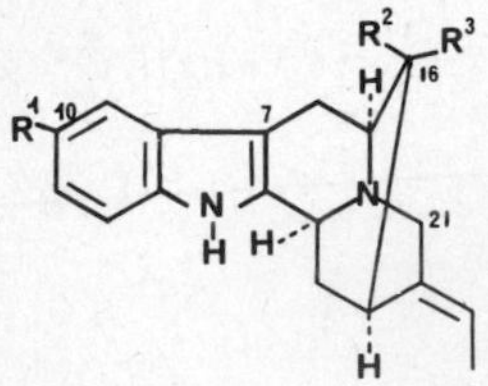

	R^1	R^2	R^3	Occurrence	Ref.
Akuammidine (syn. ervamidine)	H	$COOCH_3$	CH_2OH	*V. difformis*	[30]
				V. erecta	[8]
Vellosimine	H	H	CHO	*V. difformis*	[31]
Tombozine (syn. normacusine B)	H	H	CH_2OH	*V. erecta*	[32]
Sarpagine	OH	H	CH_2OH	*V. difformis*	[33]
10-Methoxyvellosimine	OCH_3	H	CHO	*V. major*	[34]

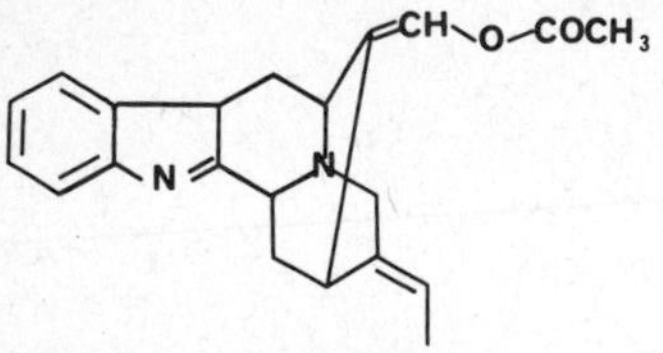

	Occurrence	Ref.
Vinorine (without stereochemistry)	*V. minor*	[35]

Group A-4.3

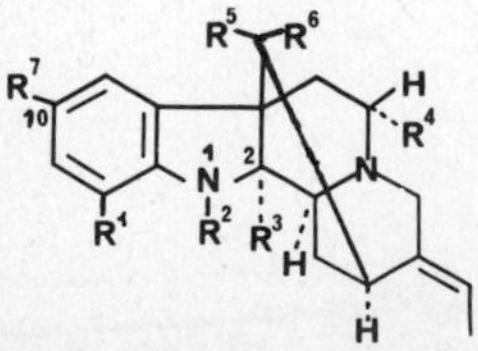

	R¹	R²	R³	R⁴	R⁵	R⁶	R⁷	Occurrence	Ref.
Picrinine (syn. vincaridine)	H	H	–O–		COOCH₃	H	H	*V. erecta* *V. libanotica*	[36] [13]
Vincaricine (structure is questionable)	OCH₃	H	–O–		COOCH₃	H	H	*V. erecta*	[37]
Vincamidine (syn. strictamine)	H	(1.2 double bond)	H	H		COOCH₃	H	*V. minor* *V. libanotica*	[38] [13]
Desacetyl akuammiline	H	(1.2 double bond)	H	CH₂OH	COOCH₃	H		*V. minor*	[39]
10-Methoxy desacetyl akuammiline	H	(1.2 double bond)	H	CH₂OH	COOCH₃	OCH₃		*V. minor*	[39]
Vincarinine (without stereochemistry)	OCH₃	H	–O–		CHO	COOCH₃	H	*V. erecta*	[40]

Group A-4.4

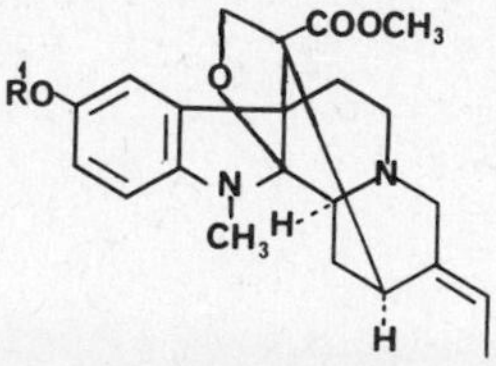

	R¹	Occurrence	Ref.
Akuammine (syn. vincamajoridine)	H	*V. major* *V. erecta* *V. herbacea*	[15] [22] [41]
O-methyl akuammine	CH₃	*V. erecta*	[42]

Group A-4.5

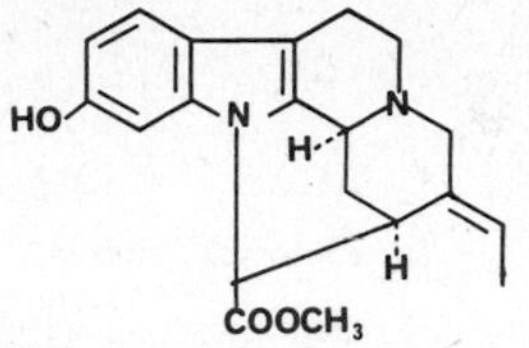

11-Hydroxypleiocarpamine   *V. erecta*   [43]

Group A-5

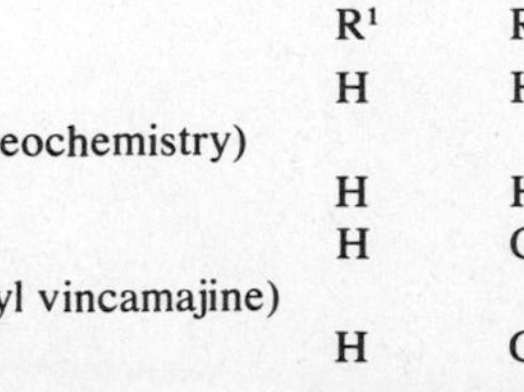

	R¹	R²	R³	R⁴	R⁵	R⁶	Occurrence	Ref.
Vincarine (without stereochemistry)	H	H	H	H	COOCH₃	H	*V. erecta* *V. herbacea*	[44] [45]
Quebrachidine	H	H	α-H	H	COOCH₃	H	*V. libanotica*	[13]
Vincamedine (syn. *O*-acetyl vincamajine)	H	CH₃	α-H	COCH₃	COOCH₃	H	*V. major* *V. difformis*	[46] [30]
Vincamajine	H	CH₃	α-H	H	COOCH₃	H	*V. major* *V. difformis* *V. libanotica*	[15] [30] [13]
Majoridine (syn. 17-*O*-acetyl-10-methoxytetraphyllicine)	OCH₃	CH₃	β-H	COCH₃	H	H	*V. major* *V. pubescens*	[47] [16]
Vincamajoreine	OCH₃	CH₃	β-H	H	H	H	*V. major*	[48]
Herbadine (structure is questionable) (C-17 has two protons)	H	H	OH	–	COOCH₃	OH	*V. herbacea* *V. libanotica*	[49] [13]

Herbamine   H CH_3 OH   – $COOCH_3$ OH   *V. herbacea*   [49]
    (structure is questionable) (C-17 has two protons)   *V. libanotica*   [13]

Group A-6

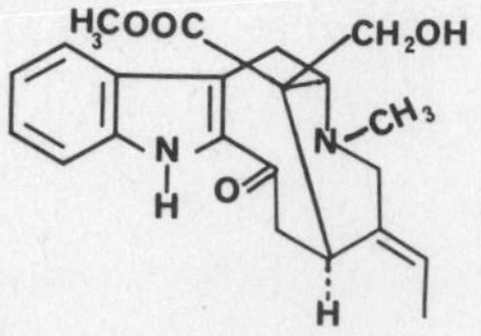

Vincadiffine   *V. difformis*   [50]

Group A-7

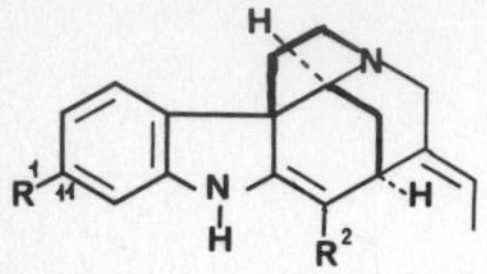

	R^1	R^2	Occurrence	Ref.
Vincanine	H	CHO	*V. herbacea*	[41]
(syn. norfluorocurarine)			*V. erecta*	[51]
Vinervidine	H	CHO	*V. erecta*	[51]
(syn. (±)-vincanine)				
Vincanidine	OH	CHO	*V. erecta*	[22]
(syn. 11-OH-norfluorocurarine)				
Vincanicine	OCH_3	CHO	*V. erecta*	[52]
Akuammicine	H	$COOCH_3$	*V. erecta*	[53]
			V. herbacea	[54]
Vinervine	OH	$COOCH_3$	*V. erecta*	[32]
(syn. 11-OH-akuammicine)				
Vinervinine	OCH_3	$COOCH_3$	*V. erecta*	[32]

Group A-8

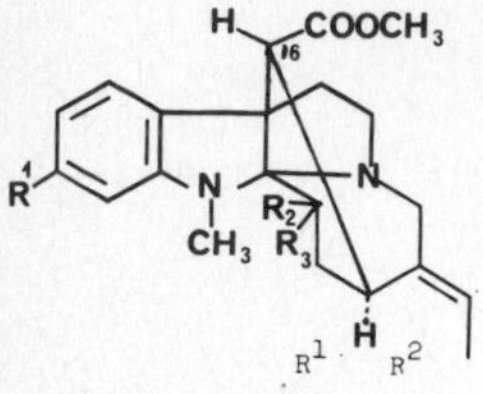

	R^1	R^2	R^3	Occurrence	Ref.
Vincoridine	H		=O	*V. minor*	[55]
(without stereochemistry at C-16)					
Vincovine	OCH_3	H	H	*V. minor*	[56]

Class B
Group B-1

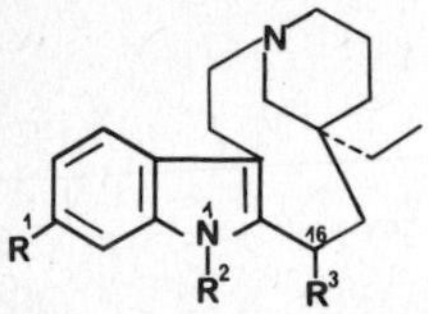

	R^1	R^2	R^3	Occurrence	Ref.
(+)-Quebrachamine	H	H	H	*V. minor*	[38]
				V. erecta	[18]
(±)-*ind-N*-Methyl-quebrachamine	H	CH_3	H	*V. minor*	[57]
Vincadine	H	H	α-$COOCH_3$	*V. minor*	[58]
(syn. 16-methoxycarbonylquebrachamine)					
Vincaminorine	H	CH_3	α-$COOCH_3$	*V. minor*	[59]
(syn. 1-methyl-16-epi-methoxy-carbonyl-					
quebrachamine)					

	R	R	R	Occurrence	Ref.
Vincaminoreine (syn. 1-methyl-16-methoxy-carbonyl-quebrachamine)	H	CH$_3$	β-COOCH$_3$	*V. minor*	[59]
Vincaminoridine	OCH$_3$	CH$_3$	α-COOCH$_3$	*V. minor*	[60]

Group B-2.1

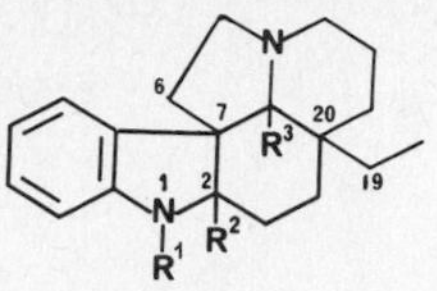

	R^1	R^2	R^3	C19(20)	C6(7)	Occurrence	Ref.
(+)-*N*-Methylaspidospermidine	CH$_3$	α-H	β-H	β	α	*V. minor*	[38]
(+)-1,2-Dehydroaspidospermidine	1,2-double bond		β-H	β	α	*V. minor*	[38]
(−)-1,2-Dehydroaspidospermidine	1,2-double bond		α-H	α	β	*V. erecta*	[18]

Group B-2.2

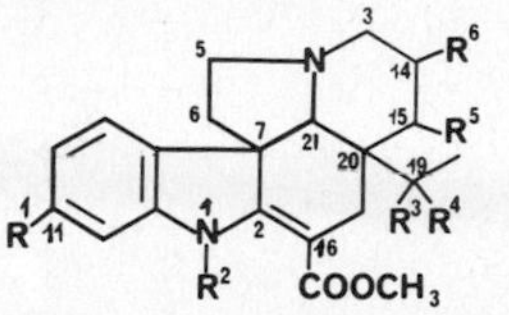

	R^1	R^2	R^3	R^4	R^5	R^6	C6(7)	C19(20)	Occurrence	Ref.
(−)-Vincadifformine (syn. ervamine)	H	H	H	H	H	H	β	α	*V. minor* / *V. erecta*	[61] / [62]
(+)-Vincadifformine	H	H	H	H	H	H	α	β	*V. difformis*	[63]
(±)-Vincadifformine	H	H	H	H	H	H	−	−	*V. minor* / *V. difformis* / *V. major*	[61] / [63] / [64]
(±)-Minovine (syn. 1-methylvincadifformine)	H	CH$_3$	H	H	H	H	−	−	*V. minor*	[61]
16-Methoxyvincadifformine (syn. ervinceine)	OCH$_3$	H	H	H	H	H	β	α	*V. minor* / *V. erecta*	[65] / [66]
(−)-Tabersonine	H	H	H	H	−	−	β	α	*V. herbacea* / *V. libanotica*	[67] / [13]
11-Methoxytabersonine	OCH$_3$	H	H	H	−	−	β	α	*V. herbacea*	[67]
Ervamicine (without stereochemistry) (14,15-double bond)	OCH$_3$	H	H	H	−	−	?	?	*V. erecta*	[66]
Minovincine (syn. 19-oxovincadifformine)	H	H	=O		H	H	β	α	*V. minor*	[68]
(−)-Minovincinine (syn. 19-hydroxyvincadifformine)	H	H	OH	H	H	H	β	α	*V. minor*	[68]
Vincesine	H	H	H	OH	H	H	β	α	*V. minor*	[69]
11-Methoxyminovincine (syn. minoriceine)	OCH$_3$	H	=O		H	H	β	α	*V. minor*	[70]
11-Methoxyminovincinine	OCH$_3$	H	OH	H	H	H	β	α	*V. minor*	[70]
Ervidinine (suspected to be 5-oxo-14,15 dihydrotabersonine)	H	H	H	H	H	H	?	?	*V. erecta*	[71]
Lochnerinine (syn. 11-methoxylochnericine)	OCH$_3$	H	H	H	–O–		β	α	*V. herbacea*	[67]
Ervinidine (syn. 7,21-dihydro,5,6-dioxo-vincadifformine) (without stereochemistry)	H	H	H	H	H	H	?	?	*V. erecta*	[71]
Ervincinine (without stereochemistry)	OCH$_3$	H	H	H	–O–		?	?	*V. erecta*	[66]
(−)-5-oxominovincine	H C$_5$=O	H	=O		H	H	α	α	*V. minor*	[72]

Group B-3.1

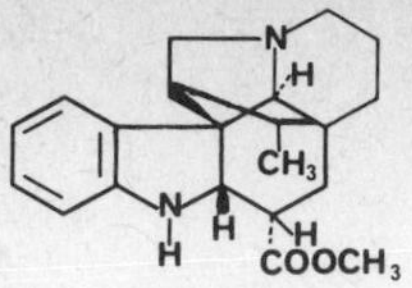

	Occurrence	Ref.
Pseudokopsinine	*V. erecta*	[22]

Group B-3.2

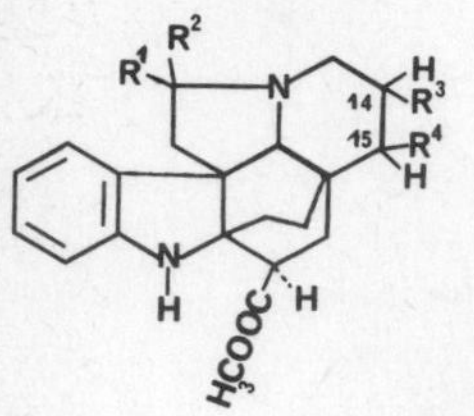

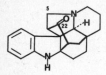

	R¹	R²	R³	R⁴	Occurrence	Ref.
Kopsinine (syn. erectine)	H	H	H	H	*V. erecta*	[22]
Kopsinilame		=O	H	H	*V. erecta*	[73]
Kopsinilamine (quaternary ammonium form of kopsinilame)		=O	H	H	*V. erecta*	[74]
Epoxykopsinine (structure is questionable)	H	H	–O–		*V. erecta*	[42]
Venalstonine	H	H	(14,15 double bond)		*V. libanotica*	[13]

Group B-3.3

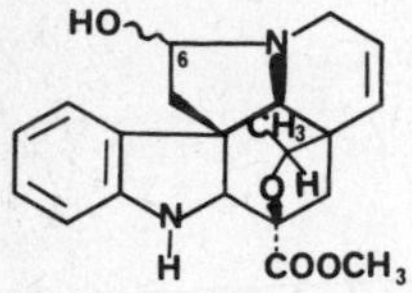

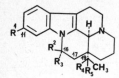

	Occurrence	Ref.
Vincoline (without stereochemistry at C-6)	*V. libanotica*	[13] [75]

Group B-4

	Occurrence	Ref.
Kopsanone (suspected to be 5,22-dioxokopsane)	*V. erecta*	[18]

Group B-5.1

	R¹	R²	R³	R⁴	R⁵	Occurrence	Ref.
Vincamine (syn. minorine)	H	OH	COOCH₃	H	H	*V. minor*	[76]
						V. major	[48]
						V. difformis	[33]
						V. herbacea	[77]
						V. erecta	[53]
16-Epivincamine (syn. isovincamine)	H	COOCH₃	OH	H	H	*V. minor*	[38]
Vincine (syn. 11-methoxy vincamine)	OCH₃	OH	COOCH₃	H	H	*V. minor*	[78]
						V. major	[48]
						V. erecta	[53]

Vincaminine	H	OH	COOCH$_3$	=O	*V. minor*	[79]
(syn. 19-oxo-vincamine, vincareine)						
Vincinine	OCH$_3$	OH	COOCH$_3$	=O	*V. minor*	[79]
(syn. 19-oxo-11-methoxy-vincamine)						
19-Hydroxyvincamine	H	OH	COOCH$_3$ OH	H	*V. minor*	[80]
(without stereochemistry)						
(−)-Eburnamonine	H	=O		H	H	*V. minor*	[38]
					V. erecta	[18]
(±)-Eburnamonine	H	=O		H	H	*V. minor*	[81]
(syn. vincanorine)						
(−)-11-Methoxyeburnamonine	OCH$_3$	=O		H	H	*V. minor*	[82]
Apovincamine	H	H	COOCH$_3$ H	H	*V. erecta*	[18]
(suspected to be artefact)	(16,17-double bond)					

Group B-5.2

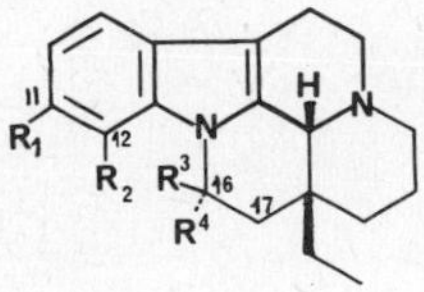

	R^1	R^2	R^3	R^4	Occurrence	Ref.
Eburnamine	H	H	H	OH	*V. minor*	[38]
(syn. pleiocarpinidine)					*V. erecta*	[18]
(±)-Eburnamine	H	H	H	OH	*V. erecta*	[18]
(suspected to be a mixture of (+) and						
(−) isomers)						
(+) Isoeburnamine	H	H	OH	H	*V. minor*	[38]
Eburnamenine	H	H	H	–	*V. minor*	[38]
	(16,17-double bond)					
11,12-Dimethoxyeburnamonine	OCH$_3$	OCH$_3$	=O		*V. minor*	[83]
(without stereochemistry)						

Group B-6

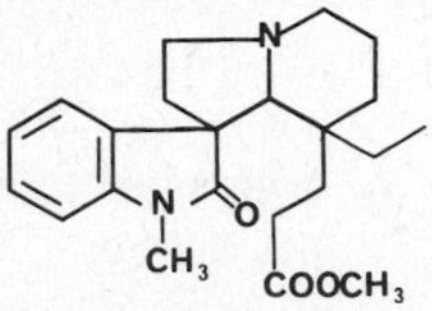

	Occurrence	Ref.
Vincatine	*V. minor*	[84]
(without stereochemistry)		

Miscellaneous

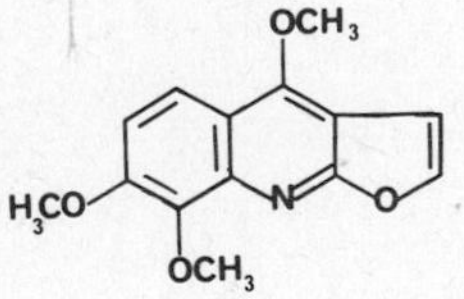

	Occurrence	Ref.
Skimmianine	*V. herbacea*	[85]
(syn. β-fagarine)		

from the point of view of occurrence of alkaloids which have a limited distribution in the plant kingdom and thus are significant from a chemotaxonomic point of view.

As was mentioned earlier, the only botanical classification of the genus *Vinca* which agrees very well with the distribution of the basic alkaloid classes A and B (table 1) is that of Novacék & Stary [8].

Keeping in mind this classification, and the different structural groups within alkaloidal classes A and B (table 2), table 3 lists the alkaloids isolated from *V. libanotica* with their corresponding group number. Table 4 lists the

Table 3. *Alkaloids isolated from Vinca libanotica with their appropriate classification*

Alkaloid	Group	Yield (mg)	Isolated previously	Ref.
Reserpinine	A-1.1	59	V. erecta	[14]
			V. major	[15]
			V. pubescens	[16]
			V. herbacea	[17]
Rauniticine	A-1.1	40	Ervine, claimed to be identical with rauniticine, from V. erecta	[12]
Picrinine	A-4.3	11	V. erecta	[36]
Vincamidine	A-4.3	74	V. minor	[38]
Vincamidine hydrate	A-4.3	229	V. minor	[38]
Quebrachidine	A-5	50	Vincarine, a stereoisomer of quebrachidine from V. erecta	[44]
			V. herbacea	[45]
Vincamajine	A-5	3923	V. major	[15]
			V. difformis	[30]
Herbamine	A-5	655	V. herbacea	[49]
Herbadine	A-5	27	V. herbacea	[49]
Tabersonine	B-2.2	18	V. herbacea	[67]
Venalstonine	B-3.2	522	Not reported from Vinca sp.	
Vincoline	B-3.3	49	Not reported from Vinca sp.	

distribution of the basic groups of alkaloids in the different species of *Vinca*.

A careful examination of table 4 shows that *V. libanotica* varies basically in its alkaloidal content not only from *V. herbacea*, but from *V. major* as well.

The major alkaloid of *V. libanotica* is vinca-majine (group A-5) which has only been reported previously from *V. major* and *V. difformis* (table 2). The presence of vincamajine in abundant amounts (about 0.05% of the dry plant material) in *V. libanotica*, together with the fact that this alkaloid has not been reported from *V. herbacea*, shows the distinctness of *V.*

Table 4. *Distribution of groups of alkaloids in Vinca species*

Class	A													
Group Section and species	1.1	1.2	2.1	2.2	3	4.1	4.2	4.3	4.4	4.5	5	6	7	8
Minor														
V. minor						1	1	3						2
Major														
V. major	1		1	1			1		1		4			
V. difformis								3			2	1		
V. pubescens	1		1								1			
Herbacea														
V. herbacea	1	2	2	2a	1	1			1		3a		2	
V. libanotica	2							3			4			
V. haussknechtii														
Erecta														
V. erecta	4		4			2	3a	2	1		1a		7	

a The structures have not been rigorously established.

libanotica, and its close relationship with *V. major.*

Also, another common feature between *V. major* and *V. libanotica* is the fact that both yield a high percentage of vincamajine type alkaloids (group A-5); 4 out of 12 alkaloids in the case of *V. major,* and 4 out of 12 in the case of *V. libanotica* (table 4).

A major difference between *V. major* and *V. libanotica* is in the fact that *V. major* has a high incidence of alkaloids oxygenated in the 10 position and belonging to groups A-4.2, A-4.4 and A-5 (majoridine, vincamajoreine, akuammicine, 10-methoxy-vellosimine) (table 2).

Comparing *V. libanotica* and *V. herbacea,* 3 common features can be found between these 2 species. First, both species contain α-methylene-indoline type alkaloids of group B-2.2 (vincadifformine family) and group A-7 (akuammicine family). Five alkaloids of these two groups have been isolated from *V. herbacea,* namely, vincanine and akuammicine (group A-7), as well as tabersonine, 11-methoxytabersonine and lochnerinine (group B-2.2). In spite of the fact that only one alkaloid of group B-2.2 has been isolated from *V. libanotica,* namely tabersonine, there are at least 3 minor alkaloids of the α-methylene-indoline type present in this plant. (Although these 3 alkaloids could not be isolated, they were detected by their chromogenic

reaction to CAS spray reagent on thin-layer chromatograms [89].) The second feature is that 2 unique alkaloids, herbamine and herbadine, belonging to group A-5, have only been reported from *V. herbacea* and *V. libanotica.* The third common feature between these 2 species is the low incidence of vincamine type alkaloids. In *V. libanotica,* vincamine and vincamine-type alkaloids (group B-5.1) are absent. This fact is an important chemotaxonomic factor, in that vincamine has been reported from all other *Vinca* species studied to date (table 2). However, of 19 alkaloids characterized from *V. herbacea,* only one, namely vincamine, belongs to group B-5.1 (vincamine family) (table 4).

On the other hand, a major difference between *V. herbacea* and *V. libanotica* is in the ability of *V. herbacea* to synthesize a high percentage of alkaloids of class A, most of which have a unique structure with a hydrogenated heterocyclic E-ring (group A-1.2 and A-2.2) (herbaline, hervine, herbaceine, herbaine). *V. herbacea* is the only source of this type of alkaloid in the genus *Vinca.*

Two alkaloids present in *V. libanotica,* namely quebrachidine and rauniticine, have not yet been reported from any *Vinca* species. However, an alkaloid with the trivial name ervine has been reported to be identical with rauniticine, the former being isolated from *V. erecta* [12].

B										
1	2.1	2.2	3.1	3.2	3.3	4	5.1	5.2	6	Total
6	2	10					9	4	1	39
		1					2			12
		2					1			9
										3
		3					1			19
		1		1	1					12
										–
1	1	6	1	4		1[a]	4[a]	2[a]		44

The presence of vincamidine, vincamidine hydrate and picrinine (group A-4.3) are also important chemotaxonomically, since this group of alkaloids has never been reported from V. herbacea or from V. major (table 4).

The isolation of venalstonine (group B-3.2) is another evidence of the uniqueness of this species, since no alkaloid of this group has been reported from any of the Vinca species, with the exception of V. erecta.

Vincoline, a new alkaloid, is the only member of a novel group (group B-3.3), and it has not been reported from any other Vinca species. Thus, the presence of vincoline and venalstonine in this plant reinforces the distinctness of this species.

The distribution of groups of alkaloids in V. libanotica is a reliable criterion to conclude that this plant represents a distinct species and consideration should be given to incorporating it into a separate section between Herbacea and Major.

Finally, at this stage it is worth mentioning that the isolation of vincoline from V. libanotica brings the number of alkaloids common to both the periwinkle genera, Vinca and Catharanthus, to 8, the first 7 being reserpinine (V. major, V. pubescens, V. herbacea, V. erecta, V. libanotica) [13–17], akuammine (V. major, V. erecta, V. herbacea) [15, 22, 41], (–)-tabersonine (V. herbacea, V. libanotica) [67], lochnerinine (V. herbacea) [67], akuammicine (V. erecta, V. herbacea) [53, 54], ajmalicine (V. erecta) [14] and reserpine (V. minor, V. herbacea) [20, 29].

Addendum

Just recently picrinine was reported to be isolated from Vinca minor [90].

References

1. Farnsworth, N R, Lloydia 1961, 24, 105.
2. Pichon, M, Mem mus hist nat 1948, XXXII, 6, 153.
3. Pichon, M, Bull mus hist nat 1951, 23, 439.
4. Pobedomova, E G, Izv akad nauk SSSR Moscow 1952, 18, 646.
5. Stearn, W T, A synopsis of the genus Vinca including its taxonomic and nomenclatural history, in the Vinca species—botany, chemistry and pharmacology (ed N R Farnsworth & W I Taylor) chap. 1. Dekker, New York, 1973.
6. Paris, R R & Moyse, H, Compt rend 1957, 245, 1265.
7. Janot, M M, Le Men, J & Garnier, J, Bull soc bot France 1965, 118.
8. Trojánek, J, Novacék, M & Stary, F, The chemotaxonomy of Vinca species, in The Vinca species—botany, chemistry and pharmacology (ed N R Farnsworth & W I Taylor) chap. 4, table 2. Dekker, New York, 1973.
9. Paris, R R & Girre, R L, Compt rend ser D 1969, 268, 62.
10. Ponert, J, Hort botan batum acad serent georgieae 1971, 82, 581.
11. Trojánek, J, Novacék, M & Stary, F, The chemotaxonomy of Vinca species, in The Vinca species—botany, chemistry and pharmacology (ed N R Farnsworth & W I Taylor) chap. 4, table 5. Dekker, New York, 1973.
12. Malikov, V M, Yuldashev, P Kh & Yunusov, S Yu, Khim prir soedin 1966, 2, 338. Through Chem abstr 66, 65684h.
13. Aynilian, G H, Farnsworth, N R & Trojánek, J, Lloydia 1973. In press.
14. Yunusov, S Yu & Yuldashev, P Kh, Dokl akad nauk uzb SSR 1958, 23. Through Chem abstr 52, 3044.
15. Kaul, J L & Trojánek, J, Lloydia 1966, 29, 26.
16. Chknikvadze, G V, Asatiani, V S, Vachnadze, V Yu & Mudzhiri, K S, Soobshch akad nauk gruz SSR 1971, 64, 345. Through Chem abstr 76, 70097e.
17. Ognyanov, I, Dalev, P, Dutscevska, H & Mollow, N, Compt rend acad bulg sci 1964, 17, 153.
18. Rhakhimov, D A, Sharipov, M R, Aripov, N Ch, Malikov, N M, Shakirov, T T & Yunusov, S Yu, Khim prir soedin 1970, 6, 713. Through Chem abstr 74, 95426h.
19. Ognyanov, I, Pyuskyulev, B, Shamma, M, Weiss, J A & Shine, R J, Chem commun 1967, 579.
20. Clauder, O, Mathe, I, Bojthe, H, Klara, G & Kocsis, A, Herba hung 1969, 8, 29. Through Chem abstr 72, 129393w.
21. Abdurkahimova, N, Kasymov, Sh E & Yunusov, S Yu, Chem nat prod 1968, 4, 116.
22. Kasymov, Sh Z, Yuldashev, P Kh & Yunusov, S Yu, Khim prir soedin akad nauk uzb SSR 1966, 260. Through Biol abstr 48, 93116.
23. Abdurakhimova, N, Yuldashev, P Kh & Yunusov, S Yu, Khim prir soedin akad nauk uzb SSR 1965, 224. Through Chem abstr, 63, 16396.
24. Ognyanov, I, Pyuskyulev, B, Kompis, I, Sticzay, T, Spiteller, G, Shamma, M & Shine, R J, Tetrahedron 1968, 24, 4641.
25. Malikov, V N, Kasymov, Sh Z & Yunusov, S Yu, Khim prir soedin 1971, 6, 640. Through Chem abstr, 74, 39172k.
26. Bhattacharyya, J & Pakrashi, S C, Tetrahedron lett 1972, 159.
27. Dzhakeli, E Z & Mudzhiri, K S, Soobshch akad

nauk gruz SSR 1970, 57, 353. Through Chem abstr 73, 25723h.

28. Ognyanov, I, Pyuskyulev, B, Bozjanov, B & Hesse M, Helv chim acta 1967, 50, 754.
29. Lyapunova, P M & Birosyak, G, Farm zh (Kiev) 1961, 16, 48. Through Chem abstr 57, 4759i.
30. Janot, M M, Le Men, J, Gosset, J & Levy, J, Bull soc chim France 1962, 1079.
31. Falco, M, Gosset-Garnier, J, Fellion, E & Le Men, J, Ann pharm franç 1964, 22, 455.
32. Abdurakhimova, N, Yuldashev, P Kh & Yunusov, S Yu, Khim prir soedin 1967, 3, 310. Through Chem abstr 68, 22113v.
33. Janot, M M, Le Men, J & Fan, C, Ann pharm franç 1957, 15, 513.
34. Potier, P, Beugelmans, R, Le Men, J & Janot, M M, Ann pharm franç 1965, 23, 61.
35. Meisel, H & Döpke, W, Tetrahedron lett 1971, 1291.
36. Kuchenkova, M A, Yuldashev, P Kh & Yunusov, S Yu, Khim prir soedin akad nauk uzb SSR 1967, 65. Through Chem abstr 68, 6205w.
37. Il'yasova, Kh T, Malikov, V M & Yunusov, S Yu, Chem nat prod 1968, 4, 327.
38. Mokry, J, Kompis, I & Spiteller, G, Collect czech chem commun 1967, 32, 2523.
39. Savaskan, S, Kompis, I, Hesse, M & Schmid, H, Helv chim acta 1972, 55, 2861.
40. Il'yasova, Kh T, Malikov, V M & Yunusov, S Yu, Khim prir soedin 1971, 7, 164. Through Chem abstr 75, 49390q.
41. Asatiani, V S & Mudzhiri, K S, Soobshch akad nauk gruz SSR 1971, 64, 341. Through Chem abstr 76, 70098f.
42. Rhakimov, D A, Sharipov, M R, Malikov, V M, & Yunusov, S Yu, Khim prir soedin 1971, 7, 677. Through Chem abstr 76, 110300z.
43. Il'yakova, Kh T, Malikov, V & Yunusov, S Yu, Khim priv soed 1970, 6, 717.
44. Yuldashev, P Kh & Yunusov, S Yu, Chem nat prod 1965, 1, 85.
45. Vachnadze, V Yu, Malikov, V M, Mudzhiri, K S & Yunusov, S Yu, Soobshch akad nauk gruz SSR 1972, 66, 97. Through Chem abstr 77, 31539b.
46. Trojánek, J & Hodkova, J, Coll czech chem commun 1962, 27, 2981.
47. Kaul, J L, Trojánek, J & Bose, A, Coll czech chem commun 1970, 35, 116.
48. Plat, M, Lemay, R, Le Men, J, Janot, M M, Djerassi, C & Budzikiewicz, H, Bull soc chim France 1965, 2497.
49. Vachnadze, V Yu, Malikov, V M, Mudzhiri, K S & Yunusov, S Yu, Khim prir soedin 1972, 341. Through Chem abstr 77, 152416t.
50. Das, B C, Garnier-Gosset, J, Le Men, J & Janot, M M, Bull soc chim France 1965, 1903.
51. Shakirov, T T & Aripov, K H N, Uzb khim zh 1969, 13, 81. Through Chem abstr 72, 103664.
52. Rhakhimov, D A, Malikov, V M & Yunusov,

S Yu, Khim prir soedin 1969, 5, 461. Through Chem abstr 72, 67165n.
53. Kuchenkova, M A, Yuldashev, P Kh & Yunusov, S Yu, Dokl akad nauk uzb SSR 1964, 11, 42. Through Chem abstr 63, 4353.
54. Aliev, A M & Babaev, N A, Farmatsiya 1969, 18, 28. Through Chem abstr 72, 15711t.
55. Kompis, I & Mokry, J, Coll czech chem commun 1968, 33, 4328.
56. Meisel, H & Döpke, W, Tetrahedron lett 1971, 1285.
57. Mokry, J & Kompis, I, Chem zvesti 1963, 17, 852. Through Chem abstr 60, 14553f.
58. Mokry, J, Dubravkova, L & Sefcovic, P, Experientia 1962, 18, 564.
59. Trojánek, J, Strouf, O, Blaha, K, Dolejs, J & Hanus, V, Collect czech chem commun 1964, 29, 1904.
60. Mokry, J & Kompis, I, Naturwiss 1963, 50, 93.
61. Mokry, J, Kompis, I, Dubravkova, L & Sefcovic, P, Experientia 1963, 19, 311.
62. Malikov, V M, Yuldashev, P Kh & Yunusov, S Yu, Dokl akad nauk uzb SSR 1963, 4, 21. Through Chem abstr 59, 11584b.
63. Djerassi, C, Budzikiewicz, H, Wilson, J M, Gosset, J, Le Men, J & Janot, M M, Tetrahedron lett 1962, 235.
64. Zsadon, B, Arvay, G & Tetenyi, P, Acta chim (Budapest) 1972, 72, 355. Through Chem abstr 77, 45490r.
65. Döpke, W & Meisel, H, Pharmazie 1968, 23, 521.
66. Rakhimov, D A, Malikov, V M, Yagudaev, M R & Yunusov, S Yu, Khim prir soedin 1970, 6, 226. Through Chem abstr 73, 45652n.
67. Pyuskyulev, B, Kompis, I, Ognyanov, I & Spiteller, G, Collect czech chem commun 1967, 32, 1289.
68. Plat, M, Le Men, J, Janot, M, M, Budzikiewicz, H, Wilson, J M, Durham, L J & Djerassi, C, Bull soc chim France 1962, 2237.
69. Döpke, W & Meisel, H, Tetrahedron lett 1971, 749.
70. — Ibid 1971, 1287.
71. Malikov, V M & Yunusov, S Yu, Khim prir soedin 1971, 640. Through Chem abstr 76, 85984r.
72. Meisel, H & Döpke, W, Pharmazie 1971, 26, 182.
73. Pachymov, D A, Malikov, V N & Yunusov, S Yu, Khim prir soedin akad nauk uzb SSR 1967, 354. Through Chem abstr 68, 3988u.
74. Malikov, V M & Yunusov, S Yu, Khim prir soedin 1971, 7, 793. Through Chem abstr 76, 110308h.
75. Weiss, S, Aynilian, G H, Abraham, D J, Farnsworth, N R & Cordell, G A, J pharm sci 1973. In press.
76. Trojánek, J, Kavkova, K, Strouf, O & Cekan, Z, Collect czech chem commun 1961, 26, 867.
77. Aliev, A M & Babaev, N A, Farmatsiya (Moscow) 1968, 17, 23.
78. Strouf, O & Trojánek, J, Chem ind (London) 1962, 2037.
79. Holubek, J, Strouf, O, Trojánek, J, Bose, A K & Malinowski, E R, Tetrahedron lett 1963, 897.

80. Mokry, J & Kompis, I, Lloydia 1964, 27, 428.
81. Döpke, W & Meisel, H, Pharmazie 1966, 21, 444.
82. Döpke, W, Meisel, H, Grundeman, E & Spiteller, G, Tetrahedron lett 1968, 1805.
83. Döpke, W, Meisel, H & Spiteller, G, Pharmazie 1968, 23, 99.
84. Döpke, W, Meisel, H & Fehlaber, H W, Tetrahedron lett 1969, 1701.
85. Vochnadze, V Yu, Malikov, V M, Mudzhiri, K S & Yunusov, S Yu, Khim prir soedin 1971, 7, 676. Through Chem abstr 76, 110298e.
86. Farnsworth, N R, The phytochemistry of *Vinca* species, in The *Vinca* species—botany, chemistry and biological activities (ed N R Farnsworth & W I Taylor) chap. 3. Dekker, New York, 1973.
87. Plouvier, V, Compt rend 1960, 251, 131.
88. Hajkova, I, Homola, V & Navratil, L, Pharmazie 1959, 9, 537.
89. Farnsworth, N R, Blomster, R N, Damratoski, D, Meer, W A & Cammarato, L V, Lloydia 1964, 27, 302.
90. Grossmann, E, Sefcovic, P & Szasz, K, Phytochemistry 1973, 12, 2058.

Discussion

Heywood: I should mention that at least part of Stearn's classification has been published in Flora Europoea, vol. 3.

As for separate matter, I should like to ask whether you chose the classification that represses the chemical relationship that you wish to have highlighted or that which best fits the general body of information available? Is the Novacék & Stary one the best classifications available irrespective of the alkaloid data? It is perfectly understandable that you should prefer one that best fits your data but this is not the same as saying the others are not feasible. I am reminded of the example of alkaloid surveys and classification in the Leguminosae-Genisteae, where the result of Faugeras and Paris supported one classification (although their results did not entirely agree with earlier alkaloid studies by Nowacki) while a survey of the polyphenols of the same tribe led to different conclusions which supported an alternative classification. There is the danger therefore that you will choose that classification which agrees with the kind of data you produce—alkaloids, terpenoids, polyphenols, etc. which is not perhaps the most scientific way to proceed.

Farnsworth: We considered only alkaloids since non alkaloid constituents of *Vinca* species are largely unknown. We applied the alkaloid data to all four classification systems. The data do not support the systems of Pichon or Pobedemova. The data do support the systems of Novacék & Stary and perhaps the system of Stearn. However not enough alkaloid data are available to accommodate the large number of species proposed by Stearn.

Hegnauer: If complex patterns of constituents are used, as taxonomic characters one should never forget that the pictures known at present are very incomplete and in fact not comparable at all because results of investigations are affected by many factors of which I will mention only chemical ones. The amount of material available affects the number of compounds identified as do extraction and separation procedures, which generally are much ameliorated during long-lasting investigations. My suggestion is to use the main constituents only for chemotaxonomic purposes until all taxa have been examined in precisely the same way.

Cronquist: The first question that occurs to a taxonomist is, how many samples of *Vinca libanotica* did you analyze, and what proportion of the geographic area of the species do they cover?

Farnsworth: Only one sample, and I do not know the total range of the species.

Cronquist: How many samples of *Vinca herbacea* did you analyze, and from what proportion of the range?

Farnsworth: None, but I rely on other reports probably based on only a few samples.

Cronquist: Then I must say that from a taxonomic standpoint your results are interesting and suggestive, but far from conclusive. For morphological characters, which are more easily studied, taxonomists have become accustomed to finding that particular specimens or local populations may appear very distinctive, but the differences fade out when the whole range of the group is studied. We must expect the same sort of problems in chemical characters.

The Quinolizidine Alkaloids

A. E. Schwarting

University of Connecticut, School of Pharmacy, Storrs, Conn. 06268, USA

The quinolizidine ring (1) is a structural unit of numerous alkaloids. It is also a component

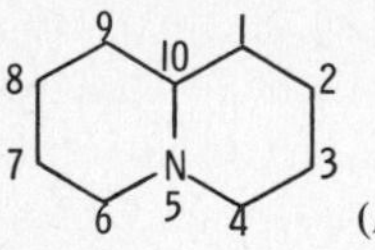

(1)

of one of the scent gland constituents of the North American beaver and of an insect metabolite. These compounds range from very simple derivatives of the bicyclic molecule to complex derivatives of the system. In some cases the quinolizidine nitrogen is shared with another, more well-known, heterocyclic ring system.

The fused ring system exists in two isomeric forms. The lone pair of nitrogen and the proton on C-10 exist either on the same side or on opposite sides of the ring plane. Both diastereomers are found among the natural alkaloids. The *trans*-isomer is lower in enthalpy and is thus the more stable isomer. There is no evidence that the pair is interconvertible through natural processes.

The ring system, unknown prior to its discovery in the lupine alkaloids (e.g. 2–4), has traditionally been associated with this large group of natural products. However, the quinolizidine unit is the nitrogen heterocycle of other alkaloids—in the small group of certain of the lythraceous alkaloids (2), in crepidine (19) one

of the orchidaceous alkaloids and in cevine (6), one of the veratrum alkaloids. In the latis virtually ignored in chemical systematics. The several other alkaloids accompanying cevine, as well as similar steroidal amines in other taxa, do not possess the quinolizidine unit. Crepidine is the only one of the known orchid alkaloids possessing this ring system. An analogous example exists in the classification of the lycopodium alkaloids. Although a great number of the alkaloids of this group possess a quinolizidine unit (e.g. 7), their proposed biogenetic character and the fact that not all of them possess this system had led to other chemical characterizations. The quinolizidine unit is also concealed in alkaloid groups classified by conventions which relate the nitrogen to another ring system component. For example the nitrogen of the protuberberine alkaloids (8) is shared on the one hand by the isoquinoline moiety but also by the quinolizidine unit. Cryptopleurine (9) is an isoquinoline alkaloid

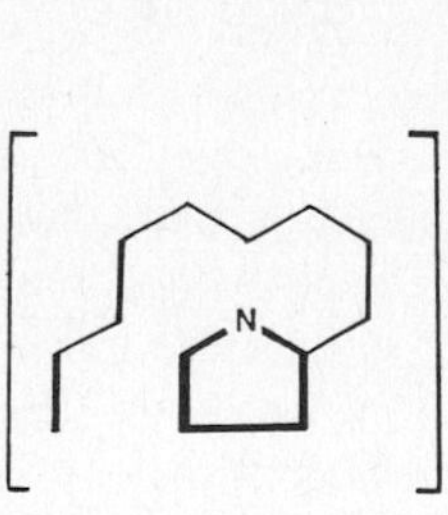

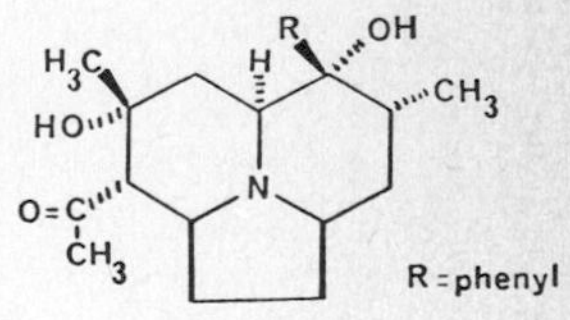

(19) CREPIDINE

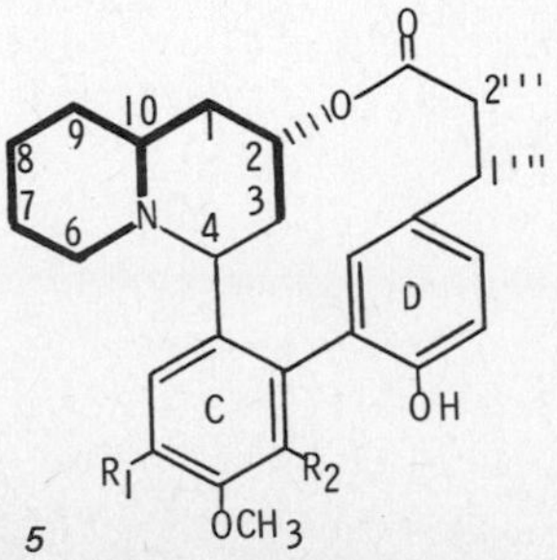

	Stereochemistry at ring junction	1'''-2'''	R₁	R₂
Lyfoline	*Trans*	–CH=CH–	–OH	–H
Cryogenine	*Cis*	–CH=CH–	–OCH₃	–H
Lythrine	*Trans*	–CH=CH–	–OCH₃	–H
Sinicuichine	*Cis*	–CH=CH–	–H	–OCH₃
Nesodine	*Trans*	–CH=CH–	–H	–OCH₃
Lythridine	*Trans*	–CHOH–CH₂	–OCH₃	–H
Heimidine	*Cis*	–CHOH–CH₂–	–OCH₃	–H
Dehydrodecodine	*Trans*	–CH=CH–	–H	–OH

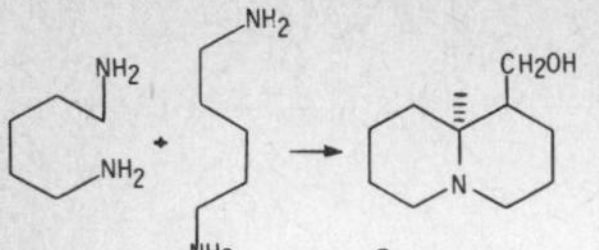

2 Lupinine

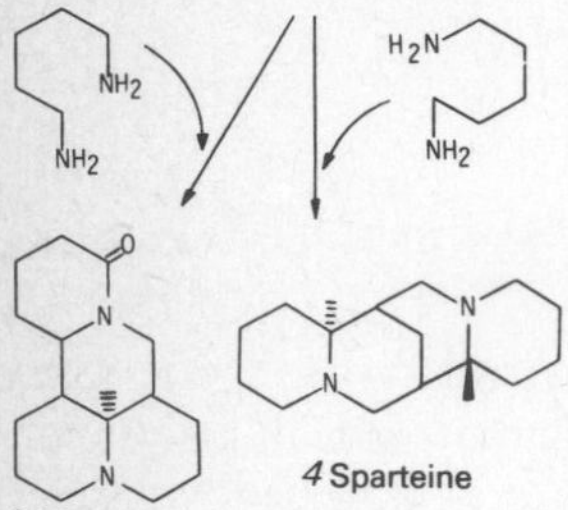

3 Matrine

4 Sparteine

by one convention and a phenanthroquinolizidine alkaloid by another.

Thus there appears to be some lack of recognition of the quinolizidine unit in the chemical classification of alkaloids. The unit is ordinarily recognized when it is the only distinguishable or recurring nitrogen heterocycle in a group. This lack of distinction can be attributed in part to a lack of unique chemical and physiological properties. However, it must be recognized that the group is small and many of the non-lupine alkaloids are of relative recent discovery.

Biosynthetic aspects

The biosynthesis of the quinolizidine unit and of the adjunct structure of several groups of alkaloids has received considerable attention. The origin of the lupine alkaloids has been studied extensively. The results of all of these studies show an extreme diverse origin of this unit. Although it is not unusual to find a dual origin for certain molecule types, it is most unusual for a unit to arise, as does the quinolizidine unit, from a medley of pathways. The known and hypothetical pathways follow.

Lysine derived alkaloids

(1) The quinolizidine unit of the lupine alkaloids is derived from lysine units via cadaverine [1, 2] through a symmetrical intermediate. Lupinine, in turn, serves as an intermediate in the biosynthesis of sparteine (4) and other $C_{15}N_2$ alkaloids. Incorporation of activity from radioactive lysine has been established for several of these.

Nobel 25 (1973) Chemistry in botanical classification

Pelletierine derived alkaloids

(1) Pelletierine (10) is an α-substituted piperidine base originating from lysine and an acetoacetate unit. It is believed that Δ^1-piperideine, the immediate precursor of the piperidine nucleus, reacts with an acetate-derived precursor to produce the pelletierine molecule. The alkylation reaction in this biosynthesis has not been demonstrated.

(2) Pelletierine or its equivalent has been demonstrated to serve as a precursor to a portion of the quinolizidine ring system of lycopodine and several other related alkaloids [6]. The dimeric nature of the hypothetical lycodine-obscurine skeleton was not observed in incorporation studies on the biosynthesis of lycopodine. Only one intact pelletierine unit is incorporated into lycopodine (7).

(3) Coccinellin (11) is a *cis*-quinolizidine N-oxide alkaloid of the beetle *Coccinella septempunctata* [7]. A hypothetical intermediate from a pelletierine precursor may be viewed as a plausible precursor. Another of the Coccinellidae, *Adalia bipunctata* L., has yielded adaline (12) [8]. Its obvious relationship to the same hypothetical precursor suggests a common origin for these two insect metabolites. The suggestions that adaline is derived from a pelletierine precursor is consistent, for example, with the proposed origin of dioscorine.

(4) Pelletierine also stands as the immediate precursor to a portion of the quinolizidine unit of certain of the lythraceous alkaloids (2, 13, 14) [9]. A condensation reaction involving an appropriate aromatic aldehyde would produce the appropriate phenylquinolizidol. We have recently isolated 2-hydroxy (a)-4-(3-hydroxy-4-methoxyphenyl) (e)-*trans* quinolizidine from *Heimia salicifolia* Link & Otto [10] and work

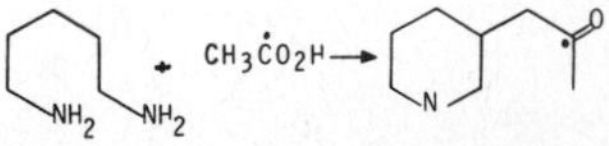

10'elletierine

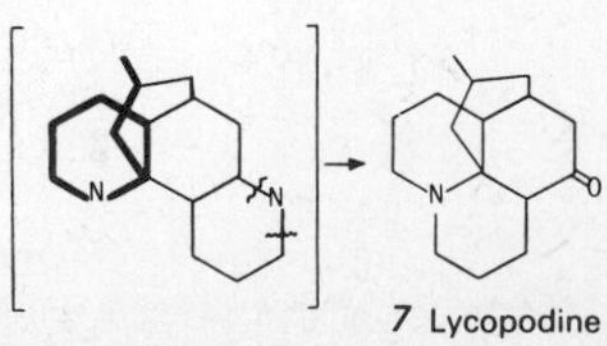

7 Lycopodine

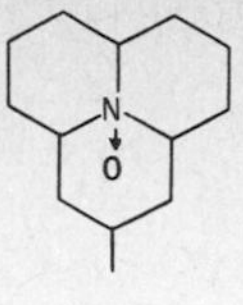

11 Coccinellin 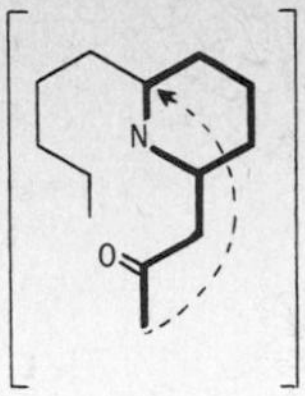 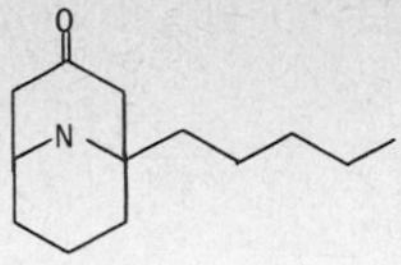   **12** Adaline

is in progress to establish the role of several benzaldehyde equivalents in the biosynthesis of the phenylquinolizidine portion of the lactonic alkaloids of this species.

Terpene derived alkaloids

(1) The carbon skeleton of the quinolizidine unit of emetine (isoquinoline alkaloid) (*15*) and ajmalacine (indole alkaloid) (*16*) originates in part from a loganin-related monoterpene unit. A Mannich reaction would account for the condensation.

(2) In cevine (*6*), castoramine (*18*) and nupharidine (*17*) the quinolizidine unit appears to arise from a terpenoid chain derived from isoprenoid units linked in the typical head-to-tail manner. Although experimental evidence is lacking, the units must be derived by way of electrophile, dimethylallyl pyrophosphate. Coupling of a molecule of dimethylallyl pyrophosphate with one of isopentenyl pyrophosphate would occur according to a well established reaction in a number of biological systems. The insertion of an atom of nitrogen is indicated, but the mechanism has not been established. The additional carbon, in the quinolizidine moiety of cevine, beyond those of the two isopentenyl groups, is translocated from another isopentenyl group according to a well known reaction in steroid biosynthesis.

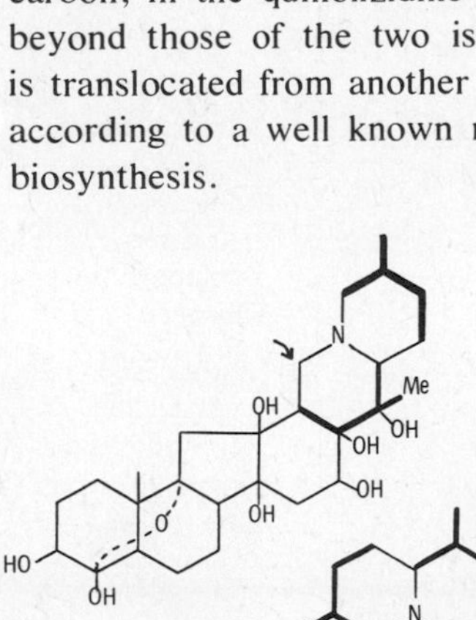

Phenylethylamine derived alkaloids

(1) In the protoberberines (e.g. *8*) the N-C bond of the quinolizidine unit, known as the berberine bridge, probably arises by an oxidative process. The precursors to these alkaloids are phenethylamine analogs with the bridge carbon representing an *N*-methyl function.

(2) Cryptopleurine, a phenanthroquinolizidine alkaloid (*9*), resembles tylophorine, a phenanthroindolizidine alkaloid. In both compounds the nitrogen is shared with an apparent isoquinoline moiety. Biosynthetic studies on tylophorine [11] indicate that this molecule is of phenethylamine (dimeric) origin. Ornithine is efficiently incorporated into tylophorine but degradation experiments have not been carried out to determine the locus of the incorporation. Comparison of results of analogous studies suggest the probability of incorporation of lysine into cryptopleurine. If this is the case, the quinolizidine ring is derived from lysine and from two molecules of a phenethylamine precursor(s).

An ornithine-acetate derived alkaloid

Crepidine (*19*) may be viewed either as an indolizidine or as a quinolizidine alkaloid. This alkaloid from *Dendrobium crepidatum* Lindl. is the only example of an alkaloid with a quinolizidine unit among the alkaloids of the Orchidaceae [12]. Inspection of the structures of other orchid alkaloids together with biogenetic considerations on crepidine and the other orchid alkaloids suggest that the fundamental precursor to this alkaloid is ornithine. Appropriate alkylation of a pyrrolidine intermediate, with an acetate- and/or propionate-derived precursor (see bracketed structure) (*19*) would provide the intermediate giving rise to the quinolizidine ring. In this context there is biogenetic analogy in the proposed origin of coccinellin and in other lysine and ornithine derived alkaloids.

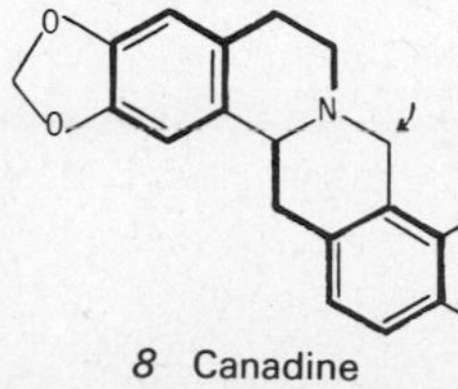

6 Cevine

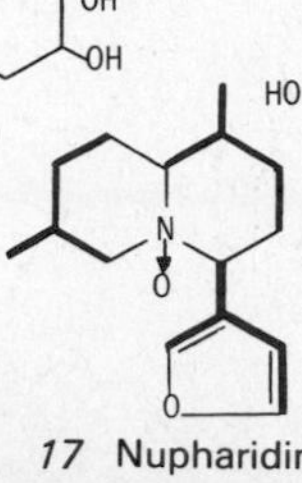

17 Nupharidine

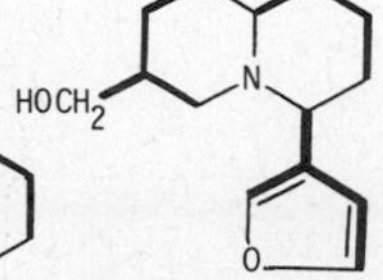

18 Castoramine

8 Canadine

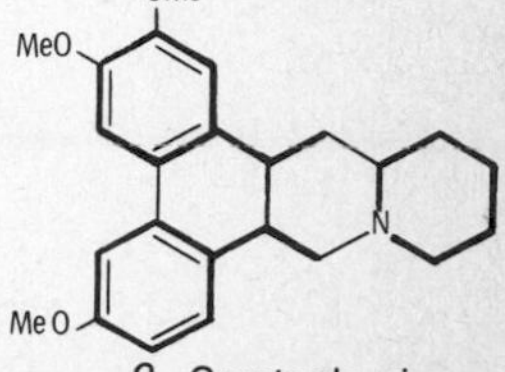

9 Cryptopleurine

Reactions

The several reactions which incorporate various precursors to give rise to the quinolizidine alkaloids are those which have been described for other alkaloids. For example, Schiff base formation leads to the condensation of the two precursor components to form the C-N bond in the lupinine alkaloids. An intramolecular Mannich reaction may be viewed as that process which creates the cyclic base. The coupling of isopentenyl units and alkylation reactions are common processes in the formation of other natural molecules. The one unique reaction among the alkaloids of this group is the radical process giving rise to the berberine bridge in the formation of the quinolizidine moiety of the protoberberine alkaloids.

Stereochemical considerations are of interest in several instances. For example, in the case of the pelletierine-derived lythraceous alkaloids, an inspection of Drieding models reveals that each of the enantiomers of pelletierine could give rise to both *cis* and *trans* phenylquinolizidones through equatorial or axial attack by a benzaldehyde equivalent.

Chemosystematics

On the basis of molecular complexity and also on the basis of the biogenesis of the fundamental ring system, the quinolizidine alkaloids are a complex and heterogenous group. This evaluation, coupled with the broad distribution of these alkaloids among quite unrelated taxa deny analysis about evolutionary relationships at almost all levels.

Among all the taxa which generate this ring system, only those producing the lupine alkaloids have been given experimental chemo-

taxonomic consideration. Hegnauer, in 1966, assembled the current knowledge [13] about the distribution of four alkaloids, lupinine, sparteine, lupanine and N-methylcytisine among eight families. The basis for the reports about the occurrence in the Monimiaceae, Ranunculaceae, Papaveraceae and Scrophulariaceae must be questioned. It would appear that valid data would restrict the occurrence to the Berberidaceae, Chenopodiaceae, Leguminosae and Solanaceae (table 1). Occurrence is among 21 genera in the Leguminosae and in one or two genera in each of the other families. Studies by Mabry, Turner & Crammer [14, 15] and others have established the occurrence in 3 tribes, Sophoreae, Podalyrieae and Genisteae of the subfamily Papilonoideae. These alkaloids

Table 1. *Distribution of the quinolizidine alkaloids*

Alkaloid type	Family distribution
Lysine derived	
Lupine-type	Berberidaceae
	Chenopodiaceae
	Leguminoseae
	Solanaceae
Lysine-pelletierine derived	
Hydrojulolidine-type	Lycopodiaceae
Lycodine-obscurine-type	Lycopodiaceae
Phenylquinolizidine-type	Lythraceae
Lythrancine-type	Lythraceae
Ornithine derived	
Crepidine type	Orchidaceae
Terpene derived	
Nuphar-type	Nymphaceae
Loganin-indole-type	Apocynaceae
	Loganiaceae
	Rubiaceae
Loganin-isoquinoline-type	Alangiaceae
	Rubiaceae
Cevine-type	Liliaceae
Phenethylamine derived	
Protoberberines	Anonaceae
	Berberidaceae
	Convolvulaceae
	Fumariaceae
	Lauraceae
	Menispermaceae
	Papaveraceae
	Ranunculaceae
	Rutaceae
Cryptopleurine-type	Lauraceae

14 Lythranidine

13 Lythrancine I

15 Emetine

16 Ajmalacine

are absent in species of the other tribes of the subfamily and are not present in other subfamilies. These authors concentrated some of their study on the genus *Baptisia*. The widespread distribution among these species did not provide useful information to establish relationships within the genus. The arguments which these authors advanced, relative to phyletic affinities, are well known and need not be repeated here.

The other taxa which produce quinolizidine alkaloids (table 1) have not been studied chemotaxonomically. The protoberberines and retro-protoberberines would surely not yield to such evaluation. There are approx. 70 different alkaloids distributed among 9 families. The Lycopodiaceae, Nymphaceae and Lythraceae have not been sufficiently investigated to provide data which will permit meaningful interpretations. A judgement may also be made on behalf of the loganin-derived quinolizidine ring system of indole and isoquinoline alkaloids; their known distribution appears restricted and the reactions which generate the monoterpene precursor are of a highly specialized nature.

Our work on the isolation and biosynthesis [16, 17] of the alkaloids of *Heimia salicifolia* and similar work [18] on the alkaloids of *Decodon verticillatus* (L.) Ell. and of a *Lagerstroemia* species has provided considerable information about a group of new alkaloids. These alkaloids may be viewed as compounds derived from a phenylquinolizidol unit and a cinnamoyl unit joined through oxidative coupling and esterification (2). The quinolizidine ring is derived in part from a pelletierine precursor; both aromatic nuclei are derived from phenethylamine precursors. All of the alkaloids possess a characteristic stereochemistry. The C-2 oxygen and the nitrogen spin pair are axial while the 4-phenyl substituent is equatorial. Both *cis* and *trans* stereoisomers occur in these

species and several alkaloids are common to both *Heimia* and *Decodon* species. There are, however, some structure differences which appear to distinguish the alkaloids of these species. All of the known *Heimia* alkaloids are of the biphenyl type, whereas in *Decodon* there are several biphenylether alkaloids. All of the alkaloids of *Heimia* are unsaturated at carbon atoms $1'''-2'''$ or they occur as the hydrate. The known *Decodon* species alkaloids are unsaturated or saturated and hydrated alkaloids have not been reported.

The alkaloids of *Lythrum anceps* Makino [19] represent a group of two additional new types of alkaloids (*13, 14*). The lythrancine I type possess a quinolizidine ring and a biphenyl unit. However, the substitution on the heterocyclic system is α–α to the nitrogen and the alkaloids are non-lactonic. The lythranidine type is an α–α substituted piperidine derivative otherwise resembling the lythrancine group. The piperidine type show structural similarity to the *Lobelia* alkaloids and to the muskopyridines of the scent gland of the musk deer.

Koehne [20], in his revision of the Lythraceae established 2 tribes, Lythreae and Nysseeae and 2 subtribes under each of these. The genus *Lythrum* is placed in the first tribe, above, with *Ammannia, Woodfordia, Cuphea, Lafoënsia* and 7 other genera. *Heimia* and *Decodon* are placed in the latter subtribe along with *Lagerstroemia, Lawsonia* and 6 other genera. In isolation and characterization studies noted earlier together with screening and isolation studies on the other lythraceous species mentioned above, certain additional information can be reported. We have not found alkaloids in leaves of species of *Ammannia, Cuphea, Lafoënsia* and *Lawsonia*. We have found alkaloids, yet to be identified, in leaves of a species of *Woodfordia*.

The Koehne treatment of the Lythraceae is quite consistent with the available alkaloid data. *Heimia* and *Decodon* alkaloids are similar and they are placed together in one subtribe of the family. The alkaloids of *Lythrum* are different chemically and the genus is in a different tribe.

Summary and conclusions

The quinolizidine ring is a component of numerous alkaloids. It is derived from a broad variety of precursors through well known biochemical reactions. The compounds are widely distributed in plants and they occur as insect and mammalian metabolites.

All of the alkaloids except the lupine alkaloids are derived from two or more fundamental precursors. These include lysine, ornithine, acetate, isoprenoid units, methyl functions, tryptophan, phenylethylamines and a benzaldehyde equivalent probably derived from the latter. Twelve alkaloid types are distinguishable.

Except for the lupine alkaloids of the Papilonoideae and possibly in the Lythraceae and the Nymphaceae can chemotaxonomic judgements be made. In the Papilonoideae the alkaloids are restricted to three tribes. The alkaloids of the closely related species of *Hemia* and *Decodon* are quite different from those of a *Lythrum* species. In the Nymphaceae only species of *Nuphar* have been shown to contain alkaloids. But sufficient work has not been done within these families to permit sensible conclusions.

It would appear that the utility of the quinolizidine ring as a key to establish botanical affinities will be restricted to subfamily, tribal or at other finite inter-generic levels. A satisfactory interpretation, in most cases, awaits further phytochemical and biosynthetic work.

Studies on the lythraceous alkaloids in our laboratory have involved a number of students and associates. I am indebted to the contributions made by these students and, in particular, to my associates Drs Ana Rother, J. Michael Edwards and J. M. Bobbitt.

References

1. Schütte, H R, Arch pharm 1960, 293, 1006.
2. — Atompraxis 1961, 7, 91.
3. Spenser, I D, Compr biochem 1968, 20, 231.
4. Keogh, M F & O'Donovan, D G, J chem soc 1970, 1972.
5. Gupta, R N & Spenser, I D, Phytochemistry 1969, 8, 1937.
6. Braekman, J-C, Gupta, R N, MacLean, D B & Spenser, I D, Can j chem 1972, 50, 2591.
7. Karlsson, R & Losman, D, Chem commun 1972, 627.
8. Tursch, B, Brackman, J C, Daloze, D, Hootele, C, Losman, D, Karlsson, R & Pasteels, J M, Tetrahedron lett 1975, 201.
9. Koo, S H, Gupta, R N, Spenser, I D & Wrobel, J T, Chem commun 1970, 396.
10. Rother, A & Schwarting, A E, Experentia. In press.
11. Mulchandani, N B, Iyer, S S & Badheka, L P, Phytochemistry 1971, 10, 1047.
12. Kierkegaard, P, Pilotti, A N & Leander, K, Acta chem Scand 1970, 24, 3757.
13. Hegnauer, R, Comparative phytochemistry of alkaloids, in Comparative phytochemistry (ed T Swain) p. 211. Academic Press, London and New York, 1966.
14. Crammer, M F & Mabry, T J, Phytochemistry 1966, 5, 1133.
15. Crammer, M F & Turner, B L, Evolution 1967, 21, 508.
16. Rother, A & Schwarting, A E, Phytochemistry 1972, 11, 2475.
17. Rother, A, Appel, H G, Kiely, J M, Schwarting, A E & Bobbitt, J M, Lloydia 1965, 28, 90 and preceding papers.
18. Ferris, J P, Boyce, C B, Briner, R C, Douglas, B, Kirkpatrick, J L & Weisbach, J A, Tetrahedron lett 1966, 3641 and preceding papers. See also ref. 9.
19. Fujita, E, Bessho, K, Fuji, K & Sumi, A, Chem pharm bull 1970, 18, 2216.
20. Koehne, E, Lythraceae in Das Pflanzenreich, Heft 17 (ed A Engler) p. 1. Engelmann, Weinheim, 1903.

Discussion

Birch: I should like to warn our botanical colleagues of a deceptive habit of the chemist. He tends to classify compounds in ways which interest him, which may not be at all helpful to the botanist. For example, the quinolizidine nucleus is probably the most interesting and chemically significant structure when it is present. However, it can arise in various ways which have little biosynthetic relations to each other. Conversely the relation of a substance to other significant ones may be obscured if it does not contain the same named structure. Botanists should not assume that the same chemical class of compound in two plants has any meaning unless the biogenetic relationship is clear. Even the term "alkaloid" is rather meaningless; it is merely an indication to the chemist that a compound containing a basic nitrogen is present.

Schwarting: Professor Birch is correct. The quinolizidine nucleus is a good example illustrating the limitations which are imposed by biogenetic considerations to systematic affinities for a given structure.

The Use of Amino Acid Sequence Data in the Classification of Higher Plants

D. Boulter

Department of Botany, University of Durham, Durham, UK

Summary

The inter-relationship of nucleic acid, protein, micromolecular and morphological characters, is presented and their relative usefulness in classification discussed. This involves a consideration of the "natural classification" and a brief discussion of the different methods used to construct phylogenies.

An affinity scheme is presented using cytochrome c amino acid sequence data from 23 higher plants; the implications of this scheme is discussed in relationship to existing phylogenetic schemes such as those of Thorne, Cronquist and Takhtajan. Preliminary serological and sequence studies of plastocyanin are also presented.

The information content of an organism is encoded exclusively in its deoxyribonucleic acid (DNA), although not all DNA is informational [1]. The information is transcribed and translated and thereby specifies the amounts and primary structures of the proteins, which are responsible for directing the chemical reactions performed by the organism (its metabolism); the visible result of this activity is the form of the organism itself (the phenotype). Over long periods of geological time, changes occur in the phenotype which are subjected to selection and other evolutionary processes, so giving rise to a series of forms which are related to each other by their evolutionary history. It might be supposed, therefore, that organisms which are closely related would have DNAs whose structures (base sequences) are more similar to each other than to those of organisms less closely related. However, the relationship is not always simple, since not all DNA is informational.

One of the major objectives of classification is to place organisms in groups, such that the greatest predictability about their characteristics is obtained; in other words, groups are formed using a few known common characters but any other characters of any member may be assumed to be common to all other members of the group. Such groups were originally called "natural groups" by Aristotle [2].

A phylogenetic classification attempts to arrange closely related organisms in groups so that the evolutionary relationships of the groups are made manifest. Phenetic classifications, on the other hand, are defined as those in which grouped members have the maximum correlation of characters [3]. Phylogenetic classifications are best established using fossil evidence, but in the absence of sufficient fossil evidence, attempts must be made to establish evolutionary relationships using other data, and Crick [4] and Zuckerkandl & Pauling [5] have suggested that nucleic acids and proteins are especially important in this connection. It is not clear, however, whether members of a phylogenetic group necessarily have the most characters, e.g. biochemical, morphological, etc., in common. It is possible that if characters such as the structure of DNA or proteins are used, a correct phylogenetic classification may be achieved but possibly it would only have predictability with regard to the characteristics of these molecules, and not to morphological and other types of characters. Some claim, therefore, that a phenetic classification is the ideal for which to strive (see [6]). However, as more characters are used to construct phenetic classifications, several classifications may be devised, each more useful for its own purpose. Another possibility is that eventually the phylogenetic and phenetic classifications may coincide, in which case the phylogenetic classification is to be preferred on the ground of parsimony. Since phylogenetic schemes based on informational macromolecules are still in their infancy and the attempt to formulate a comprehensive phenetic classification has only just started, taking a firm position on the relative merits or demerits of the two schools of thought is, at the moment, a matter of intuition and hope.

It has been suggested that for the higher plants, the task of establishing a phylogenetic classification is hopeless anyway, because there is insufficient fossil evidence [7]. I do not share this pessimistic view; I believe that the final success of such a venture will depend upon the use of various kinds of biological information, but that it will rest heavily on the use of macromolecular data. At the present time this is mainly in the form of amino acid sequences of homologous proteins; other types of macromolecular data have various limitations. With the DNA hybridization method, the limitations are mainly the technical difficulties involved in carrying out the hybridization and the interpretation of the results; with serological comparisons, the limitation is that only a small part of the protein, of the order of a 15Å^2 patch, is involved, so that other differences are neglected.

A considerable body of chemotaxonomic information involving so-called micromolecules has been accumulated [8], and some of this will contribute useful information for the establishment of a phylogenetic classification [9]. The problem is to decide which are the most useful types of molecules for this purpose, since the use of this data has limitations. For example, information is often recorded as the presence or absence of a compound in a particular organism, and the absence of a compound does not necessarily mean that the genetic information required for its production is absent also. There are many examples where genetic information is "switched off", and it is important to realise that for this reason the production of micromolecules can be affected by internal and external environmental factors. This limitation does not apply to protein sequence data since a comparison is made between amino acid residues at particular residue positions in homologous sequences. Another limitation of the micromolecular approach is that the production pathway of a compound may differ in different organisms and relationships based on the presence or absence of the compound, therefore, may be invalid.

Comparison of morphological characters will also play an important role in the establishment of a phylogenetic classification. However, it is often not possible to distinguish morphological similarities which have arisen due to convergence, from those due to common ancestry, since (1) few characters are considered; (2) there is no quantitative scale of relationship; (3) the observed similarity between two organisms may be due to the expression of different genes; and (4) morphological characters have less evolutionary stability than protein characters (amino acid residue positions in a sequence of an homologous protein set), and cannot be compared over widely differing taxa, as can the latter, e.g. the amino acid in position 1 of human cytochrome c can be compared with the amino acid in the same position in *Enteromorpha,* whereas homologous morphological characters could not be identified. Further difficulties arise due to the fact that the rates of evolutionary change of morphological characters have not remained constant. This unevenness in rate applies to the same character at different stages of evolution and also between two characters of an organism relative to one another. Thus, when using morphological characters, it is often not possible to decide whether a group diverged early with a slow rate of subsequent evolution or later at a faster rate of evolution.

Convergence has occurred during the evolution of different proteins although apparently rarely [10], but it can usually be detected by comparing the amino acid sequence of the proteins concerned, since many amino acid residue positions, i.e. characters, are considered and these are generally independent of one another.

With regard to the occurrence of uneven rates of protein evolution, Boulter [11] has given reasons for preferring the ancestral amino acid sequence method to matrix methods, for tree construction. In this method, sequences are related to computed common ancestral sequences, and these linkages should be correct even if rates of evolution are different in different parts of the tree. Along branches where faster than average rates of change have occurred, fewer positions can be used to link a sequence to the tree, but even so, it is most likely to link to its correct ancestor. With plant cyto-

Fagopyrum esculentum (Polygonales)
Spinacea oleracea (Caryophyllales)
Pastinaca sativa (Cornales)
Tropaeolum majus (Geraniales)
Guizotia abyssinica (Asterales)
Helianthus annuus (Asterales)
Arum maculatum (Arales)
Allium porrum (Liliales)
Zea mays (Poales)
Hordeum vulgare (Poales)
Triticum sp. (Poales)
Lycopersicon esculentum (Scrophulariales)
Ricinus communis (Euphorbiales)
Sesamum indicum (Scrophulariales)
Acer negundo (Sapindales)
Abutilon theophrasti (Malvales)
Gossypium barbadense (Malvales)
Sambucus nigra (Dipsacales)
Nigella damascena (Ranunculales)
Cannabis sativa (Urticales)
Phaseolus aureus (Fabales)
Cucurbita maxima (Cucurbitales)
Brassica oleracea
Brassica napus } (Capparales)
Ginkgo biloba (Ginkgoales)

Fig. 1. Phylogenetic tree relating the cytochrome *c* sequences of 25 species constructed using the ancestral sequence method.

chromes *c*, Boulter et al. [12] have recorded about 25% parallel mutations, a figure similar to that arrived at by Fitch & Margoliash [13] for animal cytochromes and in the latter case the molecular tree, apart from a few details, coincided with that established from the fossil record.

Cytochrome *c*

Amino acid sequence data

Figure 1 gives a molecular tree constructed by the ancestral sequence method, relating cytochrome *c* sequences from 19 dicotyledonous plants, 5 monocotyledonous plants and one gymnosperm. The sequences form five main groups, one containing; (1) the monocotyledons; (2) *Fagopyrum esculentum* Moench and *Spinacea oleracea* L.; (3) *Pastinaca sativa* L., *Tropaeolum majus* L., *Guizotia abyssinica* (L.f.) Cass and *Helianthus annuus* L.; with (4) *Lycopersicon esculentum* Mill., *Ricinus communis* L. and *Sesamum indicum* L.; (5) *Acer negundo* L., *Abutilon theophrasti* Med., *Gossypium barbadense* L., *Sambucus nigra* L., *Nigella damascena* L., *Cannabis sativa* L., *Phaseolus aureus* Roxb., *Cucurbita maxima*

Duchesne and *Brassica oleracea* L. and *B. napus* L. The first group to diverge from the rest is the Centrospermae group of dicotyledons, and then at approximately the same point in time, the monocotyledons and the 3 other dicotyledonous groups. The ancestral sequence method does not differentiate between the times of origin of members of the first two groups, but the last group can be sub-divided into 3 groups, (*a*) *Acer negundo, Abutilon theophrasti* and *Gossypium barbadense*; (*b*) *Sambucus nigra*; and (*c*) *Nigella damascena, Cannabis sativa, Phaseolus aureus, Cucurbita maxima* and *Brassica oleracea* and *B. napus*. The earliest point in time has been fixed at *Ginkgo biloba* L. from general fossil evidence [14].

The important question to be answered is whether this molecular tree, which relates cytochromes ancestrally, is coincident with the tree established from morphological and anatomical features of the organisms themselves and, if not, is the molecular tree more likely to give a true picture of the phylogenetic relationships? The molecular tree differs in major features from existing phylogenetic schemes, such as those of Cronquist [15], Thorne [16] and Takhtajan [17]. The molecular method is not fully discriminatory, and it would be unwise to try to establish the phylogenetic relationships of higher plants without taking into consideration other biological evidence. In the absence of sufficient fossil evidence, we have felt it necessary to seek confirmatory molecular evidence using another unrelated protein. Plastocyanin, a photosynthetic protein, has been chosen and we are carrying out both serological and amino acid sequence studies.

Plastocyanin

Serological data

The precipitin cross-reaction between *Spinacea oleracea* plastocyanin, and plastocyanin of *Lactuca sativa* L., *Vicia faba* L., *Urtica dioica* L., *Symphytum officinale* L., *Sambucus nigra, Mercurialis perennis* L. and *Cucurbita pepo* L., have been compared with the molecular tree distances between cytochrome sequences from the same or closely related genera. Except for the cross-reaction with *Vicia faba*, which was least related to *Spinacea oleracea* serologically, and least but one using cytochrome data, the order of relationship was the same by both methods. The preliminary experiments, which were carried out at Durham by Dr D. G. Wallace, corroborate the findings established from the cytochrome *c* ancestral tree.

Amino acid sequence data

Partial or complete sequences of 10 different higher plant plastocyanins have now been determined. Although these results are preliminary, certain facts already emerge. Plastocyanin of *Phaseolus aureus* and *Vicia faba* are more similar to each other than to any other sequences examined, and *Cucurbita pepo* plastocyanin is also associated with them. *Symphytum officinale* and *Lactuca sativa* are closely associated as are *Galium aparine* L. and *Sambucus nigra*. *Solanum tuberosum* L. is distinct from all these and *Mercurialis perennis* and *Urtica dioica* are also distinct from *Symphytum officinale* and *Lactuca sativa* on the one hand, and from *Cucurbita pepo*, *Vicia faba* and *Phaseolus aureus*, on the other (unpublished results of J. Ramshaw & M. Scawen).

Conclusions

Although these are only preliminary results, they tend to confirm those of the cytochrome tree. It is hoped that additional information from the other plastocyanins at present being sequenced will determine whether the molecular data give a more accurate picture of relationships than do existing schemes.

It would be premature, therefore, to try to integrate these findings with other kinds of biological information, especially as this would involve making a reappraisal of the many conflicting phylogenetic schemes which have been proposed. At this point, it is better to concentrate on those features of the tree which should remain constant, in the light of this additional information. Firstly, the monocotyledons appear as a monophyletic group. Whether they had a separate origin from the dicotyledons cannot be answered unequivocally by the data given here but the results are consistent with the suggestion of other workers that they originated from a dicotyledonous stock (see Takhtajan [17]). The results given in fig. 1 indicate

that present-day angiosperms represent a relic taxa (group) and that the five groups mentioned above have had a long and separate evolutionary history. This contrasts with the major phylogenetic schemes, i.e. those of Cronquist [15], Thorne [16] and Takhtajan [17]; Cronquist, for example, suggests that the Magnoliidae gave rise to the Rosidae and the Caryophyllidae, and that the Rosidae, in turn, gave rise to the Asteridae.

Francis & Harborne [18], on the basis of the relative complexity of the anthocyanins found in *Magnolia*, suggest that the Magnoliaceae are not as primitive as has often been suggested. Kubitzki [19] and Meeuse [20] have summarised chemotaxonomic evidence, for example the presence of the isoquinoline alkaloids in the Ranales (s.l.), which would also indicate that these were not a basic stock from which other dicotyledons arose, except possibly the Rutaceae and the Umbelliferae. This latter suggestion is not substantiated by our data which associates the Umbelliferae with the Compositae (see fig. 1); this latter association is supported by chemotaxonomic data [21].

Meeuse [20], on the basis of chemotaxonomic, phytomorphological, fossil and other data, has suggested a polyphyletic origin of the higher plants. The present work does not throw any new light on this suggestion; however, an alternative theory might be that the great diversity in the chemistry, anatomy etc. of the angiosperms, may be due to present-day groups having had a long separate evolutionary history, as is suggested in the present scheme. The cytochrome of the Centrospermae (or at least the one member we have examined), together with that of the related Polygonaceae, diverged earliest of all the dicotyledon sequences which have been determined. This affinity does not support the contention that the Ranalian-type flower is that from which all other angiosperm flowers have developed, although this is suggested by the major phylogenetic schemes and is a basis of most of the phytomorphological teaching on the flower (see, for example, [22, 23]). The present data tend to support Meeuse's suggestion that the earliest dicotyledons were diclinous and non-petaloid, and had relatively simple flowers. Melville [24, 25], in proposing the gonophyll theory of the origin of the ovary and stamen, has accumulated fossil evidence against the sporophyll concept of the flower on which the "primitiveness" of the Ranalian-type flower is based. Meeuse's views are also supported, to some extent, by the ideas of Thomas [26].

Be that as it may, there is considerable evidence that the Centrospermae are a primitive group, well separated from the rest of the dicotyledons by a long evolutionary history. For example, they normally contain betalains, rather than anthocyanins [27]; the proteinaceous bundles found in the sieve-tube plastids are of a type not found elsewhere in the dicotyledons [28], and there are primitive features in the fine-structure of the pollen walls [29], although this latter characteristic is shared by other dicotyledons including the Magnoliaceae.

I should like to thank Dr D. G. Wallace of the Carnegie Institute, Washington, D.C., for the use of his unpublished serological data, Dr J. Ramshaw, Messrs M. Scawen and D. Richardson for unpublished sequence data, and The Nuffield Foundation and Science Research Council for financial support.

References

1. Watson, J D, Molecular biology of the gene, 2nd edn. Benjamin, New York, 1970.
2. Aristotle, De partibus animalium. Tyrannion and Andronicos of Rhodes. First Century, B.C.
3. Cain, A J, Ibis 1959, 101, 302.
4. Crick, F H C, On protein synthesis, in The biological replication of macromolecules (Symp. XII, Soc exp biol) p. 138. Cambridge University Press, 1958.
5. Zuckerkandl, E & Pauling, L, J theor biol 1965, 8, 357.
6. Davis, P H & Heywood, V H, Principles of angiosperm taxonomy, p. 127. Oliver & Boyd, London, 1968.
7. Bigelow, R S, Syst zool 1958, 7, 49.
8. Hegnauer, R, Chemotaxonomie der Pflanzen. Birkhäuser, Basel, 1962–66.
9. Harborne, J B, Biochemical systematics—the use of chemistry in plant classification, in Progress in phytochemistry (ed L Reinhold & Y Liwschitz) p. 545. Interscience, New York, 1968.
10. Nolan, C & Margoliash, E, Ann rev biochem 1968, 37, 727.
11. Boulter, D, Pure & applied chem 1973, 34, 539.
12. Boulter, D, Ramshaw, J A M, Thompson, E W, Richardson, M & Brown, R H, Proc roy soc Lond B 1972, 181, 441.
13. Fitch, W & Margoliash, E, Brookhaven symp biol 1969, 21, 217.
14. Walton, J, An introduction to the study of fossil plants, 2nd edn. A & C Black, London, 1953.
15. Cronquist, A, The evolution and classification of flowering plants. Nelson, London & Edinburgh, 1968.
16. Thorne, R F, Aliso 1968, 6, 57.
17. Takhtajan, A, Flowering plants—origin and dispersal. Oliver & Boyd, Edinburgh, 1969.
18. Francis, F J & Harborne, J B, Am soc hort sci 1966, 89, 657.
19. Kubitzki, K, Taxon 1969, 18, 360.
20. Meeuse, A D J, Acta bot neerl 1970, 19, 61 and 133.
21. Hegnauer, R, Chemical evidence for the classification of some plant taxa, in Perspectives in phytochemistry (ed J B Harborne & T Swain) p. 121. Academic Press, London and New York, 1969.
22. Eames, A J, Am j bot 1931, 18, 147.
23. — Morphology of the angiosperms. New York, Toronto, London 1961.
24. Melville, R, Kew bull 1962, 16, 1.
25. — Ibid 1963, 17, 1.
26. Thomas, H Hamshaw, New phytol 1934, 33, 173.
27. Mabry, T J, Taylor, A & Turner, B L, Phytochem 1963, 2, 61.
28. Behnke, H D & Turner, B L, Taxon 1971, 20, 731.
29. Roland, F, Rev gen bot 1971, 78, 329.

Discussion

Mears: By this method of using ancestral sequences, have you been able to avoid assumption of equal probability of each amino acid substitution, including reversals?

Boulter: The assumption which has been made is that evolution has taken place by a minimum number of substitutions.

Ourisson: Are all the experimental sequences known beyond doubt? Would corrections lead to changes in the tree derived from the partially wrong data?

Boulter: Mistakes in published sequence data are known to have been made, and it is possible that there are a few errors in the data set presented here. Whether or not corrections would lead to changes in the tree would, of course, depend upon the nature of the hypothesized corrections. We have restricted our conclusions to major points, and, in my view, it is very unlikely that these would be affected by any possible errors in the experimental determination of the sequences.

Jensen: Are there two allotypes of cytochrome *c* in *Cucurbita* and how widespread are allotypes of cytochrome *c*?

Boulter: We found 2 amino acid residues at one residue position in *Cucurbita* cytochrome *c*, but could not isolate two different cytochrome *c* molecules from *Cucurbita*, although, theoretically, this should have been possible. There is little evidence of allotypes of cytochrome *c* in other higher plants or animals.

Jensen: The present affinities are based on amino acid substitutions. Have you used minimum mutation distances and, if so, are the affinities the same?

Boulter: No, but we do intend to subject data to this treatment. It is very possible that some modifications to the affinities suggested here will occur in the future, although it is unlikely that the affinity of Centrospermae relative to the dicotyledonous plants investigated will change.

Jensen: I note that the number of differences between maize and wheat is greater than that between maize and some dicotyledons.

Boulter: Yes, I would emphasize that the position of the sequences of the monocotyledons relative to those of the dicotyledons is not yet fixed firmly by the present data.

Merxmüller: I could not understand why you have chosen *Ginkgo* as the starting point in your scheme. But whatever the reason may have been, may this choice not greatly influence especially the lowest dichotomies?

Boulter: The method gives a branching topology, and *Ginkgo* is chosen as the earliest point in time and does not imply that *Ginkgo* is on the direct evolutionary line to the higher plants. The same result is given if we use the sequence of cytochrome *c* from a green alga, *Enteromorpha* (Meatyard & Boulter, unpublished). It is quite true, however, that if a sequence were used from a more closely related ancestral gymnosperm, better resolution of the lowest dicotomies might result. It is unlikely that any of the major conclusions discussed in the present paper would be affected.

The Interpretation of Comparative Serological Results

U. Jensen

Botanisches Institut, Universität Köln, 5000 Köln, W-Germany

Summary

Determinants of defined proteins may be used as tax-onomical characters. The elucidation of complete serological identity, partial identity or entirely different serological features between taxa permit the drawing of interesting conclusions, but scarcely any benefit for taxonomic purposes. This is different with multi-character systems as given by protein mixtures. Here, the serological results obtained carry tax-onomical weight and—with careful interpretation—contribute to a greater yield of taxonomical information and evidence. It can thereby be shown for the relationship levels of the Berberidaceae genera that the often postulated separation of *Nandina* must be negated and that new aspects must be taken into consideration in the attempt to reveal the na-tural system of this plant family.

During this Symposium, we have discussed primarily questions concerned with the taxonomi-cal value of secondary plant substances. Such substances are normally produced in a whole series of synthesis steps, each step being gov-erned by a certain enzyme. Thus, the characters of such substances are usually polygene-coded. Consequently, a larger number of genes provide the information determining the characters of such substances.

The situation is different with proteins, which have also been dealt with during this Sym-posium. At least the more simple proteins or polypeptide chains are each coded by one gene only; accordingly, we can speak in terms of a "one gene-one protein system" in this case. The genetic information is directly reflected in the proteins via transcription and translation.

The systematists can be interested in various aspects of the protein features, e.g. size, shape, mol. wt, primary structure, basic or acid char-acter, reactive groups, position of the isoelectric point, and so on. The form of linear arrange-ment of the amino acids, the "primary struc-ture", remains the governing factor for all these characters. The taxonomical significance of the comparison of primary structures of homologous proteins has been efficiently demon-strated by Boulter on cytochrome *c*.

The serological characters of proteins are also directly linked with the primary structure of the molecule.

This concerns certain points in the molecule ("determinants") which are capable of initiat-ing in definite cells of mammals the production of γ-globulins (immuno-globulins) which pos-sess the special bonding properties accounting for the bond to the respective protein reaction position (fig. 1).

In summarizing, we can state that the sero-logical characters of the protein reside in the determinants, located in certain confined posi-tions of the molecules. While it is still very difficult to analyse their precise structure, this is not impossible.

An idea of the size of such determinants can be formed from the results of protein fraction investigations by limited proteolytic digestion of a protein to find the smallest molecule frac-tion still showing immunological reaction. Us-ing this method, Arnon & Sela, e.g., who in-vestigated the hen egg-white lysozyme, found a determinant region extending over the amino acid sequence 60–83 forming a "loop" in the polypeptide chain [1]. In the case of the fibrous silk-fibroin the determinant region was found to comprise 8–12 amino acid residues. It must be concluded from these and similar results that the active antigenic regions of proteins are normally made up of approx. 10–20 amino acids.

The systematist is interested in the com-parison of such determinants between various taxa, of course. Thanks to the specific reac-

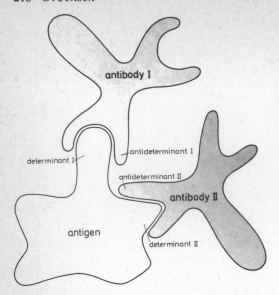

Fig. 1. The specific reactive groups of antigens and antibodies.

tion between determinants and antideterminants or, in other words, between the antigen and antibody, such comparative analyses can be accomplished without difficulty.

A priori, it would appear that great value must be attached to the significance of comparative serological investigations in revealing taxonomical relationships.

The following interpretation of the results of such investigations is based on a few examples which demonstrate the wide scope, but also the possibilities and limits of serological contributions to taxonomic evidence.

The wide field of serological studies can be indicated by means of two main research work categories of differing natures: Taxonomical comparison of (*A*) single proteins; (*B*) protein mixtures.

Taxonomical Comparison of Single Proteins

To date, comparative serological analyses of defined proteins have been carried out on a much smaller scale than could be expected when considering the claimed significance of the obtained results in revealing evolution patterns. Very little is known of "higher" plants in particular. The results of investigations on phaseolin, legumin, vicilin, some amylases, myrosi-

nase and carboxydismutase almost complete the list of available data. It could be assumed, admittedly, that the reason for this lies in the technical difficulties connected with the isolation of proteins of adequate purity and in sufficient quantity. Or is this simply due to the fact that serological evidence is considered to be too low in value for taxonomic purposes?

Before going into detail, we should attempt to estimate the extent of expectable convergences in serological characters. In the past, the extremely limited probability of convergences was emphasized as one of the main advantages of sero-taxonomical methods. If we take a polypeptide chain consisting of only 61 amino acids and take into consideration that 20 alternatives are possible for each position, we obtain a total number of conceivable combinations in the order of $20^{61}=5\times10^{79}$ compounds, almost 6 times the total number of atoms existing in the universe [2].

To-day we know that this approach is incorrect in two respects: (1) We now know that it is not the overall amino acid sequence of a protein which is the decisive factor, but only that of the determinants. For example, with 10 amino acids responsible for one determinant, the maximum number of possible alternatives is still $20^{10}=10^{13}$. However, this number of possibilities is further reduced by the fact that several positions have become invariable during evolution; any amino acid exchange would have lethal effect in this case. (2) It is not taken into account in such statistical calculations that only similarities of homologous proteins, which agree in smaller or larger parts of their primary structure, are of real interest. Particularly with the almost invariant and highly adapted proteins, such as the cytochrome *c*, a large proportion of the amino acid sequence of big plant (or animal) groups remained unchanged in evolution. Thus, the (few) changes in the variable positions could have as a pure coincidence taken place in the same manner along two separate lines of development.

Nevertheless, genuine convergences in the sphere of serological characters must still be regarded as occurring very seldom, i.e. they are not probable, but possible. For example, the existence of heterophilous antigens has been known for quite some time in the field of immunology. Although it must be assumed that these heterophilous antigens are in many cases lipo-mucopolysaccharides (like the so-called Forssman-antigen found on erythrocytes of numerous mammals and several bacteria), the possible occurrence of convergences in proteins cannot be denied. However, the "antisystematic" reactions described by Moritz and his associates [3] must be considered mainly as reactions of determinants which occur widespread and remain constant in phylogenesis, and not so much as reactions of determinants originating from convergences.

The following examples are given to outline the interpretation of comparative serological reactions of single proteins. Here, two facts will be established, that is to say: (1) the systematic ranges of determinants vary greatly. Many determinants are found over the entire plant kingdom, while others are restricted to infragenetic taxa. This explains why in some cases even homologous proteins from organisms of various classes, families and indeed genera exhibit no cross-reaction. In other words, these proteins must be regarded as being completely different in serological respect. (2) Since one protein molecule normally carries several determinants, partial identities are also feasible and can often be proved.

Closely restricted existence of plant protein determinants is found in the case of phaseolin, for example.

(*a*) Phaseolin of *Phaseolus vulgaris* L. (including the ssp. *aborigineus*), of *Ph. dumosus* MacFad., *Ph. polyanthus* Green M. and *Ph. coccineus* L. have proved to be serologically identical or virtually identical in electrophoresis [4]. My own investigations using *Ph. coccineus* antiserum have confirmed the suspected serological identity of even all other seed proteins of *Ph. vulgaris* and *Ph. coccineus* (fig. 2) by absorption tests.

Ph. cocc.~Ph. cocc.–Ph. vulg.~+*Ph. cocc.*→no reaction

Conditions

(*1*) In *Ph. cocc.'* all antideterminants against phaseolin from *Ph. coccineus* were visible, and those not visible would also react with phaseolin from *Ph. vulgaris*.
(*2*) Phaseolin from *Ph. vulgaris* does not con-

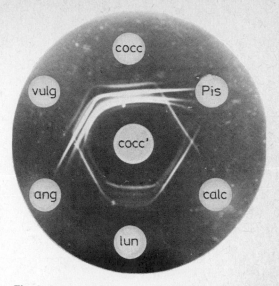

Fig. 2. Immunodiffusion reactions of a *Ph. cocc.* L. antiserum with 6 antigen systems; i.e. of the homologous *Ph. cocc.* and the heterologous *Ph. vulg.*, *Ph. angularis*, *Ph. lunatus*, *Ph. calcaratus* and *Pisum sativum* (Ouchterlony test in agar-gel).

tain additional determinants, e.g. the following would have to apply:

Phaseolin of *Ph. vulg.*~phas. Ph. vulg.–phas.–Ph. cocc.~+phaseolin *Ph. vulg.*→no reaction

(*b*) Existence of partial identity with phaseolin from *Ph. acutifolius* A. Gray (fig. 3): We conclude this from a reduced reaction strength, and consider the following conclusion of the absorption experiment as highly probable:

Phaseolin of *Ph. vulg.'*~phas. Ph. vulg.–phas.–Ph. acut.~+phaseolin *Ph. acut.*→distinctive reaction

(*c*) No phaseolin (or phaseolin determinants from *Ph. vulgaris*) is presented in the seeds of *Ph. lunatus* L. and many other species, e.g. *Ph. calcaratus* Roxb. and *Ph. angularis* (Willd.) W. F. Wight. A similar result was obtained from the Ouchterlony test with anti-*Ph. cocc.* (fig. 2): No phaseolin band with *Ph. lunatus*, *Ph. calcaratus* and *Ph. angularis*.

Barley-β-amylase (serological partial identity with wheat and rye; all classified in the same tribe Triticeae of Poaceae, sub-family Pooideae), and in particular the barley-α-amylase appear to have a wider systematic range (serological identity with the α-amylase of wheat and rye,

I Strong reaction

Ph. vulgaris ssp. *vulgaris* Ph. vulgaris ssp. *aborigineus* Ph. dumosus Ph. polyanthus Ph. cocc.	phaseolin, serologically identical or very similar

II Weak reaction

Ph. acutifolius	phaseolin, serologically similar

III No reaction

Ph. lunatus and 16 other *Phaseolus*-species Dolichos lablab Vigna sinensis Vigna catjang Glycine gracilis Glycine max Strophostyles helvola Cajanas indicus Abrus precatorius Rhynchoria phaseoloides Arachis hypogaea	no phaseolin or a phaseolin- like protein serologically quite different

Fig. 3. Grade of serological reactions between the antisystem against phaseolin obtained from *Ph. vulg.* L. ssp. *vulg.* and antigen systems of Fabaceae species; after Kloz et al. [4].

partial identity with oats and maize; maize does in fact belong to the Andropogonoideae, a further sub-family of the Poaceae).

The situation is totally different with those proteins of vital significance as enzymes of basic metabolic processes, for example carboxydismutase or cytochrome *c*. Particularly in the case of cytochrome *c*, the Boulter group was able to demonstrate with plants, too, that a high proportion of the amino acid sequences had remained unchanged during the course of evolution. This would indicate the occurrence of a great systematic range of serological cross-reactions, as proved for carboxydismutase.

Carboxydismutase ("fraction I") with at least partial agreement in serological reactivity is found in almost all green plants, from *Chlorella* to *Nicotiana* [5]. Serological identity is to be expected for very close relationships (genera, and possibly also families or classes), in the same way as for *Nicotiana* [6], see also fig. 4. In other cases at least serological partial identity is to be expected, as with *Spinacia–Nicotiana* [7] or as with *Spinacia–Chlorella*

[8]. Only with the green bacterium *Rhodospirillum rubrum* was no reaction discernible.

The systematist will now enquire about the conditions under which he can draw benefit from such results. It is obvious that proteins (enzymes) with a rigidly fixed amino acid sequence will not be selected for the investigation of the relationship level of infrageneric (and also infra-family) taxa. These could be of value, if at all, only for distinguishing between classes or divisions. In contrast, highly variable proteins, such as must be assumed for phaseolin, are naturally of value only as regards the determination of rather close relationships.

The maximum character differentiation is in all cases governed by the number of antideterminants of the protein in question, or, more precisely, by the number of antideterminants elicited by the animal subsequent to injection of the protein. In this respect, our experience

Fig. 4. Various types of serological reactions of fraction I proteins in different plants; after Kawashima et al. [6, 7], Sugiyama et al. [8].

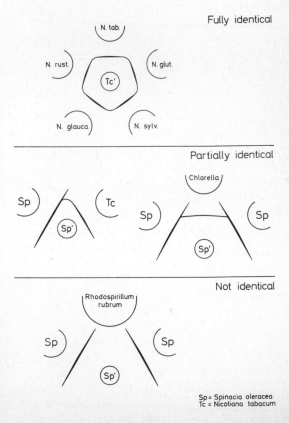

Sp = Spinacia oleracea
Tc = Nicotiana tabacum

indicates that a good result would be a count of approx. 3 antideterminants of a defined protein in the antiserum. Naturally, the knowledge to be gained from this by the systematist is of a minor order, and will in all probability scarcely exceed the trivial, well-founded taxonomic differentiations. To this extent, Heywood [9] is perfectly right in emphasizing the fact that in participation the reactions of only a limited number of genes is the drawback of (such!) serological methods. Indeed, only a few triplets of a single gene participate! Here, the question is posed as to whether the often poor informative yield of a system having few unit characters justifies the technical and preparative effort involved.

Serological Comparison of Protein Mixtures

Sero-systematic work concerned with protein mixtures form the other extreme. This applies in particular to work aimed at comparing the total protein stock of taxa (species). For technical reasons, however, the work is restricted to certain organs (often seeds, but also leaves, tubers, microspores, etc.) in definite (comparable) ontogenetic stages of individuals representing in some instances single populations only. Protein fractions (albumins, globulins) are also sometimes used, but these still represent rich protein mixtures.

In the ideal case, therefore, complete protein systems of individual taxa (e.g. species) are compared. The result is an expression of the

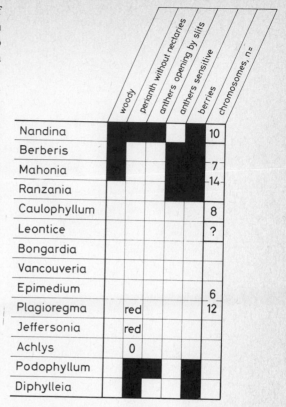

Fig. 6. The spread of important differentiating characters of the Berberidaceae genera.

serological overall similarity, to which great importance must be attached a priori in discussions concerning relationships. An ideal multi-character comparison of this type encompasses the effect of many genes, and covers characteristics of all proteins occurring in the plant, whether these be enzymes or not. Naturally, the objection made by Heywood as cited above does not apply here.

However, on a practical level an ideal system of this type can at most be approximated. The antibody producing animals in particular (mainly rabbits), which do not by any means elicit the specific antibodies against all introduced antigens, constitute a decided boundary. On the other hand the patience of the experimentator must be taken into account; he will not always perform the injections over a sufficiently long period of time, and he hesitates to employ more laborious immunisation methods which would guarantee him a higher titer of antibodies. The poorer antiserum is in "anti-

Fig. 5. Different taxonomical classifications of the Berberidaceae genera.

	Hutchinson	Takhtajan	Tischler	Syllabus Janchen
Nandina				
Mahonia, Berberis				
Ranzania				
Caulophyllum, Leontice, Bongardia, Vancouveria, Epimedium, Jeffersonia				
Achlys				
Podophyllum				
Diphylleia				

⊢—⊣ border-line of families   ⊢—⊣ border-line of tribes

⊨—⊨ border-line of subfamilies   ⊢- - -⊣ border-line of subtribes

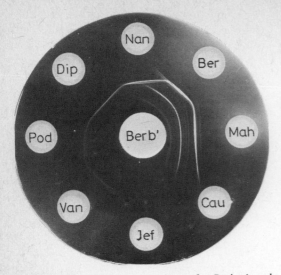

Fig. 7. Immunodiffusion reactions of a *Berberis vulgaris* L. antiserum with 8 antigen systems; i.e. of the homologous *Berberis vulgaris* antigen system and the heterologous antigen systems of *Nandina domestica* Thunb. and *Diphylleia* species, *Podophyllum emodi* Wall., *Vancouveria hexandra* Morr. et Decn. *Jeffersonia dubia* Benth et Hook f., *Caulophyllum thalictroides* Michx. and *Mahonia aquifolium*.

of perianth cycles (up to 13), morphological similarity of sepals and tepals (Hiepko) and the occurrence of 10 chromosomes (fig. 6). Depending on the significance attached to these characters, argument will be made in favour of the isolation of *Nandina* as a separate family (Nakai, Hutchinson, Takhtajan et al.) or at least as a sub-family (Hiepko et al.), or against this (Janchen et al.).

Again, many authors emphasize the special position occupied by the ligneous genera *Mahonia* and *Berberis* (e.g. Hutchinson), or separate in higher degree the bacciferous, non-ligneous genera *Podophyllum* and *Diphylleia,* which, of course, is also expressed in the chemistry by the special lignan bodies.

Results and Their Interpretation

Reactions of the Berberis vulgaris L. antisystem

The rabbits used in the experiments were immunized repeatedly (approx. 40 times) over a

characters'', the more carefully must the results be evaluated as regards a serological overall similarity.

The interpretation of comparative serological experiments with a serological multi-character system is now demonstrated by means of a concrete systematic example. This concerns the elucidation of the relationships within the Ranunculanae, in particular the Berberidaceae, which are the subject of investigations at present being carried out in our laboratory.

The examples shown in fig. 5 convey a good impression of the numerous proposals in respect of a natural classification of the Berberidaceae.

The most disputed aspect when evaluating the relationships is without doubt the position of *Nandina domestica* Thunb. Nandina shares several of the important characters used for differentiation of the family with *Berberis* and *Mahonia* (namely ligneous shoot axis; bacciferous plants) and several with *Podophyllum* and/or *Diphylleia* (namely no nectar leaves; longitudinal opening of the anthers; bacciferous plants). In the literature, however, three features always occupy a leading position to emphasize the special place of *Nandina:* the large number

Fig. 8. Results of the absorption test of a *Berberis vulgaris* L. antiserum with some Berberidaceae genera; (bottom) number of proteins (determinants or groups of determinants) involved.

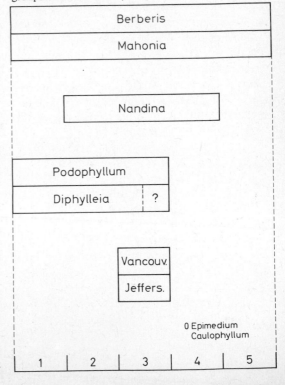

period of 2½ years, and thus developed a high number of antibodies (content of "anticharacters"). Up to 10 bands were clearly visible in the homologous reaction.

Mahonia and *Berberis:* The antigen systems of both genera reacted in an identical manner against the *Berberis* antisera (fig. 7), and without the formation of spurs in immuno-diffusion. Consequently, a serological identity exists between the two genera with respect to the antiserum employed. The value of this finding is in itself very high with respect to a general statement independent from the employed antiserum, since the reaction is based on approx. 10 precipitin lines and (assuming the average participation per protein of 2 determinant-antideterminant reactions) approx. 20 characters.

We deduce from this the justification for the assumption of a great serological similarity between *Mahonia* and *Berberis*. In our experience it is scarcely probable that a different picture

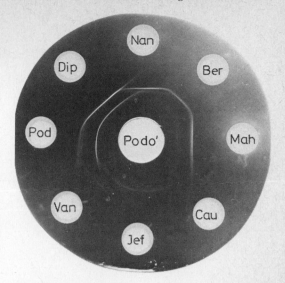

Fig. 10. Immunodiffusion reactions of a *Podophyllum emodi* Wall. antiserum with 8 antigen systems; i.e. of the homologous *Podophyllum emodi* Morr. et Decn. antigen system and the heterologous antigen systems of *Vancouveria hexandra* Wall., *Jeffersonia dubia* Benth et Hook f., *Caulophyllum thalictroides* Michx., *Mahonia aquifolium*, *Berberis vulgaris* L., *Nandina domestica* Thunb. and *Diphylleia* species; (Ouchterlony test in agar-gel).

Fig. 9. Results of the absorption test of a *Podyphyllum emodi* Wall. antiserum with some Berberidaceae genera; (bottom) number of proteins (determinants or groups of determinants) involved.

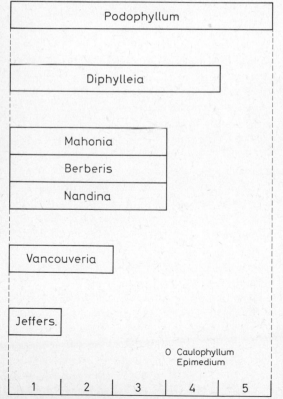

would be obtained with a *Mahonia* antiserum. Taking as a basis for comparison, similar serological identities from the even better-known family of the Ranunculaceae, the grouping together into one genus, or at least a very close relation, is suggested. Naturally, this cannot be decided solely on the basis of serological characters; but in this case the non-serological characters coincide well with this.

The serological similarities of all other Berberidaceae with *Berberis* are of a lower order.

Podophyllum and *Diphylleia: Podophyllum* and *Diphylleia* react concordantly with 2 or 3 bands against the *Berberis* antiserum. This will also indicate the possibility of a great serological similarity—or even identity—of both genera. Since too few characters take part in the reaction, this cannot decide the issue. The only way to conclusively decide this question is to use an antiserum against the antigen-system of one of the genera (or, better still, against *Podophyllum* and *Diphylleia*). Compared with the serological reactions of *Vancouveria*, *Jeffersonia* and *Caulophyllum*, those of *Podophyllum* and *Diphylleia* with *Berberis*-antiserum

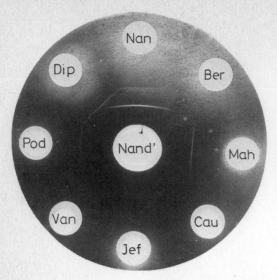

Fig. 11. Immunodiffusion reactions of a *Nandina domestica* Thunb. antiserum with 8 antigen systems; i.e. of the homologous *Nandina domestica* Wall. antigen system and the heterologous antigen systems of *Diphylleia* species, *Podophyllum emodi* Wall., *Vancouveria hexandra* Moor et Decn., *Jeffersonia dubia* Benth et Hook f., *Caulophyllum thalictroides* Michx., *Mahonia aquifolium*, and *Berberis vulgaris* L. (Ouchterlony test in agar-gel).

Vancouveria, Jeffersonia, Caulophyllum: The serological similarities of these genera with *Berberis* are of a lesser order as compared with those of *Nandina, Podophyllum* and *Diphylleia*, but do exist. In fact, the similarities of *Caulophyllum* (and *Epimedium*) with *Berberis* are so minor that only under particularly favourable test condition did a weak reaction take place (whereas with *Actaea*, for example, and even with *Nicotiana*, more pronounced reactions were noted). It would be premature to conclude on this basis that *Caulophyllum* had a more pronounced deviation as regards the serological scope of characters, but it is not impossible.

Reactions of the Podophyllum emodi Wall. antisystem

These antisera also stem from animals which were immunized over a period of $2\frac{1}{2}$ years (approx. 40 injections). In this case the not inconsiderable number of approx. 6 bands were found in the homologous immuno-diffusion reaction.

Podophyllum and *Diphylleia:* The previously suspected serological similarities of both genera have been substantially proved. Although complete identity of the serological character stock was not detected under the test conditions, the similarity was the most pronounced among all Berberidaceae genera investigated (figs 9, 10). This confirms the close relationship in the natural system.

Nandina, Berberis, Mahonia: As compared with *Podophyllum*, no factual distinction could be made between these three genera. In any case, all 3 genera were capable of mutual elimination in the absorption experiment. This is also supported by the appearance of 3 or 4 like bands in the immunodiffusion. The fact that these are only similarities with regard to the characters shared with *Podophyllum* has been revealed by the reactions of the *Berberis* antiserum, of course.

Vancouveria, Jeffersonia, and Caulophyllum: Also with respect to the reference system *Podophyllum*, these genera do not exhibit any highly substantial serological similarity (compare this also with the bands in the immuno-diffusion (fig. 10), in which *Caulophyllum* again proved to be extremely weak in reaction).

are much more pronounced; this is also reflected in the result of absorption (fig. 8). The similarity of *Podophyllum* and *Diphylleia* with *Berberis* is greater as compared with that of *Vancouveria* and *Jeffersonia* with *Berberis*. This statement is well supported by the high reactivity of the *Berberis* antiserum. According to experience gained to date, scarcely any different "relative placement series" than that of (1) *Berberis*, (2) *Diphylleia* and *Podophyllum*, (3) *Jeffersonia* and *Vancouveria* can be expected. It is possible that the reactions of both groups of these genera could branch further.

Nandina: Similar conditions apply for the reaction of *Nandina* against *Berberis*-antiserum as for *Podophyllum* and *Diphylleia*, i.e. the relevant placement series: (1) *Berberis*, (2) *Nandina*, (3) *Jeffersonia* and *Vancouveria*. The common characteristics with *Podophyllum* and *Diphylleia* are high with respect to *Berberis*, but not complete. *Nandina, Podophyllum*, and *Diphylleia* would appear to share specific identical features with *Berberis*.

Reactions of the Nandina domestica antisystem

This antiserum was relatively weak; it was generated after 16 injections over a period of only approx. 3 months. Thus, the fact that the number of precipitation bands in the homologous reaction is not yet very high (4) and the number of cross-reacting genera is also not yet high (5) is not surprising (fig. 11). Obviously, the immunized animals had only generated a few types of antideterminants. The placement series: (1) *Nandina*, (2) *Berberis, Mahonia, Podophyllum, Diphylleia* and *Jeffersonia*, and (3) other Berberidaceae, is consequently still insufficiently supported to permit any high degree of speculation.

Reactions of the Caulophyllum thalictroides antisystem

This serum reacted too species-specific to show reactions with any other genera. As a consequence, the question whether *Caulophyllum* really has very few similarities with the other Berberidaceae or not must remain fully open.

The prime information to be drawn from all discussed serological experiments: The results of the experiments show that reactions of two antisera with good cross-reacting characteristics and of two antisera having insufficient cross-reacting power (i.e. less rich antisera) permit the drawing of interesting conclusions:

1. Berberis and *Mahonia* are extremely similar from the serological aspect, and react like species of one genus.

2. Podophyllum and *Diphylleia* are relatively similar from the serological aspect (i.e. as compared with the remaining Berberidaceae).

3. Nandina features serological similarities of a high degree with *Berberis* and *Mahonia* and with *Podophyllum* and *Diphylleia*. It appears to be more closely related to *Berberis* and *Mahonia* than to the other two genera.

4. Achlys, Bongardia, Caulophyllum, Epimedium, Jeffersonia, Leontice and *Vancouveria* form one (or several) groups of genera with minor similarities to those mentioned above. Their serological similarities have not yet been clarified in detail. Out of these, *Caulophyllum* and *Leontice* are very similar in serological respects.

What conclusions must the systematist draw from this? He naturally realizes that these serotaxonomical results can make only a partial contribution towards the determination of the overall similarity of the Berberidaceae genera. However, evaluation by comparison of approx. 20–30 characters, the systematic value of which was proven in detail at the beginning of this discussion from the aspect of protein structure characteristics, will show that the serological results carry taxonomical weight in this case. Accordingly, the task is now more intensive research into other, non-serological characters with the object of substantiating the following provisional draft of a Berberidaceae system, which already gives due consideration to previous findings regarding non-serological characters:

(A) *Berberis* and *Mahonia* are again united to form a single genus.

(B) *Berberis* and *Mahonia, Nandina, Podophyllum* and *Diphylleia* form the one relationship group (sub-family?), in which connection distinction could perhaps be made between 3 tribes, the remaining genera forming the other relationship group (sub-family?).

The question posed earlier on a possible separation of *Nandina* from the family of Berberidaceae must thus be negated on the basis of the serological findings. If those arguments marshalled in favour of such a separation are then examined, they appear rather doubtful anyway:

1. The fact that the deviating chromosome number ($n = 10$) is by no means decisive is proved by the related family of Ranunculaceae, in which the Thalictreae and Coptideae also exhibit a completely different karyotypus.

2. The fact that the absence of nectar leaves need not necessarily constitute a weighty argument is confirmed for the Ranunculaceae, where *Adonis* has "already" lost the nectaria of the petals, and where *Caltha* forms only sepals, but no nectar-fertile petal-homologues at all.

3. Following Hiepko's deduction [10] that the whole perianth of *Nandina* is built up by sepals, then it is not difficult to trace back the larger number of perianth cycles to a shift of determination level [11] with simultaneous disappearance of the petals.

It would have been ideal, had we been able to test the serological similarities of the Berberidaceae with antisystems of each genus. But even the reported results permit—with careful interpretation—interesting systematic conclusions to be drawn. We would also have been in a position to characterize the proteins taking part in the precipitation, and to check the spread of the determinants. Especially interesting tasks of this nature are included in the further research programme.

References

1. Arnon, R & Sela, M, Proc natl acad sci US 1969, 62, 163.
2. Dickerson, R E & Geis, J, Struktur und Funktion der Proteine, p. 120. Verlag Chemie, Weinheim, 1971.
3. Frohne, D, Moritz, O & Jensen, U, Flora 1960, 150, 332.
4. Kloz, J, Klozova, E & Turkova, V, Preslia 1966, 38, 229.
5. Dorner, R W, Kahn, A & Wildman, S G, Biochim biophys acta 1958, 29, 240.
6. Kawashima, N, Kwok, S-Y & Wildman, S G, Biochim biophys acta 1971, 236, 578.
7. Kawashima, N & Wildman, S G, Biochim biophys acta 1971, 229, 749.
8. Sugiyama, T, Matsumoto, C & Akazawa, T, Arch biochem biophys 1969, 129, 597.
9. Heywood, V H, Taxonomie der Pflanzen, p. 112. Gustav Fischer Verlag, Stuttgart, 1971.
10. Hiepko, P, Bot Jahrb 1965, 84, 359.
11. Zimmermann, W, Ranunculaceae, in Hegi, Flora von Mitteleuropa (ed K-H Rechinger), 2nd edn. Hanser Verlag, München, 1965.

Discussion

Reichstein: May I ask what kind of material you injected in the last mentioned experiment. Have you used crude aqueous extracts? A second point: It is known, particularly for bacterial antigens but also from the work on blood group substances and others that sugar residues are most potent and very specific determining groups in antigens. The presence or absence of one single sugar on a protein will usually completely alter its antigenic specificity and power. I wonder whether glucoproteins or "glycolipids" may not be present and of primary importance for the serological reaction in such crude mixtures.

Jensen: We injected fine aqueous suspensions of seed materials in order to deliver the whole range of serological determinants to the animal. In consequence, this means that we offered not only pure polypeptides but also glycoproteins, lipoproteins etc. and even polysaccharides as antigenic material. We know, that in seed material most of the proteinic substances are soluble and are of another structure than the cell wall bounded bacterial antigens or blood group substances. Some of these proteinic substances may be glyco- or lipoproteins. We suggest that a part of the positive cross reactions with great systematical range (formerly "antisystematic reactions") may be due to common glyco-determinants. For the present they have to be considered as serosystematical characters like all the other proteinic ones.

Boulter: In the case of the legume globulin data presented by you, this protein is a glycoprotein and it is possible, since as pointed out by Dr Steinbeck, that the carbohydrate moieties are strongly antigenic, this may have affected the affinities obtained to a larger extent than the accompanying protein moiety.

Johansson: With the grasses one major problem is the protein extraction methods. One method will for instance extract 50% of the seed proteins in *Hordeum* but over 95% of the proteins in *Triticum*. No single extraction method exists solubilizing all grass seed proteins. With the grasses you therefore have to work with protein fractions, not the whole spectrum. In a few cases I have observed determinants migrating from one protein fraction to another within a genus. This migration can be caused by only one major gene. L. Munck and A. Tallberg at the Seed Association in Svalöv have found that some bands observed by disc-electrophoretic technique belong to the water-soluble fraction in one cultivar of barley but to the alcohol soluble fraction in the other sister cultivar. This difference together with a few others are caused by one major gene. Therefore you have to be very cautious making conclusions when you are aware of extraction problems.

Jensen: As I have already mentioned in the reply to Dr Reichstein, we injected fine aqueous suspensions of the seed materials. We do not know anything about the different solubility of the proteins of these suspensions in the antibody forming animals. When the accompanying substances are different, a different solubility of the same protein will be possible. Indeed, the solubility or another characteristic of a protein may be changed distinctfully by the exchange of some amino acids, which may be caused by one major gene. But there are reasons to employ methods in comparative serology, e.g. immunoelectrophoresis of crude seed material or absorption tests by these materials, which enable us to compare homologous proteins of even different solubility—if soluble at all. Fortunately, the extraction problems in other angiosperm families are not as great as in the grasses. But still we have to be aware of extraction problems.

Weimarck: Could different individuals of rabbits show different reactions?

Jensen: Usually 2–4 rabbits of one breed are used for the injections of the same antigenic material. When the injection modus was identical and all animals of good health the immunological reaction was the same or nearly the same. If not, considerable differences in the precipitation reaction in qualitative as in quantitative respect will be found. Generally, we use different kinds of immunization and choose the highest titered antiserum for the experiment.

Takhtajan: I can agree that *Podophyllum* and *Diphylleia,* as well as *Nandina* belonging to the Berberidaceae, as has been shown recently by A. Melikian and myself in a paper on the seed-coat anatomy of some berberidaceous genera published in Botanicheski Zhurnal (Leningrad). Of course the genera *Caulophyllum, Leontice* and *Gymnospermum* are not closely related to the Berberidoideae and Podophylloideae and in our joint paper with Melikian we consider them as a subfamily Leonticoideae, but without *Bongardia,* which differs considerably from these three genera in its seed-coat anatomy. What surprises me very much in a classification suggested by you on purely serological basis is a position of the genus *Epimedium.* In morphological bases *Epimedium* is rather closely related to the Berberidaceae sensu stricto and is rather far from the Leonticoideae and other herbaceous genera.

Jensen: It still remains a question, I think, whether it is justified to judge from the seed-coat anatomy only, whether *Podophyllum, Diphylleia* and *Nandina* will belong to the Berberidaceae or not. In any case it is interesting that *Nandina* belongs to the Berberidaceae even from this point of view. As regards *Epimedium* our preliminary serological studies gave no indication of mentionable connexions with the Berberidaceae s. str. To clarify this problem more comparative data are needed.

The Natural Distribution of Glucosinolates: a Uniform Group of Sulfur-Containing Glucosides

A. Kjær

Department of Organic Chemistry, Technical University of Denmark, DK-2800 Lyngby, Denmark

Summary

The chemical structures of the about 70 known, naturally occurring glucosinolates (*1*) are tabulated and briefly discussed. Their predominant appearance within the order Capparales is commented upon. The uniform, known in vivo pathway to glucosinolates is outlined and its potential significance for systematic problems adumbrated.

Amongst the vast number of well-defined secondary plant products, those displaying a discontinuous distribution pattern paired with an immediate appeal to the human senses (odour, taste, colour) have, for obvious reasons, acquired a prominent position in discussions about the merits of such products for purposes of classification and, more fundamentally, phylogenetic considerations. The glucosinolates (*1*), a structurally uniform collection of naturally occurring anions, constitute a group with some of the above characteristics.

Today, the general glucosinolate structure (*1*) encompasses about 70 naturally occurring thioglucosides, differing solely in the character of their side-chains, R. These water-soluble compounds, limited, as far as we know, to dicotyledons, uniformly undergo hydrolysis, catalyzed by endogenous plant enzymes, to glucose and energy-rich aglucones (*2*), the latter in most cases breaking down, probably spontaneously, by rearrangement and loss of sulfate, to give isothiocyanates (mustard oils) (*3*) as the major products. Nitriles (*4*), arising from fragmentation of (*2*), are alternative end-products, usually produced to a minor degree, but increasing in relative amounts under certain conditions, even to the point of dominance.

The side-chains established thus far in naturally occurring glucosinolates are summarized in table 1, divided into two major groups: aliphatic and aromatic, each subdivided according to the prevailing functionality. In the aliphatic series, the numerous hydroxylated derivates are listed next to the non-oxidized counterparts whence they formally derive.

For a detailed discussion of the individual compounds, their occurrence, chemistry, and likely biological origin, two recent reviews [1, 2] should be consulted.

Though devoid of conspicuous properties themselves, glucosinolates frequently draw attention as plant constituents by the pungency of the isothiocyanates resulting from their enzymic hydrolysis, brought about by disintegrating the plant tissues. Hence, it occasions little surprise that mustard oils and their glucosidic progenitors have been subjects of chemical curiosity for several centuries (cf [1, 3]), further spurred by the reputed utilization of many glucosinolate-producing plants as condiments, potherbs, and remedies throughout the continents and ages. Today, a wide range of powerful, modern methods are available for quick and reliable separation, identification, and quantitative determination of the individual glucosinolates and their hydrolysis products. Application of the improved techniques has vastly augmented our knowledge of the chemical character, the distribution, origin and biological function of the glucosinolates. Though far from exhaustive, our understanding of this class of natural products today compares favourably regarding detail with that of many other products. In the present context a few facets of supposed pertinence are selected for discussion.

Table 1. *Side-chains in naturally occurring glucosinolates (1)*

(1) Alkyl side-chains

Me

Et

$BzO \cdot [CH_2]_3-$

Pr^i $(R)\text{-}MeCH \cdot CH_2OH$; $(R)\text{-}MeCH \cdot CH_2 \cdot OBz$

Bu^n $MeCH(OH) \cdot [CH_2]_2-$; $HO \cdot [CH_2]_4-$

Bu^i $(Me)_2C(OH) \cdot CH_2-$

$(S)\text{-}Bu^s$ $(R)\text{-}EtCH \cdot CH_2OH$; $(R)\text{-}EtCH \cdot CH_2 \cdot OBz$

$(S)\text{-}Bu^sCH_2-$ $(S)\text{-}EtC(OH)(Me) \cdot CH_2-$

$MeO_2C \cdot [CH_2]_3-$

$EtCO \cdot [CH_2]_4-$

$PrCO \cdot [CH_2]_3-$

$PrCO \cdot [CH_2]_4-$

$MeS \cdot [CH_2]_n- (n=3\text{–}8)$ $MeS \cdot [CH_2]_n CH(OH) \cdot]CH_2]_2- (n=2, 3)$

$MeS \cdot [CH_2]_5 \cdot CO \cdot [CH_2]_2-$

$MeS \cdot CH^1{=}CH \cdot [CH_2]_2-$

$(R)\text{-}MeS(O) \cdot [CH_2]_n- (n=3\text{–}11)$ $MeS(O) \cdot [CH_2]_n \cdot CH(OH) \cdot [CH_2]_2- (n=2, 3)$

$(R)\text{-}MeS(O) \cdot [CH_2]_5 \cdot CO \cdot [CH_2]_2-$

$(R)\text{-}MeS(O) \cdot CH^1{=}CH \cdot [CH_2]_2-$

$MeSO_2 \cdot [CH_2]_n- (n=3, 4)$ $MeSO_2 \cdot [CH_2]_n \cdot CH(OH) \cdot [CH_2]_2- (n=2, 3)$

$H_2C{=}CH \cdot [CH_2]_n- (n=1\text{–}3)$ $(S)\text{-}H_2C{=}CH \cdot [CH_2]_n \cdot CH(OH) \cdot CH_2- (n=0,1)$; $(R)\text{-}H_2C{=}CH \cdot CH(OH) \cdot CH_2-$

$H_2C{=}C(Me) \cdot [CH_2]_2-$

(2) Aralkyl side-chains

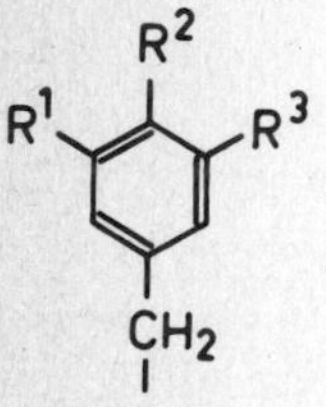

	R^1	R^2	R^3
	H	H	H
	H	OH	H
	H	OMe	H
	H	$O\text{-}(\alpha\text{-}L\text{-}$ rhamnopyranosyl)	H
	OH	H	H
	OMe	H	H
	OH	OH	H
	OMe	OMe	H
	OMe	OMe	OMe

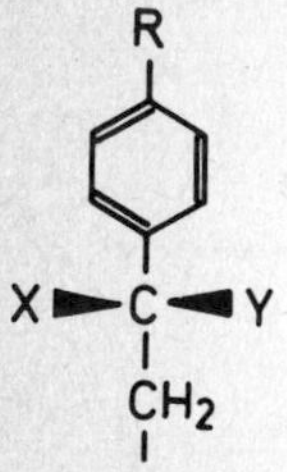

	R	X	Y
	H	H	H
	H	OH	H
	H	H	OH
	OMe	OH	H

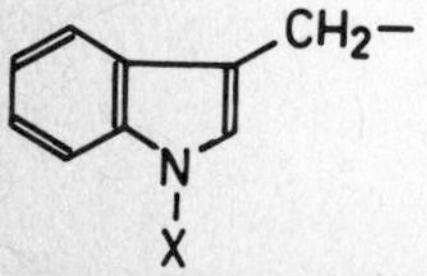

	X
	H
	MeO
	SO_3H

Table 2

Cronquist [4]	Takhtajan [5]
Tovariaceae	Tovariaceae
Capparaceae (including Koeberliniaceae and Pentadiplandraceae)	Capparaceae (including Cleomaceae and Oceanopapaver)
	Koeberliniaceae (including Canotiaceae)
	Pentadiplandraceae
Cruciferae (Brassicaceae)	Brassicaceae (Cruciferae)
Resedaceae	Resedaceae
Moringaceae	Moringaceae
	Emblingiaceae

The order Capparales as defined in the systems of Cronquist [4] and Takhtajan [5].

Natural loci of glucosinolates, as we know them today, are dicotyledons, though limited to a few families or, in some cases, even to certain genera and species. It should be pointed out here, however, that the alleged limitation in distribution, like most other reported cases of discontinuous occurrences of natural products, suffers from an obvious weakness, viz. the inherent difficulty in ascertaining the factual *absence* of a compound from a given taxon. Though in a phylogenetic context by no means void of informative value, the criterion of absence is usually less readily established by experiment than that of presence. Absence, attributable to limited sensitivity of the analytical methods employed, and absence, rooted in a genetically determined inability of the taxon to synthesize the compound in question, represent situations that are, of course, entirely different, albeit, regrettably, not always recognized as such.

Glucosinolate sources *par excellence* are families within the order Capparales, as defined by Cronquist [4] and Takhtajan [5] and set out in table 2. As far as we know, plant enzymes, myrosinases, catalyzing the hydrolysis of the thioglucosidic linkage in glucosinolates, are invariably and exclusively deposited in Nature along with the substrates. Though a great many number of characters have undoubtedly been involved in the construction of the order, as defined in table 2, the role of chemistry seems to have been a minor one, limited, in fact, to the notion that "myrosinase cells" (enzyme deposits) are a feature common to many species within the order. In fact, glucosinolates have been encountered in virtually every species studied within the families Tovariaceae, Capparaceae (sensu strictu), Cruciferae, Resedaceae, and Moringaceae. Though far from complete, the analytical coverage at the family level is substantial and sufficient to almost assure that the capacity to elaborate glucosinolates is a consistent feature throughout the major groups of table 2.

Whereas a search for glucosinolates in *Koeberlinia,* composing a family of disputed relationship, proved of no avail [1], similar studies within the monotypic or small families Canotiaceae, Oceanopapaver, Pentadiplandraceae, and Emblingiaceae (cf table 2), all of dubious affinity, are outstanding and inviting.

Though ubiquitously present within Capparales, glucosinolates are not confined to families of this order. Reliable information is on hand about their occurrence in one or more taxa belonging to the families listed in table 3 (cf [1]). Additional to the sources quoted, reference can be found to a number of taxa, even fungi, exhibiting characteristics attributed to isothiocyanates or glucosidic progenitors of these. The published evidence, however, is, at best, inadequate to support the claims [1], and we shall not discuss them here.

Of interest in the present context is the recent finding in seedlings of *Batis maritima* L. (Bataceae) of an enzyme catalyzing the hydrolysis of extraneous indolylmethylglucosinolate to products identical with those arising from a typical "myrosinase"-induced hydrolysis of this glucosinolate. However, attempts to detect glucosinolates in *Batis* plants were unsuccessful [6]. Morphological features and lack of betalains have cast doubt on the relationship of Bataceae to Caryophyllales [4]. Whereas Takhtajan [5] maintains a close phyletic relationship by placing Bataceae next to Gyrostemonaceae, inside Caryophyllales, other authors (cf [6]) have ventured on a capparidaceous alliance. It is intriguing that both Gyrostemonaceae and the order Capparales are composed of reputed glucosinolate-producing taxa.

The biosynthetic pathway to glucosinolates in intact plants is known in some detail (cf recent reviews [1, 2]). Assuming that, in principle,

Table 3. *Glucosinolate-containing families outside the order Capparales*

Family	Genera or species
Caricaceae	Several species of *Carica*
Euphorbiaceae	Species of *Putranjiva* and *Drypetes*
Tropaeolaceae	Several species of *Tropaeolum*
Limnanthaceae	*Limnanthes douglasii* R. Br.
Salvadoraceae	*Salvadora oleoides* Den.
Gyrostemonaceae	*Codonocarpus cotinifolius* (Desf.) F. Muell.

only one anabolic route is operative in Nature, the one established by experiment and summarized below, we may consider the natural distribution of glucosinolates in terms of pathways rather than final products.

The known in vivo synthesis departs from α-amino acids (5) and proceeds through consecutive oxidative steps, accompanied by decarboxylation, to the aldoximes (6). Upon additional oxidation the latter are converted into the dipolar species (7), *aci*-isomers of primary nitroparaffins (7 a), and thence, via dehydrative addition of an as yet unidentified sulfur donor, (SX), detachment of the carrier moiety, (X), introduction of the glucose residue, and, eventually, *O*-sulfonation, into the final products, the glucosinolates (1).

Viewed in this light, and allowing for unexceptional biochemical modifications of sidechains or rings along the path (oxidation, possibly followed by acylation, *O*-alkylation, and *N*-sulfonation) a fair number of the glucosinolates, listed in table 1, obviously derive from the protein amino acids alanine, valine, leucine, isoleucine, phenylalanine (tyrosine), and tryptophan. Others would require 2-aminobutyric acid and 2-aminoadipic acid as starters, both of which have been previously reported from higher plants [1].

Still others, however, stand out as homologues, again with due allowance for secondary changes such as oxidation, *O*-alkylation, and elimination, and require comments. Detailed studies (cf [1, 2]) have provided evidence for the operation, in such cases, of a chain-elongating amino acid synthesis, schematically shown below.

Only few of the homologous amino acids, required as progenitors, have been encountered as such in the appropriate species, consistent with a rapid turnover inside the metabolic pool, with little or no accumulation of amino acids.

Acceptance of the general biosynthetic scheme, in combination with the chain-elongating mechanism, renders the biosynthesis transparent, apart from details, for the great majority of known, natural glucosinolates (table 1). Notable exceptions exist, however, e.g. the origin of the oxo-alkylglucosinolates.

The capacity of converting α-amino acids (5) into oximes (6) is not confined to taxa possessing enzymic equipment for further processing of the latter into glucosinolates. Cyanogenic glycosides, $RR^1C(OSugar)CN$, with a far less restricted natural distribution, are known to derive in higher plants from α-amino acids, $RR^1CH \cdot CH(NH_2)COOH$, again, significantly, with aldoximes, $RR^1CH \cdot CH=N \cdot OH$, as obligatory biological intermediates [1, 2]. The extent to which cyanogenic glycosides and

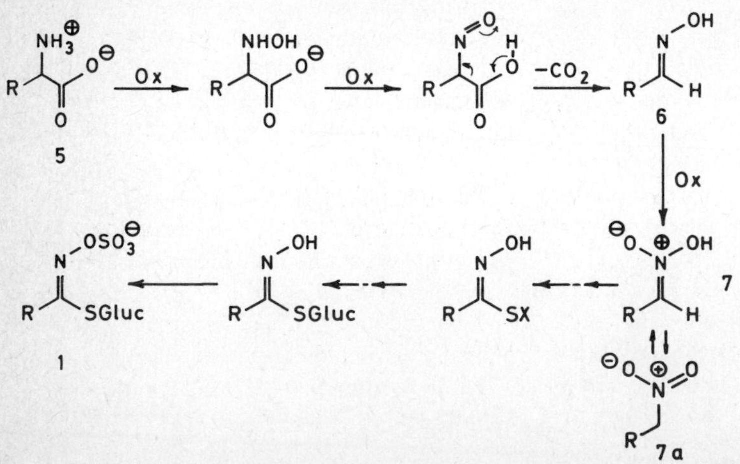

5

8

glucosinolates co-occur is unknown, but inviting for further study. Though apparently rare, cyanogenesis is not entirely unknown within, e.g., Cruciferae, but the source of hydrocyanic acid, not necessarily a cyanogenic glycoside, has not been disclosed in the reported cases. Put slightly different, evidence on hand points to the existence of an oxidative pathway, leading from protein amino acids to oximes and traversed in numerous higher plants of rather varied familial relationship. From oximes and onwards, alternative and more specialized pathways may account for the discontinuous distribution of cyanogenic glycosides, glucosinolates, and, possibly, other oxime-derived plant products. Today, the available data are insufficient for profitable speculations on the possible phyletic implications of the known in vivo pathways. Further studies along this line may, however, sharpen the contours to the point where a meaningful picture begins to emerge.

When considering the natural distribution of the individual glucosinolates, as listed in table 1, in the light of their biosynthetic origin, it follows that in taxa outside Capparales (table 3) known glucosides, with a single exception, are limited to those deriving from non-elongated amino acids. Judged on the basis of characters other than glucosinolate contents, the families of table 3, apart from Limnanthaceae and Tropaeolaceae, are hardly acceptable as a group with much internal affinity. Conceivably, the capacity of converting oximes, in their turn arising from ubiquitous amino acids, into glucosinolates may have arisen repeatedly in different evolutionary lines.

Inside Capparales (table 2), a somewhat different situation prevails insofar as Cruciferae and Resedaceae stand out as families where glucosinolates derived from elongated amino

acids abound. Capparaceae, the second largest family inside the order, is heavily dominated by the supposedly alanine-derived methylglucosinolate, never observed in Cruciferae, with only a few instances of glucosinolates suggestive of chain elongation. Known glucosinolates in the remaining, small families of Capparales are of the non-elongated type. Again, the phyletic implications of the distributional patterns are tenuous. Amino acid homologization, (5)→(8), involving a sequence of enzymes, may conceivably constitute an advanced character, in keeping with the general view on Cruciferae and Resedaceae as families with a capparidaceous ancestry [4, 5]. As always, however, "it is essential to emphasize the distinction that must be made between the evolution of individual characters and the evolution of taxa, and analogously between the degree of evolutionary advancement or primitiveness of a character and that of a taxon which has, of course, to be an average condition of those individual characters studied" [7].

The recent dissection of the collective order Rhoeadales into Papaverales and Capparales, as here defined (table 2), on i.a. morphological evidence is supported by chemistry [4, 8].

At the genus and species level knowledge of the glucosinolate pattern has, in a few cases, been put to good use as an auxiliary character in analyzing delimitations. The data available are, however, insufficient to define their specific merits for such purpose. The systematic complexity of numerous glucosinolate-producing taxa, e.g. within Cruciferae, provides a permanent challenge to our understanding of the biological position and importance of this chemically well-defined and steadily growing group of plant constituents.

234   A. Kjær

References

1. Ettlinger, M G & Kjær, A, Sulfur compounds in plants, in Recent advances in phytochemistry (ed T J Mabry, R E Alston & V C Runeckles) vol. 1, p. 58. Appleton-Century-Crofts, New York, 1968.
2. Kjær, A & Olesen Larsen, P, Biosynthesis (ed T A Geissman) vol. 2. Specialist periodical reports. The Chemical Society, London, 1973.
3. Kjær, A, Fortschr Chem org Naturstoffe 1960, 18, 122.
4. Cronquist, A, The evolution and classification of flowering plants. Nelson & Sons, London, 1968.
5. Takhtajan, A, Flowering plants. Origin and dispersal. Oliver & Boyd, Edinburgh, 1969.
6. Schraudolf, H, Schmidt, B & Weberling, F, Experientia 1971, 27, 1090.
7. Heywood, V H, The role of chemistry in plant systematics, in Chemistry in evolution and systematics (ed T Swain). Pure and appl chemistry 1973, 34, 355; p. 368.
8. Merxmüller, H & Leins, P, Botan Jahrb 1967, 86, 113.

Discussion

Mabry: Is there any evidence that the enzymes which cleave the glucosinolates present in families of the Capparales are the same as those which cleave them in families which also contain glucosinolates but belong to other orders?

Kjær: Little work has been done on glucosinolate-hydrolyzing enzymes from sources other than members of Capparales.

Turner: While you did not find glucosinates in Koeberliniaceae, did you also mean to imply that the enzymes for their hydrolysis were also not present (noting that Bataceae presumably has the enzyme but not the glucosinates)?

Kjær: In our routine examination for glucosinolates, extraneous enzyme (extracted from *Sinapis alba*) is added to ascertain whether cleavage products are formed. That also applies to *Koeberlinia*. No traces of hydrolysis products were noted. We are planning to reinvestigate the case of *Batis*.

Merxmüller: Concerning this enigmatic genus *Emblingia*, from the last investigations one fact can be stated, namely that it is to be excluded from the order Capparales for morphological, anatomical and palynological reasons. With respect to other families, a thorough investigation of Flacourtiaceae could be of interest as there are possibly rather near affinities between

this family and primitive Capparaceae. If glucosinolates would be found there too, their existence in Caricaceae should be reconsidered.

Kjær: There are, in fact, a number of statements in literature from the beginning of this century (quoted in ref [1]) to the effect that enzyme preparations from certain *Hydnocarpus* species are capable of inducing the production of allyl isothiocyanate from allylglucosinolate. Indeed, the matter deserves reinvestigation.

Lüning: In regard to the universatility of the one-carbon prolongation of the carbon chain of an amino acid, I would like to point out that our *Phalaeopsis* alkaloids are just trapping products of such malic acids as appearing in the first step of the scheme you presented. It is therefore not only the end products of the biosynthetic sequences you should look for.

Kjær: The crucial point in the biogenesis of glucosinolates is their stepwise oxidative formation from α-amino acids. The known glucosinolate side-chains encompass a considerable number, shown to be derived from elongated amino acids. These are to a very large extent limited to Cruciferae, however. It may very well be that a far greater spectrum of plants are capable of chain-elongation of α-oxo acids, including certain orchids, but the way the homologous α-oxo acids are further processed need not, and are most certainly not, the same.

Birch: Oxidations of amino acids are common in microorganisms, leading for example to N-hydroxypeptides. One of these reactions which we investigated is the oxidation of aspartic acid to β-nitropropionic acid, which does occur in some higher plants as esters with glucose. I wonder whether it is possible to extend considerations from cyanogenetic and mustard-type compounds to others such as this, which may involve N-oxidation as the initial stage?

Kjær: The glucosinolate deriving directly from aspartic acid would be carboxymethylglucosinolate, as yet an unknown compound. Conceivably, stepwise oxidation of aspartic acid to *aci*-3-nitropropionic acid takes place in a variety of higher plants. In the absence of enzymes carrying the latter into glucosinolates or cyanogenic glycosides, the nitro acid may be deposited as such, or converted into esters.

Nobel 25 (1973) Chemistry in botanical classification

The Chemistry of Resin Glycosides of the Convolvulaceae Family

H. Wagner

Institut für pharmazeutische Arzneimittellehre der Universität München, München, W-Germany

Summary

The occurrence and chemical structure of the Convolvulaceae resin glycosides and the glycosidic acids derived from them, respectively, are reviewed from a chemotaxonomical point of view.

According to Hegnauer [1] the Convolvulaceae family is divided into two subfamilies, Cuscutoideae and Convolvuloideae. There are 7 tribes belonging to the subfamily Convolvuloideae of which Convolvuleae and Ipomoeae have the most numerous and important genera. There are approx. 1100 species within 45 or 53 genera depending on the mode of classification (table 1).

Among the most striking anatomical characteristics of the family is the occurrence of rows of secretory cells with milky, resinous contents. Resin glycosides are among the most important chemical characteristics of the family. The occurrence of tropine alkaloids in Convolvulus species and lysergic acid type alkaloids in Ipomoea and Rivea species as well as a wide distribution of cinnamic acid derivatives and coumarins are also noteworthy. The last two groups of compounds are common to both the Convolvulaceae and the Solanaceae families.

Two means were used to establish the existence of resin glycosides in the family: (1) through their use as laxatives, since many resin glycosides possess a strong laxative effect; (2) through the isolation of the crude resins and the identification of the products of hydrolysis (table 2).

Conclusive evidence of the occurrence of resin glycosides is shown by the existence of characteristic short chain volatile acids, long chain hydroxy-fatty acids and sugars. On the basis of their behaviour towards acids and alkalis, a general structure scheme has been derived from Mannich & Schumann [24] for a resin glycoside from *Exogonium purga* (fig. 1).

Accordingly the resins represent complex gly-

colipids with mol. wts between 500 and 10000 or more. By means of alkaline hydrolyses the ester linkages are split and the short chain volatile fatty acids set free. Simultaneously a glycosidic acid is formed, which can be split by a further acid hydrolysis into a hydroxy-fatty acid and various sugars. The sugars occur as di, tri, tetra, penta, and hexasaccharides. In table 2 are listed the sugars and acids which have been found to date in the resin glycosides. The hydroxy fatty acids are named predominantly after the generic or species name of the plant. Exogonic acid [2] (fig. 2) deserves special mention among the esterified acids due to its exceptional structure. The differentiation between Brazilian and Mexican Jalap resin is based on this acid [3].

The ratio of ether soluble to ether insoluble portion as well as the saponification value are referred to in order to distinguish the resin of one plant from another.

The complexity of these resins and their high mol. wt prevent an isolation of pure compounds, but for systematic studies the structures of the products of alkaline hydrolysis, the volatile

Table 1. *Convolvulaceae-family*
53 genera and about 1100 species

I	Cuscutoideae (1)	
II	Convolvuloideae	
	1 Dichondreae (3)	
	2 Dicranostyleae (14)	
	3 Convolvuleae (7)	*Calystegia*
		Convolvulus
		Polymeria et al.
	4 Poraneae (3)	
	5 Erycibeae (1)	
	6 Ipomoeae (15)	*Exogonium*
		Ipomoea
		Merremia
		Operculina
		Quamoclit
		Pharbitis et al.
	7 Argyreieae (9)	

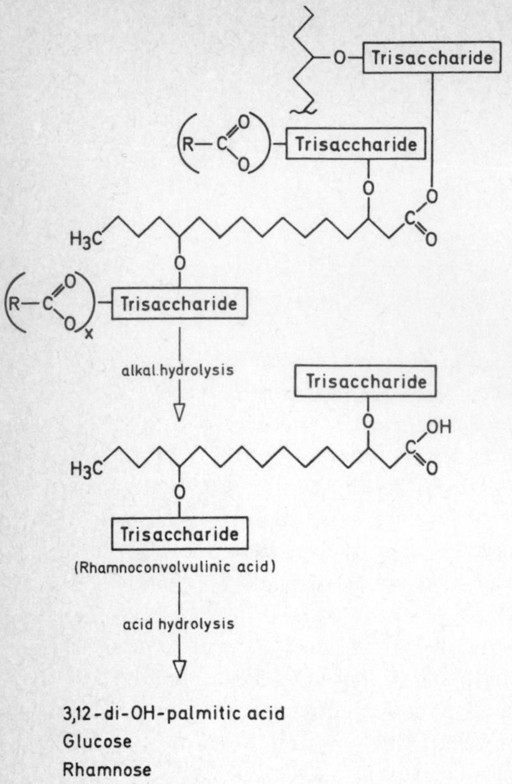

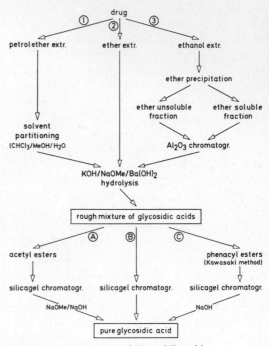

Fig. 1. Structure of the glycoresins according to Mannich & Schumann.

of the methyl sugars and their alditol acetates, by acetolysis and partial acid hydrolysis to di, tri, or tetrasaccharides. In the following section are listed the glycosidic acids of the Convolvulaceae family which have been completely or partially determined to date (fig. 2).

(1) Muricatin B

This hydroxy acid bioside, assigned the structure (+)-11-O-(4)-O-β-L-rhamnopyranosyl-β-L-rhamnopyranosyl)-hexadecanoic acid (fig. 3) was isolated by Khanna & Gupta [4] from the seed

acids and the glycosidic acids, are informative enough. Scheme 1 gives a summary of the important isolation procedures used today for glycosidic acids.

The structure determination of these glycosidic acids result from the mass spectrometric identification of the hydroxy-fatty acids, while the sequence and linkages of the sugars are determined by permethylation and identification

Scheme 1. Preparation of glycosidic acids.

Table 2. *Split products of glycosidic acids*

Volatile acids	OH-fatty acids	Sugars
Acetic acid	7-OH-C_{10}	D-glucose
Propionic acid	11-OH-C_{14} (Convolvulinolic acid)	L-rhamnose
Dimethyl-acetic acid	11-OH-C_{16} (Jalapinolic acid)	D-fucose
Methyl-ethyl-acetic acid	3,12-di-OH-C_{16} (Operculinolic acid)	D-quinovose
Iso-butyric acid	3,11-di-OH-C_{14} (Ipurolic acid)	6-deoxy-gulose
β-Methyl-β-hydroxy-butyric acid	tri-OH-C_{14} (Brasiliolic acid)	
α-Methyl-butyric acid		
n iso-valeric acid		
Tiglic acid		
4-Oxo-caprylic acid		
Exogonic acid		

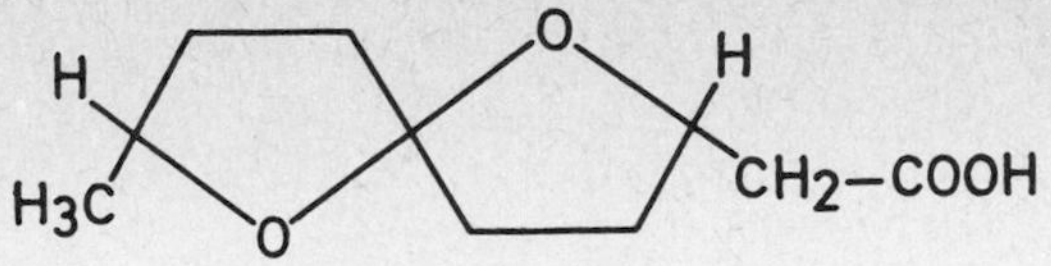

Fig. 2. Characteristic acid from *Operculina macrocarpa*. 3,6–6,9 dioxydodecanoic acid (exogonic acid).

resin of *Ipomoea muricata* Jacq. We have synthetized the anomeric compound by a modified method [5] involving condensation of α-acetobromo-α-(1→4) dirhamnose with the hydroxyfatty acid methyl ester in benzene-nitromethane as solvent (fig. 4). The data for the free hydroxy-fatty acid bioside obtained were identical with those reported by the Indian workers. However, on the basis of its optical rotation and NMR spectrum the synthetic compound is clearly an α-L, α-L-di-rhamnoside and not the β-L, β-L-bioside as postulated for the natural product.

(2) *Pharbitic acids C and D*

These two hydroxy acid–oligosides were isolated by Kawasaki [6, 7] from the seed-resin of *Pharbitis nil* Choisy. Pharbitic acids C and D are pentaglycosides and hexaglycosides respectively of Ipurolic acid in which the sugars D-glucose, L-rhamnose and D-quinovose occur in the ratio 2 : 2 : 1 and 2 : 3 : 1. The branched-chain saccharides having a terminal quinovose and rhamnose unit, have 1→2, 1→4, and 1→6 linkages (fig. 5).

(3) *Operculic acid*

This acid, isolated from the ether-insoluble root resin fraction of *Ipomoea operculata* Martin [8], contains a hexasaccharide moiety linked to 3,12-dihydroxy palmitic acid. The branched-chain hexasaccharide, composed of D-glucose and L-rhamnose, has 1→2, 1→3, 1→4, and 1→6 linkages. As in the case of the pharbitic acids only the hydroxyl groups near the methyl end of the fatty acids is glycosylated (fig. 6).

(4) *Orizabic acid A and B*

These two hydroxy acid oligosides were isolated by Kawasaki [9] from the root resin of *Ipomoea orizabensis* Ledanois. The aglycone part is once again 11-hydroxy palmitic acid. The sugar moieties are linear-chain tetrasaccharides with the sugars D-glucose, D-quinovose, L-rhamnose and D-fucose respectively. The sugars are partly 1→2 and partly 1→4 linked. The first sugar linked to the hydroxy fatty acid in orizabic acid B is D-quinovose whereas it is D-fucose in orizabic acid C. The terminal sugar in both cases is D-quinovose (fig. 7).

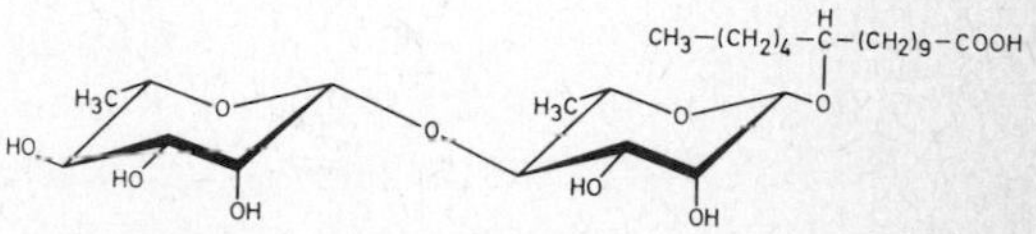

Fig. 3. Muricatin B.

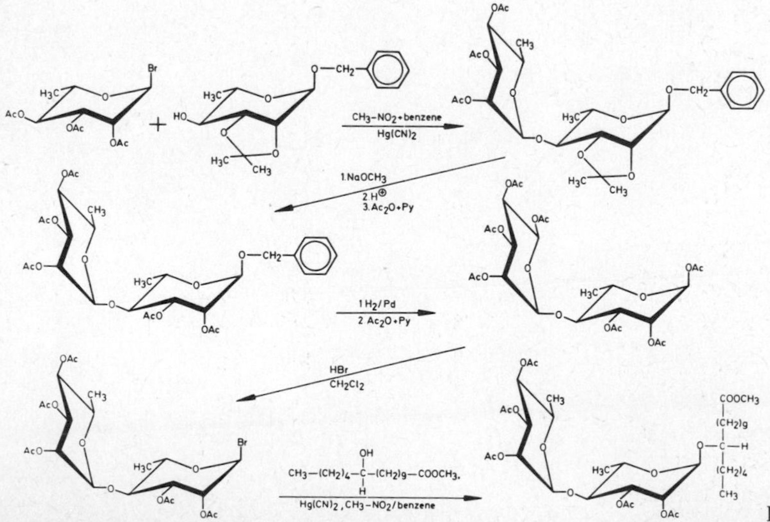

Fig. 4. Synthesis of muricatin B.

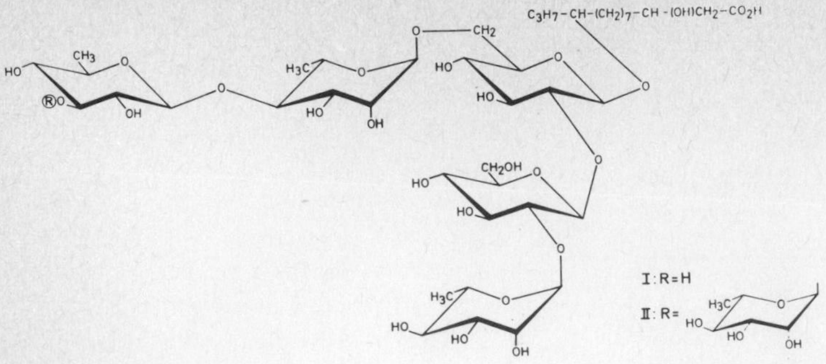

Fig. 5. I, Pharbitic acid C; II, Pharbitic acid D.

I : R = H

II : R =

(5) Microphyllic acid

This hydroxy acid glycoside was isolated as the major component of the ether-soluble part of leaf-resin from *Convolvulus microphyllus* [10]. Its aglycone is 11-hydroxy palmitic acid and the sugar units L-rhamnose, D-glucose, and as we know meanwhile D-fucose not L-fucose in the ration 3 : 1 : 1. The linkages are 1→4 and 1→6. D-Fucose is attached directly to the hydroxy-fatty acid [10] (fig. 8).

From 4 other hydroxy fatty acid glycosides from *Ipomoea* spp. only the components and the sugar ratios have been established:

(1) A hydroxy acid glycoside mixture from the petroleum ether extract of the seeds of *Ipomoea parasitica* (H.B.K.) Don has been reported [11]. Hydrolysis yields (+) 11-hydroxy palmitic acid and the sugars D-fucose, D-quino-vose and presumably 6-deoxy gulose. The absence of D-glucose and L-rhamnose in this case is of interest.

(2) A jalapinolic acid pentaglycoside, consisting of L-rhamnose and D-fucose in a ratio of 4 : 1, has been reported to occur [10] in *Ipomoea quamoclit* (fig. 9).

(3) From a Mexican Scammonium root (botanically not exactly defined) a jalapinolic acid tetraglycoside containing L-rhamnose, D-fucose and D-glucose (2 : 1 : 1) has been isolated [10].

(4) From the poisonous leaf resin glycoside of *Ipomoea fistulosa* Mart. et Chois. a mixture of hydroxy acid glycosides was isolated [12]. In this case the fatty acids were 7-hydroxy tetra-decanoic acid, 11-hydroxy hexadecanoic acid, 11-hydroxy tetradecanoic acid and 3,11-dihy-droxy tetradecanoic acid. The sugars D-glucose, D-fucose and D-quinovose+L-rhamnose were found in the ratio of about 1 : 1 : 1 (fig. 10).

As to the other resin glycosides described in the literature as having been isolated in a more or less pure form from *Convolvulus* and *Ipo-moea*-species only the products of hydrolysis have been reported. From this follows that the jalapinolic and ipurolic acid are the predominant hydroxy fatty acids and the sugars

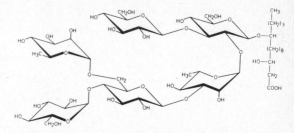

Fig. 6. Operculic acid (rhamnoconvolvulic acid).

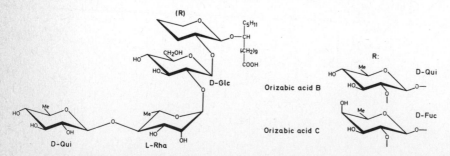

Fig. 7. Glycolipid from *Ipomoea orizabensis* Ledanois (according to ref. [9]).

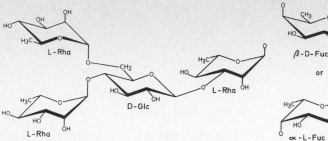

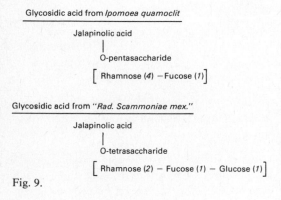

Fig. 8. Glycolipid from *Convolvulus microphyllus* Sieb. microphyllic acid [10].

D-glucose, L-rhamnose and D-fucose are the most frequent ones. The most comprehensive works carried out in the past on the resins of *Convolvulus-, Ipomoea-, Merremia-* and *Operculina* species are these of Jaretzky & Risse [13, 14], Auterhoff & Demleitner [15] and Shellard [16–18].

When one considers the results of all the chemical investigations from a taxonomic point of view one arrives at the following conclusions:

(1) The glycoside resins can definitely be looked upon as characteristic chemical features of the tribes Ipomoeae and Convolvuloideae. They are unique in their composition in the whole plant kingdom. Of similar constitution but having a smaller mol. wt and different sugar and hydroxy fatty acid moieties, are the glyco-lipids isolated from lower plants e.g. *Pseudomonas aeruginosa* [19], *Torulopsis apicola* [20], *Candida bogoriensis* and *Ustilago zeae* [21, 22, 23].

(2) The sub-family Cuscutiodeae contains, as per our investigations, only polysaccharides and no glycoside resins, thereby providing a chemical distinction of this sub-family from that of the Convolvuloideae.

(3) The hydroxy fatty acids characteristic of the two above-mentioned tribes are given in table 2. With very few exceptions these hydroxy acids have not been found anywhere else in the plant kingdom. Like the cutaneous (epidermal) hydroxy fatty acids these are saturated and have a chain length of C 16 at the most. The existence of the dihydroxy fatty acids is not confined solely to the Convolvulaceae family.

(4) The repeated occurrence of the two deoxy-sugars D-fucose and D-quinovose in the glycosidic acids seems to be of some significance. D-Quinovose has so far been found in the *Cinchona* spp. (Quinovin), in *Ladenburgia* and *Rutaceae*-species. Amino-quinovose has been found in a *Pseudomonas* species and sulfo-quinovose in bacteria, algae and the chloroplasts of higher plants. L-Fucose occurs widely in bacteria, algae, fungi and sporadically in higher plants and here particularly in excretory and cutaneous (epidermal) tissue. D-Fucose in contrast, has been found, with exception of some *cardiac glycosides* and a Streptomyces antibiotic, exclusively in the resin glycosides. Both sugars including the L-rhamnose are biosynthesized from D-glucose in several steps through reduction and epimerisation. According to the results of all the investigations so far the sugars of the D-series are β-linked while those of the L-series are α-linked (according to Klyne's rule) (fig. 11).

(5) The volatile acids, given in table 2, oc-

Glycosidic acid from *Ipomoea quamoclit*

Jalapinolic acid
|
O-pentasaccharide

[Rhamnose (*4*) − Fucose (*1*)]

Glycosidic acid from *"Rad. Scammoniae mex."*

Jalapinolic acid
|
O-tetrasaccharide

[Rhamnose (*2*) − Fucose (*1*) − Glucose (*1*)]

Fig. 9.

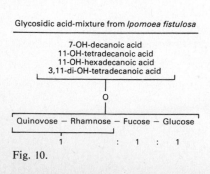

Glycosidic acid-mixture from *Ipomoea fistulosa*

7-OH-decanoic acid
11-OH-tetradecanoic acid
11-OH-hexadecanoic acid
3,11-di-OH-tetradecanoic acid

O
|
Quinovose − Rhamnose − Fucose − Glucose

1 : 1 : 1

Fig. 10.

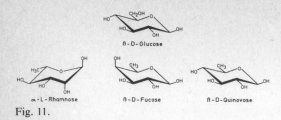

Fig. 11.

cur in varying amounts in all the glycoside resins investigated hitherto. Their repeated occurrence is noteworthy though not characteristic of the Convolvulaceae, as one also comes across such acids in resins, sap, saponins and glycosides of other families.

(6) Only conjectures can be made about the significance of these glycoside resins for the plant. As many resins display antibiotic, skin-irritant and hemolytic properties, they probably represent a defence mechanism with functions similar to those of ethereal oils or saponins. The latter two classes of chemical compounds are absent or occur in traces only in both the main tribes investigated for resin glycosides.

(7) The resin glycosides are not present in all organs of the plant in the same concentration. Therefore chemical investigations of taxonomic value have to include the seeds and roots as well as the herbs of a plant.

(8) Since the resin glycosides are of different mol. wt and chemical composition, the chemical investigations have to be performed with the alcohol-, ether- as well as the petroleum ether-soluble part of the drug.

Note added in proof: Meanwhile from *Ipomoea purga* a β-D-quinovoside of 11-hydroxytetradecanoic acid has been isolated (Singh, S & Stacey, B E, Phytochemistry 1973, 12, 1701).

References

1. Hegnauer, R, Chemotaxonomie der Pflanzen, vol. 3, p. 547. Birkhäuser, Basel and Stuttgart, 1964.
2. Graf, E & Dahlke, E, Chem Ber 1964, 97, 2785.
3. Graf, E, Dahlke, E & Voigtländer, H W, Arch Pharmaz 1965, 298, 81.
4. Khanna, S N & Gupta, P C, Phytochemistry 1967, 6, 735.
5. Liptak, A, Kazmaier, P & Wagner, H, Z Naturforsch. In press.
6. Okabe, H & Kawasaki, T, Tetrahedron lett 1970, 3123.
7. Okabe, H, Koshito, N, Tanaka, K & Kawasaki, T, Chem pharm bull 1971, 19, 2394.
8. Wagner, H & Kazmaier, P, Tetrahedron lett 1971, 3233.
9. Kawasaki, T et al., Chem pharm bull. In press.
10. Wagner, H & Schwarting, G, Z Naturforsch. In press.
11. Smith, C R Jr, Niece, L H, Zobel, H F & Wolff, I A, Phytochemistry 1964, 3, 289.
12. Legler, G, Phytochemistry 1965, 4, 29.
13. Jaretzky, R & Risse, E, Arch pharm 1940, 278, 241.
14. — Ibid 1940, 278, 379.
15. Auterhoff, H & Demleitner, H, Arzneimittelforsch 1955, 5, 402.
16. Shellard, E J, Planta med 1961, 9, 102.
17. — Ibid 1961, 9, 141.
18. — Ibid, 1961, 9, 146.
19. Jarvis, F G & Johnson, M J, J Am chem soc 1949, 71, 4124.
20. Tulloch, A P, Hill, A & Spencer, J F T, Can j chem 1968, 46, 3337.
21. Tulloch, A P, Spencer, J F T & Deinema, M H, Can j chem 1968, 46, 345.
22. Esders, T W & Light, R J, Lipid res 1972, 13, 663.
23. Lemieux, R U, Thorn, G A & Bauer, H F, Can j chem 1963, 31, 1054.
24. Mannich, C, & Schumann, P, Arch Pharm 1938, 276, 211.

Discussion

Sandberg: Is the laxative effect confined to the whole molecule or to the isolated glycosidic acids?

Wagner: The laxative activity as far as we know is bound on the intact complex resin glycosides. The glycosidic acids derived from them have shown no pharmacological activity.

Sandberg: In the case of di-hydroxy compounds are the sugars attached to one or to both hydroxy groups?

Wagner: In the glycosidic acids we and others have studied only one OH-group and in each case the OH-group near the CH_3 end was glycosylated.

Lavie: Was any other pharmacological activity detected on the compounds you have described apart from laxative or hemolytic action?

Wagner: A skin irritant activity was also reported, but no other activities.

Some Aspects of Lichen Chemotaxonomy

S. Shibata

Faculty of Pharmaceutical Sciences, University of Tokyo, Tokyo, Japan

Summary

The characteristic constituents of lichens which are noted as the basis of lichen chemotaxonomy are generally assumed to be the secondary metabolites of mycobionts of lichens. This concept has not fully been established, however.

Some lichens contain constituents identical with or similar to the metabolites of free-living fungi. Our recent experiments showed that laboratory cultures of the isolated mycobionts of some lichens produce typical lichen substances such as (+)usnic acid, zeorin and bellidiflorin.

These findings provide strong support for the above concept.

The occurrence of several new triterpenes in the lichens of *Peltigera* and *Lobaria* spp. is significant from the chemotaxonomical viewpoint.

The water-soluble polysaccharides of lichens are here discussed as a factor of lichen chemotaxonomy.

The characteristic constituents of lichens, which are known as constituting the basis of lichen chemotaxonomy, are generally recognized to be the secondary metabolites of mycobionts of lichens. This concept has not fully been established, but seems to be supported by the fact that lichens produce metabolites sometimes identical with or closely related to those of some free-living fungi [1, 2].

In our recent investigation, (+)rugulosin and skyrin, which were regarded to be unique in fungi, have been found in a lichen, *Acroscyphus sphaerophoroides* [3, 4], and the latter compound has subsequently been isolated from other lichens by other workers [5, 6]. These findings would seem to give additional support to the above concept, but more reliable support could be provided by the laboratory cultivation of symbionts of lichens.

Isolation and Cultivation of Lichen Symbionts

The isolation and cultivation of the symbionts of lichens has been studied earlier mainly of lichenological interest [7, 8], and later as a tool for biochemical investigations [9–15]. For the isolation of lichen symbionts, the phycobiont cells or the mycobiont spores are caught by a fine glass needle with a loop attached to a micromanipulator. Alternatively, the mycobiont spores dischanged from a lichen fruiting body are caught directly on the medium slant in a test tube, or else the fragments of lichen thallus, finely crushed, are cultivated in a medium.

The growth of incubated mycobionts is extremely slow in forming a solid colony of 1~15 mm in diameter from a single spore after 3–6 month incubation.

The phyco- and mycobionts of more than 60 species in 15 families of lichens have been surveyed for the laboratory cultivation [16].

The best results from the isolation and cultivation of mycobionts have so far been obtained in *Ramalina crassa* (Nyl.) Mot., *R. yasudae* Räs., and *R. subbreviuscula* Asahina.

It has been observed by several workers that ribitol is characteristically produced by *Trebouxia* spp., the green algal partner, while glucose is formed by Nostoc spp., the blue-green algal partner.

Using a ^{14}C tracer technique on the phycobionts of *Ramalina crassa* and *R. subbreviuscula*, production of ^{14}C-ribitol by the photosynthesis was revealed, and the ^{14}C-ribitol was found to be released to the fungal partner to form ^{14}C-labelled D-arabitol and D-mannitol [17].

The formation of anthraquinones, parietin, erythroglaucin, erythroglaucin carboxylic acid, emodin, fallacinal and fallacinol in the cultivated mycobiont of *Xanthoria fallax* (Hepp.) Arn. was observed. The occurrence of fragilin, 2-chloroemodin, parietin and emodin in the mycobiont of an unidentified *Caloplaca* spp.

was detected by lichen mass spectrometry [18]. These anthraquinones are also produced by the whole lichens.

By cultivation of the mycobiont of *Acroscyphus sphaerophoroides* Lév. calycin and skyrin have been obtained [20]. Pulvinic acid, pulvinic dilactone, calycin and vulpinic acid were also detected in the mycobionts of *Candelariella vitellina* (Ehrh.) Müll. Arg. by Mosbach [21].

The anthraquinonoid compounds are not unique in lichens, but are also produced by free living fungi, even by some higher plants.

However, some particular compounds which are very peculiar to lichens have also been found to be produced by the cultivated mycobionts without participation of the algal partner. On cultivation of the mycobionts of *Ramalina crassa* and *R. yasudae* at 8~20° for 5 months, (+)usnic acid was isolated in a crystalline form and in a fairly good yield [16].

A characteristic triterpenoid, zeorin, has been obtained in crystalline form from the mycobionts of *Anaptychia hypoleuca* (Ach.) Mass., and *Bombyliospora japonica* (Zahlbr.) [20], and a peculiar dark buff crystalline pigment, bellidiflorin of *Cladonia bellidiflora* (Ach.) Schaer. has been isolated along with an unidentified yellow pigment (CBF-I) and skyrin from its mycobionts, in a shaking culture [23].

The newly detected yellow pigment (CBF-I) is probably (+)*O*-diacetyl-epirugulosin whose biogenetic correlation with bellidiflorin is very noteworthy and which we are now investigating.

Until now, we have no successful means for obtaining depsides, and depsidones, the characteristic constituents of various lichens, in crystalline form from the cultured mycobionts, but the colour reactions on tlc spots have strongly suggested the formation of salazinic acid from the mycobiont of *Ramalina crassa* [22], and squamatic acid from that of *Cladonia crispata* (Ach.) Flot. [23].

The above results pertaining to the constituents peculiar to lichens from the cultured mycobionts, without participation of the phycobionts, appear to give a strong support for the fungal origin of lichen substances.

The participation of algal partners in the biosynthesis of lichen secondary metabolites has not completely been ruled out at the present

stage of knowledge, but it should be emphasized that mycobionts which compose more than 95% of the dry weight of lichen thalli also form the main chemical constituents of lichens on which chemotaxonomy of lichens mainly depends.

The Triterpenoids in *Lobaria* spp.

Asahina [24] has stated that the difference in the metabolites of the morphologically very similar lichens, *Lobaria retigera* (L.) Hoff. and *L. pulmonaria,* which might have the same fungal component, could depend on the different algal partners, since the former contains a blue-green phycobiont and the latter a green one.

Recently Yoshimura [25] proposed to classify the *Lobaria* spp. into two sections by the size and shape of spores: Lobaria section (fusiform spores) and Ricasolia section (acicular or linear spores), regardless of the kind of phycobionts. As the phenolic constituents of *Lobaria* spp., gyrophoric acid, congyrophoric acid, tenuiorin, constictic acid, scrobiculin, usnic acid, stictic acid and norstictic acid were isolated in addition to thelephoric acid, a deep violet coloured pigment in the rhizines of some species. The lichens of *Lobaria retigera* group which belong to Yoshimura's Lobaria section consist of four species, being distinguished by the presence or absence of isidia as well as by positive or negative colour reaction with *p*-phenylenediamine (PD). Yoshimura's nomenclatures are cited in parentheses.

Lobaria retigera group
Isidiate PD(−) *L. isidiosa* (Müll. Arg.) Vain.
     (*L. retigera* (Bory) Trev.)
     PD(+) *L. ididiosa* var. *subretigera* Asah.
     (*L. isidiosa* (Müll. Arg.) Vain.)

Non-isidate PD(−) *L. retigera* (Bory) Trev. (*L. kurokawae* Yoshim.)
     PD(+) *L. subretigera* Inum. (*L. pseudopulmonaria* Gyeln.)

Stictic acid and norstictic acid are responsible for the positive PD reaction.

Our recent investigation showed the presence of migrated hopane (fernane)-type triterpenes named retigeric acids A and B whose structures (*1, 2*) have been established as formulated:

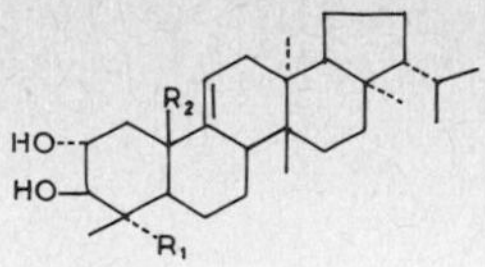

(1) Retigeric acid A
  R₁=COOH R₂=CH₃
(2) Retigeric acid B
  R₁=R₂=COOH

It is noted that the PD(+) lichens of *Lobaria retigera* group produce less of these triterpenes than do the PD(−) lichens [26]. A peculiar sesterterpene, retigeranic acid *(3)*, whose structure has recently been established by the X-ray crystallographic analysis is contained in the PD(+) lichens of *L. retigera* group [27].

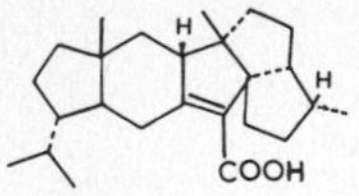

(3) Retigeranic acid

The fernane-type triterpenes are also contained in *Lobaria saccharinensis* Asahina and *L. kazawaensis* (Asah.) Yoshimura both of which possess a green algal partner, but they are now classified under Lobaria section by their fusiform spores.

The Triterpenes of *Peltigera* spp.

The lichens of *Peltigera* genus have been classified according to the difference in the phycobionts: Phlebia section consists of lichens whose phycobiont is green algae; Emprostea section includes the lichens having blue-green phycobiont and is grouped under *Peltigera malacea* group, *P. polydactyla* group and *P. canina* group.

Japanese *Peltigera* spp. are classified as shown in table 1 [28].

The lichens of *P. canina* group contain no remarkable secondary metabolites and thus give no spots on tlc. Tenuiorin *(4)* and zeorin *(5)* are contained in all the lichens of *Peltigera* spp. except those of *P. canina* group [28].

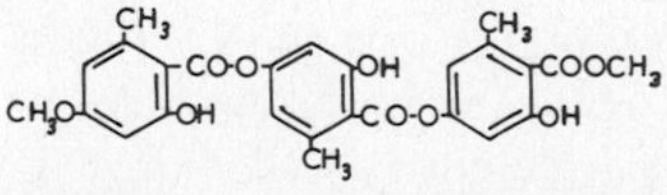

(4) Tenuiorin

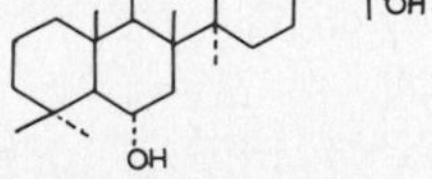

(5) Zeorin

Table 1. *Classification of the Lichens of Peltigera*

Sect. Phlebia (with green algal gonidia)
  Peltigera aphthosa (L.) Willd.
  P. nigripunctata Bitt.
  P. variolosa (Mass.) Gyel.
  P. venosa (L.) Baumg.

Sect. Emprostea (with blue-green algal gonidia)
  P. malacea group
    P. malacea (Ach.) Funck

  P. polydactyla group
    P. dolichorrhiza (Nyl.) Nyl.
    P. horizontalis (Huds.) Baumg.
    P. microphylla (And.) Gyel.
    P. polydactyla (Neck.) Hoffm.
    P. pruinosa (Gyeln.) Inum.
    P. scabrosa Th. Fr.
    P. scutata (Dicks.) Duby
    P. subscutata Gyel.

  P. canina group
    P. canina (L.) Willd.
    P. dilacerata (Gyel.) Gyel.
    P. erumpens (Gyel.) Vain.
    P. praetextata (Flörke ex Somm.) Vain.
    P. rufescens (Weiss) Humb.
    P. spuria (Ach.) DC.

Phlebic acids A and B *(6, 7)* isolated characteristically from the lichens of Phlebia section have recently been established as being hopane-type triterpenes:

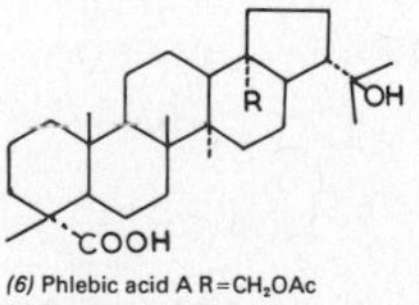

(6) Phlebic acid A R=CH₂OAc
(7) Phlebic acid B R=CH₃

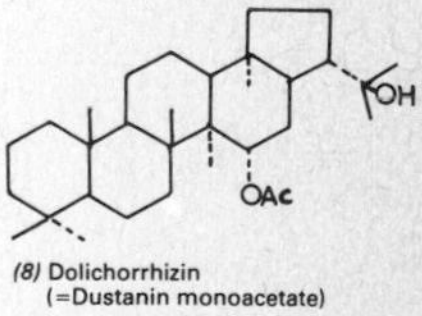

(8) Dolichorrhizin
(=Dustanin monoacetate)

Dolichorrhizin *(8)* which is identical with dustanin monoacetate is widely distributed in the lichens of both Phlebia and Emprostea section of *Peltigera* genus.

Peltigera aphthosa (L.) Willd., *P. nigripunctata* Bitt. and *P. variolosa* (Mass.) Gyel. are morphologically very similar, but can be distinguished by the absence of phlebic acid A in *P. nigripunctata*.

Peltigera aphthosa and *P. dolichorrhiza* (Nyl.) Nyl. can be distinguished chemically by the absence of phlebic acids A and B and the presence of 7-β-acetoxy-22-hydroxy-hopane *(9)* in the latter [29, 30], while ergosterol peroxide (*10*) was found in both lichens [31].

(9) 7 β-Acetoxy-22-hydroxy-hopane   (10) Ergosterol peroxide

In the chemotaxonomy of the lichens of *Peltigera* spp. the difference in phycobionts seems to agree with the different patterns of the triterpenoid metabolism, but chemotaxonomically it is not so essential, since *P. canina* group shows a quite different pattern of secondary metabolism in comparison with other *Peltigera* spp. of the same section of Emprostea having the same kind of phycobiont.

Water-soluble Polysaccharides and Their Chemotaxonomical Significance in Lichens

Among the water-soluble polysaccharides of lichens, only lichenan, isolichenan and pustulan have chemically been investigated. Lichenan =lichenin) $[\alpha]_D+9\sim18$, $\overline{DP}$ 80~400, which was initially isolated from *Cetraria islandica* (Iceland moss) was shown to be a linear homoglucan with $\beta(1\rightarrow3)$ $(1\rightarrow4)$ linkages in a ratio of 3:7 [32–34]. The sequence of linkages in lichenan was determined enzymatically as follows [35].

Glc $\beta1\rightarrow3$ Glc $\beta1\rightarrow4$ Glc $\beta1\rightarrow4$ (main unit)
Glc $\beta1\rightarrow3$ Glc $\beta1\rightarrow4$ Glc $\beta1\rightarrow4$ Glc $\beta1\rightarrow4$
Glc (subunit)

Isolichenan (=isolichenin), $[\alpha]_D+255°$ $\overline{DP}$ 34~43, occurring in lichens often being accompanied by lichenan was shown to be a linear homoglucan with $\alpha(1\rightarrow3)$ $(1\rightarrow4)$ linkages in a ratio of 3:2 [33, 36, 37].

The main part of isolichenan is Glc $\alpha1\rightarrow3$ Glc $\alpha1\rightarrow3$ Glc $\alpha1\rightarrow4$ Glc linkages. Earlier investigation and our recent studies showed the co-occurrence of lichenan and isolichenan in the following lichens:

Parmeliaceae

   Cetraria islandica (L.) Ach.
   C. nivalis (L.) Ach.
   C. islandica var. *orientalis*
   C. richardsonii Hook

Parmelia tinctorum Nyl.
P. conspersa (Ach.) Ach.
P. hypotrypella Asah. (=*Hypogymnia hypotrypella* (Asah.) Rass.)
P. nikkoensis Zahlbr. (=*Hypogymnia nikkoensis* (Zahlbr.) Rass.)

Usneaceae

Usnea barbata (L.) Wigg.
U. longissima Ach.
U. bayleyi (Stirt.) Zahlbr.

Alectoria sulcata (Lév.) Nyl.
A. sarmentosa (Ach.) Ach.

A wide distribution of the combination of lichenan and isolichenan in the lichens of *Cetraria, Usnea* and *Alectoria* spp. is presumed on the basis of foregoing results.

Pustulan $[\alpha]_D-38°C$, $\overline{DP}$ ca 120, was first reported by Drake in 1943 [38] who isolated it from *Lasallia pustulata* and *Umbilicaria hirsuta* and studied by Lindberg et al. [39] and Norman [40] to formulate it as a linear homoglucan giving $\beta(1\rightarrow6)$ linkage. Our recent investigation revealed that pustulan contains one *O*-acetyl (IR: 1735, 1250 cm⁻¹) at the 3-position of every 10 to 12 glucose units [41–43].

Pustulan is a characteristic polysaccharide in the lichens of Gyrophoraceae: *Gyrophora esculenta* Miyoshi (=*Umbilicaria esculenta* (Miyoshi) Minks.) *Umbilicaria augulata* Tuck. *U. caroliniana* Tuck., *U. polyphylla* (L.) Baumg., *Lasallia papulosa* (Ach.) Llano.

Everniin

Mićović [44] and Stefanovich [45, 46] found a cold-water insoluble polysaccharide named everniin in *Evernia prunastri,* which possessed $\alpha(1\rightarrow3)$ $(1\rightarrow4)$ linkages whose ratio is 4:1. Our recent investigation of polysaccharide of *Evernia prunastri* [47] revealed the co-occurrence of EP-3, EP-6 and EP-7.

EP-3, $[\alpha]_D+200°$ $\overline{DP}$ 70, which would correspond to everniin; EP-6, $[\alpha]_D+164°$, $\overline{DP}:160$ is an α-glucan with $(1\rightarrow3)$ $(1\rightarrow4)$ linkages (3:2) whose composition is very close with isolichenan but isolichenan can be distinguished by its specific optical rotation $[\alpha]_D+255°$ and its molecular size, $\overline{DP}:34\sim43$. EP-7 $[\alpha]_D+12°$, $\overline{DP}$ 60 is a β-$(1\rightarrow3)$ $(1\rightarrow4)$ glucan (3:1), but is different from lichenan in its proportion of $(1\rightarrow3)$ $(1\rightarrow4)$ linkages.

Table 2. *Polysaccharides of the lichens of Parmelia and the related spp.*

Lichens	$[\alpha]_D{}^a$	Sugar components in hydrolysate	Type
P. cetrarioides (Del. ex Duby) Nyl. (=Cetrelia cetrarioides (Del. ex. Duby) W. Culb. & C. Culb.	210.8	Glc	α (PC-3, isolichenan)
P. laevior Nyl.	139.7	Glc	α
P. saxatilis (L.) Ach.	98.7	Glc (Xyl. Man.)	α
P. caperata (L.) Ach.	147.0	Glc	α
P. pertusa (Schrank) Schaer (Menegazzia terebrata (Hoffm.) Mass.	79.6	Glc	α, β
P. homogenes Nyl.	98.3	Glc (Man)	α, β
P. reticulata Tayl.	55.7	Glc (15) Man (5) Gal (4)	α, β
P. conspersa (Ach.) Ach.	49.6	Glc	β, α (lichenan, isolichenan)
P. hypotrypella Asah. (Hypogymnia hypotrypella (Asah.) Rass)	44.3	Glc	β, α
P. nikkoensis Zahlbr. (H. nikkoensis (Zahlbr.) Rass.)	49.5	Glc	β, α
P. tinctorum Nyl.	43.7	Glc	β, α

^{a} $[\alpha]_D$ measured by the crude water-soluble polysaccharide fraction.

Polysaccharides of *Parmelia* spp.

Parmelia caperata contains a cold water-insoluble polysaccharide tentatively named PC-3 and a soluble fraction (PC-2), the latter of which has been proved to be identical with isolichenan. PC-3, $[\alpha]_D+201°$ (2N NaOH) IR ν_{max}^{KBr} 925 845 780 cm^{-1}, $\overline{DP}$ 100–130, has been proved to be $\alpha(1\rightarrow3)$ $(1\rightarrow4)$ glucan (1:1) [48]. It is similar to nigeran, an intracellular polysaccharide of *Aspergillus niger*, but differs from it in the molecular size (nigeran: $\overline{DP}$ 300~350). PC-3 type glucan is also occurring in *Parmelia cetrarioides*, *P. laevior* and *P. saxatilis* along with isolichenan, and in *Cladonia bellidiflora*, *Cl. alpestris*, *Cl. pacifica*, *Cl. squamosa*, *Cl. crispata*, *Cl. rangiferina* as the minor polysaccharide without coexistence with isolichenan and along with a glycopeptide which is the main component of the polysaccharide fraction.

On examining the polysaccharides of some lichens of *Parmelia* and related spp., it has been found that they are classified into 3 groups; (*a*) α-glucan type; (*b*) α-glucan dominant—β-glucan coexisting type; (*c*) β-glucan dominant—α-glucan coexisting type [49].

The Polysaccharides of the Lichens of Stictaceae

Glycopeptides are contained in the lichens of Stictaceae, *Lobaria orientalis* (Asah.) Yoshim. *L. isidiosa* (Müll. Arg.) Vain., *L. pseudopulmonaria* Gyeln., *L. linita* (Ach.) Rabenh., *L. japonica* (Zahlbr.) Asah., *Sticta gracilis* Müll. Arg., *S. wrightii* Tuck. without co-occurrence of simple glucan [50], and in some lichens of Cladoniaceae, *Cladonia bellidiflora* (Ach.), *Cl. alpestris* (L.) Rabenh., *Cl. pacifica* Ahti, *Cl. squamosa* (Scop.) Hoffm., *Cl. crispata* (Ach.) Flot., *Cl. rangiferina* (L.) G. Web. ex. Wiff. being accompanied by PC-3 type glucan.

The glycopeptides of *L. orientalis* are separated into the fraction LOF-1 $[\alpha]_D+50°$ IR ν_{max}^{KBr} 875 cm^{-1}, N% 1.01 and LOF-2 $[\alpha]_D+$ 31.5°, IR ν_{max}^{KBr} 875 cm^{-1}, N% 1.04 which we have studied recently, and LOF-1 was hydrolysed to give glucose and galactose as the main sugar components, and mannose, arabinose, xylose and rhamnose as the minor components. LOF-2 contains mainly mannose and galactose.

The carbohydrate portion of LOF-1 is linked with the peptide part by an *O*-glycosyl link-

Table 3. *Properties of polysaccharides from Ramalina crassa*

	Lichen, RC-1		Mycobiont, RC-f-1		Phycobiont, RSA-1
	RC-2	RC-3	RC-f-4	RC-f-3	RSA-2
Yield (%)	2.1	0.3	1.32	1.84	
Solubility^a	Soluble	Insoluble	Soluble	Insoluble	Soluble
Hydrolysate	Glc	Glc	Glc	Glc	Gal
$[\alpha]_D$	+31°	+140°	+36°	+136°	−85°
IR (cm^{-1})	890	780	890	800	800
		840		840	840
		920		930	866
Linkage	$\beta(1\rightarrow3)$ $(1\rightarrow4)$	$\alpha(1\rightarrow3)$ $(1\rightarrow4)$			
	3:1	5:3			
$\overline{DP}$	68	88			

^{a} In cold water.

age with serine and threonine. The results of methanolysis of permethylate of LOF-1 indicated that 1→6 glucan and 1→3 mannan are the main components of the carbohydrate portion.

Acroscyphan [47]

Acroscyphus sphaerophoroides Lév. is a very peculiar lichen occurring as one species in one genus in some very restricted places at higher altitude in the CircumPacific region from the Eastern Himalayas to the Andes mountains in South America. Our recent investigation on this lichen showed that it contains along with glycopeptide a very characteristic glucan for which we proposed the name acroscyphan.

Acroscyphan $[\alpha]_D$+176° IR: 845 cm^{-1} is insoluble in cold water and gives a very intense blue colouration with iodine. The methanolysis of acroscyphan permethylate afforded methyl 2,3,4-tri-O-methyl-, methyl 2,4,6-tri-O-methyl- and methyl 2,3,6-tri-O-methyl-D-glucopyranoside along with methyl 2,3,4,6-tetra-O-methyl-D-glycopyranoside. It has therefore been revealed that acroscyphan is a homoglucan having $\alpha(1\rightarrow3)$ $(1\rightarrow4)$ $(1\rightarrow6)$ linkages.

Polysaccharides of the Lichens of Stereocaulaceae

An $\alpha(1\rightarrow3)$ $(1\rightarrow4)$ glucan (1:4) was isolated from *Stereocaulon paschale* by Hauan & Kjølberg [51]. Our recent investigation showed that a similar glucan is also contained in other *Stereocaulon* spp. The polysaccharides of *Stereocaulon japonicum* Th. Fr., *St. sorediiferum*

Hue. and *St. vesuvianum* Pers. were studied to separate a cold-water-soluble polysaccharide, SJY-3.

The main component of SJY-2 (SJY-2-I), $[\alpha]_D$+120°, is a simple glucan with $\alpha(1\rightarrow3)$ $(1\rightarrow4)$ linkages (3:2) having some branched chains. It seems to be a characteristic polysaccharide in *Stereocaulon* spp. [52].

As previously mentioned we have provided some reliable evidences for the fungal origin of secondary metabolites of lichens. For examining the origin of lichen polysaccharides, we have made comparative studies on the polysaccharides of whole lichens, cultured mycobionts and phycobionts.

The cold-water-insoluble polysaccharide of *Parmelia caperata* (PC-3) is very similar to that separated from the cultured mycobiont, and the results obtained from the whole lichen and symbiont of *Ramalina crassa* showed that both cold-water-soluble and insoluble polysaccharides produced by the cultured phycobionts of *Parmelia caperata* and *Ramalina crassa* are apparently different in their properties from those of corresponding lichen.

Accordingly, the characteristic polysaccharides of lichens would also be the metabolites of their mycobionts [53].

Our investigations mentioned above suggested that the lichen polysaccharides represent the characteristic nature of some groups of lichens in contrast with the smaller molecular secondary metabolites which are mostly species-specific.

The lichen chemotaxonomy could be systematized on the basis of the characteristic constituents of both smaller and larger molecular size which would be recognized as the metabolites of lichen mycobionts.

I am grateful to Professor emeritus Y. Asahina, Drs S. Kurokawa, I. Yoshimura and M. Nuno for their kind advice on lichenology.

I also wish to thank my generous collaborators whose names are referred in the individual references cited below.

References

1. Culberson, Ch F, Chemical and botanical guide to lichen products. Chapel Hill, University of N Carolina Press, 1969.
2. — Ibid, suppl. Bryologist 1970, 73, 177.
3. Shibata, S, Tanaka, O, Sankawa, U, Ogihara, Y, Takahashi, R, Seo, S, Yang, D M & Iida, Y, J Jap bot 1968, 43, 335.
4. Shibata, S, Pure & appl chem 1973, 33, 109.
5. Santesson, J, Acta chem Scand 1969, 23, 3270.
6. Yosioka, I, Morimoto, K, Murata, K, Yamaguchi, H & Kitagawa, I, Chem pharm bull (Tokyo) 1971, 19, 2420.
7. Ahmadjian, V, The lichen symbiosis. Blaisdell Publ Co, Waltham, Toronto-London, 1967 (earlier papers are cited therein).
8. Thomas, E A, Beitr Kryptogamenflora Schweiz 1939, 9, 1.
9. Mosbach, K, Angew Chem (int ed) 1969, 8, 240.
10. — Acta chem Scand 1967, 21, 2331.
11. Hess, D, Z Naturforsch 1959, 14b, 345.
12. Richardson, D H S & Smith, D C, Lichenologist 1966, 3, 202.
13. Richardson, D H S, Hill, D J & Smith, D C, New phytol 1968, 67, 469.
14. Richardson, D H S & Smith, D C, New phytol 1968, 67, 69.
15. Maruo, B, Hattori, T & Takahashi, H, Agr biol chem 1965, 12, 1084.
16. Komiya, T & Shibata, S. Unpublished data.
17. — Phytochemistry 1971, 10, 695.
18. Nakano, H, Komiya, T & Shibata, S, Phytochemistry 1972, 11, 3505.
19. Santeson, J, Arkiv kemi 1968, 30, 364.
20. Nakano, H, Komiya, T & Shibata, S, Abst 92nd ann meeting of pharm soc Japan 1972 II, 254 (Osaka, Japan).
21. Mosbach, K, Acta chem Scand 1967, 21, 2331.
22. Komiya, T & Shibata, S, Chem pharm bull (Tokyo) 1969, 17, 1305.
23. Nakano, H & Shibata, S. Unpublished data.
24. Asahina, Y, J Jap bot 1933, 269, 398.
25. Yoshimura, I, J Hattori bot lab 1971, 34, 231, 364.
26. Takahashi, R, Chian, H-C, Aimi, N, Tanaka, O & Shibata, S, Phytochemistry 1972, 11, 2039.
27. Kaneda, M, Takahashi, R, Iitaka, Y & Shibata, S, Tetrahedron lett 1972, No 45, 4609.
28. Kurokawa, S, Junzenji, Y, Shibata, S & Ching, H-C, Bull nat sci mus 1966, 8, 101.
29. Takahashi, R, Tanaka, O & Shibata, S, Phytochemistry 1969, 8, 2345.
30. — Ibid 1970, 9, 2037.
31. — Ibid 1972, 11, 1850.
32. Meyer, K H & Gürtler, P, Helv chim acta 1947, 30, 751.
33. Chanda, N B, Hirst, E L & Manners, D J, J chem soc 1957, 1951.
34. Peat, S, Whelan, W J & Roberts, J G, J chem soc 1957, 3916.
35. Perlin, A S & Suzuki, S, Can j chem 1962, 40, 50.
36. Peat, S, Whelan, W J, Turvey, J R & Morgan, K, J chem soc 1961, 623.
37. Flemming, M & Manners, D J, Biochem j 1966, 100, 24.
38. Drake, B, Biochem Z 1943, 313, 388.
39. Lindberg, B & McPherson, J, Acta chem Scand 1954, 8, 985.
40. Norman, B, Acta chem Scand 1968, 22, 1623.
41. Shibata, S, Nishikawa, Y, Takeda, T & Tanaka, M, Chem pharm bull (Tokyo) 1968, 16, 2362.
42. Shibata, S, Nishikawa, Y, Takeda, T, Tanaka, M, Fukuoka, F & Nakanishi, M, Chem pharm bull (Tokyo) 1968, 16, 1639.
43. Nishikawa, Y, Tanaka, M, Shibata, S & Fukuoka, F, Chem pharm bull (Tokyo) 1970, 18, 1431.
44. Mićović, V M, Hranisavljević, M & Miljkocić-Stojanović, J, Carbohyd res 1969, 10, 525.
45. Stefanovich, Life sci 1969, 8, 122.
46. — Ph D Thesis, Faculty of Science, Univ Belgrad, 1960.
47. Takeda, T, Funatsu, M, Shibata, S & Fukuoka, F, Chem pharm bull 1972, 20, 2445.
48. Takeda, T, Nishikawa, Y & Shibata, S, Chem pharm bull (Tokyo) 1970, 18, 1074.
49. Takeda, T & Shibata, S. Unpublished data.
50. Takahashi, K, Takeda, T, Shibata, S, Inomata, M & Fukuoka, F, Chem pharm bull 1973, 21. In press.
51. Hauan, E & Kjølberg, O, Acta chem Scand 1971, 25, 2622.
52. Yokota, I, Takeda, T & Shibata, S. Unpublished data.
53. Komiya, T, Takeda, T, Nishikawa, Y & Shibata, S. Unpublished data.

Discussion

R. Santesson: The proofs you gave that lichen substances are formed by the isolated mycobiont are very important for lichen taxonomy. In 1962 Ahmadjian discussed the formation of red pigments by the isolated mycobiont of *Acarospora fuscata*. Improved culture technics will most probably prove that many more lichen substances are formed by isolated mycobionts.

Lobaria pulmonaria and *L. retigera* do not have the same mycobiont. They differ in morphological characters which are not due to different phycobionts. If two lichens have the same mycobiont they belong to one and the same species.

Many lichenologists do not regard morphologically identical chemotypes as different species. For me it is not an acceptable taxonomy to base a species on one character only, which is made by lichenologists describing "chemical species".

Shibata: I agree with you. *Lobaria pulmonaria* and *L. retigera* should have different mycobionts since they reveal different metabolic patterns.

Lavie: Should the components of the symbiosis in the lichens possibly be grown separately, one should be able to study and determine the compounds produced in the lichens by e.g. the fungi?

Shibata: As mentioned in the present talk, we are accumulating experimental evidence for the production of some very characteristic lichen substances, such as anthraquinones, (+)usnic acid, zeorin and bellidiflorin, by laboratory cultivation of mycobionts of lichens without participation of phycobionts.

Bu'Lock: I was most impressed by both the scope and the technical expertise which was displayed in your work. It was always most plausible that the phenolic lichen substances (including the polyketide quinones) would prove to be mycobiont products, and it is very pleasing to see this so abundantly confirmed. On the other hand by the same argument I would have expected the characteristic hopane-type triterpenoids to relate to the phycobiont, for so far as I know such substances have never been found in the free-living fungi but are becoming increasingly known as products from blue-green algae (and some other bacteria).

Shibata: Even hopane type triterpenes could be formed by the mycobionts of lichens, as we isolated zeorin in a crystalline form from the culture of the mycobionts of *Anaptychia hypoleuca* and *Bombyliospora japonica*.

Cronquist: I just want to point out that if it is eventually confirmed that what have been regarded as different species of lichen sometimes have the same fungal component and differ only in the algal component, some technical nomenclatural questions will be raised. The International Code of Botanical Nomenclature now states that the name of a lichen is the name of its fungal component. Under the postulated conditions one would obviously either have to change the rules or ignore the rules, or change the specific concepts.

Shibata: I am afraid, you have misunderstood this point. Opposing to Professor Asahina's earlier concept, I have demonstrated in the lichens of *Lobaria* spp. that even some morphologically similar lichens should have different mycobionts if they produce apparently different metabolites, and the difference in the phycobionts, if any, has no significant meaning in the secondary metabolism.

Geissman: To amplify Professor Shibata's response, I remind the members of his slide showing Asahina's concept of 3 different *Lobaria* (A, A', A'') associated in 3 species with the same fungus, F. Your results suggest that the fungi are indeed different, although you have not succeeded in cultivating the separate mycobionts.

Birch: I am not sure whether I have correctly understood that triterpenes and polyketides are in some cases interchangeable as metabolites? If so this might be due to the fact that terpene biosynthesis requires only acetyl coenzyme-A, whereas polyketides require malonyl coenzyme-A in addition. The latter might correlate with fatty acid production: has this been examined?

As an example of the great power of biosynthetic analysis, I might point out that despite immense differences in structure, superficially, of usnic acid and salazinic acid, these must arise from the same monomethylated polyketide precursor. In the first case, this undergoes aldol ring-closure, in the second Claisen ring-closure, and the aromatic compounds resulting undergo phenol oxidations of known type.

Shibata: In some *Lobaria* spp. the production of polyketides—depsidones—is dominant, while in others triterpenes are predominantly formed, though the correlation of the former with the fatty acid formation has not been examined.

J. Santesson: I do not fully agree with your conclusion, that the use of micromolecular lichen substances as chemical characters is restricted to the species level and below. Even at the family level the usefulness is evident—take e.g. the occurrence of parietin in virtually every member of the Teloschistaceae. Also at the generic level there are numerous examples, e.g. *Cetrelia*.

Shibata: What I mean does not restrict the smaller molecular lichen substances to species level, but I wish to emphasize that some lichen polysaccharides could be used to characterize a larger group of species, a genus or a family. The production of individual smaller size lichen metabolite is much more species dependent, though as you pointed out parietin occurs widely in Teloschistaceae. Some anthraquinones, including parietin, show a very wide distribution in fungi, lichens and even in higher plants. Some specific metabolites or a pattern of production of metabolites must be used to characterize lichen species.

Insects and Plant Chemotaxonomy

The pollination of *Ophrys* orchids

B. Kullenberg and G. Bergström

Ecological Station of Uppsala University on Öland, S-380 60 Färjestaden, Sweden

Summary

The assortative pollination of *Ophrys* species by sexually attracted aculeate Hymenoptera males is primarily based on chemical and tactile stimulation of defined species group. The volatile compounds of the flowers responsible for the specific behaviour release are mainly cyclic sesquiterpene alcohols and hydrocarbons.

The biological phenomena

The relationship between angiosperm flowers and flower-visiting insects is of the greatest importance to life on earth. This interdependence is in principle based upon the ability of flowers to offer the carbohydrate-loaded nectar and the protein-rich pollen as food to the insects, and the ability of insects to transfer pollen to the stigma, when feeding or searching for food. Flowers possess the faculty to provoke and to guide certain instinct-based behaviour in insects. Generally we are acquainted with the feeding behaviour of insects on flowers. However, certain flower-types have the ability to provoke other instinctive behaviour in insects. In the orchid genus *Ophrys*, with about 30 forms classified as species, the majority grow in the Mediterranean region. It is difficult to separate certain forms and species morphologically from each other. The flower has characteristics which arouse the inborn male copulatory behaviour of certain species of aculeate Hymenoptera [9–12]. Thus the *Ophrys* flower has the capacity to stimulate different phases of the male copulatory behaviour, from the approach flight to the attempted copulation, though accomplished copulation will never be performed. This orchid flower does not secrete nectar, neither is its pollen available to most types of insects. Female aculeate Hymenoptera are never seen to visit *Ophrys* flowers.

The male copulatory instinct is stimulated by the flower-perfume, exhaled from the labellum, one of the petals, and the approach flight as well as the descent to the flower will not be performed without this chemical stimulation. The insect's movements on the labellum, imitating the introduction of copulation, are indispensable to pollination. They depend on the superficial, epidermal structures of the labellum and of the solid construction of the labellum itself. These tactilely guided movements on the labellum cannot be released without the continued chemical, olfactory stimulation of the flower perfume. The behaviour of the male Hymenopteron has been studied by ocular observations as well as by electronic flash photography and 16 mm colour filming. During these movements, performed on the flower labellum, the males will adopt such positions and attitudes as cause the pollinia to be loosened and pollination to be accomplished.

The pollination of the various *Ophrys* types carried out by separate Hymenoptera species, or restricted species groups, is really assortative. This can be discerned in table 1. The primary delimiting factor is precisely the chemical structure of certain biologically active substances, exhaled by the flowers (the labella) of the different form groups of *Ophrys*.

The secondary form-isolating factor (figs 1, 2) of the *Ophrys* pollination has regard to the fact that in one group of *Ophrys* species the pollinia are taken by the Hymenoptera male with the tip of its abdomen, whereas in most other *Ophrys* forms the pollinia are removed by the insect's head. One species, *O. apifera*, has the ability of auto-pollination though still preserving the typical *Ophrys* attractivity for *Hymenoptera* males (fig. 3).

A third, virtually form-isolating, factor is the phenological one. Chorologically considered, the Mediterranean *Ophrys* forms do not seem to be distinctly differentiated. In fact most forms or species seem to have habitats in common. This seems to be true at least for what may be called "macro-habitats". Hans Sundermann [16] counts 31 species, but considers

Table 1. *Survey of the hitherto observed pollination of* Ophrys *by chemically and sexually excited aculeate*
Hymenoptera males

Subspecies are mentioned if pollination has been observed. Pollinators in brackets not observed by Kullenberg
***, **, *, excitation degrees in field test with highest frequency (cf [11, 12])

h and *a* after the name of the pollinator indicate that the pollinia are taken with the *head* or the tip of the *abdomen* respectively
→ scent stimulation direction

Section of the genus and species/form	Pollinators (from Kullenberg and other authors)	Attractive extracts (5 labella/1 ml hexane)
I *Orientales* Nelson		
A Aegaeae Nelson		
O. kotschyi Fl. et Soó	–	
O. cretica (Vierh.) Nelson	–	
B Reinholdianae (Renz) Nelson		
O. reinholdii Fl. ⟶	[*Melecta* sp. ♂ *h*]	
O. argolica Fl.	–	
O. lunulata Parl.	–	
II *Fuciflorae* Rchb. f.		
O. scolopax Cav. ⟶	[*Eucera* spp. I ♂♂ *h* e.g. *E. tuberculata* F., *E. longicornis* L. ⟵	*** *scolopax, tenthredinifera, apifera, bombyliflora* ** *bertolonii*
O. fuciflora (Crantz) Moench ⟶	*Eucera* spp. I ♂♂ *h*	
O. apifera Huds. ⟶	*Eucera* spp. I ♂♂ *h*	
O. bornmuelleri M. Schulze	–	
O. tenthredinifera Willd. ⟶	*Eucera* spp. I ♂♂ *h* e.g. *E. nigrilabris* Lep. ⟵	*** *scolopax, tenthredinifera, apifera, bombyliflora* ** *bertolonii, sph. atrata* * *speculum, insectifera*
III *Bombyliflorae* Rchb. f.		
O. bombyliflora Link ⟶	*Eucera* spp. II ♂♂ *h* e.g. *E. grisea* F. ⟵	*** *bombyliflora,* *** *tenthredinifera*
IV *Araneiferae* Rchb. f.		
A Basisignatae Nelson		
O. sphecodes atrata (Lindl.) ⟶ E. Mayer	*Andrena* spp. ♂♂ *h* e.g. *A. nigroaenea* K. ⟵	*** *sph. sphecodes, fusca* ** *lutea*
O. sph. sphecodes Mill. ⟶	*Andrena* spp. ♂♂ *h* e.g. *A. nigroaenea*	
O. sph. sphecodes Mill. ⟶	*Colletes cunicularius* L. ♂ *h* ⟵	*** *sph. sphecodes, sph. provincialis, arachnitiformis* ** *lutea, sph. atrata*
O. sph. litigiosa (E.G. Cam.) ⟶ Becherer	*Andrena* spp.?	
O. sph. provincialis Nelson ⟶	*Colletes cunicularius* ♂ *h*	See above
B Mediosignatae Nelson		
O. ferrum-equinum Desf.	–	
O. bertolonii Mor. ⟶	{ *Campsoscolia ciliata* F. ♂? *Eucera* spp. I ♂♂?	See below See above
V *Arachnitiformes* Nelson		
O. arachnitiformis Gren.et Phil. ⟶	{ (*Andrena* spp. II ♂♂) *Colletes cunicularius* ♂ *h*	See above
VI *Fusci-Luteae* Nelson		
O. fusca fusca Link ⟶	{ *Andrena* spp. II, III ♂♂ *a* II e.g. *A. flavipes* Pz. *Anthophora acervorum* L. ♂ *a* (one observation)	
O. f. iricolor (Desf.) O. Schwarz ⟶	*Andrena* spp. II ♂♂ *a*	
O. f. omegaifera (Fl.) Nelson	–	
O. atlantica Munby	–	
O. pallida Raf.	–	
O. lutea Cav. ⟶ *O. murbeckii* Fl.[a] (=*lutea* ssp. *subfusca* (Rchb. f.) Murb. ⟶	{ *Andrena* spp. I, III ♂♂ *a* I e.g. *Chlorandrena* spp., III e.g. *A. ovatula* K. *Andrena* spp. II ♂♂ *a*	
VII *Ophrys* L.		
O. speculum Link ⟶	*Campsoscolia ciliata*[b] ♂ *h* ⟵	*** *speculum*
O. insectifera L. ⟶	{ *Gorytes mystaceus* L. ♂ *h* *Gorytes campestris* Müll.[c] ♂ *h* } ⟵	*** *insectifera*

[a] *O. murbeckii* is regarded as a form partly separated from *O. lutea* in respect of pollination [10].

[b] The subspecies *ciliata* seems to represent the species in the western part of the Mediterranean Region.

[c] *fargei* Shuckard.

that if subspecies are created, the number of species may be reduced to 14. Erich Nelson [13] considers 21 species with numerous subspecies to be more suitable.

In order to explain the ecological relationship between the *Ophrys*-forms and their specific pollinators, as well as the differentiation of forms within the genus *Ophrys*, it seemed quite clear even from the earliest observations on its pollination that the chemistry of the flower perfumes had to be studied.

Studies on the chemistry of volatile products from *Ophrys* flowers

The first results from our work on the structure and function of the pollination excitants of *Ophrys* will be reported separately [7, 8].

We have shown that the flowers give off small amounts of simple, aliphatic compounds as well as isoprenoid substances, particularly cyclic sesquiterpene alcohols and hydrocarbons. Several such compounds have been tested in the field on *Eucera longicornis* males [12]. Electrophysiological recordings, so-called electroantennograms, have been made by Dr Ernst Priesner, Max-Planck-Institut für Verhaltensphysiologie, Seewiesen, BRD [14], on males of some *Andrena* species and of *Eucera tuberculata* and *E. longicornis*. The field tests as well as the laboratory tests show that the pollinating insects respond to specific sesquiterpene alcohols and hydrocarbons. Only compounds

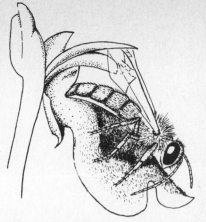

Fig. 2. Pollination of *Ophrys lutea* by a male *Andrena*. The pollinia are taken by the tip of the abdomen of the insect.

with the right stereochemistry elicit maximal behavioral and electrophysiological response.

Up till now 28 species and subspecies have been studied according to the classification of Nelson [13]. By comparison of data from, in the first place, capillary gas chromatography and mass spectrometry of compounds from *Ophrys* and of reference compounds, some types of sesquiterpenes have been identified. A large number of isomeric structures, due to positional and geometric isomerism, are possible, however, and we are currently trying to establish the absolute configuration of some of the natural compounds. The results so far already permit some conclusions regarding the taxonomical relationships between several species of *Ophrys*. These conclusions coincide, to a large extent, with those drawn from observations of pollination by aculeate Hymenoptera males [10, 11, 12].

When we looked at the distribution in *Ophrys* flowers of some types of sesquiterpene hydrocarbons, which we have tentatively called "longicyclene-type", "copaene-type" and "cadinene-type", we found that the occurrence of the copaene-type was restricted to species pollinated mainly by *Eucera* bees. This type of sesquiterpene, which may be represented by several closely related isomers in different species, has so far been found in *O. scolopax, O. apifera, O. tenthredinifera* and *O. bertolonii*. The longicyclene-type is found in *O. speculum, O. insectifera, O. lutea* and *O. fusca*, in several

Fig. 1. Pollination of *Ophrys bombyliflora* by a male *Eucera*. The pollinia are taken by the head of the insect.

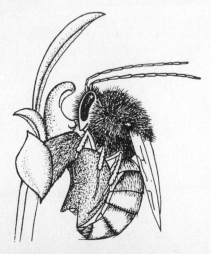

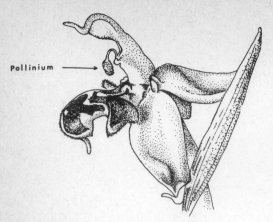

Fig. 3. Auto-pollination in *Ophrys apifera*.

forms of *O. sphecodes* and in *O. bombyliflora*. The cadinene-type is present in *O. lutea, O. fusca* and, in relatively minor amounts, in *O. bombyliflora*. This is also true for a sesquiterpene alcohol, which seems to be structurally directly related to the cadinene-type hydrocarbons. *O. lutea, O. fusca* and various forms of *O. sphecodes* are pollinated by *Andrena* males.

Typical capillary gas chromatograms from these analyses, using a splitter-free intake system [15], are shown in figs 4–6. Fig. 4 shows a chromatogram of volatile compounds from *Ophrys lutea*. They have been isolated from the labella of the flowers by an *enfleurage*-technique [1]. The regions containing sesquiterpene hydrocarbons and sesquiterpene alcohols have been marked by a larger and a smaller bracket, respectively. The component marked by an arrow was identified as *n*-octanol. *O. lutea* possesses both the longicyclene type and cadinene-type hydrocarbons as well as the

cadinol-type alcohol and is typical for the *Andrena*-pollinated group of *Ophrys*.

Fig. 5 shows a capillary gas chromatogram of *O. fuciflora*. This species is representative for those pollinated by *Eucera* males. Frequently, straight chain aliphatic hydrocarbons are added as references, permitting an accurate measurement of retention indices. An alternative and complementary method to the *enfleurage*-technique is to place a piece of a labellum directly in the splitter-free injection system of the gas chromatograph. An example of the resulting chromatogram is given in fig. 6, derived from *O. fuciflora*. Normal, long chain hydrocarbons are more pronounced in this chromatogram. The peaks corresponding to tricosane, heneicosane and nonadecane have been marked in fig. 6.

The results of the chemical analysis of the odoriferous compounds also indicate a high degree of similarity between the compositions of the volatile secretions of *O. lutea* and *O. fusca*. Considerable differences are found between the odoriferous compounds given off by forms belonging to the *O. sphecodes*-complex. By investigating these differences further, especially in relation to the variation of the analytical results within separate forms we are optimistic regarding our ability to clarify some of the complicated taxonomical interrelationships within this group.

It is highly probable that further studies will reveal that many of the volatile compounds are present in varying amounts in several, if not all, of the species and forms of *Ophrys*, and that the quantitative relation between the components of the secretions is one critical parameter for the response of potential pollinators. Chemical studies of the composition of various volatile secretions from pollinating insects are in progress. Macrocyclic lactones [3], geranyl and farnesyl esters [2, 4] and monoterpenes [6]

Fig. 4. Capillary gas chromatogram of *enfleurage* from *Ophrys lutea*. The stationary phase of the column is silicone. The temperature is programmed from +132° to +213° by 2°C/min from the right side of the chromatogram.

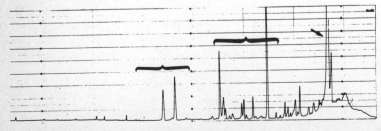

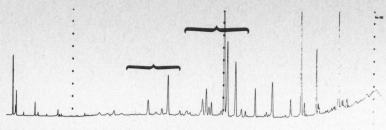

Fig. 5. Capillary gas chromatogram of *enfleurage* of *O. fuciflora*. Analytical conditions are the same as in fig. 4.

have been identified. So far, no cyclic sesquiterpenes have been found, but capillary gas chromatographic data indicate that such compounds may be present in very minute amounts [5].

Conclusions

As concerns the study of the stimuli, functioning in the assortative pollination of *Ophrys* flowers by sexually excited aculeate Hymenoptera males, the following conclusions can be drawn.

(1) The olfactory stimulation is doubtless a so-called primary key stimulus and the tactile stimulation is also of primary importance as a key stimulus. The characteristic epidermal structures on the upper side of the *Ophrys* labellum, as well as the labellum's solid construction, are indispensable for the tactilely guided movements and shiftings on it. The external morphology of the different types of hair is not of definite importance for these movements, whereas the mechanical properties and the stroking direction of the hairs, as well as

Fig. 6. Capillary gas chromatogram of a piece of one labellum of *O. fuciflora*. Analytical conditions are the same as in fig. 4.

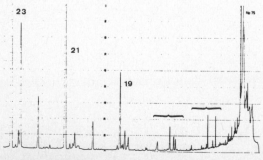

their length, are essential. Finally, it is the grouping of the hairs with different properties into fairly homogeneous pilose areas that forms the essential tactile-stimulatory basis for how the shiftings on the labellum will take place.

(2) With reference to the three main types of stimuli—olfactory, tactile and visual— and their effect on the Hymenoptera males, it seems that the olfactory stimulation may be called independent, whilst the tactile and the visual stimuli should be called dependent key stimuli, since they rely on the olfactory triggering for their ability to release. As regards the visually perceptible habitus, the labella of the *Ophrys* flowers appeal to the real fundamental of the innate releasing mechanism of a widely definable insect type, whilst, as regards the ability to give tactile and scent stimulation, they appeal to the behaviour-releasing mechanism of certain types of aculeate Hymenoptera in the male sex, and then, as regards scent stimulation, in a most specialized way. When speaking about the specialized release mechanism of the scent stimulation, one means specialization from the physiological point of view as well as from the taxonomical.

(3) The chemical compounds responsible for the attraction and stimulation of pollinators seem to be sesquiterpene hydrocarbons and alcohols mainly. Some simple aliphatic compounds may play a rôle too, perhaps in synergism with the terpenes. The large number of possible isomeric structures in these compounds, together with the high selectivity of the insect olfactory receptors, may account for the broad diversification of the plant–pollinator interrelationships. There seems to be a definite correlation between the distribution of sesquiterpenes and the pollination preferences, and it may therefore be possible to elucidate some of the taxonomical problems inside the *Ophrys*

genus by studying the sesquiterpene pattern of different forms.

(4) The morphological and chemical characteristics of the *Ophrys* flower, which distinguish it from flowers of other closely related orchid genera can, with good reason, be assumed to be adaptive, considering the regularity with which they occur within the large orchid group (Ophrydoideae) alluded to in connection with this specialized entomogamy. The *Ophrys* flower ought to be regarded as an adaptive flower. The *Ophrys* species are adapted for pollination to the male copulatory instinct of certain types of aculeate Hymenoptera.

This contribution forms part XI of the series Studies on Natural Odoriferous Compounds.

References

1. Bergström, G, Chem scripta 1973, 4, 135.
2. Bergström, G & Tengö, J, Chem scripta 1974, 5.
3. Bergström, G, Chem scripta 1974, 5.
4. Bergström, G & Tengö, J. To be published.
5. — To be published.
6. — To be published.
7. Bergström, G & Kullenberg, B. To be published.
8. Bergström, G. To be published.
9. Kullenberg, B, Zool bidrag Uppsala 1956, 31, 253.
10. — Zool bidrag Uppsala 1961, 34, 1–340.
11. — Zoon, suppl. 1, 1973, 9.
12. — Zoon, suppl. 1, 1973, 31.
13. Nelson, E, Gestaltwandel und Artbildung erörtert am Beispiel der Orchidaceen Europas und der Mittelmeerländer. Chernex-Montreux, 1962.
14. Priesner, E, Zoon, suppl. 1, 1973, 43.
15. Ställberg-Stenhagen, S, Chem scripta 1972, 2, 97.
16. Sundermann, H, Europäische und mediterrane Orchideen. Eine Bestimmungsflora. Brücke-Verlag Kurt Schmersow, Hannover, 1970.

Key-words: Aculeate Hymenoptera, Chemotaxonomy, Insect–plant relationship, Ophrys, Orchids, Pheromone, Pollination, Pseudocopulation, Sesquiterpenes

Biosynthesis and Phylogeny

Biosynthetic Pathways in Chemical Phylogeny

A. J. Birch

Research School of Chemistry, Australian National University, Canberra, Australia

Summary

Some aspects of the occurrences of related series of secondary metabolites are discussed in connection with possible "target" molecules or "target" activities. An attempt is made to arrive at methods of approaching the question of biochemical probabilities of changes in biosynthetic routes in order to assess the possible significance of a given compound as an evolutionary marker. The emphasis is on alterations of biosynthetic pathways, rather than on the structures of final molecules.

Morphological characters are determined by genetics and it has been frequenctly pointed out that the structure of a molecule is as much a morphological character as any other: indeed in cases such as colour or odour a molecular structure may represent the character in a more fundamental manner. There is, however, an added and useful dimension in considering a molecular structure: the biochemical pathway by which it has arisen. The general details of many such pathways are now being increasingly defined and understood, and in the subsequent discussion I intend to assume a knowledge of the background (e.g. [1–3]).

Three types of molecule can be broadly distinguished, from the viewpoint of biosynthetic information, although the classes merge in a number of examples. The first comprises the molecules of usually relatively high, or very high mol. wt, of metabolic function, typified by peptides which individually are "pure" compounds as usually defined. They are composed of a limited number of similar but distinct units, such as the 20 peptide amino-acids, which are necessarily joined in a defined order because of function. This order can frequently vary to some extent by replacement of one unit with another in positions where this does not eliminate a necessary function. The basis of the molecules is typically a chain or chains and the structure itself is largely sufficient to define a sequential biosynthesis. A second class, typified by many polysaccharides, contains also a string of related or possibly identical units (such as hexoses, pentoses, sugar acids, etc.) which are probably in rather a random order in the chains. The substances may consist of a mixture of molecules of different structures and different sizes. The structural aspects which can be observed to change are the proportions of the units, the numbers involved and the methods of junction, such as positions and stereochemistry with sugars, or the variety of oxidative condensations involved in lignins. Such substances are not definable as pure compounds and their functions are structural rather than catalytic.

The third class, of more interest to organic chemists, comprises the small molecules which are produced by very specific and definable sequences of chemical transformations from precursors available from metabolic pools. It includes the small units involved in the construction of the larger molecules above. While it is at this level that the most interesting and subtle information is available, it is also the most demanding in an understanding of the chemical processes involved in biosynthesis.

By considering such a biosynthetic sequence it should be possible to relate a naturally occurring compound not only to other compounds on the same sequence, but also to compounds on other related sequences. An assemblage of such information may be of use in considering the evolutionary relationships of plants. An alteration in a biosynthetic sequence, in terms of a chemical reaction or reactions, is therefore more fundamental than the structure of a substance as a morphological character per se.

Initially, rather small molecules such as encountered in "secondary" metabolites, are likely to prove useful for two reasons. One is the experimental one, that their detailed structures are more readily investigated than polymeric molecules such as peptides and nucleic acids, the other reason is the more fundamental one that they appear to mark a diversity rather than the similarity characteristic of compounds connected with basic metabolic processes. It is of course clear that variations in larger molecules such as peptide enzymes are of very great potential importance and will be increasingly used, as experimental methods become capable of examining rather small alterations in very large structures. Many such structures can of course be examined by physical and biochemical methods which yield results of great importance without necessarily defining total molecular structures. In other cases where structural requirements, rather than metabolic ones, predominate as in a number of polysaccharides, lignins, sporopollenins, etc., considerable diversity occurs and can be potentially examined by defining the nature of the units and of their junction. The emphasis however in the present discussion is on small molecules of completely known structure and probably definable biosyntheses.

Biosynthetic pathways depend on reactions which are being increasingly understood in terms of chemical mechanisms. The organic chemist has been able successfully to predict many biosynthetic pathways and processes on a chemical basis, in series such as those of alkaloids, polyketides, and C-methylated compounds. He can therefore successfully point to the fact that he has a good deal to say about the chemical probabilities of biochemical reactions. This basic mechanistic approach does not imply that enzymes are considered to be unimportant: they clearly are necessary both to make choices between alternative possible reactions and to catalyse these very specifically. Sometimes they give rise to reactions such as specific oxidations on saturated positions which the chemist would regard as mechanistically improbable. There exists therefore an action and interaction between what is chemically probable and what is biochemically known to happen: between laboratory experience which is expressed in mechanism theory, and the actual situation encountered in Nature, about which

there is still far from sufficient systematic information.

How and why do secondary metabolites arise? To begin with, there must be a biosynthetic sequence, set up for some good metabolic reason, which is capable of modification to give new products. An accidental genetic mutation affects a synthetic step in a sequence, and this is made visible by disappearance of a substance or by accumulation of an altered product. Probably the mutation involves loss, gain, blockage, or alteration of the specificity of an enzyme system. Loss of synthetic ability by mutation is presumably more common than gain or alteration, since it merely implies destruction of a system not the setting up of a new one. For example, the mutations delphinidin→cyanidin→ pelargonidin, which commonly occur, represent successive loss of ability to hydroxylate ring-B of an anthocyanin or its precursor. Whether a change will be manifest depends of course on whether the plant line continues in competition with similar ones; i.e. the alteration cannot in the long-run be deleterious and is preferably beneficial, or possibly has no significance.

The frequent occurrence of such alterations poses questions as to what functions secondary metabolites fulfil and how a plant can tolerate structural alterations in them. There are probably no universally applicable answers. We assume that in a situation in environmental equilibrium the substances must have some function, otherwise the "work" done in making them is wasted and is a drain on the resources and efficiency of the plant. It is becoming increasingly clear that many such do have functions which are usually connected with ability to compete with other plants. The simplest general assumption is that routes exist leading to "target" molecules of metabolic function, or to molecules with "target" general activities as structural materials, pigments, toxins, etc., and that modifications of these can lead to new substances which are also useful to the plant. In any case plants may not in the short run have to be as competitively efficient as animals and ecological situations are not always in long-term equilibrium. It is notable that animals do not usually produce equivalents of plant secondary metabolites.

In many cases the probable functions of secondary metabolites may be based only in a general way on structure type. For example, antioxidants could comprise almost any kind of readily oxidised phenol, and pigments could comprise any chemical class provided they absorb in the correct region of the spectrum required for the biological purpose, such as attracting insects or birds. Surface coating materials probably represent a desirable set of physical characteristics and the exact molecular composition is unimportant if these are met. For these among other reasons, mutations appear to be able to alter a number of metabolites, such as alkaloids, flavonoids, coumarins, triterpenes, waxes, etc. in a somewhat random fashion without any very notable effect on the viability of the plants.

With a knowledge of biosynthetic sequences it should be possible to determine at what point of a sequence the pathway to a compound has diverged, and what subsequent added transformations have occurred. Only rarely is a directly interrupted product of a definable pathway encountered. There may be several reasons for this. Owing to the incidence of inhibition by feed-back mechanisms or to biochemical reversibility of intermediates, some interruptions may merely cause loss of a product without observable trace, or accumulation of one much further back in the sequence than the actual break. For example, polyketide chains, probably as coenzyme-A esters, are rapidly reversible to the activated malonate-acetate units unless a stable, chemically irreversible stage is reached, usually by cyclisation or reduction. Mutation of *Penicillium islandicum,* which produces normally polyketide anthraquinones, either gives these or nothing. Reversal presumably leads to complete diversion of materials to another quite different pathway.

Many of the compounds observed are not directly on a major sequence because they are the result of further transformation processes of such intermediates. Sometimes they represent merely products of removal of a biochemically activating group, such as phosphate or coenzyme-A. Others result from rather probable chemical processes such as oxidations of readily oxidisable groups, or methylations, for example of phenols.

In a number of instances we can perceive the basic chemical reason why the synthetic sequence has been interrupted: it may be due to failure to add a required unit, lack of a cyclisation or alteration in the steric specificity of a ring-closure, or occurrence of an extra reaction leading off the original sequence. The reasons for development of additional steps may be related to the functionality of the product so obtained, depending for example on conversion of a hydrocarbon to a more polar compound, or possibly merely the necessity to dispose of the accumulating compound in a desirable manner.

When there are competing routes to different but related metabolites the picture based on considerations solely of precursor-product conversions may be too simple. As I have pointed out elsewhere [4] the genetics of production of pelargonidin and cyanidin (with the 4'- and 3',-4'-hydroxyls respectively) in the polyploid *Dahlia variabilis* seems only explicable by a consideration of the effects of dominant gene dosage, and therefore of enzyme dosage, on the relative rates of competing reactions in the two series catalysed by the same enzymes. If related, but not identical, substrates can be transformed, although at different rates, by the same enzymes, then the final products may be determined primarily by what precursors are available. In this instance there is no specific gene for one or the other anthocyanin, and the result seems to be determined by whether the withdrawal of pigment precursor is faster or slower than the rate of oxidation of the 4'-hydroxy precursor to the 3',4'-dihydroxy precursor. Unlike enzymes of metabolism which are closely tied to their single substrates, enzymes transforming secondary metabolites such as flavonoids may have much greater substrate elasticity.

It is important in attempting to assess the importance as a marker of a defined change of a biosynthetic process to decide on the degree of biochemical probability of its occurrence. Highly probable processes, for example the introduction of OH ortho or para to existing ones in a phenol, are likely to be developed repeatedly and independently and therefore to be of only minor significance. To attempt to survey some aspects of the approach to the subject, I should like to consider wider aspects of the sub-

jects of diterpenes and of flavonoids and anthocyanins. In view of the amount of work in these
well covered areas, it may seem unnecessary to
deal with them again, but they are favourite
subjects for very good reasons. Let us look first
at flavonoids in regard to related biosynthetic
areas.

General Relations of Flavonoids

As I suggested many years ago [5, 6] the carbon
skeleton arises from a cinnamoyl coenzyme-A
and 3 "acetate" units (acetyl-malonyl coenzyme-A) giving initially a chalcone. It is therefore related on the one hand to the C_6–C_3 plant
metabolites and on the other hand to the polyketides.

(1) C_6–C_3 and polyketide

The C_{15} nuclei of flavonoids such as (1) and
the C_{14} nuclei of stilbenes, merely mark a different mode of ring-closure of the same intermediate C_6–$C_3(C_2)_3$ (2), with decarboxylation
in the second case (3) (fig. 1). I also pointed out
that there are additionally a number of other
compounds with C_6–C_3–$(C_2)_n$ units with $n=2$
and upwards. Assuming the flavonoids to be the
target molecules, those with $n=2$ could be related
by loss of ability to add one C_2 unit (usually
resulting in α-pyrone derivatives such as kawain).
With $n>3$ there could be an acquired ability to
add more C_2 units. However, it is possible that
all of these routes are modifications of fatty
acid biosynthesis, with a cinnamoyl coenzyme-A
as a "starter" unit, the aryl polyketides then
being reduced or cyclised to give stable end-
products.

(2) C_6–C_3

A wide range of such structures can act as com-
petitors with different initial stages of the flavonoid route. Cinnamic acids or esters probably
result from reactions of the coenzyme-A derivative; coumarins result from direct oxidations
of the cinnamic acids; hydroxycinnamyl alcohols, which are lignin precursors, are presumably formed by reduction of the coenzyme-A
esters of the acids, and are therefore also very
directly competitive. The reduction products of
these such as eugenol, mark a much more spe-
cialised way of disposal of the cinnamyl alcohols. At a stage before the cinnamic acid,

and therefore more remote from flavonoids, are
compounds derived from phenylalanine or tyrosine, such as the isoquinoline alkaloids. Branching at a stage even before that are the ellagitannins based on gallic or on shikimic acid. A very
complex competitive network for precursors is
therefore possible. In order to obtain a meaningful picture of the result of a mutation, a wide
analysis of chemical types and their variations
is required. With the advent of new physical
tools, particularly of gas liquid chromatography
linked with mass spectrometry, such examina-
tions are increasingly feasible.

(3) Polyketides

Despite the prevalence of flavonoids in higher
plants, the incidence of pure polyketides is rare
in them compared to fungi. Some which may
have a rather close biogenetic relation to the
flavonoid route are the acylphloroglucinols or
acylresorcinols. These could represent mutations which result in the plant employing as
starter-units other available coenzyme-A esters,
such as acetyl, isobutyryl and related ones from
amino-acid degradations. In many cases the
cyclic products have undergone a further high
degree of C-methylation or C-isopentenylation,
among them humulone and lupulone in hops,
which also are mixed compounds with side-
chains related to products from valine, leucine
and isoleucine, indicating probable elasticity in
the synthetic enzymes, supported by the fact
that a similarly substituted chalcone also occurs in hops.

Modification of Flavonoid Biosynthesis

It is worthwhile to look at some biosynthetic
aspects of flavonoids which mark divergences
from the most basic pathways, and to attempt
to assess probable rarities of occurrence of alterations based on considerations of mechanisms. The substitution configuration with minimal biosynthetic steps is that of a 3,5-dihydroxy-
flavan-4-one or the related chalcone [5, 6].

Ring B

The hydroxyls in ring-B (cf 1) are almost certainly introduced in the order 4'-OH, 3'-OH,
5'-OH. The delphinidin (3',4',5'-trihydroxy)
pattern is much commoner in anthocyanins,
e.g. (3), than in other flavonoids, so that either

the enzymes involved in C-ring synthesis are highly selective of precursors, or else the third OH is introduced at a late stage. The fact already noted that mutations are almost always in the direction of loss of OH indicates that the synthetic sequence involves increasing OH, and also probably that the enzymes involved in the anthocyanin synthesis are usually sufficiently non-specific to deal with intermediates with 1, 2 or 3 OH groups in the B-ring. This conclusion does not assume that in no case is a specific gene involved for anthocyanin type; there indeed appear to be examples. Because of the rather complex mixtures often encountered with different degrees of hydroxylation and methylation it is probable that there is a widespread ability of synthetic enzymes to deal with a number of related but slightly different substrates.

An important specific oxidation of ring-B is in the 2'-position, notably of isoflavones and related rotenoids. Its introduction appears to be into a complete flavonoid rather than a precursor, and it is usually not in a position activated by another OH.

Ring A

The minimal biosynthetic pattern contains 3,5-hydroxyls, any other one requiring extra synthetic steps. The most interesting fact is that lack of one of these oxygens involves an extra reduction step, probably as I originally suggested [5, 6] in the β-polyketo-precursor of the chalcone (2), so that ring A cyclisation generates a resorcinol, rather than an orcinol derivative. The gene Y in *Dahlia* clearly determines such a reduction step. Provided that a group directly necessary for cyclisation is not involved, any carbonyl in a polyketide precursor could be reduced in this way, aromatisation occurring by dehydration rather than enolisation. Therefore 5- or 7-deoxy or 5,7-bisdeoxy flavonoid structures are possible but their synthesis involves one or two extra stages. A loss mutation under these circumstances will therefore restore the missing oxygen. The following biogenesis of flavanone itself is possible. Its oxidation product, flavone, in fact occurs in some primulas.

Further introductions of OH into one or both of the 6- and 8-positions present no mechanistic problems and mark "extra" developed stages. When introduced, such OH are frequently "cov-

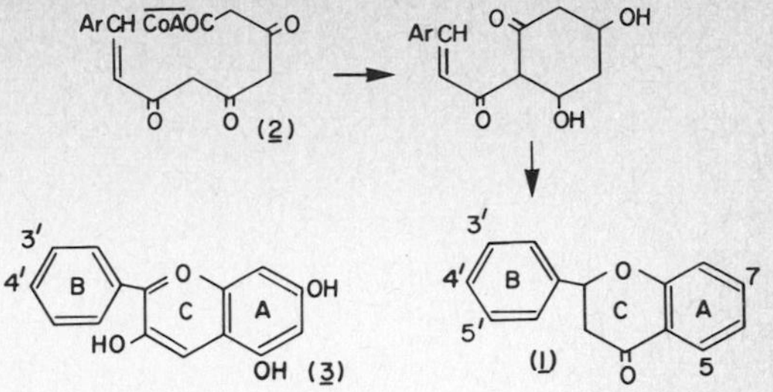

Fig. 1. Flavonoid biosynthesis.

ered up" as OMe. Other introduced groups are noted below.

Ring C

There is still some dubiety about exact sequences of oxidation-reduction processes and heterocyclic ring-closure leading to various classes of flavonoids, particularly of anthocyanins. Mechanistically, direct oxidations of chalcones can lead to the other classes [7]. Relatively independent formations appear to be flavone on the one hand and branching routes from flavanonol to flavonol or anthocyanin.

The catechin oxidation pattern is similar to the anthocyanin, but there is a higher degree of reduction. They might arise either from flavan-3,4-diols, themselves derived from flavanonols, or from anthocyanins. Although the order of events is not entirely clear in many of these cases the number and type of oxidations (oxygen-introduction or dehydrogenation) and reductions (hydrogen addition, oxygen loss) required from the initial stage is fairly clear.

A particularly interesting set of compounds are the 3-deoxyanthocyanins such as luteolinidin, where the 3-oxygenation stage characteristic of the great majority of anthocyanins has failed to develop, or been eliminated. Such anthocyanins might well be derived not from flavanonols but from flavanones through flavanols and flavones.

To summarise: in most cases at least the number of extra steps from the chalcone or its immediate precursor can be defined, if not always the order in which they occur. This information can be used to define a degree of complexity of the product, and also some idea can be gained of particularly significant alterations, either because of their mechanistic improbability or observed rarity.

Altered Skeletons

A significant and restricted process is rearrangement with transfer of ring-B to the 3-position yielding finally isoflavones or rotenoids. An even more restricted process, so far as is at present known, is the total loss of ring-B by a modification of phenol-oxidation to form the chromones with a 2-hydrogen atom [8]. A large series of significant processes involves C-alkylation either with methyl from methionine or isopentenyl (or polyisopentenyl) from the terpene pyrophosphate route. C-Methyls, when present, are invariably found attached to ring-A at the 6- and/or 8-positions, and terpene units to either ring-A or ring-B. Both types of grouping are found on oxygen as ethers, although this is far more common with the Me than C_5H_9, which is more usually attached to carbon. Terpene units are usually, although not invariably, found attached to rather highly oxygenated rings in line with the original concept of the process as an electrophilic attack by a cation.

Terpene units are also encountered in oxidised forms, or in degraded form as an unsubstituted furan ring. Some at least of these processes, especially the last, should be sufficiently rare on mechanistic grounds to form significant markers. Even rarer are the further oxidations and ring-closures leading to pterocarpans and coumestans, and the extra stages giving rise to rotenoids. On obvious biosynthetic assumptions, again the number and types of extra stages can usually be defined, although frequently not the exact order in which they occur.

Degree of Complexity

The "degree of complexity" of a given isolated substance can be defined by taking the number of stages from an arbitrary (ideally the first complete) molecule. To take a few examples, such as the rotenoid in fig. 2, taking in this case the arbitrary starting material as the tetrahydroxychalcone precursor polyketide (4), arbitrary since the stage of hydroxylation of ring-B at cinnamic acid or later is not clear.

Stages—rearrangement to isoflavone; introduction of 2 oxygens into ring-B; 3 O-methylations; oxidative ring-closure of rotenoid OMe; introduction of terpene unit from available sources; ring-closure of terpene unit. The result is 9 definable stages for the degree of

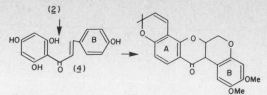

Fig. 2. The formation of deguelin.

complexity, with some doubt as to whether or not an extra reduction stage is involved producing the dihydrochromone in the rotenoid ring-closure.

Of this degree of complexity 5 steps can be classed as relatively trivial; 3 methylations, 1 (5'-)oxygen introduction and the terpene ring-closure; 2 are less trivial, the 2'-oxidation and isopentenyl introduction and 2 are outstandingly important; the isoflavonoid migration and the rotenoid ring-closure. From frequencies of occurrence however, even rather trivial processes such as methylation can be put into structural orders of probability. A 5-OMe is rarer than 7-OMe, possible because of lower reactivity due to hydrogen-bonding with a 5-carbonyl. A 3'-OMe is much more common than a 4'-OMe, possibly because a 4'-OH is frequently needed as a basis for various types of phenol oxidation [7].

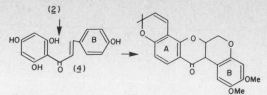

Fig. 3. Pisatin.

Pisatin (fig. 3) from the same polyketide precursor (2) requires the stages—carbonyl reduction in (2); ring-closure and aromatisation, which are probably linked; isoflavonoid migration; 2',5'-oxygenation; two methylations; methylenedioxy ring-closure; reduction of isoflavone ring; ring-closure of the furan; introduction of extra OH.

On this basis the degree of complexity is 11. It may be 12 if a reduction of isoflavone to isoflavanone is required. In terms of extra steps involved it is therefore more complex than the rotenoid above.

Of these processes, 3 are relatively trivial: 2 methylations and introduction of the 5'-OH; 3 are less trivial: 2'-oxygenation, methylenedioxy formation, furan ring-closure; others are mechanistically notable: removal of 5-OH, the isoflavonoid migration, and the processes leading to the pterocarpan nucleus with aliphatic OH.

It seems possible to gain some comparative estimates of the relative importance of at any rate related compounds by this sort of approach. Some correlations, presumably based on mechanisms, can be observed between certain types of structures. For example isoflavonoids have a 2'-oxygen more commonly than other types, so on this basis the presence of 2'-oxygen in a flavonoid would be more significant than in an isoflavonoid.

Diterpenes [9, 10]

Another rather different situation exists when there is one clearly definable route to a target molecule, contrasting with the network of routes just considered with flavonoids. "Gibberellic acid" is an example, although it is clear that a number of closely related compounds come under this title. They however differ only in the presence and type of extra oxygenated substituents, the skeleton and major structural features being the same in each case. The "extra" groups may be inserted rather late, or the synthetic enzymes may be insensitive or modifiable to the details of oxygenated substitution. Most diterpenes are related to this route.

It should be possible in such cases to define (i) the point on the sequence where the precursor of a given substance left it, and (ii) the number and types of subsequent transformations. Each departure represents an initial structural change, due to some definable change in enzymic specificity in determining a reaction. In the present instance they include alteration of absolute stereochemistry of ring-closure, failure to close a ring, the occurrence of deprotonation or of Wagner-Meerwein rearrangements in a cation reduction of a double bond, and failure to introduce an oxygen or its introduction in some other place.

It is difficult to assign any a priori degree of probability to a given mutational change, in the absence of information about how the enzymes work and how their specificity is changed. Possibly the best way to examine this is by the

statistics of occurrence. For example, in this series reversal of the absolute stereochemistry of the first mechanistically probable antiparallel A-B ring-closure seems to be frequent, whereas the change from *trans* to *cis* in this closure is never observed in the primary reaction. Total failure of a stage by destruction of an enzyme is a likely process, as already noted, but again in the absence of knowledge of the relative sensitivities of enzymes to such processes it cannot be concluded, a priori, that arrest at any one stage is more likely than at any other. Information on numerical occurrences of compounds is not readily available in the required form. If occurrences of individual structures are examined, then a rare structure counts as much as a common one, nevertheless this information is readily available and to some extent useful.

We assume that the gibberellic acid route is also required in the plant, no matter what modifications of it occur also and this can presumably be achieved by gene duplication. However, under these circumstances normal intermediates on the route are still subject to whatever biochemical control of production is necessary. It seems unlikely that any of them, except possibly the hydrocarbon (−)-kaurene, is likely to be accumulated to any extent. Such intermediates possessing correct stereochemistry would be required to undergo further metabolism to avoid control problems, although metabolism might be as chemically simple as the hydrolysis of a phosphate ester. However, the majority of bi- and tricyclic diterpene structures have the opposite absolute configuration (so called "normal" in consequence) to the gibberellic acid series, and this probably represents the reason for their formation [9, 10].

Various classifiable reactions occur with intermediates in such sequences, including carbonium-ion rearrangements of various types, and oxidations in almost every position of the molecule, with frequently subsequent ether or lactone ring-closures. Again there is no absolute method of classifying the "probability" of a given reaction, but the chemist can hazard guesses about comparative rarities by comparing examples on grounds of mechanism and occurrence.

Although there is no definable single target molecule or one unique route, somewhat similar classifications of alterations can be made in the triterpene series.

I can illustrate by several examples of a medium degree of complexity.

Manoyl oxide (fig. 4)

The normal precursor is replaced by the isomer of opposite absolute A-B configuration, branching occurring at the first stage of the process from geranyl geranyl pyrophosphate. The remaining two stages, although not inevitable are chemically highly likely and rather trivial. The hydrolysis product and ring-closure product with the final stereochemistry could indeed arise from chemical procedures not enzyme-catalysed. A slightly different possibility is that the original carbonium ion, leading to the $=CH_2$ by deprotonation, may instead be hydrated, the OH then participating in hydrolysis of the phosphate. There are probably only three stages from the open-chain precursor, with branching at the first cyclisation.

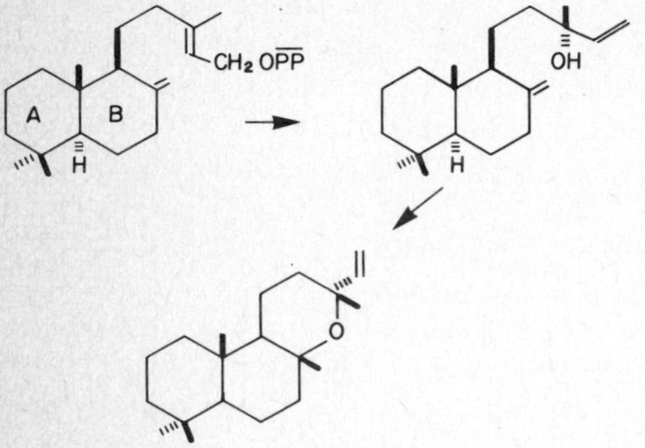

Fig. 4. The formation of manoyl oxide.

Thelepogine (fig. 5)

This alkaloid is a rather simple example of migration of Me, although not for the usual reason connected with carbonium ions generated by the A-B ring-closure or ionisation of the phosphate ester. In this case a specific oxidation of the angular CH (either to carbonium ion directly, or through OH) gives the appropriate basis for migration. It is therefore rather an unusual process leading to a cis A-B junction. The steps to produce the N-containing ring are not clear except that they involve oxidation, and probably amination via a carbonyl compound.

Although there are no automatic procedures,

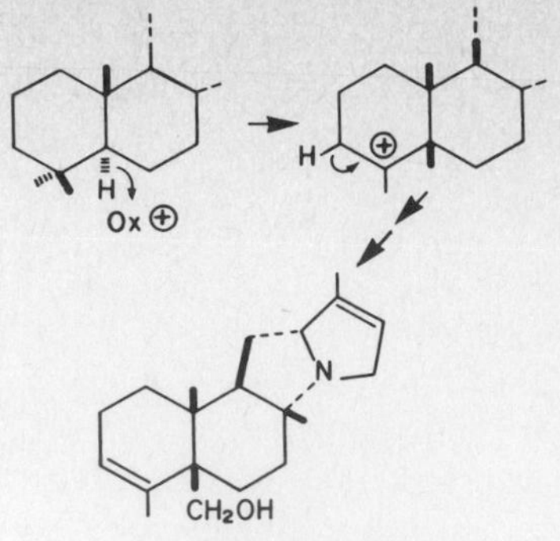

Fig. 5. Thelepogine biosynthesis.

or quantitatively definable factors, such considerations do introduce some systematisation to the superficial consideration of structures per se.

Andrographolide (fig. 6)

The initial A-B ring-closure in this case has the correct stereochemistry, and the reason for diversion must be a secondary process, although this might only be hydrolysis of the phosphate. The 3α-OH may, however, mark a different type of ring-closure via the open-chain terpene epoxide, similar to that in the triterpene series. The secondary steps, not necessarily in order, are: hydrolysis of phosphate; introductions of 4-OH, one with rearrangement of unsaturation; further oxidations, possibly via an aldehyde; ring-closure to lactone.

Oxidation of the gem-Me_2 group is fairly frequent, either axially or equatorially largely according to the absolute configuration. The other types of oxidation encountered here are rarer, although the lactone formation may be

Fig. 6. The biosynthesis of andrographolide.

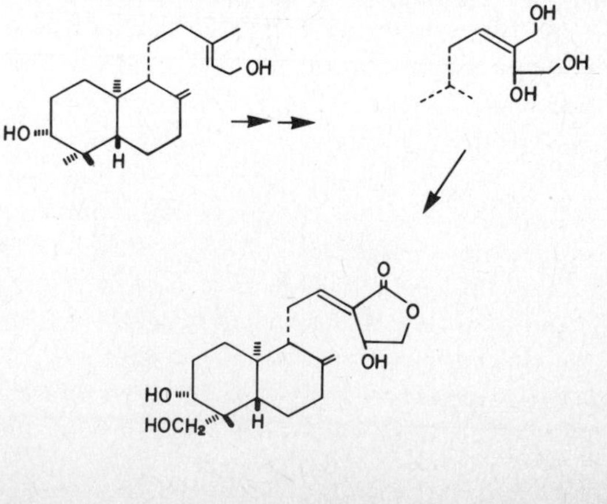

related to the type of reaction in the fairly frequent occurrence of a furan ring in this position.

Lichen Products

The lichen *Ramalina siliquosa* shows a relationship between depsidone components and growth situation and conditions [11]. The depsidones are polyketides composed of two methylorsellinic acid units in varying states of oxidation. The most primitive molecule encountered must be hypoprotocetraric acid with 4 Me, (two the residues of acetate units and two from methionine). This component is found under most favourable growth conditions. The next zone contains norstictic acid (Me, CHO, Me, CHO) which could be derived from the former by oxidation of two specific Me. In the harshest habitat occurs stictic acid, which is a mono-*O*-methylated norstictic acid, and presumably is derived from it in one stage. Two lichen types of very minor occurrence contain protocetraric acid (Me, CHO, CH_2OH, Me) derivable by oxidation of two Me of hypoprotocetraric acid, and salazinic acid (Me, CHO, CH_2OH, CHO) derivable either from protocetraric acid by oxidation of Me to CHO or from norstictic acid by oxidation of Me to CH_2OH. It seems clear at any rate that progression from mild to harsh conditions is marked by incidence, sequentially, of extra biogenetic stages in the major components of the forms encountered.

Two points of general interest emerge. A small proportion of lichens survive with no depsidone at all, showing that the presence of the latter must have only a marginal effect. Also, two strains can occur side-by-side, so that the effects are genetic and not directly due to environmental effects on biosynthesis. How far such situations occur with other organisms and other metabolites awaits further systematic investigation.

Conclusions

No automatically applicable procedures exist to assess the significance of alterations of biosynthetic pathways in connection with plant evolution. However their detailed consideration in terms of chemical probability and known occurrence may assist in defining relationships between organisms. To have much chance of success as many different pathways as possible must be examined, and far more systematic information on occurrences is desirable.

References

1. Biogenesis of natural compounds (ed P Bernfeld). Pergamon, Oxford, 1963.
2. Biosynthesis of terpenes, steroids and acetogenesis (ed J H Richards & J B Hendrickson). Benjamin, New York, 1964.
3. Biosynthetic pathways in higher plants (ed J B Pridham & T Swain). Academic Press, London, 1965.
4. Birch, A J, 17th IUPAC plenary lectures, Munich 1959, p. 73. Buttersworth, London, 1960.
5. Birch, A J & Donovan, F W, Aust j chem 1953, 6, 360.
6. Birch, A J, Fortschr Chem org Naturstoffe 1957, 14, 186.
7. Pelter, A, Bradshaw, J & Warren, R F, Phytochemistry 1971, 10, 835.
8. Birch, A J & Thompson, D J, Aust j chem 1972, 25, 2731.
9. Birch, A J, in Chemical plant taxonomy (ed T E Swain) p. 141. Academic Press, London, 1963.
10. Birch, A J, Pure and applied chem 1973, 33, 17.
11. Culberson, W L & Culberson, C F, Science 1967, 158, 1195.

Discussion

Lavie: Discussing the question of target molecules or molecules having a definite aim in the life cycle of the plant, one wonders whether the plant always produces substances that are needed for its development. We have seen at one end of the spectrum, the orchids forming patterns which mimic the female wasp or bee for its own pollination, and alternatively by a combination of enzymes due to various crossings, compounds which were not previously formed in nature are produced in the plant. The plant seems not to have a direct way to get rid of unnecessary material from its organism, may be that the continuous change of leaves for example may be one way to do that. Indeed it seems that often compounds having no value to the plant are formed and stored away and then discarded.

Birch: The problem is that plants must "work" to produce such materials, animals discard useless material taken from other organisms. However, plants may well produce compounds almost by accident to be discarded in the way discussed. Since environmental circumstances, notably light and temperature, vary rapidly, it may well be that there is frequently an imbalance between production of basic materials and their utilisation. In order to keep optimum production under minimum conditions, it may be necessary to have excess production under some conditions. The *Eucalyptus* leaf oils, for example, may exist basically to shed the effects of excessive Australian sunshine.

Ourisson: (1) By reviewing the distribution of diterpenes in the plant kingdom, rather than counting structures, one sees clearly that the "normal" series is by far the most restricted one. It was called "normal" only because it happens that the most abundant diterpenes, those of Conifers, are of these series; and they have been of course the first ones to be studied (with a few Labiatae diterpenes, also happening to be of the "normal" series). On the other hand, the *ent*-series, that of gibberellins is by far the most widely distributed one. This does not detract from any of Birch's conclusions, but I just want to correct any wrong impression.

(2) The production of huge amounts of terpenes may not be a positive evolutionary factor for the reason that it leads to energy disposal, but certainly nobody can dispute the fact that the production of terpenes from CO_2 is consuming huge amounts of solar energy. I believe this does not make it possible to accept the idea that living systems work always on the principle of maximum economy! There are certainly more economical ways to fulfill whatever role the pine oleo-resin may play, than producing α-pinene or β-carene (be it only by making limonene).

Birch: As I mentioned, a different picture may be obtained from a study of occurrences rather than of structures, but information in the former category is hard to assemble. However, in data taken from a review by Hanson, structures of bicyclic and tricyclic diterpenes show preponderantly the "normal" (nongibberellic acid) stereochemistry, whereas with almost no exceptions, tetra- and pentacyclic triterpenes are derivable from the gibberellic acid absolute configuration. I have pointed out the probable significance of this elsewhere [6].

Nobel 25 (1973) Chemistry in botanical classification

Lüning: Could you not think of the absence or presence of a certain compound as a mere presence or absence of abilities to metabolize the compound further?

Birch: The brief answer is yes, but the problem remains as to how and why it is there. The last compound in a sequence is presumably the last because it is not further changed.

Harborne: Dr Lüning's suggestion that alkaloids accumulate because plants cannot turn them over cannot apply to the flavonoids. Recent studies by Grisebach and his coworkers in parsley have shown that the enzymes for further metabolism are produced in young seedlings at the very same time as the biosynthetic enzymes are being formed.

Birch: There appears to be a considerable turnover of some secondary metabolites, as is apparent to anyone who has tried to label them biosynthetically. It is a possible speculation that some may act as storage materials; alternatively some control mechanisms may be required for compounds which may be toxic in high concentrations.

Kjaer: I would like to emphasize the irrelevancy of applying teleological arguments to biogenetic thinking.

Birch: The idea of intention and choice may not be entirely irrelevant, providing the process is clearly recognized. No conscious decision can occur, but statistically and with great wartage of undesirable types, a result may be achieved in the end, in response to a need, which is effectively similar to much a decision.

Hegnauer: Biological interpretation of characters, including chemical ones, does not mean that we imply that evolution is directed by intention. It means only that we are searching for principles and mechanisms governing selection. If, e.g., a given constituent was proved to inhibit phytopathogenic microorganisms in concentrations in which it occurs in a given plant one aspect of its presence in this plant can be understood, it is a factor of resistance against a certain number of potential pathogens. This does not mean that the plant made the constituent in order to become resistant, but only that resistance is a consequence of its presence.

Some Aspects of Organic Geochemistry

G. Ourisson

Institut de Chimie, Université Louis Pasteur, Strasbourg, France

This lecture was presented orally, replacing the review on pp. 129–134.

After describing the importance of organic matters in sediments, a few examples were given of the study of isolated, identified fossils. A general summary was given of the types of substances identifiable in sediments, and of the types of reactions they can undergo.

Newer results were given on the ubiquitous presence, and frequent prevalence, of derivatives of hopane and its 17 and 22 diastereoisomers, containing from 27 to 35 carbon atoms. Their origin was discussed, and it was proposed that they may indicate the widespread occurrence of microbial degradation of plant organic matter before inclusion into the sediments. This was taken to imply that exceptional circumstances are required for plant fossils to yield "chemical fossils" of possible use in helping with phylogenetic problems.

The work described was helped by a generous grant from ELF-ERAP, and carried out by P. Albrecht, P. Arpino, O. Sieskind, C. Spyckerelle, A. Ensminger, A. van Dorsselaer, M. Dastillung, W. Michaelis, and H. Knoche.

HOPANE

Discussion

Fredga: On behalf of the organization committee I will express our most sincere gratitude for this fascinating lecture. On an early stage of the preliminaries some botanists suggested that somebody should tell us a little about the distribution of diterpenes. Of course Professor Ourisson would be the right person to do that and so I wrote to him making a suggestion. He answered that he had intended to speak on another subject but if we wanted a lecture on diterpenes he was prepared to give it. Then I wrote that if he preferred to speak on another subject, he should of course do so. Now we have got two lectures and we are very glad for that.

Herout: I will add some facts, which speak against two rather pessimistic views of Ourisson. In our laboratories (ČSAV Praha) we recently studied two specimens of Czechoslovakian brown coal (lignites). The first one originated from Northern Bohemia (not more than 50×10^6 years old), and contained friedelin as the most typical component (with a lot of more or less aromatized derivatives). The second originated from Nováky (Slovakia, the age is higher, not exactly defined). It was characterized by quite a high content of iosene (a saturated hydrocarbon with the carbon skeleton of phyllocladene). I can summarize that in that case there exists a distinct difference, demonstrating the origin from different plant material.

Ourisson: It is indeed probable that peat for example provides an environment capable of preserving efficiently organic matter. What I have said may be general only in so far as the majority of sediments containing fossils originate probably from deposition in shallow waters. We also find friedelin in some samples, but in very small amounts, and not accompanied by friedelane. We have also found in two cases triterpenes coming most probably from higher plants (isoarborinol, arborinone).

Swain: It is unfortunate that attention has been focussed on the hydrocarbons. To throw light on higher plant evolution, it will be necessary to examine higher plant compounds and/or their degradation products, such as flavonoids, alkaloids etc. One such search by Leo and Baarghorn (Science 1969) for presumed lignin-derived aldehydes showed an interesting variation in the proportion of p-hydroxybenzaldehyde : vanillin : syringaldehyde, indicating that the older sediments contained lignin derived perhaps, from mosses.

Ourisson: (1) Triterpenes may be ubiquitous, but individual triterpenes may be good chemotaxonomic markers.

(2) We were not hoping to get triterpenes, and they were forced upon us.

(3) I agree that the evidence presented cannot be of any value for the study of evolution in plants and that it is possible that other classes of substances might give other results more useful diagnostically.

Lavie: Since until now only pentacyclic triterpenes seem to have been isolated from shales, and these compounds are known not to be degraded in nature, I wonder whether they could be regarded as a kind of "fossil compounds', i.e. compounds which remained in the plants during their evolution. The tetracyclic triterpenes are usually more easily degraded or transformed in the plant kingdom, and are by far more widespread, and therefore they could be regarded as products of subsequent developments in the evolution of plants.

Ourisson: We have isolated also sterols and 4-α-methyl sterols, which must have been derived from lanosterol or cycloartenol, both tetracyclic triterpenes. Frankly, I do not think that organic geochemical results enable one to throw light on your remark, but I also do not think you can substantiate it with facts.

Bu'Lock: Can I ask about the two extreme ends of the time-scale-organic components of pre-Cambrian deposit on the one hand and of absolutely recent sediments on the other.

Ourisson: The pre-Cambrian deposits do often contain detectable amounts of recognizable substance of biotic origin. Calvin and Eglinton are among those who have tried to go as far back as possible in time; we have studied only one old sediment, 1.5×10^9 years old and containing phytane and pristane. But the problem of contaminations becomes very dangerous.

The Bristol group are studying recent lakes and muds. A C_{31} hydrocarbon appears to be present in some samples; I do not know about higher homologs or analogs. Of course, we shall have also to check our data on recent sediments, or even on pre-sedimentary organic matter.

Natori: In the Gramineae the arborane derivatives generally coexist with the migrated hopanes, fernane derivatives. In ferns migrated hopanes again coexist generally with hopanes. Have you not ever isolated migrated hopanes?

Ourisson: No, and for instance arborane and fernane are not present in the hopane containing hydrocarbon fraction. We have also none of the methyl-ethers characteristic of Gramineae. But, in the hydrocarbon fraction, the minor peaks have not all been identified.

Application of
Combined Chemical Characters

Is the Order Centrospermae Monophyletic?

*A Review of Phylogenetically Significant
Molecular, Ultrastructural and Other Data
for Centrospermae Families*

T. J. Mabry

The Cell Research Institute and the Department of Botany, The University of Texas at Austin, Austin, Tex. 78712, USA

Summary

The Order Centrospermae as classically constituted contains both anthocyanin- and betalain-producing families. Here, the molecular and ultrastructural data are reviewed which bear upon the phylogenetic relationships of such anthocyanin families as the Caryophyllaceae and Molluginaceae to the 9 betalain families.

Possibly no better example exists in Nature where macro- and micro-molecular data are available to combine with ultrastructural, serological, embryological and morphological information to suggest evolutionary relationships for higher taxonomic categories than for the families most systematists align in the Order Centrospermae. The well known name Centrospermae is retained here in the introduction for discussion purposes; most systematists now employ the nomenclaturally more acceptable name Caryophyllales for the Order or assemble the families into two closely related Orders, the Caryophyllales and Chenopodiales.

The question of whether the Centrospermae is mono- or poly-phyletic stems in a large measure from the discovery in the last 15 years that the remarkable distribution of the red-violet and yellow betalains also represents the distribution of a unique biosynthetic pathway in higher plants [1–4]. Betalains have been detected only in the red-violet pigmented species so far examined which belong to 9 plant families (see table 1); these families appear to be evolutionarily closely related to each other and to the anthocyanin families Caryophyllaceae and Molluginaceae on the basis of several other lines of evidence. Nevertheless, the two classes of biosynthetically and structurally distinct pigments, betalains and anthocyanins, are mutually exclusive, never occurring together in the same species or even in different species of the same family [5].

It is interesting to note that the betalain pathway (fig. 1) has enabled 9 families to have pigments which correspond in their visible chromophoric properties with some of the anthocyanins (fig. 2); moreover, just as mixtures of different anthocyanins, sometimes with a co-pigment such as a flavonoid, produce a wide range of floral and leaf colorations so do mixtures of the yellow betalains (betaxanthins and betalamic acid) with the red-violet betacyanins account for a variety of violet, red, orange and yellow colors in the fruits, flowers, leaves and notably stems of members of the 9 betalain families.

Some of the 9 betalain families whose affinities were difficult to recognize on morphological grounds (e.g. Cactaceae, Didiereaceae) were allied with a greater degree of confidence with the other betalain families once their pigments were identified. For example, Cronquist in 1957 [6] placed the Didiereaceae in the Euphorbiales; later when it was known that members of this family contained betalains Cronquist and other systematists recognized its relationships to the other betalain families. Further support for this alignment and for the close relationship of all the betalain families was provided by Jensen in 1965 [7] who used antisera from the genus *Alluaudia* (Didiereaceae) to demonstrate serologically a close relationship of the Didiereaceae to all the betalain families, especially the Cactaceae and Portulacaceae (fig. 3).

The long-held view that the anthocyanin families Caryophyllaceae and Molluginaceae

Table 1. *Recent interpretations of the betalain families and those embryologically-morpho-logically-ultrastructurally related but anthocyanin-producing families*

Cronquist [8]	Takhtajan [17]	Mabry, Taylor & Turner [2] Behnke & Turner [9]
Order Caryophyllales	Order Caryophyllales	Order Chenopodiales
Aizoaceae[a]	Aizoaceae[a]	Aizoaceae[a] (including the Tetragoniaceae[a])
Amaranthaceae[a]	Amaranthaceae[a]	Amaranthaceae[a]
Basellaceae[a]	Basellaceae[a]	Basellaceae[a]
Cactaceae[a]	Cactaceae[a]	Cactaceae[a]
Chenopodiaceae[a] (including Dysphaniaceae)	Chenopodiaceae[a] (including Dysphaniaceae)	Chenopodiaceae[a] (including Dysphaniaceae if it proves to have betalains)
Didiereaceae	Didiereaceae[a]	Didiereaceae[a]
Nyctaginaceae[a]	Nyctaginaceae[a]	Nyctaginaceae[a]
Phytolaccaceae[a]	Phytolaccaceae[a]	Phytolaccaceae[a]
Portulacaceae[a]	Portulaceae[a]	Portulacaceae[a]
Caryophyllaceae[b]	Tetragoniaceae[a]	Order of Caryophyllales
Molluginaceae[b]	Caryophyllaceae[b]	Caryophyllaceae[b]
	Molluginaceae[b]	Molluginaceae[b]
	Bataceae[c]	
	Gyrostemonaceae	
	Halophytaceae[a]	
	Hectorellaceae	

[a] Known betalain families; data for Halophytaceae are not published elsewhere.
[b] Known anthocyanin families.
[c] Neither type of pigment.

are related to those which produce betalains is based upon their mutual morphological and anatomical characteristics including, notably, their distinctive embryological features [8]. Indeed, these classical data are sufficiently unique to suggest that the 9 betalain families and the Caryophyllaceae and Molluginaceae have a common ancestor. The most striking new

evidence to support a close relationship of the latter two families to the 9 which produce betalains are the ultrastructural data for sieve-tube plastids. So far as is known except for these 11 families (including the Dysphaniaceae) nowhere else in the angiosperms do the sieve-tube plastids contain ringlike inclusions composed of proteinaceous filaments [9, 10]; indeed one must look to the gymnosperms (Pinaceae) for similar ultrastructural features! The absence of these characteristic sieve-tube plastids in other families which some systematists place in the Centrospermae, including

Fig. 1. The biogenesis of betalains such as the red beet pigment betanin involves an oxidative extra-diol cleavage of L-dopa. The suspected intermediate, betalamic acid, has now been detected as a naturally occurring pigment in betalain-producing species [4].

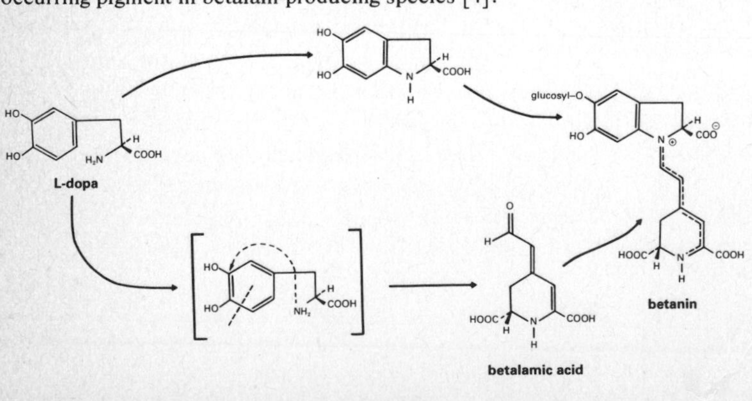

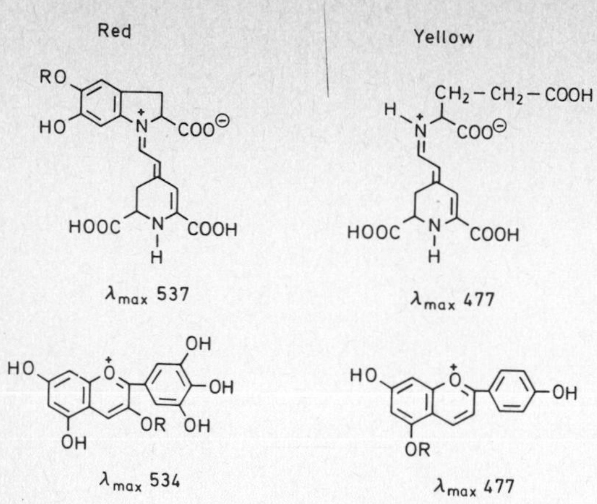

Red

λ_{max} 537

Yellow

λ_{max} 477

λ_{max} 534

λ_{max} 477

Fig. 2. The visible absorption maxima of typical red and yellow betalains (top row) are similar to certain red and yellow anthocyanins (bottom row) suggesting independent selection in different plant families for pigments with similar visible chromophoric properties.

especially two non-pigmented (i.e., neither betalains nor anthocyanins) families, Bataceae [11] (table 2) and Theligonaceae (*Theligonium*),

Fig. 3. Using antiserum from *Alluaudia procera* (Didiereaceae) Jensen [7] serologically demonstrated 100% correspondence using an antigen solution from the same species (bottom line); and notably correspondence with some species from all the betalain families. Each line on the left of the figure denotes one species which was tested versus *Alluaudia procera*; for example, 5 species from the Cactaceae were tested. (Reproduced from Botanische Jahrbücher [7] by permission of Prof. Jensen.)

and the anthocyanin-containing Polygonaceae, suggests that they are not closely related to the eleven centrospermoid families. Furthermore, the report [12] that Bataceae contains myrosin, an enzyme associated with glucosinolates, plus certain morphological features [13] suggest a relationship of this family to the Capparales. On the basis of morphological [14] and chemical [15] evidence, *Theligonium* has recently been placed in the family Rubiaceae.

The DNA-DNA and DNA-RNA hybridization data [1, 16], which we obtained in connection with our efforts to determine the extent of genetic homology among species which belong to the betalain families relative to species which are members of anthocyanin-producing families, especially the Caryophyllaceae, are of interest in connection with the present discussion.

The DNA-DNA hybridization results were somewhat surprising in that only between varieties of the same species (the red and sugar beet varieties of *Beta vulgaris*) was competition detected; that is, no differentiation between genera, let alone higher taxonomic categories, was observed [1, 16].

Next, ribosomal RNA (rRNA) was used for hybridization with DNA since it is well known that the cistrons for rRNA are relatively conserved compared with the average DNA cistrons. The results obtained for these DNA-RNA hybridizations indicated that excess rRNA from

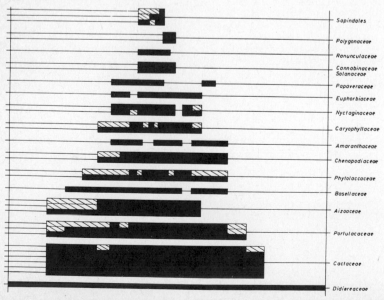

Table 2. *Sieve-tube plastids of Caryophyllidae*

 Plastids with ring-shaped bundles of filaments	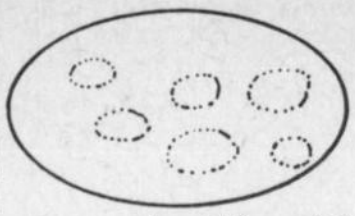 plastids with starch grains

Order 27: CARYOPHYLLALES

Phytolaccaceae	Petiveria alliacea	
	Phytolacca acinosa (3)	
	Phytolacca dioica (3)	
	Rivina humilis	
	Trichostigma peruviana (3)	
Bataceae		Batis maritima
Nyctaginaceae	Mirabilis longiflora (3)	
	Pisonia brunoniana (3)	
Molluginaceae	Mollugo verticillata	
Aizoaceae	Sceletium namaquense (3)	
	Sesuvium sp.	
Tetragoniaceae	Tetragonia expansa (1)	
Cactaceae	Pereskia aculeata	
	Pereskia grandiflora (3)	
Portulacaceae	Portulacaria afra (3)	
Basellaceae	Boussingaultia baselloides (3)	
Didiereaceae	Alluaudia humbertii (3)	
Caryophyllaceae	Cerastium biebersteinii	
	Cucubalus baccifer ⎫	
	Dianthus caryophyllus ⎬ (3)	
	Lychnis chalcedonica ⎭	
	Melandrium rubrum	
	Silene cucubalus (3)	
	Spergula arvensis	
	Vaccaria pyramidata	
Amaranthaceae	Aerva scandens ⎫ (3)	
	Celosia argentea ⎭	
Chenopodiaceae	Allenrolfea occidentalis	
	Atriplex cf. acanthocarpa	
	Atriplex reptans	
	Dysphania myriocephala	
	Beta vulgaris (2)	
	Hablitzia tamnoides (3)	
	Salicornia bigelovii	
	Salicornia virginica	
	Suaeda maritima	

Order 28: POLYGONALES

Polygonaceae		Polygonum bistortum (3)
		Polygonum fagopyrum (4)
		Rumex patientia (3)

Order 29: PLUMBAGINALES

Plumbaginaceae		Limonium arborescens
		Plumbago europaea (3)

Families in italics contain betacyanins and/or betaxanthins; key to authors that first depicted plastids: (1), Falk (1964); (2) Esau (1965); (3) Behnke (1969); (4) Arsanto (1970); without number present author.
    (Reproduced from Taxon [9] by permission of Professors Behnke and Turner; for the references referred to at the bottom of the table see [9].)

a distantly related yeast [16] could not inhibit the labeled 16S rRNA from the betalain-producing spinach (*Spinacia oleracea*, Chenopodiaceae) from hybridizing with the DNA from the same plant; on the other hand, excess rRNA from the distantly related pea, *Pisum sativum*, a member of the anthocyanin-producing Fabaceae, did reduce the homologous spinach-rRNA/spinach-DNA hybridization about 18% (fig. 4).

Some of the crucial experiments involved unlabeled rRNA from the Caryophyllaceae for competition experiments in the "spinach system". Excess rRNA from either of 3 genera from the Caryophyllaceae (*Dianthus, Cerastium* and *Stellaria*) and one species from the Bataceae reduced the spinach rRNA/spinach DNA interaction 10–15% and 17% respectively. Significantly, however, excess rRNA from several betalain-producing families reduced the spinach-rRNA/spinach-DNA hybridization in every case only 7% or less (see fig. 4). That is, the rRNA from betalain-plants showed 93.5% or more homology with the rRNA from the test system, *Spinacia oleracea* (Chenopodiaceae) [1, 16].

With regard to the initial question posed here, namely, "Is the Centrospermae monophyle-

tic?", it is the view of this author that the 11 Centrospermae families were derived from a common ancestral line from the angiosperm ancestor; this major evolutionary line subsequently gave rise to two lines prior to the origin of floral pigments (fig. 5).

Some of the research reviewed herein was supported by the NSF (Grant 29576X), the NIH (Grant HD-04488) and the Robert A. Welch Foundation (Grant F-130).

Fig. 5. The available classical, molecular and ultra-structural data suggest that the Centrospermae families were derived from a common ancestor prior to the origin of floral pigments in the Centrospermoid line.

Fig. 4. In all experiments, 0.6 μg of ³H-spinach 16S rRNA (4000 cpm/μg) were incubated with 12 μg of spinach DNA bound on a nitrocellulose filter in 0.1 ml of formamide: 4 SSC (1:1) at 40°C, in the presence of increasing amounts of rRNA from other taxa. After 38, each filter was washed with 2 ml of formamide SSC solution for 5 min at 40°C, then with 2 more ml SSC for another 5 min. Each value in this figure represents the average obtained from 3 determinations. The ratio of labeled rRNA to DNA in the hybrid in the absence of competitor rRNA was 0.33%, representing 7% binding of the input labeled rRNA. All values have been corrected for background binding using calf thymus DNA filters.

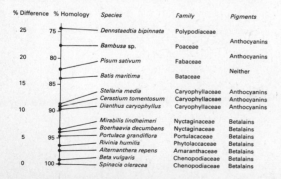

% Difference	% Homology	Species	Family	Pigments
25	75	Dennstaedtia bipinnata	Polypodiaceae	
		Bambusa sp.	Poaceae	Anthocyanins
20	80	Pisum sativum	Fabaceae	Anthocyanins
		Batis maritima	Bataceae	Neither
15	85			
		Stellaria media	Caryophyllaceae	Anthocyanins
		Cerastium tomentosum	Caryophyllaceae	Anthocyanins
10	90	Dianthus caryophyllus	Caryophyllaceae	Anthocyanins
		Mirabilis lindheimeri	Nyctaginaceae	Betalains
		Boerhaavia decumbens	Nyctaginaceae	Betalains
5	95	Portulaca grandiflora	Portulacaceae	Betalains
		Rivinia humilis	Phytolaccaceae	Betalains
		Alternanthera repens	Amaranthaceae	Betalains
		Beta vulgaris	Chenopodiaceae	Betalains
0	100	Spinacia oleracea	Chenopodiaceae	Betalains

References

1. Mabry, T J, Kimler, L & Chang, C, in Recent advances in phytochemistry (ed V C Runeckles & T C T'so) vol. 5, pp 105–134. Appleton-Century-Crofts, New York, 1973.
2. Mabry, T J, Taylor, A & Turner, B L, Phytochem 1963, 2, 61.
3. Mabry, T J, in The chemistry of alkaloids (ed S W Pelletier) pp 719–746. Reinhold, New York, 1970.
4. Kimler, L, Larson, R A, Messenger, L, Moore, J B & Mabry, T J, Chem comm 1971, 1329.
5. Kimler, L, Mears, J, Mabry T J & Rösler, H, Taxon 1970, 19, 875.
6. Cronquist, A, Bull jard bot état 1957, 27, 13.
7. Jensen, U, Bot Jahrb syst 1965, 84, 233.
8. Cronquist, A, The evolution and classification of flowering plants. Houghton Mifflin, Boston, 1968.
9. Behnke, H & Turner, B L, Taxon 1971, 20, 731.
10. Behnke, H D, Bot rev 1972, 38, 155.
11. Mabry, T J & Turner, B L, Taxon 1964, 13, 197.
12. Schraudolf, H, Schmidt, B & Weberling, F, Experientia 1972, 72, 1090.
13. Eckhardt, Th, Ber deut bot Ges 1959, 72, 411.

14. Wunderlich, R, Österr bot Z 1971, 119, 329.

15. Kooiman, P, Österr bot Z 1971, 119, 395.

16. Chang, C, Nucleic acid hybridization studies among Centrospermae. Ph.D. thesis, University of Texas, Austin, 1971; Chang, C & Mabry, T J, Biochem syst. In press.

17. Takhtajan, A, Flowering plants—origin and dispersal. Robert Cunningham & Sons, Alva, 1969.

Discussion

Heywood: So far I have not participated in the discussion on this particular problem, unlike Cronquist and Takhtajan. The solution proposed—that of a subclass containing two orders—reminds me of the title of a lecture by my predecessor in the chair of Botany at Reading, namely the "inflation of taxonomy". This situation might be termed inflation of categories! However, I should like to draw attention to the work of Tempère on specialized phytophagous beetles where representatives of two different groups—Curculionidae and Chrysomeloidae occur on a spectrum of hosts in the Centrospermae including the Caryophyllaceae. However, this no more proves the relationship between the members of the group than does any other line of evidence, chemical, fine structural or morphological. Addition of further evidence will not of itself solve the problem. The important thing is to recognize the differences in the betalain- and anthocyanin-producing groups. Whether treated as orders or suborders is not very important since decision on rank is not a precise matter. If one were to compare the various angiosperm orders in any classification in terms of degree of taxonomic differentiation, degree of evolutionary differentiation or even monophylesis, we would find they were widely different. Ranking is a matter of judgment and not susceptible to precise and rigid decision. It would be intellectually dishonest to suggest otherwise.

Mabry: I could not agree more, that the matter of rank is not the important question here. Let us not lose sight of the real questions with which we are all concerned; namely the phyletic relationships and evolutionary processes among Centrospermae families.

Hegnauer: What this means for the classification of Centrospermae is not yet clear. In this respect many views differing more or less in the interpretation of presently known facts and their impact on classification were offered during the past 10 years. It is time to stop speculations, and to extend practical work. Only after we know from all members (i.e. species and genera) of Centrospermae, especially the ones regarded to be more primitive, whether they produce betalain or anthocyanins, will speculations about the phylogenetic meaning of betalain production in this part of angiosperms become more fruitful.

Mabry: Certainly additional taxa should be examined, especially key genera and sub-families in centrospermoid families which have not been tested for their pigments. Such a project is under way presently. At the same time, we must be prepared to accept the data.

Turner: I would disagree with Professor Heywood's previous remarks in the sense that the question of rank is an intellectual one, that is, if one is attempting to show phyletic or cladistic relationships, assignment of rank is an important activity. For example, if one does not treat the Caryophyllaceae and Molluginaceae as constituting a special category in themselves (as Cronquist and Takhtajan have not done) then the user of that system (or at least the intellectual connoisseur) will not perceive the dichotomy. Most lumping and splitting by good taxonomists is done for this reason. Of course, whether they are treated as orders or suborders is not important, so long as their phyletic relationship is shown, if that is what the systematist pretends to be about.

Finally, I would like to modify Prof. Mabry's statement regarding the time of origin of betalains within the angiosperms to read as follows: "betalains probably arose amongst the angiosperms prior to the selective adaptation of anthocyanins among the flowering plants generally".

Merxmüller: I am afraid that the compromise you offered seems to me more or less useless. In the Centrospermae, however circumscribed, the Phytolaccaceae only can be considered as the basal group. As the Caryophyllaceae are relatively much more highly organized, there is, in my opinion, only one alternative left:

either the Caryophyllaceae and Molluginaceae are not at all related with the betalain-containing families (but where should they then come from?)—or we ought to accent them within this latter group as an aberrant part.

Mabry: I cannot agree, since we interpret the data to suggest that the Caryophyllaceae and Molluginaceae are phyletically close to, but distinct from, the betalain families; in any case, our view recognizes a remarkable chemical dichotomy in a group of related families. This dichotomy is not recognized by other interpretations. More important, however, is the view that our interpretation bears directly upon the evolutionary processes which occurred among centrospermoid families.

Cronquist: I am not sure it is necessary to record my remarks in the formal record of the meeting, but I am impelled to make a few comments. First, let me spit out some distasteful words that were put in my mouth. I did not say that I do not "wish to believe" that there is any connection between the centrospermous group and the Pinales. In scientific matters we are supposed to be concerned with evidence, not what we "wish to believe", and, in the vernacular "them's fightin words". I do not recall my precise phraseology, but, knowing me, I think I would have said something like this: "The likelihood of and evolutionary link between the Pinales and the Centrospermae is so remote that it can safely be dismissed". Those are still strong words, but they are based on the evidence, not on what I wish to believe.

Next let me say, in agreement with Dr Heywood, that the *rank* of a taxonomic group is not nearly so important as its *position* with respect to other taxa. At the higher taxonomic levels, in particular, the assessment of rank is highly subjective; it is not entirely immaterial, because we need to have a degree of comparability in rank for related groups of similar nature, but certainly it is not a matter of absolute right and wrong. Professor Takhtajan and I, for example, often dismiss some difference in our systems as merely reflecting the assignment of different ranks to taxa on whose evolutionary relationships we agree. In marginal cases he often splits where I lump, but both of us consider that a small matter.

If Drs Mabry and Turner had been as nearly reasonable in their earlier comments about the disposition of the Caryophyllaceae and then the Molluginaceae as they are now, not nearly so much opposition would have been raised to their proposals. From their earlier papers I gathered that they were consigning these families to some sort of taxonomic limbo, divorced from what had been considered their allies. Now they present a scheme that puts these two families collectively into a small order adjacent to the "Centrospermae", and treats these two orders collectively as forming a larger group. This larger group has exactly the same circumscription as my order Caryophyllales, so most of the argument is reduced to that of rank, and that of how the families of the mutually recognized larger group should be taxonomically organized inter se. I will here dismiss the question of rank, as being a small matter, although I see nothing to be gained from the inflation of categories to which Dr Heywood has referred.

The question of how to sort out the families of Caryophyllales (sens. lat.) is still of some importance, because it involves concepts of evolutionary relationships. Before I address myself to that question, I want to reply to some earlier comments about how and by whom the taxonomic data should be assessed. The taxonomic system is a multiple purpose system, which tries to use the data and meet the needs of as many groups of consumers as possible. But it cannot serve any of these groups absolutely without seriously compromising the needs of other groups. To take an example from another field. Those of you who have some familiarity with the constitution of the US may know that the constitution guarantees both a free press and a fair trial. Yet if either of these guarantees is enforced absolutely, it seriously compromises the other. Some sort of mutual adjustment is necessary, to meet the essential needs of both guarantees. The taxonomic system must balance many sets of data and needs, rather than only two, and the result is always likely to be less than ideal from any one standpoint. It is the taxonomist's job to evaluate all the data, from whatever source, in devising the system. Most of us come into taxonomy from the morphological side, not just because we are taxonomists, but, as has been pointed out earlier, because we are humans and are accustomed to using our eyes for evaluations. That background leaves many of us at some disadvantage in evaluating the chemical data, but evaluate it we must. The system must be synthetic, in spite of the fact that no one is equally competent in evaluating all kinds of data. I am happy to get help from the chemists, but I must still run their data through my own internal computer.

It is conceivable that the Caryophyllaceae and Molluginaceae collectively stand somewhat apart, phyletically, from the other families of the group under discussion, but I remain to be convinced. At the present time the evidence does not call for it. The serological data presented by Dr Mabry deal with only *some* of the families we are concerned with, but they support the association of the Caryophyllaceae with the other families, rather than the proposed dissociation. The Nyctaginaceae, which Dr Mabry as well as all the rest of us include in the "Centrospermae", stand at about the same distance, serologically, from his reference family (Didiereaceae) as the Caryophyllaceae, and no non-centrospermous family stands any closer to the reference family than do these two. Thus there is no support here for the separation of the Caryophyllaceae from the rest of the group.

The data on DNA hybridization likewise fail to support the separation. As I read Dr Mabry's chart, one of the genera of Nyctaginaceae shows 93% homology with *Beta,* the reference genus; and *Dianthus,* in the Caryophyllaceae, shows about 90% homology. The difference is too small to support any proposed segregation.

I am glad to note that Dr Turner thinks that the separation of the betalain families from the anthocyanin families occurred somewhat farther along in the evolutionary history of the angiosperms than Dr Mabry's chart indicates. The only way I can fit the betalain data into the rest of the picture is to assume that the betalains were substituted, phyletically, for anthocyanins in one evolutionary line that came out of the Magnoliidae.

It is conceivable that the Caryophyllaceae and Molluginaceae collectively diverged from the centrospermous line before the origin of betalains, but I think it is unlikely. The rest of the evidence suggests that the Phytolaccaceae are basal to the whole group. Thus it seems more likely to me that the Caryophyllaceae and Molluginaceae have reverted to the production of anthocyanin, more or less in correlation with the loss of betalains. We do not know how many mutations such a change would require, but there is nothing in the evidence to require more than two correlated mutations—one to interrupt the biosynthetic pathway to betalain, the other to restore the terminal link in the chain to anthocyanin. It has already been noted that many of the betalain-containing species also contain yellow flavonoids and have most of the mechanism required for the synthesis of anthocyanin. Dr Mabry's diagram shows only the terminal step for anthocyanin production to be missing in these plants.

It might possibly be argued that such a reversion is inherently unlikely, but I see no reason why it should not happen. We have heard several accounts of the independent evolutionary origin, in widely different groups, of the ability to produce certain complex chemical substances. Surely a group that has lost the ability to do something is in no worse position for getting it back, than a group that never had it is, for developing that ability.

Furthermore, if we object to such reversions in principle, we have further troubles in the subclass Caryophyllidae. The Polygonaceae and Plumbaginaceae do not have the characteristic sieve-tube plastids of the Caryophyllales (sensu meo), which are unknown in other angiosperms. Yet the evidence indicates that both of these families are derived from the Caryophyllales. The Polygonaceae, in particular, appear to be closely allied to the Caryophyllaceae, and there is a trail of Caryophyllaceous genera leading toward the Polygonaceae.

Mabry: Just as you note that our view is conceivable but you remain to be convinced so will I agree that your view is conceivable but I am not convinced; in fact, however, I believe we are much closer to the same interpretation of the data than our comments might indicate.

I do contend that a specialist in systematics such as yourself may not be equally qualified to evaluate the merits of chemical characters as phyletic markers. Just as you use an internal computer to weigh subtle differences in morphological characters so do chemists evaluate chemical characters. To me this is what this meeting is about—that we put our evaluations into a common pool in order to obtain the best possible interpretation of the data.

Most of the eminent systematists in this audience have differed on their interpretations of some of the morphological data for certain taxa placed from time to time in the Centrospermae. Is it not possible that certain interpretations by you are still in error? Perhaps the morphological features of the Caryophyllacae and Molluginaceae should be re-evaluated in view of all the data available today. Regarding the DNA-RNA data, it is of course limited but, nonetheless, it does distinguish the Caryophyllaceae from the betalain families which were tested.

I don't wish to argue whether or not the differences (based on the DNA-RNA data) are sufficient to demand a taxonomic distinction, since additional studies are clearly warranted Consideration of all the available data led us to recognize a taxonomic distinction among the eleven centrospermoid families.

Bu'Lock: Without attempting to intervene— from outside the arena—in the assessment of taxonomic niceties, I would like to come back towards what seems to me to be a basic biochemical and biochemical-genetic aspect of the present problem. This centres upon the mutual exclusiveness of betalains and anthocyanins. At first sight this would seem to necessitate the loss of one character simultaneously with the acquisition of another. Moreover at least one of this pair, the betalain sequence, must surely comprise not just one gene but a whole series of structural and regulatory genes, of such a nature that we cannot even expect them to constitute a linked group. In addition, Mabry's outline of the biosynthetic situation makes it quite clear that the mutual exclusiveness of the two processes can not have any simple biochemical explanation, since they are not at all closely linked in terms of intermediary metabolism, and any connection between them must be sought in the far more difficult area of regulatory effects.

In such a situation I feel that to speak of "two correlated mutations" or to fall back on a concept of "reversion" to which no molecular -biological meaning can be attached

is to undeservedly minimise the magnitude of the dichotomy to which the Texas school has drawn our attention. Insofar as the dichotomy is required to be bridged, in the interests of preserving the integrity of the morphological classification, it is incumbent upon the "bridgers" to supply explanations which are really meaningful in terms of biochemistry and biochemical genetics; on the other side I would say, however, that it would also be desirable if the "splitters" were to explore somewhat more enthusiastically the possible mechanisms which may underly the phenomenon of exclusiveness in this case.

Mabry: I agree completely with Professor Bu'-Lock since I am also of the opinion that the loss and gain mutations required to account for Professor Cronquist's views are sufficiently unlikely that we can at this time dismiss them. It is possible that the enzyme associated with converting a flavonoid precursor to an anthocyanin is not a member of the series of presumably related enzymes involved in other aspects of flavonoid biosynthesis. Thus, this genome may be entirely missing from the betalain families just as it is likely that the genomes which produce the enzymes associated with betalain biosynthesis are absent from the anthocyanin families.

Jensen: The serological investigation of Centrospermae taxa was instituted to throw some light on the systematical connexions of the Didiereaceae by immunological characters. The result was, as you know, the demonstration of relatively great similarities between Cactaceae, Portulacaceae and Didiereaceae. The reaction strength of all the weaker reacting taxa cannot be explained in so definite a manner, because (i) since they are weak reactions with the reference system only a relatively small part of the proteins are involved and nothing can be said of the rest of the molecules and (ii) the fact that reactions have occurred to about the same extent with Euphorbiaceae and Papaveraceae. As a consequence it cannot be said decisively that the Caryophyllaceae are connected with the betalain-containing families on this evidence.

My question is about your hybridisation

method. We know that hybridisation techniques have many limitations if used for systematical purposes. Bendich & McCarthy have pointed out the following limitation. They worked with pea RNA and DNAs of different taxa. They found that the direct binding data, a method you used too in your investigation, had astonishing differences e.g. with DNA of *Cucurbita* and yeast, the reactions with pea RNA were up to 200% that of the homologous reaction with pea DNA. They concluded that it is better to use the thermal stability profiles. My question is: Have you tested the thermal stability and if so, what results did you obtain?

Mabry: Yes, the thermal stability profiles were determined and these data were also in accord with my comments as presented here.

Ourisson: (1) The controversy about Caryophyllaceae has probably grown so intense because the conclusions of the chemists were shocking to the botanists, and because they were presented so forcefully.

Yet, it has taught us chemists a great deal, and I hope the results of the confrontation will be useful for the other cases where there are good chemical reasons to alter the accepted botanical classification: e.g. the cleavage of Euphorbiaceae into Crotonoideae and Phyllanthoideae, and to regroup Crotonoideae with Thymeleaceae.

(2) I wish to reinforce Bu'Lock's request to Mabry: please investigate in detail the biochemistry of the groups in discussion. Is *one* enzyme absent in Caryophyllaceae, or a *set* of enzymes? Can they use added intermediates? Do they form new substances by diversion of the betalain pathway? All these are questions with a potential experimental answer.

Mabry: I hope the controversy has not degenerated into one involving the personalities of the investigators; certainly the exchange of views that has occurred in this meeting is just the sort of inspiration which gives all of us reason to re-think our views, re-evaluate the data and to initiate new research projects which may ultimately reveal more precisely the phyletic relationships. Enzyme studies must be conducted; also, perhaps we can persuade Professor Boulter to sequence cytochrome *c* from centrospermoid families.

Mears: Since I worked with the Amaranthaceae (Chenopodiales) before I was a student of Professors Mabry and Turner and since I long have admired Dr Cronquists's ideas on plant family relationships, I have been trapped in the middle by this controversy, I have three remarks, one for chemists and two for the morphologists:

(1) Morphologists who consider the Chenopodiales-Caryophyllales derived from "primitive Ranalean angiosperms" through something like *Phytolacca* often believe that there has been a loss and a redevelopment of floral parts: a loss of petals with floral reduction, followed by the development of other showy floral parts. That loss and redevelopment may parallel a possible version of the betalain pigment story.

(2) Those floral parts which provide the brilliant colors of *Dianthus, Bougainvillea, Opuntia, Portulaca, Carpobrotus* and *Mollugo* apparently are not petals and probably are not homologous. Maybe some floral morphological characters would support some reorganization of the old Centrospermae.

(3) The greatest unequivocal success of chemotaxonomy at the familial and ordinal level probably has been the alliance of the Didiereaceae with the Chenopodiales. Until the formidably extensive morphological study and chemical analysis of the Didiereaceae there was no consensus on the several reported possible positions for the family. Since the study, which reported betalain pigments, I believe there has been no doubt about its inclusion in the Chenopodiales. However, although the family of *Dianthus, Alsine, Paronychia* and *Illecebrum* is heterogeneous, there was some consensus about it for a long time: the Caryophyllaceae has long been regarded as either "central" to the Centrospermae or as derived from something like the Phytolaccaceae. The straight-forward interpretation of the pigment data does not tolerate either of these ideas unless several biosynthetic reversals have occurred.

Many hope that the morphological and chemical data will agree in a thorough explanation. Such an explanation awaits a complete, morphological revaluation of the Caryophyllaceae, and indeed the Molluginaceae.

Hedberg: Although it might be of limited interest to debate the taxonomic rank of the subdivisions of Centrospermae based upon ocurrence or absence of betalains, this order is so interesting from an evolutionary point of view that it appears worth while to investigate any additional type of characters which might contribute to its understanding. One set of characters I would recommend for detailed study is provided by pollen morphology—spheroidal polyforate pollen grains are of common occurrence in most of its families and their distribution may prove taxonomically important.

Population Structure in *Circaea lutetiana, C. alpina* and *C.×intermedia* (Onagraceae) as Revealed by Thin-layer Chromatographic Patterns

G. Weimarck

Department of Plant Taxonomy, University of Lund, S-223 61 Lund, Sweden

Summary

Five collections each of *Circaea lutetiana, C. alpina* and their hybrid, *C. × intermedia*, have been studied. Variation in compounds separated by thin-layer chromatography points to a certain intrapopulational heterogeneity in all 3 taxa. The occurrence of backcrossing between *intermedia* and the parent species is suggested.

The advantages and possible sources of error of the method used are discussed.

Circaea intermedia Ehrh. is generally held to be the hybrid between *C. lutetiana* L. and *C. alpina* L., all having the chromosome number $2n=22$. Complete male sterility was found in British material of *C. intermedia* by Raven [1]. Female sterility was also absolute or almost absolute. He found evidence that allogamy prevailed in *C. lutetiana* but autogamy in *C. alpina*. With the exception of a few plants suspected to be products of back-crossing between *intermedia* and *lutetiana*, he found that plants of *lutetiana* and *alpina* had more than 90% fertile pollen.

Raven also clarified the morphological differences between the 3 taxa. He remarked that some herbarium material studied had previously been incorrectly determined.

The aims with the present study were:

(*1*) to check whether it was possible or not to recognize the 3 taxa on other than morphological criteria and whether a hybrid origin of *C. intermedia* could be confirmed in this way or not;

(*2*) to check whether populations of the 3 taxa were composed by clones, pure lines or genetically different individuals;

(*3*) to trace possible evidence of gene flow between *C. lutetiana* and *alpina* via *intermedia*.

These problems were approached mainly by the study of variation in leaf substances extractable with methanol, separated by thin-layer chromatography and made visible in UV light by a phenol reagent. A special effort was made to estimate the experimental errors and delimitations of the method used. A complementary study of the percentage of stainable pollen was made.

This paper forms part of an investigation of population structures based on chromatographically documented variation in selected plant species. A second part will be published shortly.

Material and Methods

Five collections of each taxon were cultivated. A list of localities will be given elsewhere. Voucher specimens are preserved at LD.

The plants were coded and arranged according to a random table in order to diminish possible influence of non-random environmental factors. After one year of cultivation, leaves of 10 plants from each collection were extracted in HCl-methanol (1:100). Separation was made two-dimensionally. Each spot obtained was characterised by R_F and colour after dipping in a phenol reagent. No compound was identified chemically. Two plates from each extract were examined separately from each other. This fact together with the code numbering excluded the risk of subjective biasing of the observations.

Statistical Evaluation

When using any type of phenotypic variation for the study of genotypic variation one is faced with the problem of modification and errors in observation. It is possible either to consider the entire bulk of information and estimate the size of the error, or to consider only such information as is believed to be reliable and ignore the rest.

With a view to these alternatives the evaluation of the chromatographically obtained data was performed solely on the basis of information believed to be reliable.

We are accustomed to relying on certain morphological characters when studying variation in higher plants but rejecting others on purely empirical grounds. The same procedure can be justified as regards the chromatographically obtained characters in this study. An estimation of the reliability of each character (spot) can be made thanks to the fact that the observations were made without knowledge of which plates belonged to the same plant or collection.

In the following only the absence or presence of spots will be taken into account regardless of intensity. The spots could be divided into three groups, two of them accepted for further study, one rejected.

Spots occurring in all plants of a taxon were accepted for further study, even if some of them had been occasionally found in one only of the two plates belonging to the same plant. They were considered to occur generally in the taxon concerned.

Spots occurring in some but not all plants studied in a taxon were also accepted provided there was no inconsistency at all between plates. If, for example, a given spot was found in all in two plates only of a taxon, the probability that the second one would occur in the other plate from the same plant would be 1/99, if the different observations were caused only by random variation around the perception threshold (the number of plates per taxon was 100). The same is valid if a given spot is missing in two plates only, belonging to the same plant.

The probability of finding at least one such pair in the whole material by mere chance would be about 0.85, of finding 5 pairs in the whole material 0.05 and 10 pairs 0.00001. Thirteen such pairs of present or absent spots in plate pairs from the same plant were found (table 1). However, the formula presupposes that all spots have the same tendency of inconsistent variation. This is not the case, since many spots in the material do not contribute to inconsistent variation at all.

If a given spot was found in four plates of a taxon, the probability that they would be found in 2 plants only by mere chance is about 3/10000. The probability of finding at least once in the whole material 4 plates forming 2 pairs would be about 0.06. Three such cases were actually found (table 1).

The probability that such observed differences within the material are real qualitative or quantitative ones between extracts (and, provided extraction technique was consistent and modification due to external factors negligible, between plants also) was thus considerable. It was, moreover, possible to give at least a minimum estimate of this probability.

The spots rejected were those not occurring in all plants and showing inconsistent variation between plates of the same plant. It would obviously be easy to overlook them due to an intensity close to the perception threshold. Such a spot could with a rather high degree of probability be overlooked in both plates of a plant, thus giving inadequate information.

Further numerical treatment is not performed at present. Firstly, some of the spots obviously vary simultaneously in the same way, and they may be chemically dependent one upon another. Such spots should then preferably not be considered as more than one single character. Secondly, the material cannot be regarded as randomised after some of the spots have been rejected.

Results and Discussion

Of the 54 spots listed as occurring in *Circaea lutetiana,* 21 were from the two non-rejected groups, of the 74 in *C. alpina* 33, and of the 62 in *C. intermedia* 39. Twenty-six spots did not vary between plate pairs in any of the 3 taxa, although not necessarily present in all of them. Seven of the 26 spots occurred invariably in all 3 taxa and could be ignored in this context (table 1). Seven other spots were found invariably in *alpina* and *intermedia* but not in *lutetiana,* and one spot in *lutetiana* and *intermedia* but not in *alpina.* One spot was found invariably in *lutetiana* but not in the two other taxa.

Some spots showed variation within populations. In *C. lutetiana,* two plants of collection

Table 1. *Number of plants per collection showing non-rejected spots*

Taxon, collection...	C. lutetiana					C. alpina					C. intermedia				
Spot no.	1	2	4	5	6	1	2	4	5	6	1	2	3	5	7
1–7	10	10	10	10	10	10	10	10	10	10	10	10	10	10	10
8–14	0	0	0	0	0	10	10	10	10	10	10	10	10	10	10
15	10	10	10	10	10	0	0	0	0	0	10	10	10	10	10
16	10	10	10	10	10	0	0	0	0	0	0	0	0	0	0
17	0	2	0	0	0	10	10	10	10	10	10	10	10	10	10
18	10	10	9	10	10	10	10	10	10	10	10	10	10	10	10
19	0	0	2	0	0	–					–				
20	0	0	0	0	0	10	10	10	9	10	10	10	10	10	10
21–24	0	0	0	0	0	0	0	0	1	0	0	0	0	0	0
25–26	10	10	10	10	10	0	0	0	1	0	10	10	10	10	10
27–28	–					0	0	0	1	0	10	10	10	10	10
29	–					10	10	10	9	10	–				
30	10	10	10	10	10	10	10	10	10	10	10	10	10	10	9
31	0	0	0	0	0	–					0	0	0	0	2
32	–					–					9	10	10	10	10

2 (2:2 and 2:11) showed spot 17, which was otherwise not found in the taxon although invariably found in the two others. In the plant 4:7, spot 18 was not detected although present in the rest of the material. The plants 4:7 and 4:2 showed spot 19 not found elsewhere in *lutetiana*. This spot occurred in the two other taxa but could not be taken into account because of inconsistency.

In *C. alpina*, the plant 5:4 did not show spot 20 which was consistently found in the rest of the *alpina* material and in *intermedia* but not at all in *lutetiana*. In the same plant the spot group 21–24, not found elsewhere in this study, occurred. The plant 5:16 showed spots 25–26 not found in the rest of *alpina* but invariably in the two other taxa. The same plant also showed spots 27–28 not found in the rest of *alpina*, and spot 29 could not be seen in it although consistent in all other *alpina* plants. Spots 27 and 28 could not be taken into account in *lutetiana*, but were invariably found in *intermedia*. Spot 29 could be taken into account neither in *lutetiana* nor in *intermedia*.

In *C. intermedia*, spot 30 could not be found in the plant 7:11. It was consistently present in the rest of the material studied. The plants 7:5 and 7:8 showed spot 31 not found elsewhere in *intermedia*. The plant 1:8 lacked spot 32 consistently found in the rest of the *inter-*

media material studied. Spot 31 was not found in *lutetiana* and could not be taken into account in *alpina*. Spot 32 could not be taken into account in *lutetiana* nor in *alpina*.

Thus the chromatographic patterns obtained from some plants of all 3 taxa studied deviated from the rest, and most deviations could be stated to be caused by factors other than chance with a high degree of probability. The chromatographically obtained patterns represent phenotypes. The plants showing deviating patterns had not been cultivated at the edges of the frame and did not show any conspicuous external modification. The extraction and application of the extract on the plates had been performed with the greatest care. Most differences in patterns should therefore be regarded as representing different genotypes in the material studied.

As far as the present material is concerned, *C. lutetiana* seems to have contributed to the pattern of *intermedia* with spot 15, *alpina* with spots 8–14 and with the characteristic of not showing spot 16. None of the consistently occurring spots was found in *intermedia* only.

Spots whose occurrence deviated from the rule in one taxon but which was in accordance with that in another taxon are spot 17 in *lutetiana*, 20, 25 and 26 in *alpina*. Their occurrence can be explained in two ways. Firstly, it could

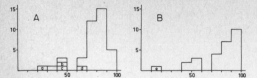

Fig. 1. *Abscissa:* % stainable pollen; *ordinate:* no. of plants. Percentage of stainable pollen in (A) 40 plants of *Circaea lutetiana*; (B) 27 plants of *C. alpina*. Plants showing chromatographic patterns indicative of hybridism are marked (a) 4:2; (b) 2:11; (c) 4:7; (d) 2:2; (e) 5:4.

be that the pure species are in these characters different in gene frequencies rather than in absolute terms. If so, this could be true also for some or all of the spots 8–16 but not shown due to the limited material studied. Secondly, it could be that the deviating plants have received genes from the other species by recent introgressive hybridization.

The percentage of stainable pollen was determined in as many plants as possible to trace possible evidence of hybridism (fig. 1). In *C. lutetiana*, the 4 plants showing deviating chromatographic patterns were among the 8 with the lowest figures of stainable pollen out of 40 plants studied with respect to this. A Wilcoxon test showed that their values differed highly significantly from the rest ($p<0.001$). Thus, they probably represent back-crossings between *lutetiana* and *intermedia* or later segregation products (they were morphologically indistinguishable from other *lutetiana* plants). *C. intermedia* was not found by me on the sites of the *lutetiana* collections 2 and 4 but has been recorded earlier from the site of collection 4. *C. intermedia* grew intermingled with *lutetiana* on the site of collection 6. The available evidence points to a minimum of one back-cross clone on the site of collection 2 and two on the site of collection 4 apart from the pure species. In addition, the 4 plants with a pollen stainability below 70% but not with chromatographic patterns indicative of hybridism may represent other segregation products.

The lowest percentage of stainable pollen in the *C. alpina* material was found in one of the two plants in collection 5 having an aberrant chromatographic pattern. The other one did not give enough pollen. A Wilcoxon test showed a rather high significance of the value ($0.01<$

$p<0.05$). It is possible that all 6 *alpina* plants with a pollen stainability of below 60% represent back-crossings or later segregation products, but other interpretations may also hold. The *alpina* plants of collection 6 grew intermingled with *lutetiana*; on the other sites neither *lutetiana* nor *intermedia* was seen in the neighbourhood.

The percentage of stainable pollen in *C. intermedia* was invariably 0. In many preparations not even empty grains could be found. The indications of back-crossing between *intermedia* and the parent species suggest, however, that sterility is not always complete.

The figures for stainable pollen in this study are in general considerably lower than those reported by Raven [1].

Conclusions

(*1*) No plant in the study would have been erroneously determined if determined on chromatographic criteria. Several spots gave good evidence of *Circaea intermedia* being the hybrid between *lutetiana* and *alpina*.

(*2*) Intrapopulational variation is indicated in all 3 taxa. In the case of *C. intermedia*, it cannot be stated whether this is caused by somatic mutation within one clone or by the co-existence of two or more clones that have arisen independently on the same site.

(*3*) In the cases of *C. lutetiana* and *alpina*, at least some intrapopulational variation is probably the result of back-crossing between *intermedia* and the parent species. A gene flow between *lutetiana* and *alpina* is therefore suggested.

Methodological Comments

The value of chromatographic data in this type of investigation largely depends upon whether the magnitude of the experimental error can be estimated and whether information can be discarded objectively if the magnitude of the error is great. Apart from the difficulty of getting an absolute reproducibility of the technique, phenotypical variation in the biological material due to the sampling of different parts of plants, plant age, environment, etc., can also contribute to the heterogeneity found [4]. This error may be rather large for some spots and may obscure actual conditions and give mis-leading results, a point which in the main has been overlooked in the literature. It is, more-over, also valid for other types of characters in different types of taxonomic work, see for example [2] for discussion. Thus, problems of a similar type are met with in the study of variation in morphological characters but are usually more easily compensated for since we have much more experience of such characters.

I am of the opinion that statistics is very valuable when testing the reliability of chro-matographical results but that sophisticated numerical indices calculated from any type of data often are dangerous and should not be used unless the effects of the formulae are com-pletely known.

Spot patterns obtained from non-hydrolyzed plant extracts often vary between individuals of the same population [4, 5]. Studies of varia-tion between different taxa based upon too few samples are therefore of limited interest and may also have contributed to a certain lack of enthusiasm shown in some quarters for chro-matographic work in connection with tax-onomy. I am convinced that evidence obtained from "spot patterns" of the type illustrated by the *Circaea* study presented here is of great value at the population level. Alone or, prefer-ably, in combination with evidence from mor-phology, cytology, pollen quality, embryology, etc., such patterns provide characters which have helped us towards a better understanding of, for instance, the degree of hybridism in *Baptisia* (classical studies by Turner and co-workers, the first of which was published as early as 1959 [3]) and the balance between apomixis, amphimixis and rhizomatous propaga-tion in *Hierochloë* [4]. Perhaps this method is of greatest use precisely in the study of apo-mictic plants or more or less sterile hybrids, in which crossings and other conventional experi-ments fail, and in cases when morphological criteria do not permit a satisfactory analysis of differences between individuals.

The point may arise that data sampling in this case involves the use of a separation tech-nique that is employed by chemists rather than the use of chemistry in botanical classification. A knowledge of the identity of the compounds concerned is of course valuable but, as far as

I can see, not indispensable in this connection and can be very laborious to attain. Studies of human populations may well be based on varia-tion in the colour of the eyes, blood groups, etc., even without a precise knowledge of the chemical background of pigments and antigens.

With reference to the subject of this sym-posium, I should like to say that chemical evi-dence in combination with other evidence enables us to classify plants also in terms of population structure and genetic system, both of which are of vital interest in plant taxonomy and the study of evolution.

Professor H. Runemark and Mr T. Karlsson, De-partment of Plant Taxonomy, Lund, have discussed the results and given constructive criticism. The numerical treatment was performed with the as-sistance of Civil Engineer H. Rootzén, Institute of Mathematical Statistics, Lund. Mrs M. Greenwood-Petersson, Lund, checked the manuscript.

The study was financed by the Nilsson-Ehle Fund of the Royal Physiographic Society, Lund, and by the University of Lund.

References

1. Raven, P H, Watsonia 1963, 5, 262.
2. Sneath, P H A & Johnson, R, J gen microbiol 1972, 72, 377.
3. Turner, B L & Alston, R, Am j bot 1959, 46, 678.
4. Weimarck, G, Bot notiser 1970, 123, 231.
5. — Taxon 1972, 21, 615.

Discussion

Turner: Do you find *C. intermedia* growing with both of its putative parents? Or even one alone? If so then your results suggest that the "hybrid" taxon is not the result of just recent hybridization, for it seems to have chemical compounds peculiar to itself.

Weimarck: C. intermedia is sometimes but not always found together with one parent or both. Raven suggested that *intermedia* could have been spread vegetatively after ancient hybridisa-tion, and my results do not conflict with this. The putative products of introgression are more likely to be of recent origin. Moreover, *C. inter-media* was not found to have any consistently occurring compounds peculiar to itself.

Grant: Do you consider it was necessary to go to the elaborate randomization of the plants in the field in order to obtain reliable chromatographic data?

Weimarck: Yes, according to my experience one may obtain a systematic error due to modification and subjective bias.

Chemosystematics in the Classification of Cultivars

W. F. Grant

Department of Biology, McGill University, Montreal, and the Genetics Laboratory, Macdonald Campus, Ste Anne de Bellevue, Quebec, Canada

Summary

Morphological descriptions of cultivars present problems in identification as clear-cut distinguishable differences between cultivars are often lacking. With the development of breeder's rights in various countries, the unequivocal identification of cultivars and the establishment of parental origins will assume increasing importance. Biochemical techniques, such as paper and thin-layer chromatography, electrophoresis, and serology offer ways for the identification of cultivars. More than one biochemical system (different enzymes) or the use of different plant parts (leaves, roots) may be necessary to completely distinguish all cultivars of a species due to limited intercultivar genetic variability. Intracultivar genetic heterozygosity may also be diagnosed through such methods.

Secondary phenolic compounds from leaf samples of over 900 cultivars of *Manihot esculenta* Crantz have been used for cultivar classification by means of two-dimensional thin-layer chromatography. Intercultivar differences for 55 fluorescent spots have been compared by a cluster analysis computer program in which the cultivars have been grouped at different S values. By means of chromatography and spectrophotometry flavonoids have been identified as the glycosides of quercetin and luteolin and the cinnamic acids as chlorogenic acid, esters of *p*-coumaric, caffeic, ferulic, and sinapic acids, and the glycosides of caffeic and ferulic acids.

A wide range of chemical compounds have been used in recent years as characters, or markers, in the fields of evolutionary genetics and systematics to aid in the study of genetic and taxonomic variation. These compounds range from alkaloids, flavonoids, phenolics and terpenoids to macromolecules such as lipids, enzymes and DNA. A major emphasis in chemosystematic studies has been placed on the flavonoids since their occurrence in the plant kingdom is widespread. Both paper and thin-layer chromatographic techniques have been developed for ease in their analysis. At the same time, electrophoretical and serological techniques have been developed to study the phylogenetic relationships of organisms based on differences in enzymes or other proteins. Studies using the chemosystematic approach have been undertaken by various investigators to help in resolving such problems as species relationships [29, 31, 32, 34], interspecific hybridity [9, 18], polyploidy [34, 35, 38], and ecological [12] and geographical [15] differences between taxa.

Considerable morphological variation exists in some cultivated varieties of plants which is not unlike that for species in certain genera. At the same time, morphological descriptions of cultivars present problems in identification as clear-cut distinguishable morphological differences between cultivars are often lacking for positive identification. In addition, chromosome number differences are not present at the cultivar level which may aid in differentiating taxa at the species level. With the development of the Plant Variety Protection Act of 1970 in the United States, the adoption of breeder's rights by the Swedish Official Committee of Cultivars in Sweden, and active discussions concerning breeder's rights in Canada and possibly other countries, the unequivocal identification of cultivars and the establishment of parental origins will assume increasing importance. In general, biochemical markers have never been consciously selected by the plant breeder, and therefore, it is expected that they will be useful in combination with morphological traits to resolve cases of disputed parentage just as similar methods have been used in man and other animals.

Techniques borrowed from the repertoire of the biochemists have been welcomed by plant breeders and geneticists who have been faced with the problem of cultivar identification. In addition to individual biochemical tests, chromatography, electrophoresis and serology have been found useful, in many cases, to distinguish cultivars. It has also been of considerable interest to geneticists to try to associate specific biochemical markers with either morphological features and/or resistance to disease and insect pests inherent with the organisms under study. It may be expected that the use of biochemical markers in cultivar recognition will be extended in the very near future to specific chromosomes, or chromosome regions, with the techniques which have been developed to show chromosome banding patterns by the use of quinacrine mustard, and various staining procedures, such as that of G- and C-banding for the identification of different human syndromes [42].

In recent years several schematic representations have been devised to show phenetic relationships [44, 47, 48]. Recently, Baum & Lefkovitch [6, 7] have described the few instances of the application of numerical taxonomic techniques to infraspecific taxa and have dealt with the problems of classifying large numbers of cultivars by taximetric methods. They have grouped some 5000 oat (*Avena sativa* L.) cultivars into clusters based on morphological criteria. Using grain yield per plant, Hanson & Moll [30] have assessed the genetic relationships between cultivars by comparing the weighted differences in gametic frequencies between the populations based on non-additive gene effects (dominance and epistasis) using a diallel mating design.

The use of chemosystematics in cultivar identification has received only minimal attention to date but as biochemical techniques become better known by those engaged in cultivar studies it is expected that considerably more attention will be focused on this aspect. A brief review of some of the uses of biochemical markers in the classification of cultivars will be presented here as well as some of the studies in progress in my laboratory on the classification and identification of phenolics in cultivars of cassava (*Manihot esculenta* Crantz).

Biochemical markers in classification

Anthocyanin pigmentation. Blank [11] in his review on anthocyanin pigments of plants has enumerated different authors who have used anthocyanin formation to differentiate cultivars including those of *Hordeum vulgare* L., *Theobroma cacao* L. and *Triticum vulgare* Vill. In a paper chromatographic investigation of the anthocyanin pigments found in the apiculi of glumes of 20 cultivated rice (*Oryza sativa* L.) varieties, Mizushima et al. [39] showed that plants of certain cultivars which contained chrysanthemin and keracyanin could readily be differentiated from those cultivars of other genotypes where these compounds were remarkably decreased and sometimes undetectable.

Grasses. Cultivar identification of creeping bentgrass (*Agrostis palustris* Huds.) and Kentucky bluegrass (*Poa pratensis* L.) is a major problem for turf growers, breeders, and seed producers when distinguishable morphological differences are lacking for positive identification. Wilkinson & Beard [50] have used acrylamide gel electrophoresis to distinguish the cultivars of these two species. Of the 6 bentgrass cultivars, all could be readily separated. Of the 10 bluegrass cultivars, 6 could be placed into 3 groups of 2. Two of the cultivars could be identified singly. Two others exhibited no characteristic banding pattern which was considered possibly due to the lack of genetically uniform plants within these cultivars. The authors considered acrylamide gel electrophoresis to have potential in cultivar identification of these grasses.

Secale cereale L. A thin-layer chromatographic study of 15 inbred lines of the diploid Swedish rye (*S. cereale*) cultivar, steel-rye, has been carried out by Fröst [24]. For comparison he studied 18 plants taken from the general population of this cultivar. He found significant differences in the number of spots between the inbred lines and in the population plants. The variation in the number of spots between the inbred lines was greater than that between the population plants. The mean number of spots was higher in the population plants than in the inbred lines possibly indicating fewer spot differences with increased homozygosity. Since the plants were all grown under the same environmental conditions, Fröst considered these differences to be genetically conditioned.

Avena sativa L. By means of thin-layer chromatography, Dhesi et al. [19] were able to separate 4 out of 6 oat (*A. sativa*) cultivars from seed extracts.

Recently, Asker & Fröst [5] have reported on a thin-layer chromatographic study of 22 biotypes of oats (*Avena* sp.) including 15 cultivated varieties of *A. sativa*. In chromatograms from leaves of young plants they observed 20 different spots of which 11 occurred in 16 cultivars and 4 were common to all. Only small differences were noted between cultivars. In chromatograms from mature plants 26 spots were observed with 12 occurring in 20 cultivars. Differences in chromatogram pattern between cultivars of mature plants was not as great as the intracultivar differences between young and mature plants. The number of spots observed in these *Avena* chromatograms was fewer than in *Hordeum* [25] and was considered as one reason that a clear discrimination between all the cultivars was not obtained.

Williamson et al. [51] carried out a study to determine the extent and nature of electrophoretic variation of esterase isozymes within commercial cultivars of oats. Three esterase banding patterns were found for the cultivar Putman 61:

Banding patterns	1	2	3
Frequency	183	3	4

The majority of the 190 plants fell into one class. In the cultivar Orbit, 21 banding patterns were found with the highest frequency being 69 plants out of 176 in one type. Six different patterns were found among 189 plants of Nodaway—the largest group consisting of 132 plants. The intracultivar variation was considered partly as a consequence of mutation, outcrossing and procedural differences. In Nodaway segregation of heterozygous loci was considered the major source for the cause of variation. The results suggested that if isozyme variation reflects genetic variation, then it might be related to the adaptability of a cultivar.

In a study of three oat cultivars, Almgård & Norman [4] were able to separate one of the cultivars from the other two by the electrophoretic

pattern of the peroxidases from leaf homogenates.

Smith & Frey [46] have used serological techniques to define the genotypic relationships of 6 oat cultivars to a seventh. They compared antigens from pollen, leaf, whole grain, scutellum and embryo but found those of whole grain and embryo agreed best with the genotypic relationships based on genetic parameters. With the common antigens present in closely related plants, the authors considered it necessary to make reciprocal comparisons.

Hordeum vulgare L. By means of thin-layer chromatography, Fröst & Holm [25] studied the differences in phenolic compounds from leaves of 20 cultivated varieties of barley (*H. vulgare*). They observed as many as 50 different spots but included in their comparisons only 19 which were consistent with respect to position and size. The 20 varieties separated into two main groups which they designated A (15 varieties) and B (5 varieties). From an analysis of the lineage of these cultivars, the spot patterns were observed to be inherited as blocks and it was possible to associate particular patterns from hybrid originating cultivars with their parental species. In a further study, Fröst & Holm [26] analyzed an additional 17 cultivars using 18 marker spots. As in the first study, these cultivars separated into the group A and B patterns. With two exceptions, the classification of the cultivars into these two groups was as expected on the basis of information on their parentage.

In a study of four barley cultivars, Almgård & Norman [4] observed that one of the 4 cultivars could be distinguished from the other 3 by an esterase system using leaf homogenates; another cultivar could be distinguished from the other 3 by using a catalase system from leaf homogenates, and electrophoresis of peroxidases in root homogenates made it possible to separate the 4 cultivars into 2 groups of 2. In these 3 systems no intracultivar variation was noted.

Frydenberg et al. [27] have shown that heatstable α-amylase isozymes are readily detected electrophoretically in extracts from germinated barley seeds. They found three different zymotypes and classified 117 cultivars of European

Table 1. *Frequency distribution (%) of α-amylase zymotypes and DDT responses in Canadian and European barley cultivars*

Cultivars	Amylase zymotypes					DDT response		
	DC	DC_1	DC & DC_1	DC_2C	DC_2C & DC	Res.	Susc.	Res. & Susc.
Canadian[a]	74.5	7.3	1.8	–	16.4	–	100.0	–
European[b]	53.8	37.6	5.1	3.4[c]	–	52.1	42.7	5.2

[a] From Fedak & Rajhathy [22]; [b] From Frydenberg et al. [27]; [c] Cultivars from US.

ancestry for their α-amylases and their response to foliar application of DDT (table 1). They found old cultivars to be often polymorphic, whereas monomorphism was more characteristic among newer but well-established ones. Among very recent cultivars, or incipient cultivars not yet named, polymorphism was often observed.

By agar gel electrophoresis Fedak & Rajhathy [22] studied the same 3 α-amylase zymotypes from germinated seeds of 55 Canadian cultivars of barley and also classified them for their response to foliar applications of DDT. The most striking difference between the European and the Canadian cultivars was the much lower frequency of the DC_1 isozyme and the presence of DC_2C in the Canadian cultivars (table 1). The authors noted that the distribution of the α-amylase zymotypes reflected the origin of the parental lines and the subsequent breeding procedures in Europe and Canada.

All of the Canadian barley cultivars tested were susceptible to DDT. Over half of the European cultivars, which are predominantly 2-rowed in contrast to 6-rowed in Canada, were resistant (table 1). A few European cultivars were polymorphic for DDT response.

Also, Fedak & Rajhathy [23] analyzed the plumule tissues of these same 55 cultivars by means of starch gel electrophoresis to determine these esterase isozyme patterns. In this study 10 alleles at 4 loci were used to separate the different cultivars. Including the 3 α-amylase alleles previously described [22], 13 alleles can be used for barley cultivar identification in addition to the *ddt* locus. Because of a great deal of similarity in pedigrees, many barley cultivars cannot be distinguished by using one or both of these enzyme systems. However, the authors considered that the use of

a combination of several enzyme systems could conceivably increase the usefulness of enzyme markers in the identification of barley cultivars. Nielsen & Frydenberg [41] have provided further information on the distribution of esterase and α-amylase isozymes and DDT reactions in 107 European barley cultivars.

Other biochemical characters which have been used to differentiate cultivars of barley are protein, amylase and amylose content and pericarp phenolic staining reaction (for references refer to [25]).

Triticum vulgare Vill. Almgård [1] by means of starch gel electrophoresis has studied several different enzyme systems in 12 Swedish wheat (*T. vulgare*) cultivars and was able to separate the cultivars into 6 groups. In 2 cases, plants of a cultivar could be divided into 2 distinct esterase patterns.

In a thin-layer chromatographic study of glycoflavones of 3 cultivars of hard red spring wheat (*T. aestivum* L. em. Thell.), their extracted AABB tetraploids, and 2 cultivars of durum wheat (*T. turgidum* L. var. *durum*), almost identical chromatograms were obtained for 11 compounds [17]. The cultivar Prelude could be distinguished from Thatcher & Rescue mainly on quantitative differences, while the later two could not be distinguished from each other. The durum cultivars differed slightly from the hexaploids, however, almost no qualitative or quantitative differences were observed between the hexaploids and their extracted tetraploid counterparts, indicating considerable duplication of flavonoid-synthesizing genes.

It has also been reported that T. Dabrowska, Institute of Plant Genetics, Polish Academy of Sciences, Poznan, Poland, has initiated comparative disc electrophoretic research with isoenzymes extracted from barley and wheat cultivars [21].

Citrullus, Cucumis and *Phaseolus*. Singh & Thompson [45] carried out a paper chromatographic study of flavonoids in stem and leaves of 10 cultivars of water melon (*Citrullus vulgaris* Schrad.), 8 of cucumber (*Cucumis sativus* L.) and 13 of beans (*Phaseolus vulgaris* L.). From the differences in chromatographic patterns, it was possible to separate the water melon cultivars at 10 days of age into 5 groups. All the cucumber cultivars were individually identified. Four cultivars of beans were identified individually, while the remaining 9 were put into two groups of 5 and 4 each. The authors suggested that this method offers considerable help in simplicity, speed, and convenience to breeders, seed producers and regulatory agencies, in the identification of cultivars.

Pisum sativum L. The peroxidase banding patterns in four yellow and two fodder pea (*P. arvense*) cultivars were studied by Almgård [2] in order to determine whether this method was useful for pea cultivar identification since these cultivars cannot be safely separated by visual inspection of their seeds. Chromatograms of all 4 of the yellow-seeded cultivars showed different patterns but the patterns were not uniform in plants of 3 cultivars. Two separate patterns were noted in a 50 plant sample of one cultivar, one pattern represented by 9 plants. The pattern of the 41 other plants was found in 7 plants of two of the other cultivars but by applying esterase coloring the plants could largely be relegated to the cultivar to which they belonged. The 50 plant progenies of the two fodder cultivars were uniform within cultivars and clearly different between cultivars.

Cucumus sativus L. Brown et al. [14] undertook a study to determine whether the chromatographic patterns produced by flavonoids of snap bean (*C. sativus*) leaves could be associated with cultivar characteristics and therefore could be used as a tool in snap bean breeding. In comparing spot patterns with morphology, 17 of the 43 cultivars had patterns similar to the cultivar Contender and 9 to the cultivar Harvester. The pole or runner type cultivars

were distinguished from the bush type by the absence of a single spot. The results indicated that it was possible within limits to classify snap bean cultivars by chromatographic spot patterns. While a few cultivars had widely differing patterns, the relatively similar patterns in the majority of the cultivars was considered the result of breeders choice of parents rather than a lack of genetic variability. The studies with snap beans suggested the possibility of screening segregating populations chromatographically and discarding undesirable types at an early stage.

Glycine max (L.) Merr. Dhesi et al. [19] have used thin-layer chromatography to differentiate between 4 cultivars of soybeans (*G. max*) from seedling leaf extracts.

Larsen [36] carried out a disc electrophoresis study of 61 soybean cultivars to determine if differences in seed proteins exist which might be used to supplement morphological characters in cultivar identification. Previously, Hilty & Schmitthenner [33] did not find any differences in leaf proteins in a study which involved only 2 cultivars. Larsen [36] found the main differences between cultivars to be the presence of 2 proteins (A and B) which separated the cultivars into 2 groups. Of the 61 cultivars, 13 contained only the A protein and 48 contained only the B protein. All cultivars with black hilums possessed the B protein. The uneven distribution of the two proteins among the cultivars was attributed to the predominance of the B protein in the original parents from which the cultivars were derived. In a later study, Larsen & Caldwell [37] were able to trace back the lineage of the A protein to an introduced cultivar Mandarin.

By starch gel electrophoresis, Brim et al. [13] compared the isoperoxidase banding patterns for the cotyledons and radicles of dry and germinating seed of 4 cultivars of soybeans. Two of the cultivars had the same banding pattern in both cotyledons and roots and therefore, while distinct from the other 2 cultivars, they could not be distinguished from each other by this enzyme system for these plant parts.

In a survey of 447 cultivars and strains of soybeans, Buttery & Buzzell [16] have analyzed for differences in leaf flavonoids and studied their inheritance with the idea of obtaining information that would aid in cultivar identification. By means of thin-layer chromatography and spectrophotometry they were able to group the cultivars according to the presence of flavonol glycosides and they obtained some evidence for geographic differences in the groupings. Thirteen accessions of *Glycine soja* Sieb. & Zucc., a wild species, all contained quercetin glycosides and appeared to be different in flavonols than the brown pubescent cultivars of *G. max*.

Other biochemical characteristics to supplement flavonoid data in the identification of soybean cultivars are the peroxidases, ureases, and a trypsin inhibitor (for references see [16]).

Lotus corniculatus L. Cyanogenesis has been of interest in forage plants for many years on account of its toxic effects to cattle, as a marker in genetic studies [40] and for its use in systematics where its presence or absence in different taxa has served as a distinguishing character [28]. Tests by the author for the presence of HCN in different cultivars of *Lotus corniculatus* have shown both quantitative and qualitative differences between plants and that it is possible to select HCN free plants within a cultivar.

Medicago sativa L. By means of serological tests of seed proteins of alfalfa (*M. sativa*) Esposito et al. [20] were able to distinguish 9 cultivars. They considered immunodiffusion techniques to be sufficiently definitive to separate unknown seed lots of these 9 cultivars from each other.

By gel electrophoresis, Rommann et al. [43] studied the soluble proteins in 4 cultivars of alfalfa. They did not find any qualitative differences in banding patterns between the cultivars. However, after staining the lipoproteins with Sudan Black B different patterns were observed for each of the cultivars. Bingham & Yeh [10] have also identified the seed protein phenotypes from 30 cultivars of *M. sativa* by means of gel electrophoresis. The protein was qualitatively similar in 12 out of 13 bands. On the basis of the presence and intensity of the single band difference, the cultivars could be classified into 4 groups. In addition, within

these 4 groups it was possible to separate most of the cultivars on the basis of visual density differences in other bands. Since most cultivars differed more in electrophoretic patterns than they did in gross morphology, they considered there were sufficient differences to warrant further investigation of the usefulness of electrophoresis in the identification of alfalfa cultivars.

Berrang et al. [8] noted both qualitative and quantitative differences in saponins between two cultivars (DuPuits and Lahontan) of alfalfa. The cultivar DuPuits contained about 1.5 times the total saponin content of Lahontan. The non-acidic soysapogenols A and B were the major aglycones found in Lahontan in contrast to medicagenic acid and oxymedicagenic acid which were the predominant aglycones in DuPuits.

Mangifera indica L. A paper chromatography study of fluorescent compounds from the leaves of 21 cultivars of mango (*M. indica*) has been carried out [49]. Seven cultivars could be distinguished from all the others and the remaining 14 fell into groups of 2 to 5 cultivars each.

Manihot esculenta Crantz. Studies are in progress on the classification of a world wide collection of approx. 2 500 cultivars of cassava (*M. esculenta*) which has recently been established at the Centro Internacional de Agricultura Tropical, Cali, Colombia, South America. This center has initiated research on all aspects of cassava and is specializing in this crop as its major undertaking. The cassava collection has been assembled from experimental stations throughout the tropics and considerable duplicate material is suspected to have been gathered. A comparison of the secondary phenolic compounds from leaves by means of thin-layer chromatography appeared to be a way to readily identify duplicate accessions.

A total of 55 fluorescent compounds have been observed in cultivars of cassava from North and South America. A master chromatogram showing the location of the spots and their R_F values are given in fig. 1. A cluster analysis program has been used for grouping the cultivars. At the time of writing, 993 cultivars including a number of duplicate acces-

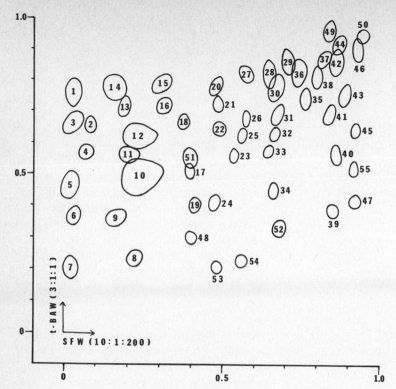

Fig. 1. Master chromatogram showing the location of the 55 fluorescent spots found in leaf samples of North and South American cultivars of *Manihot esculenta* on cellulose two-dimensional thin-layer chromatographic plates.

sions have been classified at the 60 to 95% S value levels. The clustering was as follows:

At the 60% level the 993 cultivars separated into 9 clusters and at the 70% level into 29 clusters. At the 80% level 962 cultivars separated into 107 clusters; 31 cultivars were dropped since they did not have an S value of 80% with any other cultivar. At the 90% level 769 cultivars separated into 239 clusters; 224 cultivars were dropped. At the 95% level 337 cultivars separated into 142 clusters; 656 cultivars were dropped.

By aid of computer programming the cultivars have been clustered specifically in relation to one particular cultivar, Llanera, which is considered to be a very desirable economic type. The evaluation of our classification with morphological and other attributes is still to be completed.

By means of chromatography, spectrophotometry and the use of authentic compounds, a

number of flavonoids and cinnamic acid derivatives have been identified (fig. 2). The flavonoids have been identified as glycosides of quercetin and luteolin. For the cinnamic acid derivatives, alkaline and acid hydrolysis indicated the presence of chlorogenic acid, esters of *p*-coumaric, caffeic, ferulic, and sinapic acids, and the glycosides of caffeic and ferulic acids. The identification of additional compounds is in progress.

Discussion

From a brief review of the literature it is clear that biochemical techniques offer considerable scope for use in the identification of cultivars. With the development of breeder's rights whereby a new cultivar must be distinguishable from cultivars already appearing on approved lists, biochemical markers should be of tremendous value. Almgård [3], a leader in this biochemical approach to cultivar identification, has already been able to have approved a new fodder pea (*Pisum arvense*) cultivar (cv. Timo, derived from the cross Hero×Parvus) which was morphologically indistinguishable from one of its

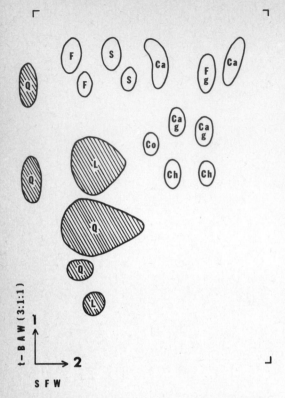

Fig. 2. Phenolic compounds identified in leaf samples of *Manihot esculenta* cultivars by two-dimensional thin-layer chromatography on cellulose plates. Flavonoids (crosshatched): Q, quercetin glycoside; L, luteolin glycoside. Cinnamic acids: Ca, caffeic acid ester; Ca g, caffeic acid glycoside; Ch, chlorogenic acid ester; Co, *p*-coumaric acid ester; F, ferulic acid ester; Fg, ferulic acid glycoside; S, sinapic acid ester.

parents (Parvus) through the use of electrophoresis. Almgård found an esterase marker which revealed a uniform intercultivar difference in an examination of 100 seedlings of each cultivar.

The ease of biochemical intraspecific intercultivar identification would appear to reflect the degree of morphological distinctness which in turn reflects the genetic make-up of the plants concerned. For example, the grouping of the barley cultivars by means of α-amylase electrophoretic patterns clearly reflects the limited genetic background from which the cultivars were derived, and hence, the difficulty in differentiating the cultivars from each other.

At the same time it has been amply demonstrated that even in genotypically closely related organisms the use of additional biochemical markers or systems (such as the use of different enzymes) are likely to differentiate cultivars from one another. A corollary is that it may be necessary to use multiple systems for some time in order to differentiate organisms with a limited genetic background.

An alternative approach is to analyze and compare different tissues or organs using the same technical procedure. For example, Brim et al. [13] have shown genotypic differences in peroxidase isozyme banding patterns of the epicotyl and seed coat in cultivars of *Glycine max*. Similarly, Smith & Frey [46] have found whole grain and embryo to give the best results in their serological studies of oat cultivars. A combination of different enzyme systems and both leaf and root homogenates has been used successfully to differentiate cultivars of *Hordeum vulgare* [4].

Part of the difficulty of precise cultivar identification by biochemical techniques will be the discovery of a certain amount of genetic heterogeneity which may develop in such material and which will be biochemically diagnosable. A number of factors, such as outcrossing and mutation, could lead to genetic differences in plants of a cultivar. Thus unrecognized genetic variability of cultivars may be determined. Such polymorphisms in plants of a cultivar have been reported in the studies referred to here [2, 23, 27, 51].

The correlation of biochemical markers with morphological features and/or resistance to disease and insect pests offers considerable practical use in detecting desirable genotypes. In certain cases it may be possible to eliminate undesirable plants at an early age of growth [14]. In *Manihot esculenta* we are comparing the chromatographic patterns between plants of a cultivar which are susceptible to virus and have a strong mosaic leaf pattern with those of plants of a cultivar which are described as having field immunity. In the chromatograms from the plants of the resistant cultivar there are several spots present which are absent in those of the susceptible cultivar. Further work is in progress to fully confirm these findings.

The chromatographic studies reported by the author on *Manihot esculenta* have been carried out with the capable technician assistance of Dr B. H. Soma-roo and Mr M. L. Thakur. Financial assistance to Macdonald College for studies on *Manihot* has been provided by the International Development Research Centre, Ottawa. The author's studies reported on *Lotus* and the purchase of a Unicam Ultraviolet Recording Spectrophotometer were made possible through grants from the National Research Council of Canada which are gratefully acknowledged.

References

1. Almgård, G, Hereditas 1969, 63, 444.
2. — Pisum newsl 1970, 2, 9.
3. — Hereditas 1971, 69, 287.
4. Almgård, G & Norman, T, Agri hort genet 1970, 28, 117.
5. Asker, S & Frost, S, Hereditas 1973, 73, 17.
6. Baum, B R & Lefkovitch, L P, Can j bot 1972, 50, 121.
7. — Ibid 1972, 50, 131.
8. Berrang, B, Davis, K H Jr & Wall, M E, 1972. Personal commun.
9. Bhatia, C R, Buiatti, M & Smith, H H, Am j bot 1967, 54, 1227.
10. Bingham, E T & Yeh, K J, Crop sci 1971, 11, 58.
11. Blank, F, Bot rev 1947, 13, 241.
12. Bragg, L H & McMillan, C, Am j bot 1966, 53, 893.
13. Brim, C A, Usanis, S A & Tester, C F, Crop sci 1969, 9, 843.
14. Brown, G B, Deakin, J R & Hoffman, J C, J Am soc hort sci 1971, 96, 477.
15. Brunsberg, K, Bot notiser 1965, 118, 377.
16. Buttery, B R & Buzzell, R I, Crop sci 1973, 13, 103.
17. Dedio, W & Kaltsikes, P J, Crop sci 1972, 12, 219.
18. Desborough, S & Pcloquin, S J, Phytochemistry 1966, 5, 727.
19. Dhesi, N S, Desormeaux, R W & Pauksens, J, Proc assoc off seed analysts 1967, 57, 120.
20. Esposito, V M, Ulrich, V & Burrell, R G, Crop sci 1966, 6, 489.
21. Fairbrothers, D E, Serolog mus bull 1972, 48, 8.
22. Fedak, G & Rajhathy, T, Can j pl sci 1971, 51, 353.
23. — Ibid 1972, 52, 507.
24. Fröst, S, Hereditas 1966, 55, 68.
25. Fröst, S & Holm, G, Hereditas 1971, 69, 25.
26. — Ibid 1972, 70, 259.
27. Frydenberg, O, Nielsen, G & Sandfaer, J, Z Pflanzenzücht 1969, 61, 201.
28. Grant, W F & Sidhu, B S, Can j bot 1967, 45, 639.
29. Grant, W F & Zandstra, I I, Can j bot 1968, 46, 584.
30. Hanson, W D & Moll, R H, Genetics 1973, 74, 133.
31. Harney, P M & Grant, W F, Am j bot 1964, 51, 621.
32. — Can j genet cytol 1965, 7, 40.
33. Hilty, J W & Schmitthenner, A F, Phytopathology 1966, 56, 287.
34. Iiyama, K & Grant, W F, Can j bot 1972, 50, 1529.
35. Johnson, B L, Barnhart, D & Hall, O, Am j bot 1967, 54, 1089.
36. Larsen, A L, Crop sci 1967, 7, 311.
37. Larsen, A L & Caldwell, B E, Crop sci 1969, 9, 385.
38. Mitra, R & Bhatia, C R, Genet res 1971, 18, 57.
39. Mizushima, U, Kondo, A & Konno, N, Jap j breed 1963, 13, 88.
40. Nelson, O E Jr, Ann rev genet 1967, 1, 245.
41. Nielsen, G & Frydenberg, O, Z Pflanzüchtg 1972, 68, 213.
42. Pearson, P, J med genet 1972, 9, 264.
43. Rommann, L M, Gerloff, E D & Moore, R A, Crop sci 1971, 11, 792.
44. Scora, R W, Am j bot 1967, 54, 446.
45. Singh, K & Thompson, B D, Proc Am soc hort sci 1961, 77, 520.
46. Smith, R L & Frey, K J, Euphytica 1970, 19, 447.
47. Sokal, R R, Sci am 1966, Dec 106.
48. Sokal, R R & Sneath, P H A, Principles of numerical taxonomy. W H Freeman & Co, San Francisco, 1963.
49. Teas, H J, Winters, H F & Almeyda, N, Proc assoc south agr workers 1959, 56, 223.
50. Wilkinson, J F & Beard, J B, Crop sci 1972, 12, 833.
51. Williamson, J A, Kleese, R A & Snyder, J R, Nature 1968, 220, 1134.

Discussion

Reichstein: Is there a possibility of identifying the compounds responsible for the additional spots in the virus resistant forms and of checking whether these compounds have any direct protective activity?

Grant: The results which I have reported indicating definite differences in the number of spots on chromatograms between resistant and susceptible plants have been obtained only in the last few weeks. We are now in the process of identifying the particular compounds which appear to differentiate these two types and further work will be necessary to determine if they have protective activity.

Birch: An increasing knowledge of phytoalexins shows that compounds may not be present until infection occurs. The pattern of components may be very different in infected to non-infected material, or alternatively there may be no obvious differences between resistant and non-resistant strains until they are exposed to infection.

Heywood: One problem I have come across in working in the cultivars, especially horticultural ones, is the variation from country to country between what is ostensibly the same cultivar. This is evident in our work at Reading on carrot (*Daucus carota*) cultivars in association with the Institut National Agronomique at Versailles, France. The same cultivar from USSR, Germany, France, Japan, Great Britain, etc. may vary even in the micromorphology of the mericarp surface details (using scanning electron microscopy) which we have found to be the most useful means of identification. The part you eat—the carrot root—does not lend itself to easy cultivar recognition except in terms of general shape and colour. The chemical work done by Dr Harborne in our group at Reading tends to confirm the variation.

As regards the cassava cultivars, I wonder if you could explain how you knew your material actually belonged to a particular cultivar before characterizing the material chemically if you did not use morphological features for recognizing it? In other words, if the cultivars can only be characterized chemically how does one know that one has a particular cultivar before analysing it? There seems to be some circularity involved here.

Grant: The International Center of Tropical Agriculture at Cali, Columbia, has been assembling cultivars from different countries throughout the tropics. At present, I can only assume that I am receiving cultivars unless they designate otherwise. They have informed me that they are interested in knowing of plants which have properties similar to what they consider to be a very desirable cultivar called Llanera. For example, they have sent me material of several species for comparative purposes. I would expect that some of the collections do not represent cultivars but rather plants of potential economic value.

Biological and Chemical Screening of Plant Materials

F. Sandberg

Department of Pharmacognosy, Faculty of Pharmacy, Lindhagensgatan 128, S-112 51 Stockholm, Sweden

Summary

Many different ways of approach have been taken in the search for natural products with biological activity. Phytochemical and biological screening procedures are described, and examples of different methods are given. Some screening results are discussed, and some aspects of field work are presented. The importance of plant identification is pointed out.

In this review I want to briefly present some methods used in the search for plants and plant products of pharmacological and chemical interest [30]. The prospects of such investigations are promising, since it has been estimated that only 5–6% of the world's flora has been studied chemically in any detail. The main purpose of the various screening methods is thus the selection of plant material for further study, but the results may also be of interest to taxonomists, biochemists and ethnobotanists.

Basically, there are two ways of looking for new biologically active compounds. We can either search for (1) the compound (phytochemical screening) or (2) for the effect it produces (biological screening). An extremely thorough and extensive review of various screening procedures has been published by Farnsworth [11], with more than 800 references to the literature. Before embarking on our main theme, we would like to cite Farnsworth [11] regarding the personal training and background of investigators in this truly interdisciplinary field. "A great deal of common sense, a broad background in the medical sciences, and some knowledge of plant constituents and of chemotaxonomic relationships are all necessary for one to select the most promising plants for study" ([11], p. 227).

Biological Screening

Pharmaceutical natural products research deals with pharmacologically active substances occurring in plants and also in animals. The first question the pharmacognosist asks is therefore: "Does this or that plant contain any substance with a biological effect?" The second question is: "What structure does the biologically active substance have?" It should be pointed out that both therapeutic and toxic effects are equally interesting at this stage of the investigation.

In order to test the biological effects of plant material simple screening methods have been worked out using mice and rats [29]. This pharmacological screening can be performed in the field or, more usually, in the home laboratory.

Pharmacological screening as part of the field work was carried out by a research group headed by D. B. Taylor. Their procedure in the field was as follows: From plants growing in a region outside Tingo Maria small amounts of leaves, bark and/or root were collected at random. The taxonomist in the group collected sufficient herbarium material for reference. The locality was indicated as accurately as possible. In the simple field laboratory a crude extract of the collected plant material was prepared and injected into mice for the screening procedure. If the screening showed any effect of interest, the Taylor group returned to the place where the plant was found in order to collect a large amount of plant material for further phytochemical and pharmacological research in the home laboratory in Los Angeles. This type of field work requires basic laboratory facilities for extraction and screening and mice must be flown out to the field base.

Nobel 25 (1973) Chemistry in botanical classification

Another way of establishing the pharmacological effects of plants is to carry out the screening procedure on rats in the home laboratory. Our own experience is limited to folklore medicine from two regions in Equatorial Africa: The southwestern part of the Central African Republic (Berberati region) and the northern part of Congo/Brazzaville (Ouesso Region) [28]. With the aid of Swedish Baptist missionaries, who were well acquainted with the regions investigated, suitable persons were interviewed regarding plants used in native medicine, their mode of use and the vernacular names. This information had to be crosschecked in order to ensure that the data were reliable. The man who supplied the information on the use of a certain plant was asked to point out that particular plant in the field. One to 3 kg of the plant part used was collected, dried and sent by air to our laboratory in Stockholm where screening was performed on rats. A positive screening result confirms the value of the local use of a plant. From 200 species about 10% showed pronounced effects, which motivated further investigations.

Another type of pharmacologic screening has been carried out in our department by Samuelsson [26, 27]. Parasitic Loranthaceae species have been screened for their content of basic toxic proteins, tested by intraperitoneal injection in mice. The intraperitoneal toxicity method was also used to follow the purification of the basic proteins of viscotoxin type.

In all the above-mentioned examples the therapeutic and/or toxic effects on man and laboratory animals have been observed. Observations on the effects on domestic animals can also be used. They react in two ways: either they avoid certain plants or they eat the toxic plants and get a specific disease: mouldy *Melilotus* gives "sweet clover disease" (internal bleeding), *Senecio* species produce severe destruction of the liver and so on. Recent examples have come from New Zealand and Colombia.

Under the auspices of the National Cancer Institute in Bethesda, Md, USA, a survey of plants used to combat cancer has recently been carried out by Hartwell [14].

The empirical knowledge of toxic and medicinal plants gained by our ancestors was passed on from generation to generation by oral tradition, and was eventually annotated in the form of herbals and therapeutic manuals, as well as in herbarium notes and ethnobotanical studies. The retrieval of information from these sources can be both cumbersome and laborious, but the method has been used in varying degrees by several workers.

The unequivocal identification of plants mentioned in the *old herbals* can however sometimes be accomplished through a careful study of text and illustrations.

The Mexican Badianus manuscript is an outstanding example. Written by an Aztec physician, Martín de La Cruz, in 1552 and with detailed colour illustrations, it still provides problems and furnishes possibilities for drug plant research. A newly richly commented edition of this book has recently appeared [9].

The advantage of securing information from herbarium notes is obvious. The data are first hand and there is usually no problem in identifying the plant. Altschul [2] has collected 6000 notes of medical interest from an examination of the Harvard University herbaria.

Schultes [37] has discussed the role of ethnobotany in the search for psychoactive drugs.

A very illustrative recent example connected with our own work is as follows: A Norwegian missionary—a physician (not a botanist!)—working in Zaire (Congo ex-Belge) learned of a plant, the leaves of which were used to facilitate childbirth. The taxonomist found the plant to be *Oldenlandia affinis* (Roem et Scult) DC. (Rubiaceae). In the investigation from the Central African Republic ([28], p. 27) the child-birth facilitating effects of this plant had already been established. Phytochemical and pharmacological investigations in Oslo have demonstrated the presence of a polypeptide with oxytoxic effect.

The screening of African species of *Strychnos* for alkaloids in this laboratory began with the information that the root bark of *Strychnos icaja* Baill. was used in arrow poisons and the root as an ordeal poison [28].

Continued pharmacologic screening of various fractions during the work-up of *S. icaja* led to the first isolation of strychnine from an African *Strychnos* species, and to the new alkaloid 4-hydroxy-strychnine [35].

Phytochemical and pharmacological screening was carried out in our department [34, 36] on a unique plant material of African *Strychnos* (well determined by Dr A. J. M. Leeuwenberg, Wageningen). A semi-quantitative estimation of the convulsant and muscle-relaxant effects was also made.

Leeuwenberg [17] has published a botanical revision of the African *Strychnos* species, and also reviewed their use in Central African ordeal and arrow poisons [5]. An excellent summary of the ethnobotany of these plants was compiled by Bisset [4], and was followed by a thorough discussion of the alkaloids, including the screening of 180 herbarium samples [6].

Phytochemical Screening

The natural products that have received the greatest attention with regard to possible future medicinal applications are the alkaloids and saponins. Numerous papers have been published describing the screening for these two types of compounds.

The methods used to detect these compounds must be simple and rapid, but they should also be selective and easy to use. A field test for alkaloids has been described by Raffauf [22], and found by us to be very useful. This test is based on a not completely specific reaction, which means that some compounds other than alkaloids will also give a positive result. Such false-positive reactions will probably have to be accepted in field screening, but by applying several different tests their number can be reduced. Herbarium material has also been tested with some success [24]. The decomposition of alkaloids with time may however affect the results.

To avoid duplication of data, reference publications are available enumerating the plants known to contain certain compounds (e.g. [23, 38]).

Steroidal and triterpenoid saponins and sapogenins are important economic plant products. The possibility of using steroidal sapogenins for the synthesis of cortisone and related substances has lead to their extensive use in the pharmaceutical industry. Today, the synthesis of the above-mentioned drugs and the even more important anticonceptives depends almost entirely on plant sapogenins such as diosgenin (from *Dioscorea* species), stigmasterol (from soy beans) and hecogenin (from *Agave* species). The most important plant sources at present are the Mexican *Dioscorea* species, *D. composita* Hemsl. and *D. floribunda* Mart. et Gal., both of which have a high sapogenin content in the root tubers.

However, some of the *Dioscorea* species have probably been collected to the point of extinction (e.g. *D. sylvatica* Spreng. of southern Africa), or are over-exploited as commercial sources of diosgenin [12, 20].

Many surveys have been carried out to detect saponins in plant material. Simple tests for saponins include the testing of the hemolytic activity of plant extracts as well as the formation of persisting froth when plant material is shaken up with water. These properties of the saponins are, however, common to both steroidal and triterpenoid saponins. A search for one of the two groups must utilize other methods, such as infrared spectrophotometry analysis of the crude, isolated sapogenins to bring about a differentiation. Hardman & Fazli [13] have screened various *Trigonella* species for the presence of steroidal sapogenins in their seeds. The aim of their research is to find a plant rich in diosgenin or related compounds and that lends itself more easily to cultivation than the *Dioscorea* species. *Trigonella foenum-graecum* L. seeds are known to contain diosgenin and in the current investigation *T. coerulea* Ser in DC., *T. corniculata* Sibt et Sm. and *T. cretica* Boiss. were found to contain steroidal sapogenins, including diosgenin. The methods for assaying the seeds were blood haemolysis, colour reaction, infrared spectrophotometry, and thin layer chromatography. The authors conclude that *Trigonella* "may prove to be a good source of steroidal sapogenins˜for the steroid industry and screening is continuing" [13].

In the Soviet Union, great problems have been encountered in supplying these raw materials for the steroid industry. *Dioscorea* does not occur in this country, and the interest of the Soviet research workers has therefore been focused on the screening of *Solanum* species as well as other plants. Some *Solanum* species contain the steroidal alkaloid solasodine, closely related to diosgenin. In a recent paper, Rabinowich [21] has discussed the introduction and cultivation of possible new drug crops of this type in the Soviet Union. *Solanum laciniatum* Ait. is already being cultivated for the isolation of solasodine. Other promising plants are *Dioscorea deltoidea* Wall. of Indian origin and and *Yucca gloriosa* L., which are now being tested as presumptive crops in southern Russia. From an economic point of view the cultivation of *D. deltoidea* and *Y. gloriosa* was deemed profitable [21].

The role of alkaloid screening in the development of a folk medicine into a drug of therapeutic importance is exemplified by *Rauwolfia serpentina* (L.) Benth. ex Kurz. The root of this plant has been used for centuries in Indian medicine as a cure for insanity, epilepsy and high blood-pressure. In the 1930s Indian workers started an investigation of the plant and isolated several alkaloids. However, these alkaloids could not account for all of the interesting effects found in the plant extracts. The search for active principles continued, but not until 1952 was the alkaloid reserpine isolated and shown to possess all the main pharmacological properties of the root.

This discovery initiated an intensive screening for this therapeutically valuable alkaloid in the genus *Rauwolfia*, and subsequently also in the other, closely related genera of the Apocynaceae. *Rauwolfia* is a genus of about 100 species, which are widely spread in the tropics, and alkaloid screening revealed the presence of reserpine in several of these. On the basis of this screening, *R. vomitoria* Afz. of Africa (Congo ex-Belge) and *R. tetraphylla* L. of America (the Caribbean) are now used for the commerical extraction of reserpine.

Reserpine was also found in more distantly related species, e.g. *Alstonia constricta* F. Muell., and the search continues. The pharmacological properties of reserpine were also looked for in other *Rauwolfia* alkaloids and were found in rescinnamine and deserpidine.

In these *Rauwolfia* studies botanists, pharmacognostists, chemists and pharmacologists cooperated to a very high degree as ably shown in the *Rauwolfia* monograph prepared by Woodson et al. [39], and in the more popular report by Kreig [16].

In contrast to the search for reserpine sources,

the phytochemical screening for the occurrence of isothiocyanates and their parent glucosides in the families Cruciferae and Capparidaceae [15] has no immediate therapeutic aspects. These compounds have however been discussed as a cause of poisoning in cattle. The restriction of this type of screening to one type of substance (not a single substance) within two families, gave within a reasonable period of time two results: (1) an excellent chemotaxonomic contribution to critical genera within these families; (2) a decisive criterium in the choice of forage plants.

By checking the avaible literature and by interviewing botanists and chemists in the Middle East it was found that most of the plants from the arid zones here were almost unknown phytochemically. Professor Vivi Täckholm, Cairo, invited the author to start a phytochemical screening of desert plants in Egypt. With the help of the resources of the Desert Insitute, Matariah, Cairo, the occurrence of alkaloids and saponins in desert plants was studied. This project was started in 1956 and has resulted in several important discoveries [32]. The family Chenopodiaceae was found to be promising: triterpenoid saponins were found in *Anabasis* species [31], and alkaloids of new type in *Haloxylon* species [8, 33].

Working within a single family, Lüning [19] has screened more than one thousand species of Orchidaceae for their alkaloid content. A systematic treatment of these alkaloid analyses was attempted but the distribution of alkaloids in Orchidaceae is rather complicated, different types of compounds being found in closely related species. The only way of coming to grips with this problem is to investigate the biosynthesis of these alkaloids. This also applies to other groups of plants, e.g. the Cactaceae, where the two main alkaloid groups, the phenethylamines and the tetrahydroisoquinolines, have been shown to have a common biosynthetic origin [1, 7].

A screening procedure is not necessarily limited to the finding of and the preliminary characterization of given compounds. In a screening process for Cactaceae alkaloids and their biosynthetic intermediates, Agurell [1] and De Vries et al. [10] have used the combination of gas chromatography and mass spectrometry for

Table 1. *Possibly fruitful families for alkaloid study*

	Number of species			
	Total in family	Containing only unnamed alkaloids	Containing named alkaloids	Unnamed / Named
Verbenaceae	2600	41	3	14
Caryophyllaceae	2100	33	2	16
Labiatae	3200	106	23	5
Scrophulariaceae	3000	65	9	6
Amaranthaceae	800	15	1	15
Umbelliferae	2900	37	9	4
Asclepiadaceae	1800	59	13	4
Compositae	20000	371	129	3
Eupherbiaceae	7300	117	41	3
Apocynaceae	1300	171	480	0.4
Papaveraceae	675	2	204	0.1

the rapid identification of already-known compounds in alkaloid extracts. This sensitive technique also makes it easy to observe new alkaloids and obtain information regarding their structures. Such a rapid screening process can presumably present valuable chemotaxonomic information that may be useful in the classification of this taxonomically difficult group.

The continued search for further alkaloid-bearing plants will be directed towards "non-classical" alkaloid families, illustrated in table 1 [18].

This table arranges 9 families in decreasing order of probable fruitfulness in the search.

The families Apocynaceae and Papaveraceae, which have been well searched, are included for comparison.

References

1. Agurell, S, Lloydia 1969, 32, 206.
2. Altschul, S von Reis, Lloydia 1967, 30, 192.
3. Bandoni, A L, Mendiondo, M E, Rondina, R V D & Coussio, J D, Lloydia 1972, 35, 69.
4. Bisset, N G, Lloydia 1970, 33, 201.
5. Bisset, N G & Leeuwenbert, A J M, Lloydia 1968, 31, 208.
6. Bisset, N G & Phillipson, J D, Lloydia 1971, 34, 1.
7. Bruhn, J G, Lundström, J & Svensson, U, Lloydia 1971, 34, 183.
8. Carling, C & Sandberg, F, Alkaloids of *Haloxylon articulatum*. Acta pharm suec 1970, 7, 285.
9. Del Pozo, E C, in Ethnopharmacologic search for psychoactive drugs. USPHS no. 1645, pp 59–76. Washington D.C., 1967.
10. De Vries, J X, Moyna, P, Diaz, V, Agurell, S & Bruhn, J G, Rev Latinoam Quim 1971, 2, 21.
11. Farnsworth, N R, J pharm sci 1966, 55, 225.
12. Hardman, R, Tropical sci 1969, 11, 196.
13. Hardman, R & Fazli, F R Y, Planta med 1972, 21, 131.
14. Hartwell, J L, Lloydia 30–34. (Last installment, with index, Lloydia 34, 386–425, 1967–71).
15. Kjaer, A & Thomsen, J, Phytochemistry 1963, 2, 29.
16. Kreig, M B, Green medicine. Rand McNally Co., New York, 1964.
17. Leeuwenberg, A J M, The Loganiaceae of Africa VIII. Strychnos III. Med Landb Hoogsch Wageningen 1969, 69, 1–316.
18. Li, H L & Willaman, J J, Econ bot 1972, 26, 61.
19. Lüning, B, Phytochemistry 1967, 6, 857.
20. Martin, F W, Econ bot 1969, 23, 373.
21. Rabinowich, I M, Rastitelnye Resursy 1972, 8, 321. (In Russian.)
22. Raffauf, R F, Econ bot 1962, 16, 171.
23. — A handbook of alkaloids and alkaloid-containing plants. Wiley-Interscience, New York, 1970.
24. Raffauf, R F & Altschul, S von Reis, Econ bot 1968, 22, 267.
25. Rusby, H H, J Am pharm ass 1918, 7, 770.
26. Samuelsson, G, Acta pharm suec 1966, 3, 353.
27. — Ibid 1969, 6, 441.
28. Sandberg, F, Cahiers de la Maboké 1965, III, 5–49.
29. — Proc int symp medicinal plants, Kandy 1964, pp 187–201. Colombo, 1966.
30. Sandberg, F & Bruhn, J G, Bot notiser 1972, 125, 370.
31. Sandberg, F & Shalaby, A F, Sv farm tidskr 1960, 64, 677.

32. Sandberg, F, Michel, K-H, Staf B & Tjernberg-Nelson, M, Acta pharm suec 1967, 4, 51.
33. Sandberg, F, Haglid, F & Norin, T, Acta pharm suec 1967, 4, 97.
34. Sandberg, F, Lunell, E & Ryrberg, K J, Acta pharm suec 1969, 6, 79.
35. Sandberg, F, Roos, K, Ryrberg, K J & Kristiansson, K, Acta pharm suec 1969, 6, 103.
36. Sandberg, F, Verpoorte, R & Cronlund, A, Acta pharm suec 1971, 8, 341.
37. Schultes, R E, in Ethnopharmacologic search for psychoactive drugs. USPHS no. 1645, pp 33–57. Washington, 1967.
38. Willaman, J J & Li, K L, Lloydia 1970, 33, suppl. 3A.
39. Woodson, R E, Jr, Youngken, H W, Schlitter, E & Schneider, J A, Rauwolfia: Botany pharmacogonsy, chemistry & pharmacology. Little, Brown & Co., Boston, 1957.

Discussion

Farnsworth: To point out the importance and role of natural products in the American prescription market we are carrying out an analysis of new and refilled prescriptions over a 12 year period (1959–1970). Only prescriptions containing one or more natural products as active ingredients were considered. It was found that about 47% of the prescriptions contained natural products (plant, microbial or animal origin), this percentage remaining constant over the survey period. Higher plant products account for about 22% of all prescriptions (in 1967 about 250 million prescriptions) and this percentage did not change appreciably from 1959 to 1970 (the total number, however, increased dramatically). The total value (in $) of higher plant products in the American prescription market is in excess of 1 billion annually at the consumer level. Of the many plant products represented in the survey (±100), only 3 are produced commercially by synthesis, i.e. caffeine, ephedrine isomers and papaverine, all of the others are still extracted from plants.

Professor Sandberg has mentioned that it is often difficult to enlist the aid of a botanist to collect quantities of plant material for phytochemical investigation. I agree and propose that the phytochemist should add a qualified botanist to his staff to solve the problem. Further, it is even more difficult to enlist the aid of a pharmacologist to screen plant extracts,

and pure compounds. One of the reasons why I left the University of Pittsburgh to assume a similar position at the University of Illinois is that in Illinois I would have 7 Ph.D. pharmacologists on my staff. They have been most cooperative in aiding our phytochemical program and we have now set up an antibiotic screen, an analgesic screen, an anti-inflammatory screen, and will soon have a CNS screen (Hippocratic), as well as several in vitro enzyme inhibition screens, i.e. tyrosine hydroxylase inhibition to predict hypotensive activity and acrosine inhibition to predict anti-fertility activity.

We are now prepared to offer these screening services to each of you and to other natural product investigators, at no charge, within reasonable limits. About 5–10 g of crude extract would be desirable or 25–100 mg of any pure compound.

Professor Sandberg has lectured on the use and value of phytochemical and pharmacological screening in the search for new drugs in plants. Although, at one time I was in agreement with his remarks, today I do not feel that phytochemical screening is of much value in this respect. Random selection-mass pharmacological screening of large numbers of plants is out of the question because of the expense involved, except in a few cases. We are currently investigating new approaches to finding biologically active plants for phytochemical studies. Our first effort will be published soon (J pharm sci), and it involves central nervous system depressant plants. To begin with, we utilized our computorized natural product data bank. We asked the computor for the names of all plants which (a) had a folkloric reputation as being used as sedatives; (b) those whose extracts had been shown to have some degree of CNS depressant activity in laboratory tests. About 575 names were obtained. We scored these plants from 10 to −5 on the following criteria: (a) Laboratory CNS depressant activity; (b) folklore CNS depressant activity; (c) presence of alkaloids; (d) the plant is related to plants known to yield active CNS depressant compounds; (e) the plant has been relatively uninvestigated or is in an obscure genus or family. If the plant was reported as toxic, a value of −5 point was assigned. A maximum score was +30 points, and the lowest score was −5 points for any of the 575 plants. We dropped all plants from consideration that were scored as +14 or less. This left 75 plants on the list. From this list we deleted those plants such as *Rauwolfia serpentina, Papaver somniferum,* etc. that were known sources of sedative compounds. This left a list of 20 plants. We obtained 17 of the 20 plants and evaluated extracts of them in activity cage studies (measuring decreases in spontaneous motor activity) using mice as test animals. Activity of the extracts was compared with that of pentobarbital sodium. It was shown that 9 of the 17 plant extracts were equal to or greater in sedative activity than pure pentobarbital sodium.

Thus, we were able to select 17 plants from a list of 575 suspected to have sedative activity, with more than 50% of the 17 proving to be candidates for phytochemical studies with the hope of finding new sedative compounds.

Because this appears to be a useful method for selecting CNS depressant plants for phytochemical studies we intend to extend our work to other pharmacological categories, i.e. analgesic, anti-inflammatory, etc. These are in progress.

Birch: I should like to draw attention to the CSIRO Phytochemical Survey in Australia as a model collecting organisation for the chemist. Material, in large quantities, is collected by a qualified botanist employed by CSIRO. This is an admirable matter for the chemist and authenticates the material used. The survey also carried out spot tests for alkaloids and other components. These indications are available to assist choice of material. A very few chemists have therefore carried out a great deal of work. Although carried out primarily for chemical reasons, the resulting information is available for botanists in Australia.

General discussion

Takhtajan: I should like to make some general remarks, because as Professor Cronquist has said, we are both "generalists", although we are not generals. There is only one systematics, and this essentially involves correlation and synthesis of all available data from all available sources of information. There is, therefore, properly no "chemotaxonomy", no "cytotaxonomy", and so on; these are jargon words for sources of information, not for methods. One thing we must avoid is the amateurish use of chemical data. Many systematists are insufficiently trained in chemistry, and many chemists do not sufficiently understand systematics. Thus we often see unprofessional conclusions based on chemical data. We need team work to help to integrate chemical information into the traditional classification of organisms. In seeking to make new classifications, therefore, chemists should take into account those founded on broad comparative-morphological bases. In spite of these criticisms, I think we all realise that at this Symposium we have begun an important and useful dialogue between chemists and systematists, and I hope very much that we will be able to continue this at the XIIth International Botanical Congress in Leningrad in 1975.

Harborne: I would like to stress that the chemical groupings of families related to grasses were presented in my paper (p. 103) without suggesting that these represented a *taxonomic* reclassification. The evidence, in the case of the grasses and palms, I regard as being strong, since there are 5 chemical characters uniting the two families. I would add that a numerical analysis of many morphological characters has supported this view and it seems probable that the grasses and palms should be placed nearer to each other than present systems allow.

Turner: Professor Takhtajan has re-emphasized the statement by Professor Cronquist that systematists should be generalists. The trouble is that in the area of higher plant phylogeny, generalists are variable as regards their capacity to generalize. I would like to make the analogy that such a generalist is like a commander in charge of an army. He enters the fray with a certain kind of deployment. But history will show that most battles are won by the general who develops the best weaponry. In fact, weapons make *all* the difference between modern armies. It is my opinion, then, that the systematic general who wishes to emerge "victorious" from the phyletic battlefield must recognise *and use* the superior weaponry. Thus, comparative protein chemistry is a king of armament that heretofore has not been available to the generalist. He simply cannot treat this as just another private in uniform or, indeed, a modified cannon. And I think that Professor Cronquist and perhaps Professor Takhtajan both recognize the potency of such weapons. But, alas, they already have their forces deployed and their battle lines drawn. Perhaps had they had our present knowledge of cytochrome *c* before they constructed their respective systems, they would have deployed their forces (taxa) differently. In any case a good general engaged in such activity, today, should not only use the new weaponry, he should also understand its potency and limitations. I believe most such generals can understand the data being produced by comparative chemistry and the conceptual reasons for its generally superior resolving power, if they take the trouble to study it carefully. If they cannot they should at least designate an appropriate military aide to advise them accordingly.

On the other side of the coin, I would like to point out that many chemists working in the area of chemosystematics are woefully ignorant of what most evolutionary biologists are about, as indicated by some of the discussions following some papers presented here. Heywood has aptly remarked that many of the questions smack of arguments resolved by biologists nearly a century ago. It is surprising that chemists seem to discover again and again many principles and conceptions elucidated by plant systematists of earlier generations. For example, Zuckerkandl & Pauling in their delightful paper on the new field of "paleogenetics" published some years ago,[1] also in a Swedish journal, seemed to have discovered anew, in their proposal for the reconstruction of ancestral proteins by a comparative study of extant proteins, what has long been a pastime for plant taxonomists: the reconstruction of "primitive type" flowers in their minds eye by studying the floral morphology of numerous extant flowers. Indeed, the early American systematists Bessey did this: as have, no doubt, Professors Cronquist and Takhtajan in the construction of their systems. Hence, these workers tend to think of the *Magnolia*-type flower as a representative of the "primitive type" flower. The kind of subjectivity with which such floral reconstruction is fraught is perhaps best understood when one considers the phyletic arrangements accorded the angiosperms by the late John Hutchinson. Using the same kinds of data available to most plant taxonomists, he came up with quite different interpretations from those workers cited above. And that is the problem with morphological data; it is open to so many interpretations.

Compare this state of affairs with the present knowledge of cytochrome *c*. The distinction between primitive and derived types is not under serious controversy, and most biochemically-oriented workers readily agree as to what its ancestral amino-acid sequence might have been. If similar data were assembled for other proteins we would be sure to have an increasingly "one-model" phyletic tree constructed, albeit using characters from every discipline.

I can best conclude this sermon by an appeal to plant systematists to become more familiar with the conceptual bases for the interpretation of chemical data (emphasized by Mabry and others) but with an equal appeal to those chemists working in the area of plant systematics to become familiar with evolutionary concepts. The successful generalist must have both kinds of insights.

Heywood: Might I suggest as an aphorism that in classification a character is only as good as

its correlative value. This applies to any kind of character, including chemical, and explains perhaps why systematists are often reluctant to accept chemical evidence which leads to disruption of existing classifications. Another dictum is in fact that classification must make sense.

I believe that in meetings of this kind involving botany and chemistry, we are faced with problems of presentation and organisation as regards our data. The chemist, on the one hand, can state clearly and unambiguously what his results are, and although he does not go so far (at least in most cases) as to say "these are the chemical facts, now let us see yours", he is certainly in a stronger position as regards marshalling his evidence. The systematist, on the other hand, is unable to set out, detail by detail, organ by organ, system by system, what his evidence is, since it is derived at so many levels from so many sources that it would require many hours or pages to go through it. Often the information has never been brought together unless the group in question has been the subject of a recent monograph. One is therefore frequently attempting to assess precise chemical data against an uncorrelated or unorganized systematic background. The same applies to systems of classification, such as those of Professors Cronquist, Takhtajan and others where the full documentation is not published by them. In the case of the Hutchinson system, the author himself stated that only morphological information was used, apart from anatomical data in some cases supplied by Dr Metcalfe and pollen data provided by Professor G. Erdtman.

Systematics can be likened to a mincing machine into which data of all sources is fed and processed to form a series of "sausages" encapsulating an unspecified amount of data, and recognisable only by a few characters. The basic recipe for the construction of these "sausages" is usually secret, yet it is such encapsulated pieces of information that we have to work with and communicate with. Systematics can also be regarded as a waste disposal machine although what one regards as waste depends on the judgment of the individual worker.

[1] J theor biol 1965, 8, 357.

Hedberg: Utilizing the metaphor suggested by Professor Heywood, I suggest that botanists as "sausage-makers" must make a compromise between the fabrication of large quantities of low-grade "sausage", containing mainly coarse morphological fragments, and smaller amounts of high quality "sausage" spiced with chemotaxonomical, embryological, and other ingredients. Most taxonomists certainly prefer to concentrate on the time-consuming spicy grade, but for obvious reasons we have to work largely with the quick service coarse variety. Large areas of the globe are still insufficiently known botanically, and thousands of species are being exterminated before they can be classified. For our high-grade "sausage"making or detailed taxonomic revisions, we should try to utilize characters from as many independent fields as possible, such as gross morphology, anatomy, embryology, cytology, chemotaxonomy and palynology. However, when attempting to approach a phyletic classification we should utilize only such characters which disclose the direction of evolutionary change, and in this connection the amino acid sequence data presented to us by Professor Boulter seems to be very promising. Another set of characters where the direction of change can often be traced is found in pollen morphology.

Integration of all the types of data mentioned in this Symposium and several others into a balanced system requires a wide knowledge on the part of the botanist. Botanists with sufficiently multifarious training to carry out such tasks are extremely scarce. Indeed, even the supply of taxonomists capable of adequate manufacture of the coarse variety of taxonomic "sausage" corresponding to the writing of local floras and determination keys is insufficient, and so is the funding for taxonomic research. I would, therefore, like to put in a plea for stronger support for taxonomic research and training of taxonomic botanists.

Grant: I wish to re-emphasize the fact that many biosystematists have entered the area of chemosystematics simply to obtain additional characters to help resolve relationships between the taxa under study. For example, when the biosystematist is confronted with a complex genus, or subgenus, where the species have the same chromosome number—as in the situation of the cultivars which I have discussed—or where chromosome pairing relationships between taxa have not provided satisfactory information. Chemical data are analogous to the use of chromosome banding techniques and nuclear DNA content to elucidate species relationships.

Tétényi: We must first underline the equivalence of chemical characters with any other kind of character used for the classification of living organisms. The various codices of plant nomenclature accept this situation totally. We have to take into account, however, that different plants have their variation in form and function, and this means a variation in their morphological and their physiological attributes which is due ultimately to differences in their chemical make up. We should remember that the stability of chemical characters are generally determined genetically, although this cannot be taken for granted and must be proved by studies over several generations, and by taking into account the h^2-value, which enables one to distinguish between environmental and hereditary factors.

The appearance of a given chemical character can be taken as general for the taxa concerned but we must recognise that quantitative differences exist. The variability of chemical characters can be arranged roughly at three biochemical levels; DNA-RNA, enzyme-protein, and finally metabolic activity. These correspond to the cytological, anatomical and morphological levels of plant structure. The determination of characters is dependent on this three-level hierarchy being more and more difficult as one delves closer to the genome. Usually, however, the chemical characters and morphological structure are more or less interdependent.

The most important facet of the use of chemical characters is the homology of biosynthetic routes. An examination of this homology can often give us a real answer to taxonomic problems and sometimes an insight into phylogeny.

Runemark: One important difference between morphological and chemical characters is as follows. Both types of character are governed by complex genetic systems which can easily change by mutations. The system controlling morphological characters, however, is comparatively stable since genes with additive effects play the most important role. Mutations, therefore, rarely give a drastic morphological effect. Mutations affecting chemical systems, on the other hand, have relatively often considerable effect on the structure of the chemical compounds produced, because a single gene may control an important enzyme on any given biosynthetic route.

Reichstein: It should be noted that it is often found that the presence or absence of some particular chemical compounds may be taxonomically significant at the species or the genus level, but not on higher levels. A good example is the cardiac glycosides. In some genera of the Apocynaceae, e.g. *Strophanthus,* all known species are rich in cardenolides, other genera in the family contain none. In *Acokanthera,* only the representatives of the subgenus Acokanthera are rich in cardenolides, those of the subgenus Carissa contain only small amounts. We must also accept the fact that families as far apart from the Apocynaceae as Scrophulariaceae, Cruciferae, Euphorbiaceae, Liliaceae, etc. do contain cardenolides, some again only in a few genera. In the Cruciferae, only the members of *Cheiranthes, Erysimum* and a few other genera contain cardenolides; the same is true for the Euphorbiaceae.

Bu'Lock: There are two separate points which I would like to offer for discussion, which emerge from biosynthetic and biochemical considerations. First I believe that taxonomists should apply the same criteria of the validity to chemical characters as they do to botanical characters, the difference being that the application of these criteria in the former category requires a different expertise. Some characters are false because they are heterogeneous, but realization of this may be slow. The unqualified use of such invalid categories as "tannins" or "alkaloids" is disappearing in the face of their very apparent heterogeneity, but seemingly better defined categories such as "anthraquinones" may be equally heterogeneous. This was pointed out by chemists many years

ago, but even now, when the biosynthetic heterogeneity of such groups is well-established, it is still liable to be misunderstood by less critically informed botanists. The second point is one which follows from the possibility of interspecific transfers of genetic material which is now well-established even for higher organisms. In all phylogenetic arguments it is assumed that characters are only derived ancestrally, and though we should still accept this for the overwhelming majority of transvers, occasional interspecific transmissions have almost certainly occurred. They are, I suppose, most probable between plants and infective fungi, with backtransmission to other plant hosts, and I believe that if we look at the plant and fungal distribution of, for instance, the ergot alkaloids, gibberellic acid, or flavonoids, we shall be persuaded to take such mechanisms into account. Far from disrupting any structure which chemosystematics may be able to assemble, I believe such mechanisms may ease our task by allowing us to take certain anomalies rather less seriously than, we would at first sight be inclined to do.

Mears: I had noted that the process of genomic transfer among micro-organisms *via* virus vectors could occur in plants and had suggested that such a process may have occurred in vascular plants. During the preparation of my paper for the AIBS-ASPT symposium on the "Amentiferae" (Minneapolis, 1972), it became obvious to me that the verified distribution of the structurally moderately-complicated tylophorine alkaloids in *Tylophora* (Asclepiadaceae) and *Ficus* (Moraceae), two groups usually never considered closely related, either reflects an atrocious error in plant sampling and identification or a transfer of genetic material through non-reproductive processes.

Sørensen: I belong myself to the group of natural product chemists who have only worked with the so called micromolecular compounds. Summing up my experience, I think that we chemists should be very cautious in trying to assess the taxonomic importance of our work. The amount of material we have studied is quite insufficient, and the way we have handled the materials often very improper. For example, consider the stilbenes, since the concept of chemotaxonomy in Scandinavia starts with Holger Erdtman's work on the stilbenes from pine heartwoods. It was somewhat of a shock when about 3 years ago a number of similar stilbenes were isolated from spruce needles and this finding reminded us of the need to investigate all parts of a plant before making any statement about what it can synthesize or not. If we see what plants have been analysed for all known classes of compounds in all parts of the plant, the data dwindles to relative insignificance.

However, in spite of my general pessimism, I have found some promising lines to follow, and I am particularly grateful to Dr Bu'Lock and some of his biochemical colleagues in this respect. The joke that "all characters are equal" has been repeated often during this meeting. I quite agree that this is true if you are a plant breeder or the like and only have the need for looking for some simple character which differentiates varieties. But if you have anything more profound to consider, like phylogeny, I must repeat one need which a chemical character has to possess, which I have stressed before: it has to be conservative. What Bu'Lock and his colleagues have shown is that in fatty acid biochemistry, there occur certain fundamental steps which are responsible for the development of whole classes of secondary products. These steps are seemingly all very old in an evolutionary sense and very conservative. The synthetic ramifications from these steps, however, are of importance. For example, the cyclopentene acids of the Flacourtiaceae and the cyclopropene acids of the Sterculiaceae must have deviated early; again two different acetylenic acid varieties, the Santalaceae-type and the crepenynic acid-type which leads to all the hundreds of acetylenes of the Basidiomycetes and the Compositae-Umbelliferae, show a fundamental difference.

The optimism aroused by these biochemical insights is reinforced by chemotaxonomic results from animals. Dr Boulter selected the living fossil *Gingko* as his reference species yesterday. In fishes and reptiles we have a large number of living fossils, and I have personally been very much impressed by the results of bile acid research. This gives a convincing demonstration of conservation through some 200–300 million years of biochemistry. As we slowly increase our knowledge of biochemical steps of equal importance, I suppose we will then have valuable information bearing on the phylogeny of plants.

Mears: There is a difference between complicated morphological and complicated chemical characters. Since all described plant species must have been examined at least once, it is likely that most of the very unusual morphological characters (such as the peculiar leaves of *Nepenthes* or the pollenia of orchids) would have been found and noted wherever they occur among plants. However, that is not the case with many different kinds of chemical compounds. I find it inconceivable that unrelated plant taxa could produce an alkaloid as complex as the daphniphylline, but I have to admit that I do not indeed know the distribution of these compounds. Very often, then, chemists find only what they search for and that, to a great extent, distinguishes the known distribution of chemical characters from the known distribution of morphological characters.

Hegnauer: Characters are made use of in plant systematics according to their correlation with other characters. If the correlations are unsatisfactory, either the character or the taxa might have been misinterpreted or, alternatively, the character may be useless at the taxonomic rank concerned. If reinvestigation of the character proves that it was misinterpreted, its taxonomic meaning has to be re-evaluated. For example, one can consider anthraquinones in Rhamnaceae and Rubiaceae. After it was shown that anthraquinones are elaborated along different pathways in these two families, this character can no longer be used in estimating similarity between them. Nevertheless, anthraquinones remain perfectly good characters at the tribal and generic levels in both families. The same applies to morphological characters, e.g., when the pseudosyncarpic nature of a gynaeceum has been demonstrated, it will no longer be confused with a true syncarpic one.

Throughout the plant kingdom, present knowledge of the true nature of characters is very unevenly distributed. Systematic botany has to classify plants on the basis of present knowledge. Improvement of classification will go on as long as our knowledge about characters and their occurrence in taxa increases.

Birch: It is perhaps worthwhile to note that I made the division into "polyketide" and "other" types of anthroquinones in 1957. I was not able at the time to define the route to the "other" type, as has been done since. Nevertheless, structural considerations did make the division and if this had been believed it could have been used. I think this is more generally true, and it is clearly difficult to prove every theory biochemically before attempting to make use of it.

In connection with biosynthetic control mechanisms, it seems clear that one reason why secondary metabolites sometimes accumulate in large quantities is that control mechanisms may be minimal, unlike the close control necessary with general metabolic pathways. The genetic data in *Dahlia* are very instructive. In this polyploid the quantity of flavonoid formed increases up to a limit which is presumably determined by available precursor. After a certain genetic dosage, increased genes for a given character increase the competitive power, and effect the composition of pigment mixtures without altering the total quantities. This kind of picture may be much more general if all biogenetically related substances in adjacent sequences are taken into account.

Swain: Besides the structural chemical data, whether coming from micro- or macromolecules, one should also consider the important information which has been obtained from an electrophoretic and serological examination of relatively crude protein mixtures. In many cases, as Professors Jensen and Fairbrother have shown, these can lead to important taxonomic discrimination. However, there are a number of anomalies which need to be taken note of.

Jensen: The significance of serological findings varies according to whether one uses single (pure) proteins or mixtures of proteins as antigenic systems. When using and comparing a

pure protein, we have to remember that it has only a limited number of determinants, the range of which in different taxa *can* lead to irregular "nonsystematic" results. When using mixtures of many proteins we get comparative data from many different molecules and we are unlikely to be misled in this way.

Ourisson: I would very much like to see botanists put chemists to test by listing problems of interest to them, and for which they feel that some additional information could be valuable. As Professor Cronquist mentioned that he could provide a list of such problems, I suggest this list should be published with the proceedings.[1]

Sørensen: I have always found that botanists list critical species, genera or sections in a very complete way in their publications. The difficulty as far as I am concerned has always been to get authentic material. In the Compositae, for example, I listed some 25–30 cases which needed chemical examination. I have, through Hortus Bergianus, only obtained one single example, all my efforts to get the rest have been in vain.

Swain: It is obvious from the wide-ranging nature of this discussion that we have come a long way in the last ten years since the first publications of Hegnauer and Alston & Turner. It is certain that the next ten years will see an ever increasing interest in the application of chemical and biochemical data to taxonomic problems, and all who have contributed to the success of this Symposium should be happy in the knowledge that it has undoubtedly instituted a necessary dialogue between botanists and chemists. I am sure that all look forward to the continuation of our discussions in the near future.

---

[1] See p. 316.

A list of some angiosperm families of controversial or doubtful affinities
on which chemical data might appropriately be brought to bear

Bataceae – related to Caryophyllales, or what?

Buddlejaceae – related to Loganiaceae or Scrophulariaceae, or what?

Callitrichaceae – related to Boragniaceae–Verbenaceae, or what?

Calyceraceae – better with Dipsacales or Campanulales,?

Caprifoliaceae – related to Rubiales, Cornales, or what?

Coriariaceae – related to Ranunculales, or to Sapindales–Rutales, or what?

Corynocarpaceae – related to Ranunculales, or Celastrales, or what?

Crossossomataceae – related to Paeoniaceae or Rosaceae or what?

Elaeagnaceae and Proteaceae – related to each other, or not, and closer to Rosales or Myrtales?

Euphorbiaceae – related to Celastrales or Malvales or what?

Fouquieriaceae – related to families of Violales, or to Polemoniaceae, or what?

Haloragaceae and Gunneraceae – really related to Myrtales, or what?

Hippuridaceae – related to Haloragaceae or to Scrophulariales, or what?

Hydrostachyaceae – related to Podostemonaceae. or Scrophulariales, or what?

Juglandaceae – related to Hamamelidales–Fagales–Myricales, or to Anacardiaceae?

Lecythidaceae – related to Myrtales, or Malvales or what?

Nelumbonaceae – related to Nymphaeaceae or not?

Oleaceae – related to Loganiaceae, or Scrophulariaceae, or what?

Petrosaviaceae – related to Triuridaceae, or Liliaceae, or what?

Rafflesiaceae cum Hydnoraceae and Mitrastemonaceae – related to Santalales, or what?

Rhizophoraceae – better in Cornales or Myrtales?

Sabiaceae – related to Ranunculales, or Sapindales, or what?

Triuridaceae – related to what?

Typhales – related to what?

Urticales – related to Hamamelidales, or Malvales?

A. Cronquist

Concluding remarks

Human beings are biased, and in making these concluding remarks, I lay claim to considerable humanity.

The primary objective of this meeting was to build communication bridges between taxonomists and chemists, and then to use these to promote fruitful discussions. Both the Organising Committee and the participants can be congratulated in that this objective was fully realised; it became clear during the meeting that the points of view not only of the chemists and taxonomists, but also of the cytologist, geneticist and ecologist, were being interwoven.

Taxonomists at Work

Cronquist, Heywood and Takhtajan explained that taxonomists construct taxonomic systems by considering the sum total of the similarities and dissimilarities among organisms, and that this totality of information should be reflected in the final system adopted. However, until now, morphological characters have formed the major source of information on which taxonomic schemes have been based. As Cronquist pointed out, this is due partly to the relative ease of obtaining this information and also because man is psychologically adapted to the use of visual pattern data. Turner suggested that whilst we must accept morphologically based schemes as a historical fact, it would have been most interesting if the data had been available, to have constructed schemes solely using chemical data and then fit the morphological data to them.

The method of organising information into a coherent taxonomic system, is subjective in nature—the brain processes the information, balancing and weighing the many considerations in a way not fully understood. In presenting their systems, taxonomists note down relatively few of the characters they have used and even if pressed, would be unable to say clearly exactly how they arrived at their final scheme.

We are fortunate in having Cronquist and Takhtajan with us, since they are the authors

responsible for the two most widely accepted systems. Although these were arrived at independently they are very similar; but it should be borne in mind that there are many other systems based mainly on morphology and constructed using this intuitive process, which are contradictory to them and to each other.

Major Problems of Constructing Phyletic Schemes

The question of the relative merits and demerits of phyletic versus phenetic classifications was not discussed, but many speakers and discussants tacitly accepted the desirability of making a phyletic classification. However, this view would not be held by all the general taxonomists attending, and some have not yet made up their minds one way or the other.

Two major difficulties arise when using comparisons of features of present-day organisms to construct phyletic schemes, one is to identify homologous structures, i.e. to recognize convergence when it occurs, and the second, to avoid distortions due to differing rates of evolution.

There is clear evidence that these difficulties cause problems with schemes based mainly on morphological characters. The use of amino acid sequence data, on the other hand, largely avoids the difficulties. I pointed out that although the data are sparse, there is little evidence for the occurrence of convergence of protein structures during evolution as judged by the amino acid sequences from animals, or to a lesser extent, from plants. Furthermore, use of the ancestral sequence method of tree construction largely avoids distortions due to differing rates of evolution.

The position with respect to microchemical data is less clear. Many speakers, Birch, Bu'-Lock, Hegnauer, Herout, Lavie and Tétényi, to mention a few, gave examples where recording just the presence or absence of particular compounds was misleading; the whole biosynthetic pathway has to be considered. Schwarting discussed the situation with the quinolizidine alkaloids, which originate from several primary biochemical systems, in the Leguminosae, Lycopodiaceae, Lythraceae and Nymphaeaceae. Whereas the existence of separate biosynthetic routes of a group of compounds is of considerable use in distinguishing between related groups of organisms, comparisons of the occurrence of compounds made between organisms with different biosynthetic routes for them, could be misleading. Whilst these speakers saw lack of biosynthetic information as a serious limitation, Bate-Smith felt that we must work with what is at present available, since we might have to wait a long time for the biosynthetic pathways to be elucidated in some cases.

The ability of the organic chemist to predict biosynthetic pathways from a consideration of the structures of related compounds was clearly demonstrated in Birch's lecture. As he put it, "Nature is a good organic chemist". In considering reasons for the occurrence of the vast range of secondary compounds, Birch developed the idea of "target" molecules and target activities. Gibberellic acid was given as an example of a target molecule relative to diterpenes, and colour as that of a target activity with reference to betalains or anthocyanins.

Although much can be established about biosynthetic pathways from a consideration of the chemical structure of their intermediates, Reichstein stressed the importance of knowing about the enzymes involved. Here there are large gaps in our knowledge and biochemists have shown a surprising reluctance to work on the enzymes of the metabolism of secondary compounds, a term objected to by some.

Chemists at Work

A feature of the symposium was the considerable amount of elegant chemical data presented. Limitations of space allow me to mention, at this point, only the work of Kjær with glucosinolates, Wagner with glycolipid resins and Shibata with the triterpenes of lichens.

Many speakers felt that the usefulness of many groups of micromolecules as chemical characters in taxonomy was at the generic level or below. Thus, Lavie has shown in *Withania* the existence of several chemotypes differing in

the substitution pattern and contents of withanolides, but having identical morphology. Two further types were also identified, but in the latter case certain morphological differences exist. Cross-breeding experiments and determination of the products produced in the offspring has led to the understanding of the biogenetic pathways leading to specific substitution patterns; chemical groupings in specific locations of the compounds were found to be of dominant character, whereas others were recessive. A further elegant study of this type was that of Schantz & Forsén, on the ethereal oils of *Chrysanthemum vulgare*.

Turner & Flake, in a contribution on volatile constituents and their utility and potential as taxonomic characters, concluded that the qualitative and quantitative expression of these compounds, especially terpenes, has a genetic base. Although environmental factors cause variability, considerable insight about speciation or adaptational processes, especially at the specific level or lower, can be obtained by using populations. Their conclusions were based particularly on chemical data from numerous populations of *Juniperus virginiana*. Population structure in some *Circaea* spp. was discussed by Weimarck, who showed that variations in chemical compounds indicate inter-populational heterogeneity. As pointed out by Turner many years ago, chemical characters are particularly useful in deciphering the occurrence of hybrids in populations of closely related species. Aynilian, Trojánek & Farnsworth, on the basis of a phytochemical examination of the alkaloids of *Vinca* species, concluded that *Vinca libanotica* should be classified as a distinct species. Lastly, Grant, describing the use of chemosystematics in the classification of cultivars gave examples where these methods were now of economic importance.

However, some molecules, such as the sesquiterpenes of *Artemisia* spp., the pseudoguaianolides of the Compositae and the flavanoids of the Gesneriaceae, are accepted by some taxonomists as capable of suggesting solutions to problems above the generic level. Geissman's studies of the sesquiterpene constituents of *Artemisia* spp. belonging to the limited section Tridentatae, showed that these compounds provided characters of use in recognising alliances and distinctions between four groups of species. However, he clearly pointed out that crossovers occur not only between groups within the section but also with species in other sections and, in some cases, other tribes of the family. Herz showed that the pseudoguaianolides and modified pseudoguaianolides found in the Compositae, can be divided into two groups. On the evidence gathered so far, one group are characteristic secondary metabolites of the Ambrosiinae and *Parthenium* (Melampodiinae), and the other typical of certain Heleniinae, although occasionally encountered elsewhere. One may also mention in this connection, Harborne's use of flavonoids as systematic markers in the family of Gesneriaceae, and that of Natori et al., with white and brown rot species of the Polyporaceae. Nevertheless, even for these compounds, many participants would strongly object to such generalization above the species level. In contrast, Swain, Bu'Lock, Bate-Smith and others, believe that relationships between even higher taxa may be indicated by comparisons of suitable compounds, e.g. isoprenoids, flavonoids. An example was given by Fredga, Bendz & Apell, in the occurrence of 3-desoxyanthocyanidins in mosses and ferns but only rarely in angiosperms. This compound does occur in several species of *Juncus* and *Luzula* they investigated, and if the pigment pattern found here is representative of the whole family, the Juncaceae would be of special interest among the angiosperms.

Ourisson surveyed the evidence from fossil chemistry and concluded that, as yet, it could not play an important role in chemotaxonomy.

The Great Betalain Battle

One of the focal points of the conference was the difference in viewpoint between, in particular, Cronquist and Takhtajan, on the one hand, and Turner and Mabry, on the other, as to the significance to be placed upon chemical data in relationship to the morphology in the Centrospermoid families (Caryophyllales). Of the twelve families normally referred to this order, i.e. excluding the Bataceae, two, Caryophyllaceae and Molluginaceae, do not contain betacyanins or betaxanthins. The restriction of the betacyanins and betaxanthins (betalains) to the other families of the Caryophyllales was

first reported by Mabry, Taylor and Turner in 1963, and it had become generally thought that these authors advocated the removal of the Caryophyllaceae and Molluginaceae from the Caryophyllales. During the course of the conference it became clear that this was a misintelligence and one which had led to some confusion on both sides. More recently, Behnke & Turner reported that all the families of the Caryophyllales, including Caryophyllaceae and Molluginaceae, contain specific sieve-tube plastid inclusions, not found in other dicotyledonous families, including those of the related orders, Polygonales and Plumbaginales. This reinforced Cronquist and Takhtajan's refusal to accept what they considered insufficient grounds for setting Caryophyllaceae aside, particularly in view of considerable morphological evidence upon which the families of the order were originally circumscribed. Bu'-Lock, in support of Turner, pointed out that the biosynthetic pathways leading to the betalains are biogenetically very different to that of the anthocyanins, and that an "evolutionary reversal" was not a simple matter in this case. The present position is that Turner and Mabry would split the order Caryophyllales into two orders or sub-orders, one with anthocyanins present and betalains absent, containing the Caryophyllaceae and Molluginaceae, and the other with betalains present and anthocyanins absent, containing the ten betalain families. Turner and Mabry's view is that the Caryophyllaceae and Molluginaceae are related to the other families of the Caryophyllales but that they should be treated as a separate phyletic group. Cronquist and Takhtajan, however, remained unconvinced that these chemical characters have the phyletic significance afforded them by Turner and Mabry. Merxmüller emphasized that in view of the morphological evidence from the order as a whole, a compromise positioning of the two non-betalain families was not possible. The whole discussion illustrated the problem of weighing the significance of chemical evidence when it conflicts with morphological data. The betalain question was not resolved and, no doubt, both sides will hope to accumulate further ammunition for a renewal of the discussion (battle?) at Leningrad, in 1975.

The Present Situation in Chemosystematics

I felt that there is still some reluctance on the part of classical taxonomists to consider some of the chemical data now available. I think this stems from a feeling on their part that they cannot always evaluate these data correctly. Mabry has suggested that they could avoid this difficulty by co-operating more closely with the chemist and accepting his evaluation. Whilst, in the past, chemists have sometimes suggested changes in classification on the basis of relatively few, albeit precise, chemical characters, I detected that most of the chemists here were prepared to present their data and leave it to the taxonomist to evaluate taxonomically. The same problem exists for the "classical" taxonomist in the interpretation of comparative serological results, as presented by Jensen. There are considerable difficulties, both technical and interpretative, in the use of these methods and as a result most of the data have been collected without reference to systematic questions. The earlier results obtained with plants have been severely criticised and this has led, in my view, to some neglect of this approach. Hopefully, more serological studies will now be undertaken, bearing in mind the very real warning given by Jensen as to the difficulties involved. In his discussion of the results obtained with the Ranunculaceae and Fabaceae, he concluded that antigens consisting of many proteins were more useful than those consisting of a single purified protein. However, several participants felt that this conclusion was not of general applicability and pointed out that it is important to know something about the structure of the protein or proteins which are being injected as antigens.

It also became clear during the course of the meeting, that chemists are much more aware of the relative usefulness of different types of chemical compounds and reaction pathways in different taxonomic situations; both Sandberg and Kjær stressed the importance of correct plant identification. As we learn more about the apparent anomalies of some of the chemical findings, for example Bu'Lock pointed out that control mechanisms may be extremely complicated in higher organisms, so the usefulness of chemical data for taxonomy increases.

It is on this optimistic note that I have chosen to end. However, before doing so I would like, on behalf of the participants, to thank our hosts for arranging this fruitful Symposium, and to extend to the Organising Committee, in particular Professors Fredga and Bendz; the Sponsors, The Nobel Foundation and its Nobel Symposium Committee financed from the Tri-Centennial Fund of the Bank of Sweden and the Nobel Institute for Chemistry; the Editorial Committee and Secretariat, and the staff at Södergarn, our sincere appreciation of their help.

Donald Boulter

Nobel Symposia 1–22

Publishers:
Almqvist & Wiksell, Stockholm, Sweden
John Wiley & Sons, New York, USA

Nobelsymposium 1, 1966
  Muscular afferents and motor control
  (ed Ragnar Granit)

Nobelsymposium 2, 1967
  Prostaglandins
  (ed Sune Bergström & Bengt Samuelsson)

Nobelsymposium 3, 1967
  Gamma globulins
  (ed Johan Killander)

Nobelsymposium 4, 1968
  Current problems of lower vertebrate phylogeny
  (ed Tor Ørvig)

Nobelsymposium 5, 1967
  Fast reactions and primary processes in chemical kinetics
  (ed Stig Claesson)

Nobelsymposium 6, 1968
  Problems of international literary understanding (ed Karl Ragnar Gierow)

Nobelsymposium 7, 1968
  International protection of human rights
  (ed Asbjörn Eide & August Schou)

Nobelsymposium 8, 1968
  Elementary particle theory
  (ed Nils Svartholm)

Nobelsymposium 9, 1968
  Mass motions in solar flares and related phenomena
  (ed Yngve Öhman)

Nobelsymposium 10, 1969
  Disorders of the skull base region
  (ed Carl-Axel Hamberger & Jan Wersäll)

Nobelsymposium 11, 1969
  Symmetry and function of biological systems at the macromolecular level
  (ed Arne Engström & Bror Strandberg)

Nobelsymposium 12, 1970
  Radiocarbon variations and absolute chronology
  (ed Ingrid U Olsson)

Nobelsymposium 13, 1970
  Pathogenesis of diabetes mellitus
  (ed Erol Cerasi & Rolf Luft)

Nobelsymposium 14, 1970
  The place of value in a world of facts
  (ed Arne Tiselius & Sam Nilsson)

Nobelsymposium 15, 1971
  Control of human fertility
  (ed Egon Diczfalusy & Ulf Borell)

Nobelsymposium 16, 1973
  Frontiers in gastrointestinal hormone research
  (ed Sven Andersson)

Nobelsymposium 17, 1971
  Small states in international relations
  (ed August Schou & Arne Olav Brundtland)

Nobelsymposia 18 and 19 (Postponed)

Nobelsymposium 20, 1972
  The changing chemistry of the oceans
  (ed David Dyrssen & Daniel Jagner)

Nobelsymposium 21, 1972
  From plasma to planet
  (ed Aina Elvius)

Nobelsymposium 22, 1972

ESR applications to polymer research
  (ed Per-Olof Kinell & Bengt Rånby)

Nobel Symposia 23–25

Publishers
Nobel Foundation, Stockholm, Sweden
Academic Press, New York, USA

Nobel Symposium 23, 1972
Chromosome identification—technique and applications in biology and medicine
  (ed Torbjörn Caspersson & Lore Zech)

Nobel Symposium 24, 1973
  Collective properties of physical systems
  (ed Bengt Lundqvist & Stig Lundqvist)

Nobel Symposium 25, 1973
  Chemistry in botanical classification
  (ed Gerd Bendz & Johan Santesson)